Novel Antioxidants for Animal Nutrition

Novel Antioxidants for Animal Nutrition

Editors

Luciana Rossi
Matteo Dell'Anno

MDPI

Basel • Beijing • Wuhan • Barcelona • Belgrade • Novi Sad • Cluj • Manchester

Editors

Luciana Rossi
Department of Veterinary Medicine and Animal Sciences—DIVAS
Università degi Studi di Milano
Lodi
Italy

Matteo Dell'Anno
Department of Veterinary Medicine and Animal Sciences—DIVAS
Università degli Studi di Milano
Lodi
Italy

Editorial Office

MDPI
St. Alban-Anlage 66
4052 Basel, Switzerland

This is a reprint of articles from the Special Issue published online in the open access journal *Antioxidants* (ISSN 2076-3921) (available at: www.mdpi.com/journal/antioxidants/special_issues/Novel_Antioxidants_Animal_Nutrition).

For citation purposes, cite each article independently as indicated on the article page online and as indicated below:

Lastname, A.A.; Lastname, B.B. Article Title. *Journal Name* **Year**, *Volume Number*, Page Range.

ISBN 978-3-7258-1158-8 (Hbk)
ISBN 978-3-7258-1157-1 (PDF)
doi.org/10.3390/books978-3-7258-1157-1

Contents

Preface

It am enthused to announce that the printed version of our successful Special Issue, which delves into the realm of innovative antioxidants for animal nutrition, is now available. Our journey into this fascinating field of novel antioxidants for animal nutrition has been marked by insights, discoveries, and the collective efforts of important researchers.

As we navigate the intricate pathways of nutrition, it becomes increasingly evident that antioxidants play a pivotal role in shaping the health and well-being of both humans and animals. In the design of functional diets, their importance cannot be neglected.

I would like to take this opportunity to extend my sincerest gratitude to Prof. Luciana Rossi for her unwavering dedication and unparalleled expertise. Prof. Rossi's invaluable awareness has been fundamental in shaping the trajectory of this Special Issue, and her guidance has been a beacon of light throughout my journey. It is with deep appreciation that I acknowledge her contributions and the immeasurable impact she has had on our work.

Together, we have crafted a collection of papers that I am confident will contribute to the future of animal nutrition research.

Luciana Rossi and Matteo Dell'Anno
Editors

antioxidants

Editorial

Novel Antioxidants for Animal Nutrition

Luciana Rossi and Matteo Dell'Anno *

Department of Veterinary Medicine and Animal Sciences—DIVAS, University of Milan, 29600 Lodi, Italy; luciana.rossi@unimi.it

* Correspondence: matteo.dellanno@unimi.it

In recent years, the importance of nutrition has notably escalated, with antioxidants emerging as crucial ingredients in the formulation of functional diets pivotal for promoting animal health and preventing diseases. Antioxidants can be obtained from a variety of natural sources or can be chemically synthesized in laboratory settings. Oxidative balance refers to the equilibrium between oxidants and antioxidants, and it is crucial for overall health. On the contrary, oxidative stress, resulting from a disparity favoring oxidants, can lead to cellular damage and contribute to the development of pathological states. The role of dietary antioxidants has also gained renewed interest in cancer research.

Antioxidants offer multifaceted benefits in animal diets, including supporting animal health, reducing disease incidence, and ultimately contributing to the reduction of antibiotics use in livestock. By promoting overall health and bolstering the immune system, antioxidants help to decrease the need for antimicrobials. Additionally, sustainable farming practices encourage the inclusion of antioxidants, aligning with the efforts to mitigate antimicrobial resistance, and supporting public health and food safety initiatives.

The bioaccessibility of antioxidant compounds can be influenced by various factors that are not only related to the differences in the digestive systems of ruminants and monogastrics, but also by the enzymatic portfolio, feed composition, microbiota contributors, and many other species-specific variables. Over the recent decade, a wide range of natural extracts and compounds have been placed on the market as functional ingredients or additives. Nevertheless, their heterogeneity and varying dosages have resulted in contradictory findings in the literature, necessitating further in-depth investigations to thoroughly characterize each antioxidant for optimal inclusion in animal feed.

check for updates

Citation: Rossi, L.; Dell'Anno, M. Novel Antioxidants for Animal Nutrition. *Antioxidants* **2024**, *13*, 438. https://doi.org/10.3390/antiox13040438

Received: 29 March 2024
Accepted: 3 April 2024
Published: 7 April 2024

Beslo et al. (Contribution 1) reviewed the use of plant polyphenols in animal nutrition, proposing their use in animal nutrition for a sustainable approach and underlining the importance of understanding the intracellular mechanisms of polyphenols' antioxidant activity to maximize their potential. Cell models offer a valuable tool for evaluating antioxidants by providing a controlled experimental system to study their protective effects against oxidative stress and their potential applications in promoting cellular health and mitigating intestinal disease processes. Cell models also contribute to the reduction of using animals in research by providing an alternative approach to study biological processes. In this context, the contribution of Xu et al. (Contribution 2) represents an intriguing approach focused on the IPEC-J2 cell model to evaluate stevioside, a natural sweetener. It registered protective effects on intestinal cells under induced oxidative stress, indicating its potential application in promoting gut health and mitigating inflammation and apoptosis. Dabrowska et al. (Contribution 3) evaluated the chlorogenic acid derived from green coffee extract on peripheral blood mononuclear cells from racehorses' blood, suggesting its potential as a natural supplement for improving immune function and oxidative stress management in equine species.

Antioxidant supplementation in animal nutrition could enhance animal welfare, the quality of products, and reduce stress and the use of antimicrobials, which is in line with One Health principles. This is particularly relevant in young animals due to their susceptibility to stressors and immune system immaturity, highlighting the importance of proactive measures in promoting their health and welfare. In particular, piglets face numerous challenges during the weaning period that can compromise their immune system, increasing the high risk of pathogenic invasions (Contribution 4). Cai et al. (Contribution 5) investigated the protective role of silybin, derived from *Silybum marianum*, against a hepatic oxidative injury induced by paraquat in weaned piglets. Results demonstrated that a dietary supplementation of 400 mg/kg silybin alleviated liver damage by enhancing antioxidant capacity, inhibiting inflammation, and improving mitochondrial function. Pastorelli et al. (Contribution 6) evaluated the partial replacement of synthetic vitamin E by natural polyphenols in post-weaning piglets, revealing encouraging results in maintaining growth performance, antioxidant status, and immunity. Their findings suggest that the replacement of an equivalent of 87.9 mg/kg vitamin E with polyphenols extracted from different plants could be viable alternatives to conventional vitamin E commonly used in animal nutrition. Furthermore, the addition of 3% thyme to diets for fattening pigs showed positive effects on redox status and lipid metabolism, highlighting the potential of this plant extract as a functional feed additive for adult pigs (Contribution 7).

Furthermore, Laviano et al. (Contribution 8) assessed the effect of hydroxytyrosol supplementation in the colostrum and milk in Iberian sows during late pregnancy. Their study revealed that hydroxytyrosol supplementation increased polyunsaturated fatty acids in the colostrum and exhibited a lower desaturase capacity in 20-day milk. These modifications in milk lipid composition could potentially benefit the oxidative status and gut health of piglets during the early stages of life.

In broilers, selenium nanoparticles and organic selenium supplementation (0.5 mg/kg) showed positive effects in the physico-chemical quality of the breast, health status, and the antioxidant potential in muscle and liver (Contribution 9). Particularly, the nano-sized Se confirmed the protective activity against mitochondrial damage in hepatocytes without showing any toxic effects. The use of myricetin, a natural flavonoid, supplemented at 600 mg/kg in broilers showed a significant improvement in growth performance and antioxidant capacity of meat, and reduced the interleukin relative expression, severity of *Eimeria* spp. lesions, and oocyst shedding in challenged broiler chickens (Contribution 10).

In calves, in-milk dietary supplementation of 6 g/day of chestnut and quebracho tannins reduced the occurrence of neonatal calf diarrhea without affecting the protein digestibility of milk. Thus, this confirmed the role of natural polyphenols in the reduction of pathology occurrence and antibiotic treatments (Contribution 11).

In dairy ewes, a mixture of natural antioxidants derived from thyme, anise, and olive improved milk yield, oxidative stability, and somatic cell count, without influencing the milk composition. This underscored the potential of plant bioactive compounds to enhance milk quality and udder health (Contribution 12). Dalaka et al. (Contribution 13) investigated the antioxidant capacity of sweet whey derived from cheeses obtained from bovine, ovine, caprine, or a mixture of ovine/caprine milk before and after in vitro digestion. Particularly, ovine sweet whey exhibited higher antioxidant activity in both non-digested and digested cheese, suggesting ovine whey as a valuable source of antioxidants for both human and animal nutrition.

Mavrommatis et al. (Contribution 14) focused on microalgae as renewable and sustainable sources of bioactive compounds. The antioxidant activity of microalgae significantly varies between species and depends on growth conditions. The use of microalgae in animal nutrition revealed variable findings that were highly dependent on the composition of species and their percentage of inclusion. The use of carbohydrate-active enzymes can increase nutrient bioavailability because of recalcitrant microalgae cell wall degradation, making it a promising strategy for monogastric nutrition and for improving livestock productivity. The main limitations of microalgae lie in their cost-effectiveness and their cultivation technology, which should be improved by reducing production costs, thus increasing profitability.

Collectively, the published studies underlined the importance of harnessing the potential of natural antioxidants in animal nutrition. By exploring a wide array of bioactive compounds and their mechanisms of action, researchers are plotting the way for sustainable approaches in livestock farming. As we strive for a more holistic and ethical approach in livestock production, the supplementation of natural antioxidants into animal feed represents a promising frontier for enhancing health, welfare, and productivity in livestock.

Conflicts of Interest: The authors declare no conflicts of interest.

List of Contributions

1. Bešlo, D.; Golubić, N.; Rastija, V.; Agić, D.; Karnaš, M.; Šubarić, D.; Lučić, B. Antioxidant activity, metabolism, and bioavailability of polyphenols in the diet of animals. *Antioxidants* **2023**, *12*, 1141.
2. Xu, Q.; Liu, M.; Chao, X.; Zhang, C.; Yang, H.; Chen, J.; Zhou, B. Stevioside improves antioxidant capacity and intestinal barrier function while attenuating inflammation and apoptosis by regulating the NF-κB/MAPK pathways in diquat-induced oxidative stress of IPEC-J2 cells. *Antioxidants* **2023**, *12*, 1070.
3. Dąbrowska, I.; Grzędzicka, J.; Niedzielska, A.; Witkowska-Piłaszewicz, O. Impact of Chlorogenic Acid on Peripheral Blood Mononuclear Cell Proliferation, Oxidative Stress, and Inflammatory Responses in Racehorses during Exercise. *Antioxidants* **2023**, *12*, 1924.
4. Duarte, M.E.; Stahl, C.H.; Kim, S.W. Intestinal damages by F18+ Escherichia coli and its amelioration with an antibacterial bacitracin fed to nursery pigs. *Antioxidants* **2023**, *12*, 1040.
5. Cai, L.; Ming, D.; Chen, W.; Zhao, Y.; Li, Y.; Sun, W.; Pi, Y.; Jiang, X.; Li, X. Silybin Alleviated Hepatic Injury by Regulating Redox Balance, Inflammatory Response, and Mitochondrial Function in Weaned Piglets under Paraquat-Induced Oxidative Stress. *Antioxidants* **2024**, *13*, 324.
6. Pastorelli, G.; Benamri, R.; Faustini, M.; De Bellis, R.; Serra, V.; Turin, L.; Haumont, M.; Durand, P.; Bianchessi, L.; Prost-Camus, E. Partial Replacement of Synthetic Vitamin E by Polyphenols in Post-Weaning Piglets. *Antioxidants* **2023**, *12*, 1752.
7. Klimiuk, K.; Sembratowicz, I.; Tutaj, K.; Czech, A. Effect of thyme (*Thymus vulgaris* L.) used in diets with extruded flaxseed on the antioxidant and lipid profile of the blood and tissues of fattening pigs. *Antioxidants* **2023**, *12*, 1045.
8. Laviano, H.D.; Gómez, G.; Muñoz, M.; García-Casco, J.M.; Nuñez, Y.; Escudero, R.; Molina, A.H.; González-Bulnes, A.; Óvilo, C.; López-Bote, C. Dietary Vitamin E and/or Hydroxytyrosol Supplementation to Sows during Late Pregnancy and Lactation Modifies the Lipid Composition of Colostrum and Milk. *Antioxidants* **2023**, *12*, 1039.
9. Bień, D.; Michalczuk, M.; Łysek-Gładysińska, M.; Jóźwik, A.; Wieczorek, A.; Matuszewski, A.; Kinsner, M.; Konieczka, P. Nano-sized selenium maintains performance and improves health status and antioxidant potential while not compromising ultrastructure of breast muscle and liver in chickens. *Antioxidants* **2023**, *12*, 905.
10. El-Ghareeb, W.R.; Kishawy, A.T.; Anter, R.G.; Aboelabbas Gouda, A.; Abdelaziz, W.S.; Alhawas, B.; Meligy, A.M.; Abdel-Raheem, S.M.; Ismail, H.; Ibrahim, D. Novel antioxidant insights of myricetin on the performance of broiler chickens and alleviating experimental infection with Eimeria spp.: Crosstalk between oxidative stress and inflammation. *Antioxidants* **2023**, *12*, 1026.
11. Dell'Anno, M.; Frazzini, S.; Ferri, I.; Tuberti, S.; Bonaldo, E.; Botti, B.; Grossi, S.; Sgoifo Rossi, C.A.; Rossi, L. Effect of Dietary Supplementation of Chestnut and Quebracho Tannin Supplementation on Neonatal Diarrhoea in Preweaning Calves. *Antioxidants* **2024**, *13*, 237.
12. Karageorgou, A.; Tsafou, M.; Goliomytis, M.; Hager-Theodorides, A.; Politi, K.; Simitzis, P. Effect of Dietary Supplementation with a Mixture of Natural Antioxidants on Milk Yield, Composition, Oxidation Stability and Udder Health in Dairy Ewes. *Antioxidants* **2023**, *12*, 1571.

13. Dalaka, E.; Politis, I.; Theodorou, G. Antioxidant Activity of Sweet Whey Derived from Bovine, Ovine and Caprine Milk Obtained from Various Small-Scale Cheese Plants in Greece before and after in vitro Simulated Gastrointestinal Digestion. *Antioxidants* **2023**, *12*, 1676.
14. Mavrommatis, A.; Tsiplakou, E.; Zerva, A.; Pantiora, P.D.; Georgakis, N.D.; Tsintzou, G.P.; Madesis, P.; Labrou, N.E. Microalgae as a sustainable source of antioxidants in animal nutrition, health and livestock development. *Antioxidants* **2023**, *12*, 1882.

antioxidants

MDPI

Article

Fucoidan Supplementation Improves Antioxidant Capacity via Regulating the *Keap1/Nrf2* Signaling Pathway and Mitochondrial Function in Low-Weaning Weight Piglets

Chenggang Yin [1,†], Qingyue Bi [1,2,†], Wenning Chen [1,†], Chengwei Wang [3,*], Bianca Castiglioni [4], Yanpin Li [1], Wenjuan Sun [1], Yu Pi [1], Valentino Bontempo [5], Xilong Li [1] and Xianren Jiang [1,*]

1 Key Laboratory of Feed Biotechnology of Ministry of Agriculture and Rural Affairs, Institute of Feed Research, Chinese Academy of Agricultural Sciences, Beijing 100081, China; 82101231287@caas.cn (C.Y.); biqingyue981106@126.com (Q.B.); 82101215363@caas.cn (W.C.); liyanpin@caas.cn (Y.L.); sunwenjuan@caas.cn (W.S.); lixilong@caas.cn (X.L.)
2 College of Agriculture, Yanbian University, Yanji 133000, China
3 College of Life Science, Jiangxi Science and Technology Normal University, Nanchang 330013, China
4 Institute of Agricultural Biology and Biotechnology (IBBA-CNR), Via Einstein, 26900 Lodi, Italy; casti@ibba.cnr.it
5 Department of Veterinary Medicine and Animal Science (DIVAS), University of Milan, 26900 Lodi, Italy; valentino.bontempo@unimi.it
* Correspondence: 1020130201@jxstnu.edu.cn (C.W.); jiangxianren@caas.cn (X.J.); Tel.: +86-10-8210-8134 (C.W. & X.J.)
† These authors contributed equally to this work.

Abstract: Fucoidan (FC) is known for its antioxidant properties, but it has unclear effects and mechanisms on weaned piglets. Two experiments were conducted to determine the optimal FC dosage in piglet diets and its protective effect against lipopolysaccharide (LPS)-induced oxidative stress. In experiment one, 24 low weight weaned piglets were randomly assigned to four dietary treatments: a basal diet (FC 0), or a diet supplemented with 150 (FC 150), 300 (FC 300), or 600 mg/kg FC (FC 600). In experiment two, 72 low-weaning weight piglets were randomly allocated into four treatments: a basal diet (CON), or 300 mg/kg of fucoidan added to a basal diet challenged with LPS (100 μg LPS/kg body weight) or not. The results showed that FC treatments increased the G:F ratio, and dietary FC 300 reduced the diarrhea incidence and increased the plasma IGF-1 concentrations. In addition, FC 300 and FC 600 supplementation increased the plasma SOD activity and reduced the plasma MDA concentration. LPS challenge triggered a strong systemic redox imbalance and mitochondrial dysfunction. However, dietary FC (300 mg/kg) supplementation increased the activity of antioxidant enzymes, including SOD, decreased the MDA concentration in the plasma and liver, down-regulated *Keap1* gene expression, and up-regulated *Nrf2*, *CAT*, *MFN2*, *SDHA*, and *UQCRB* gene expression in the liver. These results indicated that dietary fucoidan (300 mg/kg) supplementation improved the growth performance and antioxidant capacity of low-weaning weight piglets, which might be attributed to the modulation of the *Keap1/Nrf2* signaling pathway and the mitochondrial function in the liver.

Keywords: fucoidan; lipopolysaccharide; liver; oxidative stress; low-weaning weight piglets

check for updates

Citation: Yin, C.; Bi, Q.; Chen, W.; Wang, C.; Castiglioni, B.; Li, Y.; Sun, W.; Pi, Y.; Bontempo, V.; Li, X.; et al. Fucoidan Supplementation Improves Antioxidant Capacity via Regulating the *Keap1/Nrf2* Signaling Pathway and Mitochondrial Function in Low-Weaning Weight Piglets. *Antioxidants* **2024**, *13*, 407. https://doi.org/10.3390/antiox13040407

Academic Editor: Dong Uk Ahn

Received: 8 February 2024
Revised: 13 March 2024
Accepted: 26 March 2024
Published: 28 March 2024

1. Introduction

In current intensive farming, the employment of high-yielding sows coupled with early weaning strategies has become a prevalent practice, helping to enhance sow productivity and bolster the economic yield of farms [1,2]. However, a consequence of this practice is the surge in the number of low-weaning weight piglets. The digestive system of these low-weaning weight piglets is not fully mature, the growth rate is relatively slow, the resistance to external stimuli is poor, and the diarrhea incidence and mortality rate are often high [3–5]. Early weaning would trigger a series of complex physiological and psychological reactions in piglets, including changing eating habits and adapting to the new social environment, which may lead to their physiological and immune function abnormalities, causing serious oxidative stress [6,7]. Oxidative stress is due to the overproduction of reactive oxygen species (ROS) that exceeds the processing capacity of the antioxidant system, so maintaining the redox balance within cells is crucial for maintaining intestinal homeostasis [8,9]. In addition, the early weaning may lead to oxidative stress and mitochondrial dysfunction in the liver, which in turn affects its redox status and function [10]. It has been found that weaning stress can cause oxidative stress and oxidative damage in the liver of piglets, activate the *MAPK* pathway, increase the apoptosis of liver cells, and affect liver function [11,12]. A disturbance in the balance of oxidation reduction can interfere with the biological processes of mitochondria, leading to mitochondrial dysfunction [13]. Although mitochondria play a crucial role in energy production, they are also one of the main organelles that produce ROS [14]. Mitochondrial dysfunction can lead to excessive ROS production and affect liver function [15].

At present, research focusing on enhancing the body's antioxidant capacity through the inclusion of functional additives in the diet is gaining increasing attention. Previous studies in our laboratory have found that functional feed additives can enhance the antioxidant capacity of piglets [7,8], improve mitochondrial function [9], and promote the growth of piglets. Recent research in our laboratory has revealed that functional feed additives can improve the antioxidant capacity of weaned piglets by activating the *Nrf2* signaling pathway and optimize their mitochondrial function, thus effectively reducing the oxidative damage of paraquat to the liver [16]. The *Nrf2* signaling pathway plays a central regulatory role in the oxidative stress response in vivo, and it controls the expression of a variety of genes or enzymes related to antioxidants [17]. In the non-stressed state, *Keap1* forms ubiquitin E3 ligase complexes with CULLIN3 (CUL3), resulting in the polyubiquitination of *Nrf2*, which is then rapidly degraded by the proteasome system. However, under electrophilic or ROS stress, the active cysteine residue of *Keap1* is directly modified, which reduces the ubiquitin E3 ligase activity of the *Keap1–CUL3* complex, thereby stabilizing *Nrf2* [18,19]. Therefore, the regulation of *Nrf2* by *Keap1* plays a crucial role in cell resistance to oxidative stress and maintenance of cell homeostasis.

Fucoidan is a natural polysaccharide compound derived from kelp and brown algae, first identified in 1913 by the Swedish scientist Professor Kylin, who ultimately christened it "Fucoidan" [20]. As a substance with significant antioxidant properties, Fucoidan has been the subject of extensive research in recent years. Previous in vitro and in vivo studies found that fucoidan could improve cell viability and protect cells from oxidative damage induced by 2,2′-Azobis (2-methylpropionamidine) dihydrochloride (AAPH) by clearing intracellular ROS and inhibiting cell apoptosis. Fucoidan improves the survival rate of zebrafish by eliminating ROS, inhibiting lipid peroxidation, inhibiting cell death, and thus inhibiting oxidative stress [21]. A further study demonstrated that the oral intake of fucoidan decreased the concentrations of reactive oxygen species (ROS) and malondialdehyde (MDA) in mouse serum, while simultaneously boosting the activity of glutathione peroxidase (GSH-Px) and superoxide dismutase (SOD), increasing the production of adenosine triphosphate (ATP), and restoring the concentrations of mitochondrial respiratory chain complexes in cardiac tissue, thereby reducing oxidative stress and preventing mitochondrial functional damage [22]. Furthermore, fucoidan demonstrated the ability to decrease lipid peroxide (LPO) and MDA concentrations in the liver. However, it is noteworthy that higher doses of fucoidan (2000 mg/kg) may potentially trigger inflammation and metabolic disorders [23]. Drawing from the findings of prior studies, we found that fucoidan has potential as a new antioxidant, but there is a need to pay attention to the dosage in use.

To our knowledge, the literature offers limited insights into the impact of fucoidan on weaned piglets, particularly with regard to the effects of the dosage of fucoidan and antioxidant capacity of weaned piglets. In this study, we focused on piglets with lower weaning weights, who are more prone to oxidative stress. We designed two experiments with the aim of preliminarily determining the optimal dosage of fucoidan and its effect on the plasma antioxidant status of low-weaning weight piglets. Subsequently, an oxidative stress model was established by stimulating low-weaning weight piglets with lipopolysaccharide (LPS). This model was utilized to further investigate the mechanisms by which fucoidan regulates oxidative damage in low-weaning weight piglets.

2. Materials and Methods

2.1. Animal Ethics Approval

The trial was conducted from July to August 2022 and September to October 2023 at the Tianpeng Experimental Farm located in Langfang, Hebei province. The animal protocol in this study was approved by the Animal Care and Use Committee of the Institute of Feed Research of the Chinese Academy of Agricultural Sciences (IFR-CAAS20221010 for experiment one and IFR-CAAS20230825 for experiment two).

2.2. Animals and Treatment

Experiment one: In reference to the previous selection scheme [5,24], this study selected 72 healthy weaned piglets (Duroc × Landrace × Yorkshire), with an average body weight (BW, 6.62 ± 0.13 kg) and age (25 ± 1 days). These weaned piglets were categorized into high, medium, and low weight groups based on their body weight. A total of 24 weaned piglets with low-weaning body weight (BW, 5.81 ± 0.05 kg) were randomly divided into 4 treatment groups with 6 replicates per group and 1 piglet per replicate. The dietary treatments were as follows: basal diet without FC (FC 0), FC 0 + 150 mg/kg FC (FC 150), FC 0 + 300 mg/kg FC (FC 300), or FC 0 + 600 mg/kg FC (FC 600). The experiment lasted for 21 days and plasma samples were collected (Figure 1A). The Fucoidan used in this study was purchased from Zhenlu Biotechnology Co., Ltd. (Xi'an, China), purity ≥ 98% and derived from kelp.

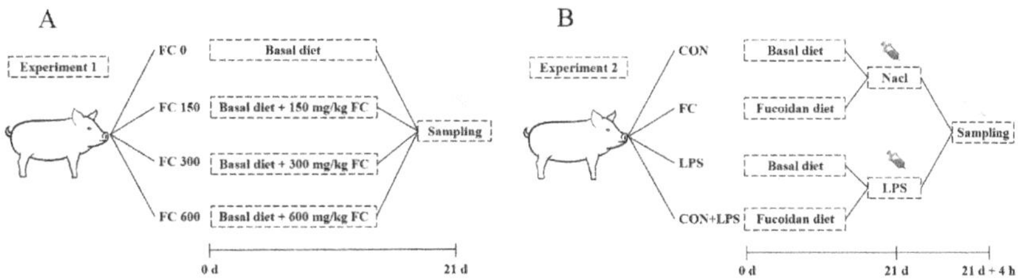

Figure 1. Schematic diagram of experimental design. (**A**) experiment one; (**B**) experiment two.

Experiment two: According to the selection protocols of experiment one, this study selected 72 healthy low weight weaned piglets (BW, 6.01 ± 0.32 kg) from 216 healthy weaned piglets (Duroc × Landrace × Yorkshire) with an average body weight (BW, 8.14 ± 0.15 kg) and age (28 ± 1 day). They were randomly allocated into 4 treatments with 6 replicates per treatment and 3 piglets per pen: a basal diet (CON) or 300 mg/kg of fucoidan added to a basal diet (FC) challenged with LPS or not. On day 21, one piglet was selected from each pen. Piglets in the challenged groups were intraperitoneally injected with 1 mL of LPS (*Escherichia coli* O55:B5, Sigma Chemical, Burlington, MA, USA) at 100 µg/kg BW, and the other piglets were intraperitoneally injected with the same amount of sterile saline (0.9% NaCl). The selection of LPS dosage was based on previous studies on weaned piglets [25]. All the selected piglets were euthanized 4 h later, and plasma and liver samples were collected (Figure 1B).

Animals were purchased from a Langfang commercial farm and housed in a nursery room. During the experiment, the diet for the piglets was formulated meeting the National Research Council (2012) nutrient requirements (Appendix A: Tables A1 and A2).

The experiment one and experiment two composition and nutrient levels of the basal diet are shown in Table A1. The basal diet did not contain any antibiotic growth promoters and the form of diet was mash. Piglets were accommodated in slatted floor pens (1.7 m × 1.5 m) with unrestricted access to feed and water. The room's initial temperature was 28 °C and was gradually reduced to 26 °C. The room was lit naturally and artificially, with ventilation provided by speed-controlled fans. Each pen had two drinking fountains and an adjustable trough. Standard farm procedures were followed for disinfection and vaccination.

2.3. Sample Collection

Experiment one: Body weight (BW) was recorded individually at the beginning and the end of the trial. Any culling or mortality was recorded daily and feed consumption was corrected for accordingly. Growth performance was evaluated by calculating the average daily gain (ADG), average daily feed intake (ADFI), and gain-to-feed ratio (G:F) for each pen. To determine the incidence of diarrhea, fecal scores were monitored daily by visually appraising each subject using the following five-point fecal consistency scoring system: 1 = hard, dry pellet; 2 = firm, formed stool; 3 = soft, moist stool that retains its shape; 4 = soft, unformed stool; and 5 = watery liquid that can be poured. A liquid consistency (score 4–5) was considered indicative of diarrhea [26]. The incidence of diarrhea (%) was calculated as a percentage of the number of piglets with diarrhea divided by the total number of piglets in each treatment. At the end of the experiment (day 21), blood was taken from each piglet through the jugular vein to the heparin tube, left for 30 min, and centrifuged at 3000 rpm for 10 min. The plasma was separated and stored at −20 °C for analysis.

Experiment two: At the end of the experiment (day 21), the plasma of the selected piglets after LPS challenge (4 h) was collected. The specific procedure was the same as in experiment one. After LPS challenge for 4 h, the piglets were stunned by a portable electrical stunner (the output voltage is 220 V) and bled quickly to be euthanized. Liver samples were collected and placed in cryogenic vials (Corning Incorporated, New York, NY, USA), frozen in liquid nitrogen, and stored at −80 °C for analysis.

2.4. Assay of Antioxidant Indices

The antioxidant indicators in both the plasma and liver, including the activities of SOD, catalase (CAT), and GSH-Px, as well as the concentrations of MDA, and the plasma growth hormone concentrations of insulin-like growth factor 1 (IGF-1), were determined using commercial assay kits as instructed (Enzyme-linked Biotechnology Co., Ltd., Shanghai, China). The concentrations of SOD, MDA, CAT, and GSH-Px were determined by micromethod according to the instructions. The absorbance was determined by microplate spectrophotometer (Bio Tek Instruments, Inc, Shanghai, China) at the appropriate wavelength, and the sample concentration was calculated. The superoxide anion (O^{2-}) was generated via the reaction system of xanthine and xanthine oxidase. This anion can interact with WST-8 to yield a water-soluble dye, formazan, which exhibits absorption at 450 nm. The activity of SOD, which can eliminate O^{2-}, thereby inhibiting the formation of formazan, can thus be measured. MDA reacts with thiobarbituric acid (TBA) to produce a red product that has a maximum absorption peak at 532 nm. The content of lipids containing peroxides can be estimated by colorimetry. Concurrently, the absorbance at 600 nm was measured, and the difference in absorbance at these two wavelengths was used to calculate the MDA content. CAT has the ability to decompose H_2O_2, which has a characteristic absorption peak at 240 nm. Consequently, the absorbance of the reaction solution at 240 nm decreased over time. The activity of CAT can be calculated based on the rate of change in absorbance. GSH-Px catalyzes the oxidation of glutathione (GSH) by H_2O_2 to produce glutathione disulfide (GSSG). Glutathione reductase (GR) then catalyzes the reduction of GSSG by nicotinamide adenine dinucleotide phosphate (NADPH) to regenerate GSH, oxidizing NADPH to NADP+ in the process. NADPH has a characteristic absorption peak at 340 nm, while NADP+ does not. Therefore, the activity of GSH-Px can be calculated by measuring the rate of decrease in absorption at 340 nm. The IGF-1 concentration was determined using an ELISA kit, samples were measured using microplates pre-coated with porcine IGF-1-trapping antibodies, and color was developed using TMB substrates through incubation and thorough cleaning. TMB turned blue under peroxidase and eventually yellow under acid. Finally, the absorbance was measured at a wavelength of 450 nm using an enzyme labeler to calculate the concentration of the sample. The protein concentration of the liver crude enzyme fluid was determined using the BCA protein quantitative kit, and the specific steps were carried out according to the instructions (Beijing Huaxing Bochuang gene Technology Co., Ltd., Beijing, China).

2.5. Real-Time Quantitative PCR Analysis (qPCR)

The hepatic RNA extraction and quantitative polymerase chain reaction (qPCR) procedure follows the method outlined by Cai et al. [8]. In succinct terms, for liver samples, the total RNA extraction employed Trizol reagent (Beijing AidLab Biotechnology Co., Ltd., Beijing, China) in accordance with the manufacturer's stipulations. Firstly, we took 50 mg of liver tissue and added it to 1 mL of lysis buffer for tissue homogenization, then incubated it at room temperature for 5 min. Next, we added 0.2 mL of chloroform, vigorously shacked it for 15 s, and then incubated it at room temperature for 3 min. Afterwards, the sample was centrifuged at 12,000 rpm at 4 °C for 10 min using a centrifuge (Hitachi Koki Co., Ltd., Tokyo, Japan). Then, we took the supernatant, added anhydrous ethanol equal to half its volume, mixed well, and transferred it into an RA adsorption column, then centrifuged it at 12,000 rpm for 45 s. Subsequently, we added 500 μL of RE deproteinization solution, centrifuged it at 12,000 rpm for 45 s, and then discarded the waste liquid. Next, we added 500 μL of RW wash solution, centrifuged it at 12,000 rpm for 45 s, and repeated this step once. Then, we centrifuged the sample at 13,000 rpm for 2 min, added 60 μL of Rnase-free water, let it stand at room temperature for 2 min, centrifuged it at 12,000 rpm for 1 min, and finally obtained the liver RNA. The concentration and quality of RNA were scrutinized using a Nano Drop™ One/One Cmicro UV-Vis spectrophotometer (Thermo Fisher Scientific, Inc., Boston, MA, USA). The instrument calibration was performed using Rnase-free water prior to detection. Following this, complementary deoxyribonucleic acid (cDNA) was synthesized by a two-step reverse transcriptional procedure using the appropriate concentration of the reagent according to the quantitative concentration of liver RNA and the kit instructions (Beijing Takara Biomedical Technology Co., Ltd., Beijing, China). The cDNA was diluted with Rnase-free water at the appropriate concentration, packaged, and stored for further detection. The qPCR analysis was executed utilizing a CFX96 Touch real-time fluorescent qPCR system (Bio-Rad Laboratories Inc., Berkeley, CA, USA). The relative expression of the target gene was determined by employing the $2^{-\Delta\Delta CT}$ method, wherein glyceraldehyde-3-phosphate dehydrogenase (GAPDH) functioned as the designated housekeeping gene. The specific primer sequences utilized in the qPCR assay can be found in Appendix B: Table A3.

2.6. Statistical Analysis

SPSS 19 (IBM, Armonk, NY, USA) was used for the statistical analysis. The univariate analysis of variance (ANOVA) and Tukey's honest significance difference test were used for the statistical analysis. In addition, the Kruskal–Wallis test for non-normally distributed data sets was used to determine statistical significance. Orthogonal polynomial comparison tests with linear and quadratic effects were used to evaluate the effects of different doses of FC supplementation. Data are expressed as the mean of the standard error (SE). The difference was considered to be significant when $p < 0.05$, and the difference was considered to have a trend when $0.05 \leq p < 0.10$.

3. Results

3.1. Growth Performance and Diarrhea Incidence

The effects of different doses of FC supplementation on the growth performance of low-weaning weight piglets are presented in Figure 2A–E. Compared with the FC 0 group, dietary FC 150, FC 300, and FC 600 significantly increased the G:F ratio of low-weaning weight piglets ($p < 0.05$), and dietary FC 300 and FC 600 tended to increase the final BW ($p = 0.059$) and ADG ($p = 0.057$). There were no significant differences in the ADFI among experimental groups ($p > 0.05$). In addition, FC supplementation linearly increased the final BW, ADG, and G:F ratio ($p < 0.05$), and tended to linearly increase the ADFI ($p = 0.097$). In addition, there was a quadratic effect of FC supplementation on the G:F ratio ($p < 0.05$).

The result of different doses of FC supplementation on the diarrhea incidence of low-weaning weight piglets in experiment one is shown in Figure 2F. Compared with the FC 0 group, FC 300 supplementation significantly reduced the diarrhea incidence of low-weaning weight piglets from day 0 to 21 ($p < 0.05$). However, there was no significant difference in diarrhea incidence among FC 0, FC 150, and FC 600 groups ($p > 0.05$).

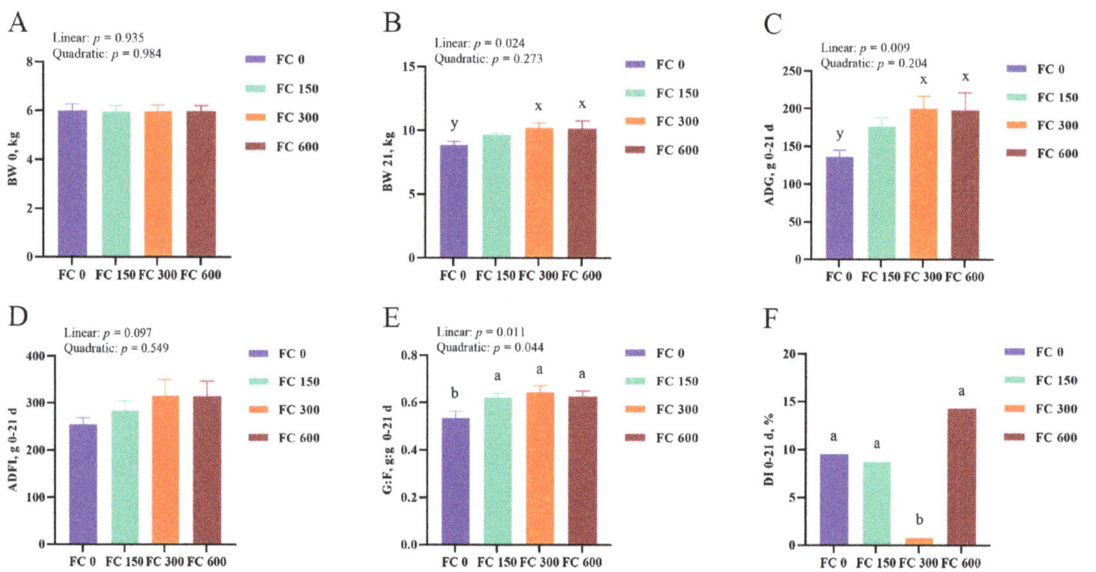

Figure 2. The effects of different doses of fucoidan (FC) supplementation on growth performance and diarrhea incidence of low-weaning weight piglets are presented at 0 to 21 days. (**A**,**B**) Body weight of piglets on day 0 (**A**) and 21 (**B**). (**C**–**E**) Average daily gain (**C**), average daily feed intake (**D**), and ratio of average daily gain to average daily feed intake (**E**) from 0 to 21 days. (**F**) Diarrhea incidence (DI) from day 0 to 21. Data were expressed as mean with their standard errors represented by vertical bars, ($n = 6$). Orthogonal polynomials were used to evaluate linear and quadratic responses to the concentrations of FC treatment. FC 0: basal diet, FC 150, FC 300, and FC 600 group, basal diet adding 150, 300, and 600 mg/kg FC, respectively. ^{a,b} Means listed in the same row with different superscripts are significantly different ($p < 0.05$). ^{x,y} Means listed in the same row with different superscripts showed a tendency to be different ($0.05 \leq p < 0.10$).

3.2. *Plasma IGF-1 Concentrations*

The effects of different doses of FC supplementation on the plasma IGF-1 concentrations of low-weaning weight piglets are shown in Figure 3E. Compared with the FC 0 group, the plasma IGF-1 concentrations in the FC 300 group were significantly increased ($p < 0.05$). In addition, the plasma IGF-1 concentrations linearly increased with the increase in FC supplemental concentrations ($p < 0.05$). There was no quadratic effect of FC supplementation on the plasma IGF-1 concentrations ($p > 0.05$).

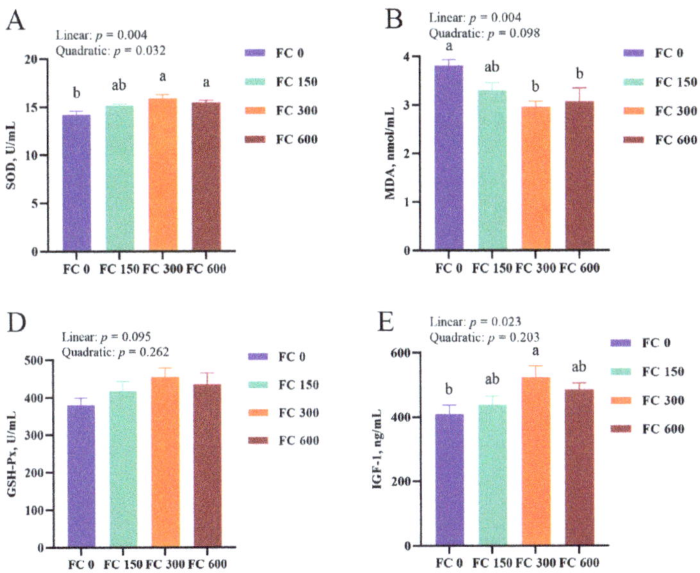

Figure 3. The effects of different doses of FC supplementation on plasma antioxidant enzyme activity and IGF-1 concentrations of low-weaning weight piglets is presented at day 21. (**A–D**) The plasma concentrations of indicators related to antioxidants (SOD, MDA, CAT, GSH-Px). (**E**) The plasma concentrations of IGF-1. Data were expressed as mean with their standard errors represented by vertical bars, (n = 6). Orthogonal polynomials were used to evaluate linear and quadratic responses to the concentrations of FC treatment. FC 0: basal diet, FC 150, FC 300, and FC 600 group, basal diet adding 150, 300, and 600 mg/kg FC, respectively. ^{a,b} Means listed in the same row with different superscripts are significantly different ($p < 0.05$).

3.3. Plasma Antioxidant Enzyme Activity

The effects of different doses of FC supplementation on the plasma antioxidant enzyme activity of low-weaning weight piglets are shown in Figure 3A–D. FC 300 and FC 600 supplementation significantly increased the plasma SOD activity and reduced the plasma MDA concentration compared to the FC 0 group ($p < 0.05$). There were no significant effects on plasma GSH-Px and CAT activities in all treatment groups ($p > 0.05$). In addition, FC supplementation linearly and quadratically increased the plasma SOD activity ($p < 0.05$) and linearly reduced the plasma MDA concentration ($p < 0.05$).

In addition, the effects of FC supplementation on plasma antioxidant enzyme activity of low-weaning weight piglets under LPS challenge are shown in Figure 4A–D. LPS challenge decreased the plasma SOD activity and increased the MDA concentration of low-weaning weight piglets ($p < 0.05$). Compared with the LPS group, the plasma SOD activity in the LPS + FC group was increased ($p < 0.05$), and the MDA concentration tended to be decreased in the LPS + FC group ($p = 0.091$). In addition, dietary FC significantly increased the plasma SOD activity and decreased the plasma MDA content compared to the CON group ($p < 0.05$).

3.4. Hepatic Antioxidant Enzyme Activity

The effects of FC supplementation on hepatic antioxidant enzyme activity of low-weaning weight piglets under the LPS challenge are shown in Figure 4E–H. Compared with the CON group, LPS reduced SOD activity and increased MDA concentration in the liver of low-weaning weight piglets ($p < 0.05$). Compared with the LPS group, the activity of CAT in the liver of LPS + FC piglets was significantly increased ($p < 0.05$), and the concentration of MDA was decreased in the LPS + FC group ($p < 0.05$), and dietary FC to LPS-challenged piglets tended to increase the activity of SOD ($p = 0.087$). In addition, hepatic CAT activity in the FC group was significantly increased compared with CON group ($p < 0.05$).

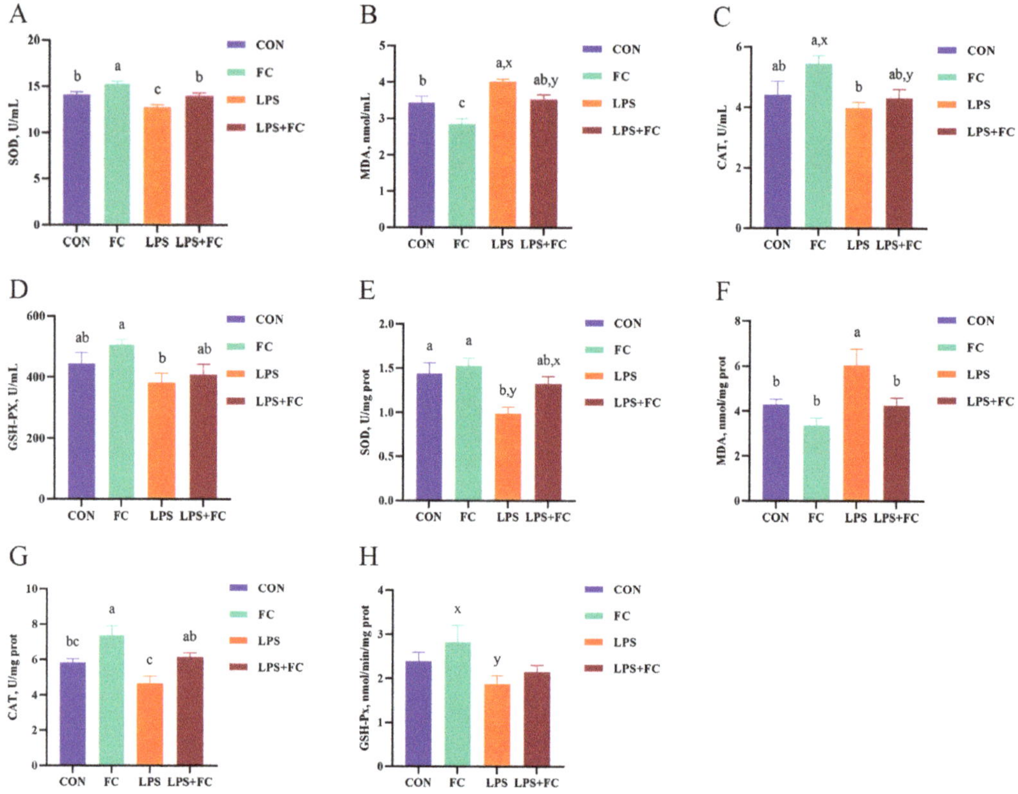

Figure 4. The effects of FC supplementation on the antioxidant enzyme activity of plasma and liver in LPS-challenged low-weaning weight piglets. (**A–D**) The plasma concentrations of indicators related to antioxidants (SOD, MDA, CAT, GSH-Px). (**E–H**) The liver concentrations of indicators related to antioxidants (SOD, MDA, CAT, GSH-Px). Data were expressed as mean with their standard errors represented by vertical bars, ($n = 6$). CON: basal diet, FC: basal diet + 300 mg/kg FC. $^{a–c}$ Means listed in the same row with different superscripts are significantly different ($p < 0.05$). ^{x,y} Means listed in the same row with different superscripts showed a tendency to be different ($0.05 \leq p < 0.10$).

3.5. Hepatic Antioxidant Genes mRNA Expression

The effects of FC supplementation on the hepatic antioxidant gene mRNA expression in low-weaning weight piglets under LPS challenge are shown in Figure 5A–I. Compared with the CON group, LPS challenge down-regulated *GCLC* mRNA expression in the liver of low-weaning weight piglets ($p < 0.05$). Compared with the LPS group, administering dietary FC to LPS-challenged piglets up-regulated *CAT* mRNA expression ($p < 0.05$) and down-regulated *Keap1* mRNA expression ($p < 0.05$) in the liver. The mRNA expression of *Nrf2* ($p = 0.072$), *SOD1* ($p = 0.091$), and *GCLM* ($p = 0.094$) in the LPS + FC group tended to be up-regulated compared to the LPS group. In addition, FC supplementation up-regulated the mRNA expressions of *Nrf2*, *SOD1*, and *GCLM* and down-regulated the mRNA expression of *Keap1* compared to the CON group ($p < 0.05$). In addition, all treatment groups had no significant effects on liver *HO1* and *NQO1* gene expression ($p > 0.05$).

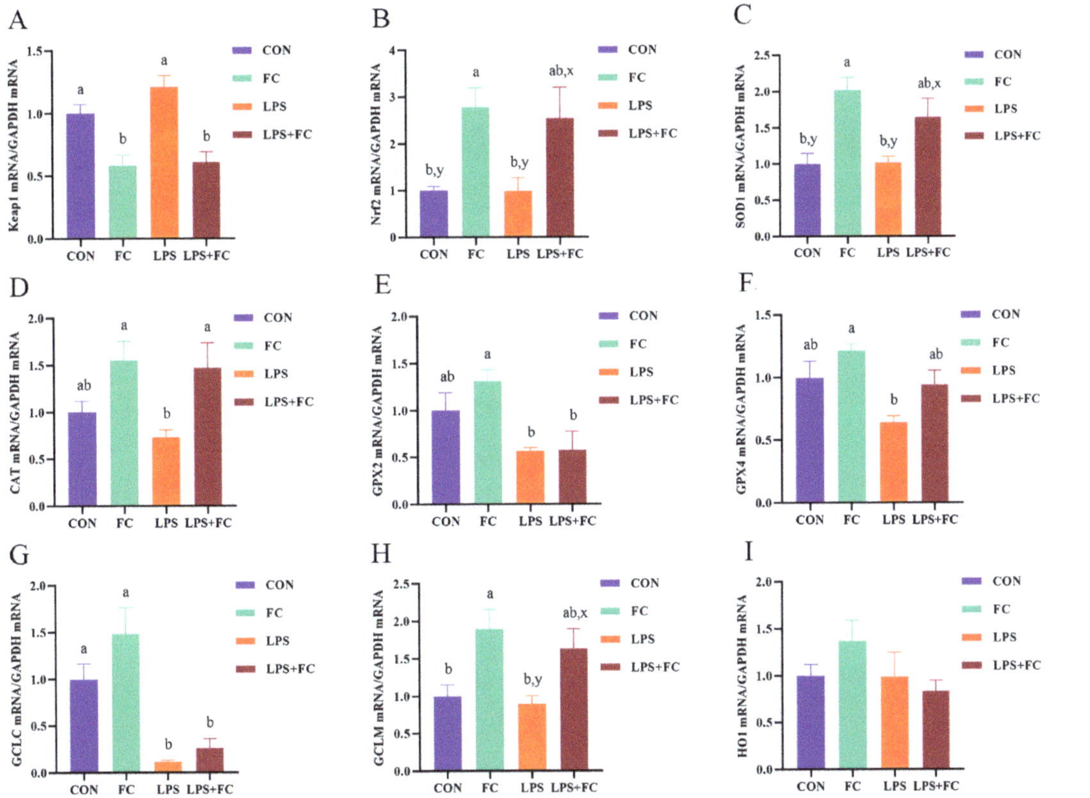

Figure 5. The effect of FC regulation and LPS challenge on liver antioxidant gene mRNA expression in low-weaning weight piglets. (**A–I**) Relative mRNA expression of *Keap1*, *Nrf2*, *SOD1*, *CAT*, *GPX2*, *GPX4*, *GCLC*, *GCLM*, and *HO-1*. Data were expressed as mean with their standard errors represented by vertical bars, ($n = 6$). CON: basal diet, FC: basal diet + 300 mg/kg FC. ^{a,b} Means listed in the same row with different superscripts are significantly different ($p < 0.05$). ^{x,y} Means listed in the same row with different superscripts showed a tendency to be different ($0.05 \leq p < 0.10$).

3.6. Hepatic Mitochondrial Genes mRNA Expression

The effects of FC supplementation on hepatic mitochondrial gene mRNA expression in low-weaning weight piglets under LPS challenge are shown in Figure 6A–J. Compared with the CON group, LPS challenge down-regulated the expression of the mitochondrial fusion gene *MFN2* and division gene *FIS1* mRNA ($p < 0.05$) and tended to the decrease the division gene DRP1 ($p = 0.051$) in the liver of low-weaning weight piglets. Conversely, supplementation with FC partially improved mitochondrial biogenesis gene expression in the liver (Figure 6A–E). Compared with the LPS group, administering dietary FC to LPS-challenged piglets up-regulated the expression of the mitochondrial fusion gene *MFN2* mRNA in the liver ($p < 0.05$). In addition, the analysis of the expression of genes related to mitochondrial respiratory chain membrane proteins (Figure 6F–J) showed that supplementation with FC significantly up-regulated the expression of *SDHA* and *UQCRB* genes in the liver of weaned piglets stimulated by LPS ($p < 0.05$). The Spearman correlation analysis found significant correlations among *Keap1/Nrf2* signaling pathway genes, antioxidant enzyme-related genes, and mitochondrial respiratory chain membrane protein-related genes (Figure 6K).

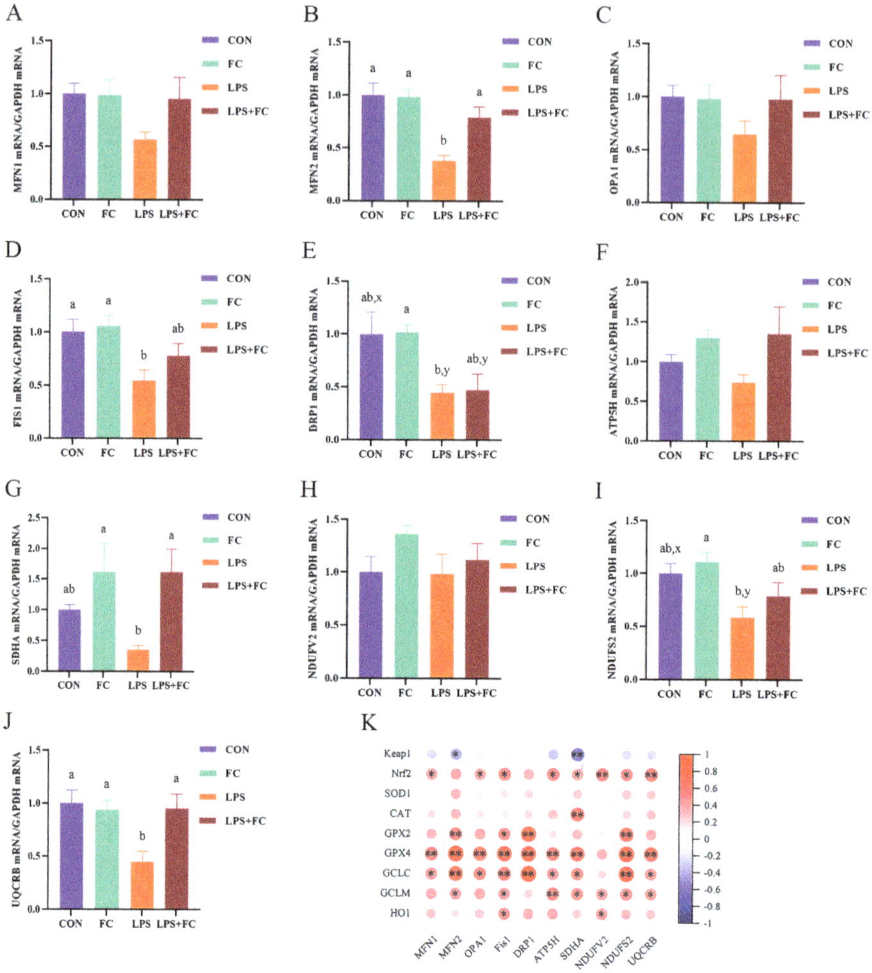

Figure 6. The effect of FC regulation and LPS challenge on liver mitochondrial biogenesis and respiratory chain membrane protein-related gene mRNA expression in low-weaning weight piglets. (**A–J**) Relative mRNA expression of *MFN1*, *MFN2*, *OPA1*, *FIS1*, *DRP1*, *ATP5H*, *SDHA*, *NDUFV2*, *NDUFS2*, and *UQCRB*. (**K**) The heatmap of Spearman's correlation between the expression of mitochondrial function-related genes and *Keap1/Nrf2* signaling pathway genes. Data were expressed as mean with their standard errors represented by vertical bars, (n = 6). CON: basal diet, FC: basal diet + 300 mg/kg FC. ^{a,b} Means listed in the same row with different superscripts are significantly different ($p < 0.05$). ^{x,y} Means listed in the same row with different superscripts showed a tendency to be different ($0.05 \leq p < 0.10$). * $p < 0.05$, ** $p < 0.01$. (**A**) MFN1 = mitofusin 1, (**B**) MFN2 = mitofusin 2, (**C**) OPA1 = optic atrophy 1, (**D**) FIS1 = mitochondrial fission protein 1, (**E**) DRP1 = dynamin-related protein 1, (**F**) ATP5H = ATP synthase, H+ transporting, mitochondrial F0 complex, subunit, (**G**) SDHA = succinate dehydrogenase complex flavoprotein subunit A, (**H**) NDUFV2 = NADH ubiquinone oxidoreductase core subunit V2, (**I**) NDUFS2 = NADH ubiquinone oxidoreductase core subunit S2, and (**J**) UQCRB = ubiquinol cytochrome c reductase binding protein. (**K**) Keap1 = kelch-like ECH-associated protein 1, Nrf2 = nuclear factor-erythroid 2-related factor 2, SOD1 = superoxide dismutase 1, CAT = catalase, GPX = glutathione peroxidase, GCLC = glutamate-cysteine ligase catalytic subunit, GCLM = glutamate-cysteine ligase modifier subunit, HO1 = heme oxygenase 1.

4. Discussion

Weaning usually induces an increase of ROS in piglets, particularly in low-weaning weight piglets; consequently, weaning can disrupt the balance of oxidation reduction, reduce antioxidant capacity, cause oxidative stress and oxidative damage in the tissues and intestine, and potentially lead to diarrhea, inhibited growth, and even death [5,7,27–29]. It is well understood that piglets with a lower weaning weight are more susceptible to post-weaning challenges than pigs with a heavier weaning weight [5,30]. Therefore, piglets with low-weaning weight need antioxidants with nutritional regulation functions to improve the antioxidant capacity of weaned piglets, reduce diarrhea, and promote growth post weaning.

Our results in experiment one demonstrated that fucoidan (FC) supplementation could linearly improve the growth performance of low-weaning weight piglets and the optimal dose of FC was 300 mg/kg. The findings of this study were in accordance with previous research, which found that supplementations with FC could increase feed conversion and improve the growth performance of weaned piglets [31]. Moreover, the supplementation of various forms and doses of FC can enhance the growth rate and health status of young chicks [32], weaned kids [33], fish [34], and Penaeus monodon [35]. Previous studies demonstrated that FC could reduce the diarrhea incidence of weaned piglets [36,37]. In this study, FC supplementation at varying concentrations in low-weaning weight piglets revealed a significant reduction in the incidence of diarrhea at a concentration of 300 mg/kg. Walsh et al. [37] reported that adding 240 mg/kg FC to the diet of weaned piglets can reduce the incidence of diarrhea. On the other hand, Rattigan et al. [38] found that adding 250 mg/kg FC effectively improved the fecal consistency of weaned piglets. Based on these results, we speculated that the source and processing techniques of FC might determine its optimal dosage and method of use. These findings provide us with valuable references, helping us to better understand and utilize FC.

The free radical metabolism and antioxidant systems of piglets may undergo disruption after weaning, thereby instigating oxidative stress responses, which have the potential to interfere with the host's antioxidant system and disrupt the cellular redox equilibrium [7,8,38,39]. The antioxidant defense system, which primarily comprises antioxidant enzymes such as SOD, CAT, and GSH-Px, along with other non-enzymatic antioxidants, can eliminate ROS [40,41]. MDA, the end product of lipid peroxidation, is frequently used as a marker for oxidative damage [42]. The role of functional nutritional supplements in improving the antioxidant capacity of piglets is widely recognized [43,44]; however, the effectiveness would vary due to differences in the type, source, and dosage of supplements. Our previous studies showed that supplementation with yeast hydrolysate from *Kluyveromyces fragilis* at 10 g/kg could significantly increase the activity of SOD and reduce the concentration of MDA in the plasma of weaned piglets, thereby enhancing the antioxidant capacity [7]. Furthermore, our studies revealed that a supplementation with 400 mg/kg of silybin could efficaciously mitigate the redox imbalance in weaned piglets and counteract the growth retardation induced by paraquat [8]. Our results in experiment one showed that supplementation with 300 and 600 mg/kg FC could enhance the activity of SOD and decrease the concentration of MDA in the plasma of piglets with a low-weaning weight. Consistent with previous studies, the use of FC or brown algae extracts to improve the antioxidant capacity of animals and inhibit oxidative stress has been verified in a variety of animal models [21,22,32,33,35]. Moreover, our study observed that the supplementation of FC at a dosage of 300 mg/kg notably elevated the concentrations of IGF-1 in the plasma of piglets with a low-weaning weight; the increase in IGF-1 concentrations exhibited a positive correlation with the piglets' growth performance [45], thereby substantiating the growth-enhancing impact of FC supplementation. This is a finding that further corroborates the beneficial role of FC in promoting growth.

Although our preliminary investigation suggested that FC has the potential to augment the antioxidant and growth-enhancing capabilities of low-weaning weight piglets, this is far from sufficient to fully elucidate the mechanism of action of FC on oxidative stress in piglets. Therefore, we further designed an experiment and established an oxidative stress model using LPS-challenged low-weaning weight piglets to further study the alleviating effect of FC on oxidative damage. LPS has been substantiated as an effective inducer of oxidative stress in experimental animals, thereby validating its use in the establishment of oxidative stress models [25,46]. Our research findings indicate that LPS challenge triggers a pronounced systemic redox imbalance in low-weaning weight piglets, as evidenced by a decrease in SOD activity and an increase in MDA concentration in the plasma and liver. The challenge model in our study was similar to previous results, verifying the successful establishment of our experimental animal model [25,46,47]. The liver, being the primary organ for metabolism and detoxification, plays a crucial role in defending against severe infections and exogenous stimuli, and in tissue repair [48]. *Keap1* acts as a negative regulator of *Nrf2*, inhibiting its nuclear translocation. The *Nrf2* signaling pathway serves as the central regulator of oxidative stress responses within biological organisms, controlling the expression of various antioxidant response-related genes or enzymes [17]. When cells are damaged, *Nrf2* is up-regulated and activates phase II enzymes, enhancing the cells' tolerance to oxidative stress [49]. Therefore, the *Keap1* regulation of *Nrf2* plays a crucial role in cellular resistance to oxidative stress and the maintenance of cellular homeostasis. In experiment two, our observations revealed that the supplementation with FC could down-regulate the expression of the *Keap1* gene, up-regulate the activation of the *Nrf2* pathway, and stimulate the expression of *CAT* antioxidant-related genes. This regulatory mechanism effectively mitigates oxidative stress, thereby offering substantial protection against liver damage induced by LPS to low-weaning weight piglets. This dual regulatory effect further underscores the potential of FC supplementation as a potent modulator of gene expression in the context of antioxidant defense mechanisms.

The mitochondrion, an essential organelle, is instrumental in controlling redox processes and lipid metabolism within liver cells [50]. Fission and fusion are crucial for maintaining mitochondrial balance by isolating and eliminating impaired components. When mitochondria malfunction, this may hinder fusion or stimulate fission to stop the integration of damaged parts into the healthy mitochondrial system [51]. The genes *FIS1* and *DRP1* play a regulatory role in mitochondrial division, while *MFN1*, *MFN2*, and *OPA1* primarily oversee the regulation of mitochondrial fusion [52]. Our research findings indicate that LPS challenges mitochondrial function by down-regulating the mRNA expression of *MFN2*, *FIS1*, and *DRP1*. Dietary supplementation with FC can partially reverse these adverse effects. Specifically, FC can alleviate LPS challenge-induced mitochondrial damage in the liver of low-weaning weight piglets by regulating the mitochondrial *MFN2* gene in the liver. Previous studies demonstrated that FC could prevent mitochondrial functional damage [22,53]. The mitochondrial oxidative phosphorylation system, which is crucial to cellular metabolism, consists of five enzyme complexes, including NADH dehydrogenase (Complex I), succinate dehydrogenase (Complex II), ubiquinol cytochrome c oxidoreductase (Complex III), cyanide sensitive oxidase (Complex IV), and ATP synthase (Complex V) [54,55]. Our findings demonstrated that the addition of FC counteracted the reduction in the expression of *SDHA* (Complex IV) and *UQCRB* (Complex II) genes in the liver of LPS-challenged low-weaning weight piglets. Thus, our results suggest that FC may enhance the activity of the mitochondrial oxidative phosphorylation system, preventing mitochondrial function damage.

5. Conclusions

The supplementation with an optimal dose of FC (300 mg/kg) exhibits benefits in the antioxidant capacity and growth performance of low-weaning weight piglets. The antioxidant function of FC might be attributed to the inhibited *Keap1* expression, controlled nuclear migration of *Nrf2*, enhanced CAT activity, various antioxidant enzymes, and improved mitochondrial function. Further, the antioxidant function may have eventually played a protective role against liver oxidative stress damage. Thus, the optimal dose of FC used in this study could provide a theoretical reference for the application of FC as a novel and natural antioxidant in swine production.

Author Contributions: Conceptualization, C.W. and X.J.; methodology, C.W. and X.J.; validation, X.L. and X.J.; formal analysis, C.Y. and Q.B.; investigation, C.Y. and W.C.; data curation, C.Y. and Y.L.; writing—original draft preparation, C.Y.; writing—review and editing, B.C., V.B. and X.J.; supervision, W.S. and Y.P.; project administration, X.J.; funding acquisition, C.W. and X.J. All authors have read and agreed to the published version of the manuscript.

Funding: This research was funded by the National Natural Science Foundation of China (32260850), the China Postdoctoral Science Foundation (2023M730594), and the Agricultural Science and Technology Innovation Program of the Chinese Academy of Agricultural Sciences (CAAS-ZDRW202305).

Institutional Review Board Statement: The animal study protocol was approved by the Institutional Review Board (or Ethics Committee) of the Animal Care and Use Committee of the Institute of Feed Research of the Chinese Academy of Agricultural Sciences (IFR-CAAS20221010 and IFR-CAAS20230825).

Data Availability Statement: All data is included in the article.

Conflicts of Interest: The authors declare no conflicts of interest.

Appendix A

Table A1. The composition of basal diet and nutrient levels in experiment one (as-fed basis, %).

Items	Experiment Treatments			
	FC 0	FC 150	FC 300	FC 600
Ingredients				
Corn	16.45	16.45	16.45	16.45
Extruded corn	32.00	32.00	32.00	32.00
Soybean meal, 46%CP	14.00	14.00	14.00	14.00
Extruded soybean	11.50	11.50	11.50	11.50
Fish meal	5.60	5.60	5.60	5.60
Whey	15.00	15.00	15.00	15.00
Soybean oil	1.00	1.00	1.00	1.00
Dicalcium phosphate	0.40	0.40	0.40	0.40
Limestone ($CaCO_3$)	0.75	0.75	0.75	0.75
Salt	0.30	0.30	0.30	0.30
Choline chloride (60%)	0.05	0.05	0.05	0.05
L-Lysine HCl	1.20	1.20	1.20	1.20
DL-Methionine	0.09	0.09	0.09	0.09
Threonine	0.27	0.27	0.27	0.27
Tryptophan	0.02	0.02	0.02	0.02
Phytase	0.02	0.02	0.02	0.02
Acidifier	0.35	0.35	0.35	0.35
Zinc oxide	0.20	0.20	0.20	0.20
Vitamin and mineral premix [1]	0.80	0.80	0.80	0.80
Total	100.00	100.00	100.00	100.00
Nutrition composition (Analyzed value)				
GE, KJ/g	16.84	16.80	16.85	16.84
Crude protein	19.45	19.48	19.45	19.46
Calcium	0.77	0.76	0.78	0.77
Phosphorus	0.65	0.66	0.64	0.65
Ether extract	4.22	4.21	4.21	4.22
Crude Ash	6.21	6.19	6.18	6.21
Nutrition composition (Calculated value)				
ME, MJ/kg	14.23	14.23	14.23	14.23
Lysine	1.30	1.30	1.30	1.30
Methionine	0.38	0.38	0.38	0.38
Threonine	0.76	0.76	0.76	0.76
Tryptophan	0.21	0.21	0.21	0.21

[1] Premix supplied per kg of diet: niacin, 38.4 mg; calcium pantothenate, 25 mg; folic acid, 1.68 mg; biotin, 0.16 mg; vitamin A, 35.2 mg; vitamin B_1, 4 mg; vitamin B_2, 12 mg; vitamin B_6, 8.32 mg; vitamin B_{12}, 4.8 mg; vitamin D_3, 7.68 mg; vitamin E, 128 mg; vitamin K_3, 8.16 mg; zinc ($ZnSO_4 \cdot H_2O$), 110 mg; copper ($CuSO_4 \cdot 5H_2O$), 125 mg; selenium (Na_2SeO_3), 0.19 mg; iron ($FeSO_4 \cdot H_2O$), 171 mg; cobalt ($CoCl_2$), 0.19 mg; manganese ($MnSO_4 \cdot H_2O$), 42.31 mg; iodine ($Ca(IO_3)_2$), 0.54 mg.

Table A2. The composition of basal diet and nutrient levels in experiment two (as-fed basis, %).

Items	Experiment Treatments			
	CON	FC	LPS	FC + LPS
Ingredients				
Corn	16.45	16.45	16.45	16.45
Extruded corn	32.00	32.00	32.00	32.00
Soybean meal, 46%CP	14.00	14.00	14.00	14.00
Extruded soybean	11.50	11.50	11.50	11.50
Fish meal	5.60	5.60	5.60	5.60
Whey	15.00	15.00	15.00	15.00
Soybean oil	1.00	1.00	1.00	1.00
Dicalcium phosphate	0.40	0.40	0.40	0.40
Limestone ($CaCO_3$)	0.75	0.75	0.75	0.75
Salt	0.30	0.30	0.30	0.30
Choline chloride (60%)	0.05	0.05	0.05	0.05
L-Lysine HCl	1.20	1.20	1.20	1.20
DL-Methionine	0.09	0.09	0.09	0.09
Threonine	0.27	0.27	0.27	0.27
Tryptophan	0.02	0.02	0.02	0.02
Phytase	0.02	0.02	0.02	0.02
Acidifier	0.35	0.35	0.35	0.35
Zinc oxide	0.20	0.20	0.20	0.20
Vitamin and mineral premix [1]	0.80	0.80	0.80	0.80
Total	100.00	100.00	100.00	100.00
Nutrition composition (Analyzed value)				
GE, KJ/g	17.00	16.89	17.00	16.89
Crude protein	19.29	19.31	19.29	19.31
Calcium	0.79	0.78	0.79	0.78
Phosphorus	0.67	0.65	0.67	0.65
Ether extract	4.32	4.29	4.32	4.29
Crude Ash	6.23	6.13	6.23	6.13
Nutrition composition (Calculated value)				
ME, MJ/kg	14.23	14.23	14.23	14.23
Lysine	1.30	1.30	1.30	1.30
Methionine	0.38	0.38	0.38	0.38
Threonine	0.76	0.76	0.76	0.76
Tryptophan	0.21	0.21	0.21	0.21

[1] Premix supplied per kg of diet: niacin, 38.4 mg; calcium pantothenate, 25 mg; folic acid, 1.68 mg; biotin, 0.16 mg; vitamin A, 35.2 mg; vitamin B_1, 4 mg; vitamin B_2, 12 mg; vitamin B_6, 8.32 mg; vitamin B_{12}, 4.8 mg; vitamin D_3, 7.68 mg; vitamin E, 128 mg; vitamin K_3, 8.16 mg; zinc ($ZnSO_4 \cdot H_2O$), 110 mg; copper ($CuSO_4 \cdot 5H_2O$), 125 mg; selenium (Na_2SeO_3), 0.19 mg; iron ($FeSO_4 \cdot H_2O$), 171 mg; cobalt ($CoCl_2$), 0.19 mg; manganese ($MnSO_4 \cdot H_2O$), 42.31 mg; iodine ($Ca(IO_3)_2$), 0.54 mg.

Appendix B

Table A3. Primer sequences used for RT-qPCR.

Gene	Primer Sequence (5′→3′)	Product Length, bp	Accession No.
ATP5H	F: CATTGACTGGGTAGCCTTTG R: CTTCTCAGGTAGAGCAGCCA	115	XM_021066093.1
CAT	F: CCTGCAACGTTCTGTAAGGC R: GCTTCATCTGGTCACTGGCT	72	NM_214301.2
DRP1	F:GTAAACCGAAGCCAGAAGGACA R: CAAGTGGCGATAGGAAGGGTGG	102	XM_021069575.1
FIS1	F:CCAAAGGGAGCAAAGAGGAGCA R: CCTGGTTGTTCTGTGGCTCTGT	132	XM_021086263.1
GAPDH	F: GCTTGTCATCAATGGAAAGG R: CATACGTAGCACCAGCATCA	86	NM_001206359.1
GCLC	F: GGAGAGGGGAGAAAGTTGTC R: GCCTTCGCTGCTTCATCATC	103	XM_021098556.1
GCLM	F: GCTTCGAGACTGTATCCAAA R: CTTTCATCGGGATTTATTTT	132	XM_001926378.4
GPX2	F: TCTCCAGTGTGTCGCAATGA R: TCGATGGTCAGAAAGCGACG	104	NM_214201.1
GPX4	F: GATTCTGGCCTTCCCTTGC R: TCCCCTTGGGCTGGACTTT	173	NM_214407.1
HO1	F: GAGAAGGCTTTAAGCTGGTG R: GTTGTGCTCAATCTCCTCCT	74	NM_001004027.1
Keap1	F: AGCTGGGATGCCTCAGTGTT R: AGGCAAGTTCTCCCAGACATTC	100	NM_001114671.1

Table A3. *Cont.*

Gene	Primer Sequence (5′→3′)	Product Length, bp	Accession No.
MFN1	F:CAATAGAAGAGAGGGAAGACC R: TATTTGCCACCTCCTCTGTAA	117	NM_001315732.1
MFN2	F:AGAGGAGAAGAGGAGCGTCAAGA R: ACATCACACTCACCAGGCTGC	95	XM_021095370.1
NDUFS2	F: CTAAACGCGCAGAGATGAAGA R: CCTCAATGGCAGTGTATGTGG	108	XM_005663166.3
NDUFV2	F:CCCAGATACTCCATTTGATTTCA R: AATTTCTGCCACCTTGTTCATG	169	NM_001097475.2
Nrf2	F: GAGAAGGCTTTAAGCTGGTG R: GTTGTGCTCAATCTCCTCCT	103	XM_005671981.3
OPA1	F:CAGAGGATGGTGCTTGTTGAC R: AGTATGATGGCGTTGGGATTC	128	XM_021070065.1
SDHA	F:TCTCTGAGGCCGGGTTTAACACA R: CACCTCCAGTTGTCCTCCTCCAT	124	XM_021076930.1
SOD1	F: GAAGACAGTGTTAGTAACGG R: CAGCCTTGTGTATTATCTCC	93	NM_001190422.1
UQCRB	F: GGATGACGATGTAAAAGAAGCCA R: TCCTCCTCATATTTTGTCCACTG	141	NM_001185172.1

ATP5H: ATP synthase, H+ transporting, mitochondrial Fo complex, subunit, *CAT*: catalase, *DRP1*: dynamin related protein 1, *FIS1*: mitochondrial fission protein 1, *GAPDH*: glyceraldehyde-3-phosphate dehydrogenase, *GCLC*: glutamate-cysteine ligase catalytic subunit, *GCLM*: glutamate-cysteine ligase modifier subunit, *GPX2*: glutathione peroxidase 2, *GPX4*: glutathione peroxidase 4, *HO1*: heme oxygenase 1, *Keap1*: kelch-like ech-associated protein 1, *MFN1*: mitofusin 1, *MFN2*: mitofusin 2, *NDUFS2*: NADH ubiquinone oxidoreductase core subunit S2, *NDUFV2*: NADH ubiquinone oxidoreductase core subunit V2, *OPA1*: optic atrophy 1, *SDHA*: succinate dehydrogenase complex flavoprotein subunit A, *Nrf2*: nuclear factor erythroid2-related factor 2, *SOD1*: superoxide dismutase 1, *UQCRB*: ubiquinol cytochrome c reductase binding protein.

References

1. Wang, L.; Zhang, S.; Johnston, L.; Levesque, C.; Yin, J.; Dong, B. A systematic review and meta-analysis of dietary fat effects on reproductive performance of sows and growth performance of piglets. *J. Anim. Sci. Biotechnol.* **2022**, *13*, 12. [CrossRef] [PubMed]
2. Mao, X.; Peng, J.; Cui, C.; Gu, Q.; Zhou, L.; Chao, W.; Sun, H.; Peng, J.; Wei, H. Effect of gestation dietary methionine-to-lysine ratio on methionine metabolism and antioxidant ability of high-prolific sows. *Anim. Nutr.* **2021**, *7*, 849–858.
3. Craig, A.; Muns, R.; Gordon, A.; Magowan, E. Extended nursing and/or increased starter diet allowances for low weaning weight pigs. *Asian-Australas. J. Anim. Sci.* **2020**, *33*, 1301–1309. [CrossRef] [PubMed]
4. Magowan, E.; Ball, M.; McCracken, K.; Beattie, V.; Bradford, R.; Robinson, M.; Scott, M.; Gordon, F.; Mayne, C. The performance response of pigs of different wean weights to 'high' or 'low' input dietary regimes between weaning and 20 weeks of age. *Livest. Sci.* **2011**, *136*, 232–239. [CrossRef]
5. Collins, C.; Pluske, J.; Morrison, R.; McDonald, T.; Smits, R.; Henman, D.; Stensland, I.; Dunshea, F. Post-weaning and whole-of-life performance of pigs is determined by live weight at weaning and the complexity of the diet fed after weaning. *Anim. Nutr.* **2017**, *3*, 372–379. [CrossRef]
6. St-Pierre, B.; Palencia, J.; Samuel, R. Impact of early weaning on development of the swine gut microbiome. *Microorganisms* **2023**, *11*, 1753. [CrossRef]
7. Yin, C.; Comi, M.; Cai, L.; Chen, W.; Perricone, V.; Xiao, J.; Agazzi, A.; Li, X.; Jiang, X. Hydrolysed yeast from *Kluyveromyces fragilis* improves plasma antioxidant efficiency and immunoglobulin concentration, and faecal microbiota of weaned piglets. *Ital. J. Anim. Sci.* **2023**, *22*, 578–588. [CrossRef]
8. Cai, L.; Gao, G.; Yin, C.; Bai, R.; Li, Y.; Sun, W.; Pi, Y.; Jiang, X.; Li, X. The effects of dietary silybin supplementation on the growth performance and regulation of intestinal oxidative injury and microflora dysbiosis in weaned piglets. *Antioxidants* **2023**, *12*, 1975. [CrossRef] [PubMed]
9. Li, Y.; Jiang, X.; Cai, L.; Zhang, Y.; Ding, H.; Yin, J.; Li, X. Effects of daidzein on antioxidant capacity in weaned pigs and IPEC-J2 cells. *Anim. Nutr.* **2022**, *11*, 48–59. [CrossRef]
10. Novais, A.; Deschene, K.; Martel-Kennes, Y.; Roy, C.; Laforest, J.; Lessard, M.; Matte, J.; Lapointe, J. Weaning differentially affects mitochondrial function, oxidative stress, inflammation and apoptosis in normal and low birth weight piglets. *PLoS ONE* **2021**, *16*, e0247188. [CrossRef]
11. Yu, L.; Li, H.; Peng, Z.; Ge, Y.; Liu, J.; Wang, T.; Wang, H.; Dong, L. Early weaning affects liver antioxidant function in piglets. *Animals* **2021**, *11*, 2679. [CrossRef] [PubMed]
12. Luo, Z.; Zhu, W.; Guo, Q.; Luo, W.; Zhang, J.; Xu, W.; Xu, J. Weaning induced hepatic oxidative stress, apoptosis, and aminotransferases through MAPK signaling pathways in piglets. *Oxid. Med. Cell. Longev.* **2016**, *2016*, 4768541. [CrossRef] [PubMed]
13. Guerbette, T.; Boudry, G.; Lan, A. Mitochondrial function in intestinal epithelium homeostasis and modulation in diet-induced obesity. *Mol. Metab.* **2022**, *63*, 101546. [CrossRef] [PubMed]

14. Valera-Alberni, M.; Canto, C. Mitochondrial stress management: A dynamic journey. *Cell Stress* **2018**, *2*, 253–274. [CrossRef] [PubMed]
15. Zheng, J.; Yuan, S.; Wu, C.; Lv, Z. Acute exposure to waterborne cadmium induced oxidative stress and immunotoxicity in the brain, ovary and liver of zebrafish. *Aquat. Toxicol.* **2016**, *180*, 36–44. [CrossRef]
16. Cai, L.; Ming, D.; Chen, W.; Zhao, Y.; Li, Y.; Sun, W.; Pi, Y.; Jiang, X.; Li, X. Silybin alleviated hepatic injury by regulating redox balance, inflammatory response, and mitochondrial function in weaned piglets under paraquat-induced oxidative stress. *Antioxidants* **2024**, *13*, 324. [CrossRef]
17. Zhou, W.; He, H.; Wei, Q.; Che, L.; Zhao, X.; Liu, W.; Yan, Y.; Hu, L.; Du, Y.; Yin, Z.; et al. Puerarin protects against acetaminophen-induced oxidative damage in liver through activation of the *Keap1/Nrf2* signaling pathway. *Food Sci. Nutr.* **2023**, *11*, 6604–6615. [CrossRef]
18. Kobayashi, M.; Yamamoto, M. *Nrf2-Keap1* regulation of cellular defense mechanisms against electrophiles and reactive oxygen species. *Adv. Enzym. Regul.* **2006**, *46*, 113–140. [CrossRef] [PubMed]
19. Wang, Y.; Liu, B.; Wu, P.; Chu, Y.; Gui, S.; Zheng, Y.; Chen, X. Dietary selenium alleviated mouse liver oxidative stress and NAFLD induced by obesity by regulating the Keap1/Nrf2 Pathway. *Antioxidants* **2022**, *11*, 349. [CrossRef]
20. Zhao, Y.; Zheng, Y.; Wang, J.; Ma, S.; Yu, Y.; White, W.; Yang, S.; Yang, F.; Lu, J. Fucoidan extracted from undaria pinnatifida: Source for nutraceuticals/functional foods. *Mar. Drugs* **2018**, *16*, 321. [CrossRef]
21. Wang, L.; Jayawardena, T.; Hyun, J.; Wang, K.; Fu, X.; Xu, J.; Gao, X.; Park, Y.; Jeon, Y. Antioxidant and anti-photoaging effects of a fucoidan isolated from *Turbinaria ornata*. *Int. J. Biol. Macromol.* **2023**, *225*, 1021–1027. [CrossRef]
22. Ji, Y.; Jin, D.; Qi, J.; Wang, X.; Zhang, C.; An, P.; Luo, Y.; Luo, J. Fucoidan protects against doxorubicin-induced cardiotoxicity by reducing oxidative stress and preventing mitochondrial function injury. *Int. J. Mol. Sci.* **2022**, *23*, 10685. [CrossRef] [PubMed]
23. Fang, L.; Sun, H.; Yang, L.; He, D.; Ren, C.; Zhu, C.; Lv, G. Effects of fucoidan on growth performance, immunity, antioxidant ability, digestive enzyme activity, and hepatic morphology in juvenile common carp (*Cyprinus carpio*). *Front. Mar. Sci.* **2023**, *10*, 1167400.
24. Van Nieuwamerongen, S.; Bolhuis, J.; Van, P.; Kemp, B.; Soede, N. Effects of pre-weaning housing in a multi-suckling system on performance and carbohydrate absorption of relatively light and heavy piglets around weaning. *Animal* **2018**, *12*, 802–809. [CrossRef]
25. Chen, F.; Wang, H.; Chen, J.; Liu, Y.; Wen, W.; Li, H.; Huang, X. *Lactobacillus delbrueckii* ameliorates intestinal integrity and antioxidant ability in weaned piglets after a lipopolysaccharide challenge. *Oxid. Med. Cell. Longev.* **2020**, *2020*, 6028606. [CrossRef] [PubMed]
26. Jiang, X.; Awati, A.; Agazzi, A.; Vitari, F.; Ferrari, A.; Bento, H.; Crestani, M.; Domeneghini, C.; Bontempo, V. Effects of a blend of essential oils and an enzyme combination on nutrient digestibility, ileum histology and expression of inflammatory mediators in weaned piglets. *Animal* **2015**, *9*, 417–426. [CrossRef]
27. Tang, X.; Liu, X.; Fang, R. Role of epidermal growth factor in regulation of intestinal health of weaned piglets. *Chin. J. Anim. Nutr.* **2021**, *33*, 614–621.
28. Feng, Y.; An, Z.; Chen, H.; He, X.; Wang, W.; Zhang, H.; Li, F.; Liu, D. Ulva prolifera extract alleviates intestinal oxidative stress via *Nrf2* signaling in weaned piglets challenged with hydrogen peroxide. *Front. Immunol.* **2020**, *11*, 599735. [CrossRef] [PubMed]
29. Bergeron, N.; Robert, C.; Guay, F. Feed supplementation with arginine and zinc on antioxidant status and inflammatory response in challenged weanling piglets. *Anim. Nutr.* **2017**, *3*, 236–246. [CrossRef]
30. Larriestra, A.; Wattanaphansak, S.; Neumann, E.; Bradford, J.; Morrison, R.; Deen, J. Pig characteristics associated with mortality and light exit weight for the nursery phase. *Can. Vet. J.* **2006**, *47*, 560–566.
31. O'Shea, C.; McAlpine, P.; Sweeney, T.; Varley, P.; O'Dohert, J. Effect of the interaction of seaweed extracts containing laminarin and fucoidan with zinc oxide on the growth performance, digestibility and faecal characteristics of growing piglets. *Brit. J. Nutr.* **2014**, *111*, 798–807. [CrossRef] [PubMed]
32. Sweeney, T.; Meredith, H.; Vigors, S.; McDonnell, M.J.; Ryan, M.; Thornton, K.; O'Doherty, J. Extracts of laminarin and laminarin/fucoidan from the marine macroalgal species laminaria digitata improved growth rate and intestinal structure in young chicks, but does not influence *Campylobacter jejuni* colonisation. *Anim. Feed Sci. Technol.* **2017**, *232*, 71–79. [CrossRef]
33. Yang, W.; Chen, J.; Guo, G.; Wang, S.; Peng, S.; Gao, Z.; Zhao, Z.; Lan, R.; Yin, F. The effects of fucoidan dietary supplementation on growth performance, serum antioxidant capacity, immune function indices and intestinal morphology in weaned kids. *Animals* **2022**, *12*, 574. [CrossRef]
34. Tuller, J.; De, S.; Jerry, D. Dietary influence of fucoidan supplementation on growth of lates calcarifer. *Aquac. Res.* **2014**, *45*, 749–754. [CrossRef]
35. Sivagnanavelmurugan, M.; Thaddaeus, B.; Palavesam, A.; Immanuel, G. Dietary effect of *Sargassum wightii* fucoidan to enhance growth, prophenoloxidase gene expression of *Penaeus monodon* and immune resistance to *Vibrio parahaemolyticus*. *Fish Shellfish Immun.* **2014**, *39*, 439–449. [CrossRef]
36. Rattigan, R.; Sweeney, T.; Vigors, S.; Thornton, K.; Rajauria, G.; O'Doherty, J. The effect of increasing inclusion levels of a fucoidan-rich extract derived from *Ascophyllum nodosum* on growth performance and aspects of intestinal health of pigs post-weaning. *Mar. Drugs* **2019**, *17*, 680. [CrossRef] [PubMed]
37. Walsh, A.; Sweeney, T.; O'Shea, C.; Doyle, D.; O'Doherty, J. Effect of dietary laminarin and fucoidan on selected microbiota, intestinal morphology and immune status of the newly weaned pig. *Br. J. Nutr.* **2013**, *110*, 1630–1638. [CrossRef] [PubMed]

38. Tang, X.; Xiong, K.; Fang, R.; Li, M. Weaning stress and intestinal health of piglets: A review. *Front. Immunol.* **2022**, *13*, 1042778. [CrossRef] [PubMed]
39. Xu, X.; Wei, Y.; Hua, H.; Jing, X.; Zhu, H.; Xiao, K.; Zhao, J.; Liu, Y. Polyphenols sourced from *Ilex latifolia* thunb. relieve intestinal injury via modulating ferroptosis in weanling piglets under oxidative stress. *Antioxidants* **2022**, *11*, 966. [CrossRef]
40. Pardo, Z.; Fernandez-Figares, I.; Lachica, M.; Lara, L.; Nieto, R.; Seiquer, I. Impact of heat stress on meat quality and antioxidant markers in iberian pigs. *Antioxidants* **2021**, *10*, 1911. [CrossRef]
41. Kumar, S.; Wang, M.; Liu, Y.; Zhu, Z.; Fahad, S.; Qayyum, A.; Zhu, G. Vanadium stress alters sweet potato (*Ipomoea batatas* L.) growth, ros accumulation, antioxidant defense system, stomatal traits, and vanadium uptake. *Antioxidants* **2022**, *11*, 2407. [CrossRef] [PubMed]
42. Vega, C.; Reyes-Castro, L.; Rodriguez-Gonzalez, G.; Bautista, C.; Vazquez-Martinez, M.; Larrea, F.; Chamorro-Cevallos, G.; Nathanielsz, P.; Zambrano, E. Resveratrol partially prevents oxidative stress and metabolic dysfunction in pregnant rats fed a low protein diet and their offspring. *J. Physiol.* **2016**, *594*, 1483–1499. [CrossRef] [PubMed]
43. Yang, M.; Yin, Y.; Wang, F.; Bao, X.; Long, L.; Tan, B.; Yin, Y.; Chen, J. Effects of dietary rosemary extract supplementation on growth performance, nutrient digestibility, antioxidant capacity, intestinal morphology, and microbiota of weaning pigs. *J. Anim. Sci.* **2021**, *99*, skab237. [CrossRef] [PubMed]
44. Wang, Y.; Chen, X.; Huang, Z.; Chen, D.; Yu, B.; Yu, J.; Chen, H.; He, J.; Luo, Y.; Zheng, P. Dietary ferulic acid supplementation improves antioxidant capacity and lipid metabolism in weaned piglets. *Nutrients* **2020**, *12*, 3811. [CrossRef] [PubMed]
45. Han, T.; Bjrkman, S.; Soede, N.; Oliviero, C.; Peltoniemi, O. IGF-1 concentrations after weaning in young sows fed different pre-mating diets are positively associated with piglet mean birth weight at subsequent farrowing. *Animal* **2021**, *15*, 100029. [CrossRef] [PubMed]
46. Xia, B.; Meng, Q.; Feng, X.; Tang, X.; Jia, A.; Feng, J.; Zhang, H. Probing the molecular regulation of lipopolysaccharide stress in piglet liver by comparative proteomics analysis. *Electrophoresis* **2018**, *39*, 2321–2331. [CrossRef] [PubMed]
47. Zhang, Y.; Deng, Z.; He, M.; Pastor, J.; Tedo, G.; Liu, J.; Wang, H. Olive oil cake extract stabilizes the physiological condition of lipopolysaccharide-challenged piglets by reducing oxidative stress and inflammatory responses and modulating the ileal microbiome. *Food Funct.* **2021**, *12*, 10171–10183. [CrossRef] [PubMed]
48. Rishi, G.; Subramaniam, V. The liver in regulation of iron homeostasis. *Am. J. Physiol. Gastr. L.* **2017**, *313*, 157–165. [CrossRef]
49. Song, Z.; Cheng, K.; Zheng, X.; Ahmad, H.; Zhang, L.; Wang, T. Effects of dietary supplementation with enzymatically treated artemisia annua on growth performance, intestinal morphology, digestive enzyme activities, immunity, and antioxidant capacity of heat-stressed broilers. *Poult. Sci.* **2018**, *97*, 430–437. [CrossRef]
50. Ma, X.; McKeen, T.; Zhang, J.; Ding, W. Role and mechanisms of mitophagy in liver diseases. *Cells* **2020**, *9*, 837. [CrossRef]
51. Twig, G.; Elorza, A.; Molina, A.; Mohamed, H.; Wikstrom, J.; Walzer, G.; Stiles, L.; Haigh, S.; Katz, S.; Las, G.; et al. Fission and selective fusion govern mitochondrial segregation and elimination by autophagy. *Embo J.* **2008**, *27*, 433–446. [CrossRef] [PubMed]
52. Guan, L.; Che, Z.; Meng, X.; Yu, Y.; Li, M.; Yu, Z.; Shi, H.; Yang, D.; Yu, M. MCU up-regulation contributes to myocardial ischemia-reperfusion injury through *calpain/OPA-1*-mediated mitochondrial fusion/mitophagy inhibition. *Int. J. Implant. Dent.* **2019**, *5*, 7830–7843. [CrossRef] [PubMed]
53. Xing, M.; Li, G.; Liu, Y.; Yang, L.; Zhang, Y.; Zhang, Y.; Ding, J.; Lu, M.; Yu, G.; Hu, G. Fucoidan from *Fucus vesiculosus* prevents the loss of dopaminergic neurons by alleviating mitochondrial dysfunction through targeting ATP5F1a. *Carbohydr. Polym.* **2023**, *303*, 120470. [CrossRef] [PubMed]
54. Vercellino, I.; Sazanov, L. The assembly, regulation and function of the mitochondrial respiratory chain. *Nat. Rev. Mol. Cell Biol.* **2021**, *23*, 141–161. [CrossRef]
55. Protasoni, M.; Perez-Perez, R.; Lobo-Jarne, T.; Harbour, M.; Ding, S.; Penas, A.; Diaz, F.; Moraes, C.; Fearnley, I.; Zeviani, M.; et al. Respiratory supercomplexes act as a platform for complex III-mediated maturation of human mitochondrial complexes I and IV. *Embo J.* **2020**, *39*, e102817. [CrossRef]

antioxidants

Article

Silybin Alleviated Hepatic Injury by Regulating Redox Balance, Inflammatory Response, and Mitochondrial Function in Weaned Piglets under Paraquat-Induced Oxidative Stress

Long Cai [1,†], Dongxu Ming [1,†], Wenning Chen [1,†], Ying Zhao [1,2], Yanpin Li [1], Wenjuan Sun [1], Yu Pi [1], Xianren Jiang [1,*] and Xilong Li [1,*]

1 Key Laboratory of Feed Biotechnology of the Ministry of Agriculture and Rural Affairs, Institute of Feed Research, Chinese Academy of Agricultural Sciences, Beijing 100081, China; 82101211211@caas.cn (L.C.); mdx920825@163.com (D.M.); 82101215363@caas.cn (W.C.); ying.zhao@student.uliege.be (Y.Z.); liyanpin@caas.cn (Y.L.); sunwenjuan@caas.cn (W.S.); piyu@caas.cn (Y.P.)

2 Precision Livestock and Nutrition Unit, TERRA Teaching and Research Centre, Gembloux Agro-Bio Tech University of Liege, 5030 Gembloux, Belgium

* Correspondence: jiangxianren@caas.cn (X.J.); lixilong@caas.cn (X.L.); Tel.: +86-010-8210-8134 (X.L.)

† These authors contributed equally to this work.

Abstract: Silybin (Si) is the main element of silymarin isolated from the seeds of *Silybum marianum* L. Gaernt., which has superior antioxidant properties. However, the protective role of Si in maintaining liver health under oxidative stress remains ambiguous. This study aimed to investigate the underlying mechanism of the beneficial effect of dietary Si against hepatic oxidative injury induced by paraquat (PQ) in weaned piglets. A total of 24 piglets were randomly allocated to four treatments with six replicates per treatment and 1 piglet per replicate: the control group; Si group; PQ group; and Si + PQ group. Piglets in the control group and PQ group were given a basal diet, while piglets in the Si and Si + PQ groups were given a Si-supplemented diet. On the 18th day, the pigs in the PQ treatment group received an intraperitoneal injection of PQ, and the others were intraperitoneally injected with the same volume of saline. All piglets were sacrificed on day 21 for plasma and liver sample collection. The results showed that dietary Si supplementation mitigated PQ-induced liver damage, as proven by the reduction in liver pathological changes and plasma activity of alanine transaminase and aspartate transaminase. Si also improved superoxide dismutase and glutathione peroxidase activities and total antioxidant capacity, as well as decreased malondialdehyde and hydrogen peroxide concentration in the liver, which were closely related to the activation of the nuclear factor-erythroid 2-related factor 2 signaling pathway. Meanwhile, Si reduced tumor necrosis factor-α and interleukin-8 production and their transcript levels as well as abrogated the overactivation of nuclear factor-κB induced by PQ. Importantly, Si improved mitochondrial function by maintaining mitochondrial energetics and mitochondrial dynamics, which was indicated by the elevated activity of mitochondrial complexes I and V and adenosine triphosphate content, decreased expression of dynamin 1 protein, and increased expression of mitofusin 2 protein. Moreover, Si inhibited excessive hepatic apoptosis by regulating the B-cell lymphoma-2 (Bcl-2)/Bcl-2-associated-X-protein signaling pathway. Taken together, these results indicated that Si potentially mitigated PQ-induced hepatic oxidative insults by improving antioxidant capacity and mitochondrial function and inhibiting inflammation and cell apoptosis in weaned piglets.

Keywords: silybin; hepatic damage; inflammation; antioxidation; mitochondrial function; apoptosis; weaned piglets

Citation: Cai, L.; Ming, D.; Chen, W.; Zhao, Y.; Li, Y.; Sun, W.; Pi, Y.; Jiang, X.; Li, X. Silybin Alleviated Hepatic Injury by Regulating Redox Balance, Inflammatory Response, and Mitochondrial Function in Weaned Piglets under Paraquat-Induced Oxidative Stress. *Antioxidants* **2024**, *13*, 324. https://doi.org/10.3390/antiox13030324

Academic Editors: Luciana Rossi and Matteo Dell'Anno

Received: 8 February 2024
Revised: 2 March 2024
Accepted: 5 March 2024
Published: 6 March 2024

1. Introduction

The liver is the principal organ responsible for metabolism and detoxification. It plays a crucial role in nutrient storage, xenobiotic metabolism, antioxidation, and immunoregulation within the body [1]. However, young animals are susceptible to liver injury induced by various stressors, including dietary and environmental stress, due to their physiological immaturity [2]. Oxidative stress is recognized as a vital pathological mechanism underlying the initiation and progression of liver injury [3]. Excessive reactive oxygen species (ROS) induce lipid peroxidation, protein oxidation, and DNA damage [4–6]. More importantly, free radicals interfere with the process of mitochondrial dynamics and bioenergetics, and then cause mitochondrial structure and function disorders [7], subsequently aggravating liver intracellular oxidative stress and triggering endogenous apoptotic pathways, ultimately leading to liver injury and dysfunction [8,9]. Therefore, it is necessary to develop effective nutrition intervention strategies to protect animals against liver damage injury from oxidative stress.

In recent years, significant attempts have been made to improve liver health through dietary antioxidants such as phytonutrients. Silybin, as a kind of flavonoid lignan, is the main active constituent of silymarin extracted from milk thistle, and plays an essential role in eliminating ROS and regulating antioxidant capacity [10]. In addition, silybin possesses a wide range of pharmacological activities, such as hepatoprotective, anti-inflammatory, and antibacterial activities [11,12], and improves mitochondrial function by regulating biogenesis and bioenergetics [13,14]. Dietary silymarin/silybin has been reported to improve growth performance and exert a promoting effect on systemic health in pigs and poultries [15–18]. Consistently, our previous research also found that dietary silybin decreased diarrhea incidence and improved growth performance in weaned piglets; moreover, silybin supplementation effectively alleviated intestinal injury by improving the antioxidant properties, mitochondrial function, and microbial community composition of weaned piglets [19]. However, the beneficial effects of silybin on liver health in weaned piglets under oxidative stress, and their underlying mechanisms, are still unknown.

Therefore, the purpose of this present study was to explore the potential of silybin in attenuating liver oxidative injury challenged with paraquat (PQ). PQ is a well-known redox imbalance inducer that can produce ROS continuously by interfering with the electron transmission on the mitochondrial respiratory chain, thereby inducing damage to different organs, including the liver, gastrointestinal tract, and other organs [20–23]. Thus, it is widely used to establish an oxidative stress model [24,25]. We measured the effect of dietary silybin on liver pathological changes, plasma biochemical parameters, hepatic enzyme activity, and inflammatory cytokine content. Subsequently, the activation of the Nrf2/Keap1 and NF-κB signaling pathways, mitochondrial function-related gene and protein expression, and initiation of hepatocytic apoptosis were further analyzed. Our findings can provide a novel nutritional intervention strategy for improving the liver health of weaned piglets and lay a theoretical foundation for the incorporation and utilization of silybin in animal feed.

2. Materials and Methods

2.1. Animal Ethics Approval

All animal procedures in this study complied with the ARRIVE guidelines and were approved by the Animal Care and Use Committee of the Institute of Feed Research of the Chinese Academy of Agricultural Sciences (IFR-CAAS20220428) [26].

2.2. Animals and Treatment

A total of 24 Duroc × (Landrace × Yorkshire) weanling piglets of similar age (28 ± 1 d) and initial body weight (7.68 ± 0.37 kg) were randomly allocated to 4 treatments with 6 replicates (pens) per treatment, ensuring an equal distribution of initial body weight and sex (half male and half female). The treatment included the Ctrl group (basal diet); Si group (basal diet supplemented with 400 mg/kg silybin); PQ group (basal diet); and Si + PQ group (basal diet supplemented with 400 mg/kg silybin). On the 18th day, the pigs in the PQ group and Si + PQ group received an intraperitoneal injection of 4 mg/kg BW paraquat (PQ) (methyl viologen hydrate, Huaxia Chemical Reagent Co., Ltd., Chengdu, China), and other groups received an intraperitoneal injection of saline. The dose of PQ to induce oxidative stress in the liver of the piglets was determined according to a previous report [25]. The administered level of silybin (purity > 97%, Panjin Tianyuan Pharmaceutical Co., Ltd., Panjin, China) in this study was determined based on a previous study [19]. The experiment was conducted at the Tianpeng experimental farm and lasted for a total of 21 days. The ingredient and nutrient levels of the basal diet met the nutritional requirements according to the National Research Council (2012) and have been described in our previous report [19,27]. Piglets were given ad libitum access to feed and fresh water in pens with slatted floors.

2.3. Sample Collection

At the end of the trial (day 21), approximately 6 mL blood samples were collected via the jugular veins of the piglets and centrifuged at 3000× *g* for 4 °C at 10 min to obtain plasma. Subsequently, all piglets were euthanized after being stunned by a portable electrical stunner (the output voltage was 220 V). The liver samples from the right medial lobe were collected and snap-frozen in liquid nitrogen and stored at −80 °C for further analysis. In addition, portions of the liver samples from the right medial lobe were fixed in fresh 4% paraformaldehyde for hematoxylin and eosin (H&E) staining analysis.

2.4. Plasma Biochemical Analysis

The activities of alanine transaminase (ALT), aspartate transaminase (AST), and alkaline phosphatase (ALP) in plasma were measured using a corresponding commercial kit (Jiangsu Meimian Industrial Co., Ltd., Yancheng, China) as described by the manufacturer's instructions.

2.5. Determination of Inflammatory Cytokine Content

The frozen liver was homogenized in chilled saline using an automatic homogenizer. Then, the homogenate was centrifuged at 3000× *g* for 15 min at 4 °C to obtain the supernatant, and the total protein concentration in the liver supernatant was subsequently detected. The content of tumor necrosis factor-α (TNF-α) and interleukin (IL)-6, IL-8, and IL-10 in the liver homogenate were measured using porcine-specific enzyme-linked immunoassay kits (Shanghai Enzyme-linked Biotechnology Co., Ltd., Shanghai, China) according to the manufacturer's protocols.

2.6. Determination of Enzyme Activity

Following the instructions of the protocols, the activities of superoxide dismutase (SOD), catalase (CAT), and glutathione peroxidase (GSH-Px); the total antioxidant capacity (T-AOC) level; and the content of malondialdehyde (MDA) and hydrogen peroxide (H_2O_2) in the liver supernatant were detected using kits purchased from Nanjing Jiancheng Biotechnology Co., Ltd. (Nanjing, China). The activities of mitochondrial complex I (COX I) and complex V (COX V) in the supernatant were measured using commercially available kits (Beijing Solarbio Science & Technology Co., Ltd., Beijing, China). After a similar homogenate step, the activity of caspase 3 and caspase 9 in the supernatant was determined using a specific kit purchased from Beyotime Biochem. Co., Ltd. (Shanghai, China). The assay steps were performed according to the instructions provided by the manufacturer.

2.7. Adenosine Triphosphate Content Assay

The liver homogenate was boiled on an electric hot plate for 10 min, cooled down on ice, and subsequently centrifuged at 3500× *g* for 10 min at 4 °C to obtain the supernatant. The content of adenosine triphosphate (ATP) in the supernatant was determined by phosphomolybdate colorimetry using a commercial kit (Nanjing Jiancheng Bioengineering Institute, Nanjing, China).

2.8. Histopathological Staining

The liver tissues were fixed with fresh 4% paraformaldehyde for more than 24 h, then embedded in paraffin after dehydration. The slices were stained with hematoxylin and eosin (H&E) for histopathological examination with an upright optical microscope (Niko, Tokyo, Japan). The histological score was assessed as described in a previous report [28].

2.9. Real-Time Quantitative PCR Analysis (RT-qPCR)

RNA extraction, reverse transcription, and qPCR of liver tissues were performed according to the procedure described in a previous report [29]. Briefly, total RNA was isolated using Trizol reagent (Thermo Fisher Scientific, Inc., Boston, MA, USA) according to the manufacturer's instructions. Then, 1 μg RNA was used to produce cDNA by reverse transcription using the PrimeScript RT reagent kit (Takara Biotechnology Co., Ltd., Dalian, China). The real-time quantitative PCR was conducted using SYBR Green reagent (Thermo Fisher Scientific, MA, USA) on a CFX96 Touch real-time PCR instrument (Bio-Rad Laboratories Inc., Berkeley, CA, USA). The mRNA expression of the target gene relative to the housekeeping gene (*GAPDH*) was calculated using the $2^{-\Delta\Delta CT}$ method. The primer sequences used in this study are listed in Table 1.

Table 1. The primers for quantitative real-time PCR.

Gene [1]	Accession Number	Primer Sequence (5′-3′)	Product Size (bp)
GAPDH	NM_001206359.1	F: GCTTGTCATCAATGGAAAGG R: CATACGTAGCACCAGCATCA	86
TNF-α	NM_214022.1	F: CTCACGTCCTTCTGGTTTAG R: CCCTGATTTCTAAGTGTTGC	96
IL-6	NM_214399.1	F: AATGTCGAGGCTGTGCAGATT R: TGGTGGCTTTGTCTGGATTCT	82
IL-8	NM_213867.1	F: CCGTGTCAACATGACTTCCAA R: GCCTCACAGAGAGCTGCAGAA	75
IL-10	NM_214041.1	F: GACGATGAAGATGAGGAAGA R: AGGTTTTTCTTTGGTTTCCC	54
CAT	NM_214301.2	F: CCTGCAACGTTCTGTAAGGC R: GCTTCATCTGGTCACTGGCT	72
SOD1	NM_001190422.1	F: GAAGACAGTGTTAGTAACGG R: CAGCCTTGTGTATTATCTCC	93
GPX1	NM_214201.1	F: TCTCCAGTGTGTCGCAATGA R: TCGATGGTCAGAAAGCGACG	104
GPX4	NM_214407.1	F: GATTCTGGCCTTCCCTTGC R: TCCCCTTGGGCTGGACTTT	173
NDUFS2	XM_005663166.3	F: CTAAACGCGCAGAGATGAAGA R: CCTCAATGGCAGTGTATGTGG	108
NDUFV2	NM_001097475.2	F: CCCAGATACTCCATTTGATTTCA R: AATTTCTGCCACCTTGTTCATG	169
SDHA	XM_021076930.1	F: TCTCTGAGGCCGGGTTTAACACA R: CACCTCCAGTTGTCCTCCTCCAT	124
UQCRB	NM_001185172.1	F: GGATGACGATGTAAAAGAAGCCA R: TCCTCCTCATATTTTGTCCACTG	141
ATP5H	XM_021066093.1	F: CATTGACTGGGTAGCCTTTG R: CTTCTCAGGTAGAGCAGCCA	115

[1] *GAPDH*, glyceraldehyde-3-phosphate dehydrogenase; *TNF-α*, tumor necrosis factor-α; *IL*, interleukin; *CAT*, catalase; *SOD*, superoxide dismutase; *GPX*, glutathione peroxidase; *NDUFS2*, NADH ubiquinone oxidoreductase core subunit S2; *NDUFV2*, NADH ubiquinone oxidoreductase core subunit V2; *SDHA*, succinate dehydrogenase complex flavoprotein subunit A; *UQCRB*, ubiquinol-cytochrome c reductase binding protein; *ATP5H*, ATP synthase subunit d.

2.10. Western Blotting Analysis

The protein from liver tissues was extracted using an ice-cold RIPA buffer supplement with 1% phosphatase and protease inhibitors (Huaxing Biotechnology, Beijing, China). The total protein concentration of the extracting solution was measured with the BCA assay kit (Huaxing Biotechnology, Beijing, China). Then, the expression of the target proteins was detected by immunoblotting assays, and the relative abundance of protein expression was normalized to GAPDH. The specific antibodies were used as follows: Nrf2 (1:1000, Abcam, Cambridge, UK), Keap1 (1:1000, Abcam, Cambridge, UK), NF-κB p65 (1:1000, Cell Signaling Technology, Boston, MA, USA), Phospho-NF-κB p65 (1:500, Cell Signaling Technology, Boston, MA, USA), Drp1 (1:1000, Affinity Biosciences, Cincinnati, OH, USA), Mfn2 (1:1000, Affinity Biosciences, Cincinnati, OH, USA), Cleaved caspase 3 (1:1000, Affinity Biosciences, Cincinnati, OH, USA), Bax (1:1000, Affinity Biosciences, Cincinnati, OH, USA), Bcl-2 (1:1000, Abcam, Cambridge, UK), GAPDH (1:2000, Cell Signaling Technology, Boston, MA, USA), HRP-linked anti-rabbit IgG (1:2000, Cell Signaling Technology, Boston, MA, USA), and HRP-linked anti-mouse IgG (1:2000, Cell Signaling Technology, Boston, MA, USA).

2.11. Statistical Analysis

The data analysis was conducted by two-way ANOVA with SPSS statistical software version 19 (IBM, Armonk, NY, USA). One-way ANOVA was adopted to analyze the differences among all groups when the interaction was significant, followed by Tukey's honest significant difference test for multiple comparisons. All data are expressed as mean with standard error (SE). A significance level of $p < 0.05$ was considered statistically significant, while a significance level of $0.05 \leq p < 0.1$ indicated a significant trend.

3. Results

3.1. Effects of Dietary Silybin Supplementation on Liver Injury in Piglets Challenged with PQ

As shown in Figure 1A, H&E staining showed that PQ challenge led to swelling and vacuolar degeneration of liver cells in the piglets, along with hepatic karyopyknosis and karyolysis. Conversely, dietary silybin supplementation alleviated the abovementioned pathological changes. Consistently, compared with the PQ group, dietary silybin administration significantly decreased the histological score of liver lesions in the PQ-challenged piglets ($p < 0.05$) (Figure 1B). Then, we further measured the activities of AST, ALT, and ALP in plasma and found that PQ challenge significantly increased ALT activity ($p < 0.05$). However, dietary silybin supplementation obviously reduced ALT activity ($p < 0.05$) and tended to decrease AST activity ($p = 0.07$) in the PQ-challenged piglets (Figure 1C,D). In addition, neither diet nor PQ treatment affected ALP activity in the plasma of the piglets ($p > 0.05$) (Figure 1E).

3.2. Effects of Dietary Silybin on Hepatic Oxidative Stress in Piglets Challenged with PQ

As shown in Figure 2, compared with the control group, the piglets challenged with PQ had significantly decreased GSH-Px activity ($p < 0.05$) and increased H_2O_2 levels ($p < 0.05$), and tended to have a lower T-AOC level ($p = 0.10$) and higher MDA content ($p = 0.08$) (Figure 2C–E). In contrast, dietary silybin significantly increased the T-AOC level and decreased the concentration of MDA and H_2O_2 ($p < 0.05$), and tended to increase the activities of SOD ($p = 0.06$) as well as GSH-Px ($p = 0.06$) (Figure 2B–E). Moreover, there was no significant difference in CAT activity among all groups ($p > 0.05$) (Figure 2A).

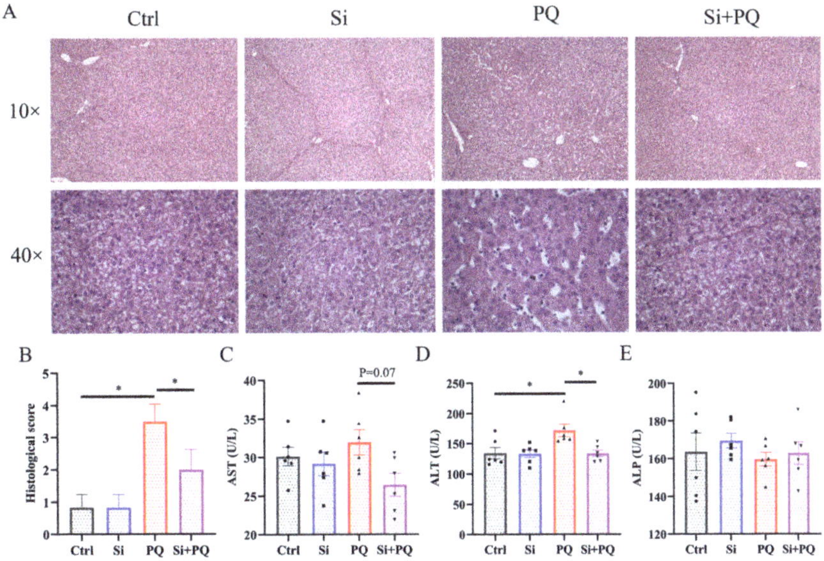

Figure 1. Dietary silybin supplementation alleviated liver injury induced by paraquat in piglets. (**A**) Representative H&E staining of liver tissues (captured at 100× or 400× magnification). (**B**) Histopathology scores of liver tissues. (**C–E**) Plasma activities of AST, ALT, and ALP. Ctrl, piglets were given a basal diet and were challenged with saline; Si, piglets were given a silybin-supplemented diet and were challenged with saline; PQ, piglets were given a basal diet and were challenged with paraquat; Si + PQ, piglets were given a silybin-supplemented diet and were challenged with paraquat; AST, aspartate transaminase; ALT, alanine transaminase; ALP, alkaline phosphatase. Data are expressed as mean ± standard error (n = 6). * $p < 0.05$.

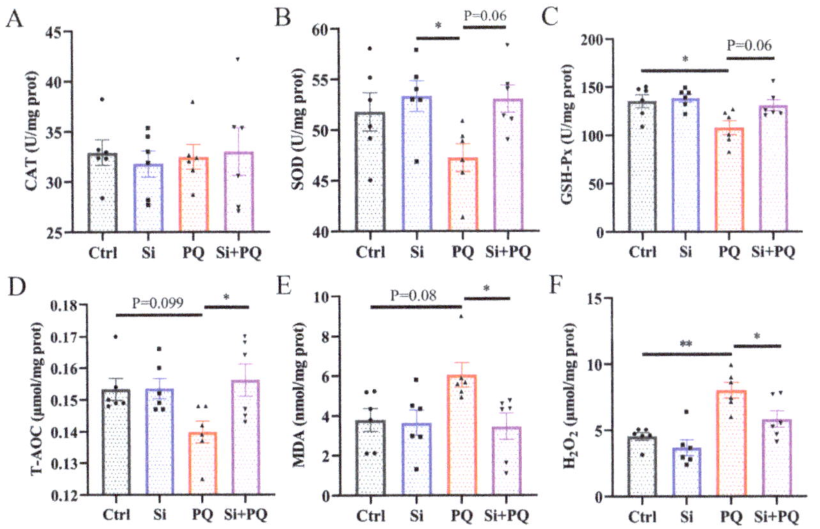

Figure 2. Dietary silybin administration ameliorated PQ-induced oxidative stress in the liver of piglets. The activities of CAT (**A**), SOD (**B**), and GSH-Px (**C**) and the level of T-AOC (**D**), MDA content (**E**), and H_2O_2 level (**F**) in the liver. Ctrl, piglets were given a basal diet and were challenged with saline; Si, piglets were given a silybin-supplemented diet and were challenged with saline; PQ, piglets were given a basal

diet and were challenged with paraquat; Si + PQ, piglets were given a silybin-supplemented diet and were challenged with paraquat; CAT, catalase; SOD, superoxide dismutase; GSH-Px, glutathione peroxidase; T-AOC, the total antioxidant capacity; MDA, malondialdehyde; H_2O_2, hydrogen peroxide. Data are expressed as mean ± standard error (n = 6). * $p < 0.05$, ** $p < 0.01$.

3.3. Effects of Dietary Silybin on Hepatic Activation of Nrf2 Signaling Pathway in Piglets Challenged with PQ

As shown in Figure 3, we found that PQ challenge significantly decreased the expression level of the *GPX4* gene ($p < 0.05$), but not the expression levels of the *SOD1* and *GPX1* genes ($p > 0.05$) (Figure 3B–D). However, dietary silybin supplementation significantly elevated the mRNA expression of *SOD1* ($p < 0.05$) and tended to enhance the expression levels of *GPX1* ($p = 0.097$) and *GPX4* ($p = 0.099$) in the liver of the PQ-challenged piglets (Figure 3B–D). Neither silybin supplementation nor PQ treatment affected *CAT* mRNA expression in the liver ($p > 0.05$) (Figure 3A). Interestingly, the Western blotting results showed that silybin supplementation increased the nuclear factor-erythroid 2-related factor 2 (Nrf2) protein expression in the presence or absence of PQ challenge ($p < 0.05$) (Figure 3E–G).

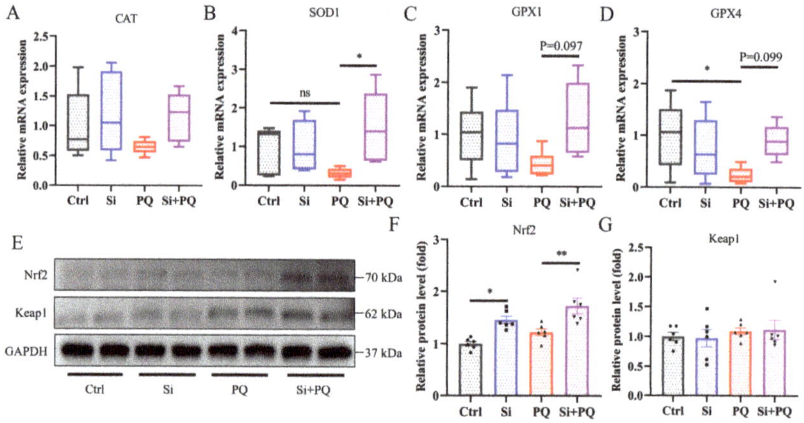

Figure 3. Silybin enhanced hepatic antioxidant capacity by activating the Nrf2 signaling pathway. The expression of Nrf2 signaling pathway genes including *CAT* (**A**), *SOD1* (**B**), *GPX1* (**C**), and *GPX4* (**D**). (**E–G**) The relative protein expression level of Nrf2 and Keap1. Ctrl, piglets were given a basal diet and were challenged with saline; Si, piglets were given a silybin-supplemented diet and were challenged with saline; PQ, piglets were given a basal diet and were challenged with paraquat; Si + PQ, piglets were given a silybin-supplemented diet and were challenged with paraquat; *CAT*, catalase; *SOD*, superoxide dismutase; *GPx*, glutathione peroxidase; *Nrf2*, nuclear factor-erythroid 2-related factor 2; *Keap1*, kelch-like ECH-associated protein 1. Data are expressed as mean ± standard error (n = 6). * $p < 0.05$, ** $p < 0.01$.

3.4. Effects of Dietary Silybin on Inflammatory Response in Piglets Challenged with PQ

As shown in Figure 4, compared with the control group, PQ challenge caused an increase in the levels of TNF-α and IL-8 ($p = 0.05$ and $p < 0.05$, respectively) (Figure 4A,C). In contrast to the PQ group, silybin dramatically reversed these trends and increased IL-10 content in the liver ($p = 0.08$) (Figure 4A,C,D). Consistently, the relative mRNA abundance of *TNF-α* and *IL-8* was upregulated by PQ treatment ($p = 0.07$ and $p = 0.06$, respectively) (Figure 4E,G), compared with the control group. However, it was dramatically downregulated by dietary silybin supplementation ($p = 0.06$ and $p < 0.05$, respectively). In addition, dietary silybin significantly reduced the increase in protein abundance of nuclear factor-kB (NF-κB) and phosphorylated NF-κB induced by PQ treatment ($p < 0.05$) (Figure 4I,J).

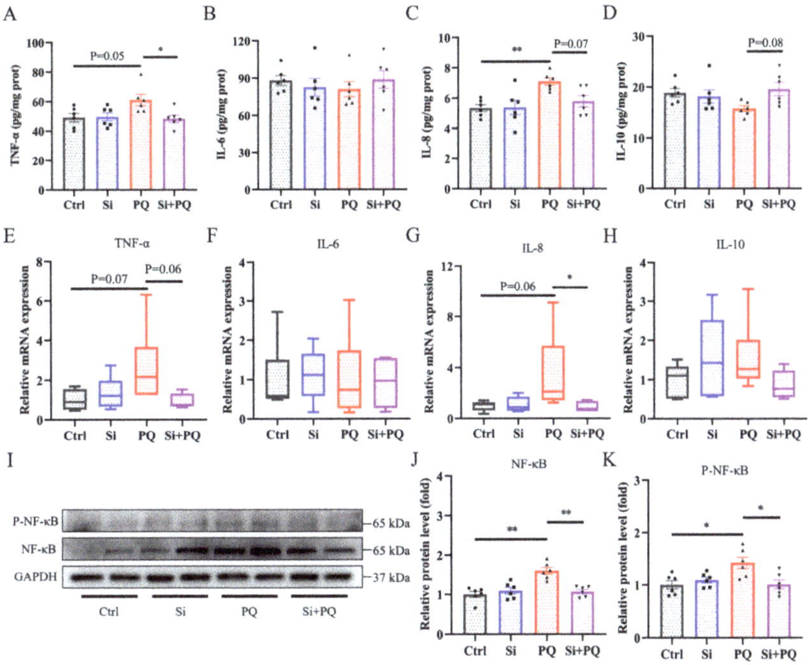

Figure 4. Dietary silybin supplementation alleviated hepatic inflammation induced by paraquat in piglets. The concentration of TNF-α (**A**), IL-6 (**B**), IL-8 (**C**), and IL-10 (**D**). The mRNA expression level of *TNF-α* (**E**), *IL-6* (**F**), *IL-8* (**G**), and *IL-10* (**H**). (**I–K**) The relative protein expression level of P-NF-κB and NF-κB. Ctrl, piglets were given a basal diet and were challenged with saline; Si, piglets were given a silybin-supplemented diet and were challenged with saline; PQ, piglets were given a basal diet and were challenged with paraquat; Si + PQ, piglets were given a silybin-supplemented diet and were challenged with paraquat; TNF-α, tumor necrosis factor-α; IL, interleukin; NF-κB, nuclear factor-kB; P-NF-κB, phosphorylated NF-κB. Data are expressed as mean ± standard error (n = 6). * $p < 0.05$, ** $p < 0.01$.

3.5. Effects of Dietary Silybin on Mitochondrial Function in Piglets Challenged with PQ

As shown in Figure 5, PQ challenge decreased the activities of mitochondrial complexes I and V and subsequently reduced the ATP content in the liver ($p < 0.05$) (Figure 5A–C). However, dietary silybin supplementation alleviated the PQ-induced decrease in mitochondrial complex I and V activity compared to the PQ group, thereby enhancing the production of ATP ($p < 0.05$). Consistently, we found that PQ treatment downregulated the expression level of NADH ubiquinone oxidoreductase core subunit V2 (*NDUFV2*) ($p < 0.05$) and tended to decrease ATP synthase subunit d (*ATP5H*) ($p = 0.05$) compared to the control group (Figure 5D), while silybin supplementation elevated the abovementioned gene expression in the liver of the PQ-challenged piglets ($p = 0.07$ and $p = 0.06$, respectively). In addition, dietary silybin also upregulated the mRNA expression of *NADH* ubiquinone oxidoreductase core subunit S2 (*NDUFS2*) in response to PQ challenge ($p = 0.06$) (Figure 5D). Moreover, PQ challenge increased the protein abundance of dynamin 1 (Drp1) and decreased the protein abundance of mitofusin 2 (Mfn2), while silybin supplementation significantly reversed the abovementioned trend ($p < 0.05$) (Figure 5E–G).

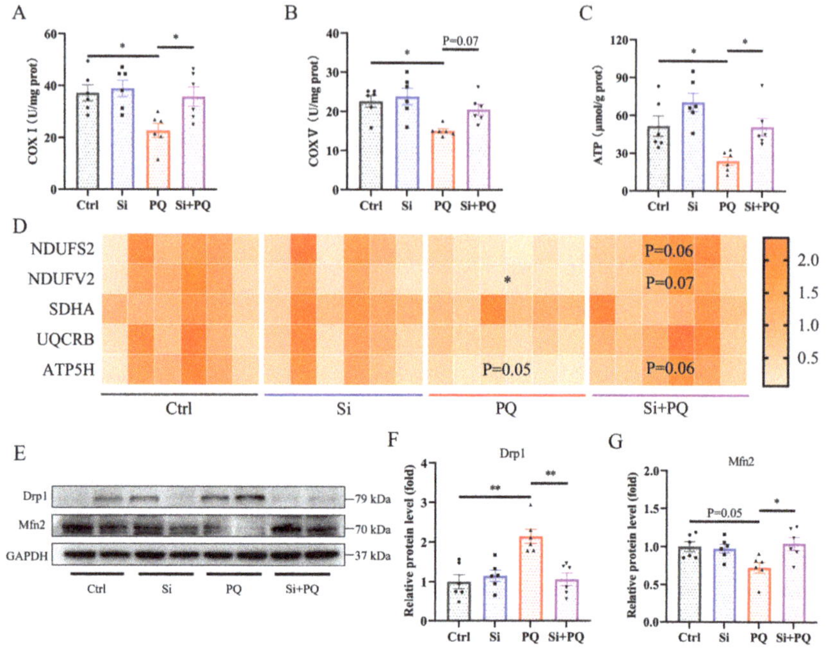

Figure 5. Dietary silybin supplementation protected against PQ-induced mitochondrial dysfunction. (**A**,**B**) The activities of COX I and COX V in the liver. (**C**) The level of ATP. (**D**) The expression of mitochondrial respiratory chain protein complex genes including *NDUFS2*, *NDUFV2*, *SDHA*, *UQCRB*, and *ATP5H*. (**E**–**G**) The relative protein expression level of Drp1 and Mfn2. Ctrl, piglets were given a basal diet and were challenged with saline; Si, piglets were given a silybin-supplemented diet and were challenged with saline; PQ, piglets were given a basal diet and were challenged with paraquat; Si + PQ, piglets were given a silybin-supplemented diet and were challenged with paraquat; COX, mitochondrial complex; ATP, adenosine triphosphate. *NDUFS2*, NADH ubiquinone oxidoreductase core subunit S2; *NDUFV2*, NADH ubiquinone oxidoreductase core subunit V2; *SDHA*, succinate dehydrogenase complex flavoprotein subunit A; *UQCRB*, ubiquinol-cytochrome c reductase binding protein; *ATP5H*, ATP synthase subunit d; Drp1, dynamin 1; Mfn2, mitofusin 2. Data are expressed as mean ± standard error (n = 6). * $p < 0.05$, ** $p < 0.01$.

3.6. Effects of Dietary Silybin Supplementation on Hepatocyte Apoptosis in Piglets Challenged with PQ

As shown in Figure 6, PQ challenge significantly increased the activities of caspase 3 and caspase 9 in the liver, which was reversed by silybin supplementation ($p < 0.05$) (Figure 5A,B). Interestingly, dietary silybin tended to decrease the activity of caspase 3 and caspase 9 in the non-challenged piglets ($p = 0.07$ and $p = 0.09$, respectively). Western blot analysis showed that PQ challenge significantly enhanced the protein expression of the Cleaved caspase 3 in the liver compared with the control group ($p < 0.01$) (Figure 5D). Conversely, silybin supplementation markedly downregulated Cleaved caspase 3 protein expression, upregulated B-cell lymphoma-2 (Bcl-2) protein expression, and increased the ratio of Bcl-2 to Bcl-2-associated-X-protein (Bax) in piglets upon PQ challenge ($p < 0.05$) (Figure 5D,E,G). Furthermore, silybin also increased Bcl-2 protein expression in the non-challenged piglets ($p < 0.01$) (Figure 5E). There was no significant difference in the Bax protein expression among all groups ($p > 0.05$) (Figure 5F).

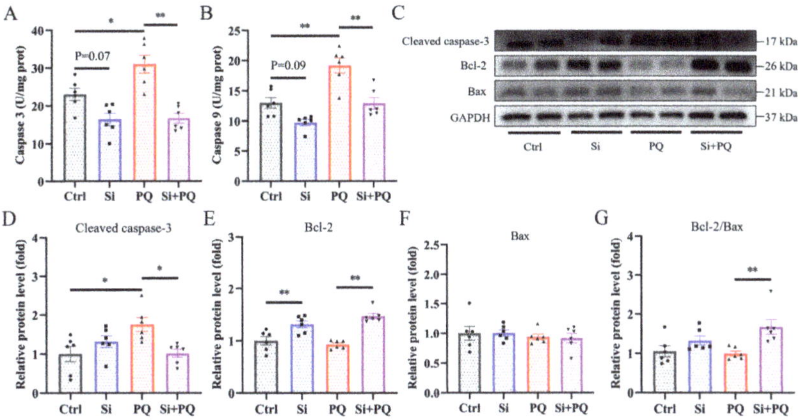

Figure 6. Dietary silybin addition suppressed hepatocyte apoptosis. (**A**,**B**) The activities of caspase 3 and caspase 9 in the liver tissues. (**C**) Representative image of Western blot. The protein expression levels of Cleaved caspase 3 (**D**), Bcl-2 (**E**), Bax (**F**), and the ratio of Bcl-2 to Bax (**G**). Ctrl, piglets were given a basal diet and were challenged with saline; Si, piglets were given a silybin-supplemented diet and were challenged with saline; PQ, piglets were given a basal diet and were challenged with paraquat; Si + PQ, piglets were given a silybin-supplemented diet and were challenged with paraquat; Bcl-2, B-cell lymphoma-2; Bax, Bcl-2-associated-X-protein; Bcl-2/Bax, the ratio of Bcl-2 to Bax. Data are expressed as mean ± standard error (n = 6). * $p < 0.05$, ** $p < 0.01$.

4. Discussion

The liver is a highly metabolically active organ in mammals that plays an important role in maintaining systemic health but is extremely susceptible to oxidative stress caused by various factors. Herein, we investigated the protective effect of silybin on liver injury induced by PQ challenge in weaned piglets and its potential molecular mechanism. Consistent with previous reports [30–32], histopathological staining analyses showed that PQ challenge caused noticeable pathological changes in the liver, evidenced by hepatic karyopyknosis, karyolysis, and vacuolation of hepatocytes. However, silybin supplementation alleviated hepatic morphological injury induced by PQ in piglets, which was proved by a reduced pathological score. A previous study also suggested that silymarin administration mitigated fibrosis, granulomatosis, and cytolytic necrosis in the liver of CCl_4-challenged broiler chickens [33]. Plasma biochemical indices such as AST, ALT, and ALP are the most sensitive markers of the degree of liver damage, and the increased activity of these enzymes in plasma indicates increased liver damage [34]. In the present study, dietary silybin supplementation lowered plasma ALT and AST activity in the PQ-challenged piglets, suggesting that silybin may have protective effects against liver injury induced by PQ treatment. These results were in agreement with those observed by Zhang et al. (2021) [35], which verified that micelle silymarin supplementation significantly decreased serum AST activity in sows on day 21 postpartum. In addition, silybin addition decreased the serum activities of ALT and γ-glutamyl transpeptidase in Peking ducks [18]. Taken together, these results indicate that dietary silybin administration has a protective function against the hepatic injury induced by PQ in piglets.

It is known that maintaining a proper redox balance is an effective strategy for alleviating liver injury [36]. The antioxidant enzymes secreted by hepatocytes are the primary line of defense against oxidative stress. In particular, SOD is an essential component of the antioxidant enzyme system, catalyzing the disproportionation of superoxide anion radicals into oxygen and hydrogen peroxide; the latter is further decomposed into water and oxygen by CAT [37]. In addition, GSH-Px also has a strong ability to scavenge free radicals, and T-AOC can reflect the overall antioxidant capacity of the body [38,39]. A previous study suggested that PQ challenge triggered severe oxidative stress in piglets [25]. Consistent with this result, our data also showed that PQ treatment decreased GSH-Px activity and T-AOC levels and augmented MDA and H_2O_2 levels in the liver. More importantly, dietary silybin supplementation has been shown to enhance SOD and GSH-Px activity and T-AOC levels, and subsequently lower H_2O_2 content and the level of lipid peroxidation. Another study also corroborated that silymarin addition protected against oxidative stress in gilts by reducing protein carbonyl contents in hepatic tissue [40]. Jiang et al. (2020) [15] reported that silymarin supplementation alleviated oxidative stress caused by pregnancy through enhancing serum activities of CAT and GSH-Px in sows. These results indicated that dietary silybin supplementation relieved oxidative stress by regulating the function of the antioxidant defense system.

As mentioned above, the hepatoprotective effects of silybin may be intimately related to its regulation of antioxidant capacity. Thus, the effect of silybin on the activation of the Nrf2 signaling pathway in the liver was further investigated. Nrf2/Keap1 signaling plays a vital role in resisting oxidative injury. Impaired oxidative redox status induces the dissociation of Nrf2 from Keap1 and translocation into the nucleus, which then binds to the antioxidant response element to regulate the expression of downstream antioxidant genes, such as CAT, SOD, and GPx, thus coping with cells from oxidative-stress-induced cellular injury [41,42]. In the present study, we found that Nrf2 protein expression significantly increased in response to silybin supplementation, and the mRNA expression of *SOD1*, *GPX1*, and *GPX4* was also in parallel with the alteration trend of Nrf2 protein expression, suggesting that dietary silybin enhanced antioxidant capacity in the PQ-challenged piglets. Coherently, our previous research found that dietary silybin improved intestinal antioxidant capacity in weaned piglets through regulating the Nrf2 signaling pathway [19]. Another study indicated that silymarin administration alleviated thioacetamide-induced acute liver injury by activating the Nrf2/Keap1 pathway [43]. An in vitro study also confirmed that silybin addition attenuated H_2O_2-induced oxidative stress by modulating Nrf2 signaling [44]. Taken together, these results demonstrated that dietary silybin supplementation mitigated liver damage through regulating the Nrf2 signaling pathway and subsequently enhancing antioxidant capacity in piglets.

Oxidative stress can trigger an inflammatory response, which in turn can directly exacerbate redox imbalance [45]. The increased content of proinflammatory cytokines is tightly related to the activated inflammatory response [46]. In the present study, PQ challenge elevated the secretion of TNF-α and IL-8 and their gene expression in the liver compared with the control group. However, dietary silybin dramatically reversed these trends and increased IL-10 concentration in the PQ-challenged piglets, which indicated that silybin attenuated the PQ-induced excessive inflammatory response by diminishing the generation of proinflammatory cytokines. Consistent with our results, dietary silybin dose-dependently ameliorated triptolide-induced liver injury in rats by suppressing proinflammatory cytokine TNF-α, IL-6, and IL-1β production [47]. Similarly, silymarin supplementation decreased serum TNF-α and IL-1β concentration in lactating sows [15,48]. It is well known that the NF-κB signaling pathway plays an important role in inflammatory responses, which can promote the expression of downstream target genes and the secretion of inflammatory mediators when activated by external stimuli [49,50]. A previous study has observed that PQ challenge triggers an inflammatory response by elevating the phosphorylation of NF-κB in the liver of weaned piglets [25]. In the current study, increased protein abundance of P-NF-κB and NF-κB was also found after PQ treatment, but dietary silybin significantly suppressed the protein expression of P-NF-κB and NF-κB compared with PQ-challenged piglets. The passivation of the NF-κB signaling further reduced the expression of proinflammatory cytokines, which was consistent with the alteration trend of cytokine levels in the liver. In line with our results, some studies have also reported that silybin inhibited the NF-κB signaling cascade to alleviate hepatic inflammation and steatohepatitis [51,52]. Taken together, these results indicated that dietary silybin supplementation mitigated hepatic inflammation signaling by abrogating the activation of the NF-κB pathway to reduce the expression of inflammatory cytokines.

The liver contains a large number of mitochondria, which play an essential role in cellular bioenergetics, regulating redox balance and apoptosis [53]. The mitochondrial respiratory chain consists of five enzymatic complexes, which are mainly responsible for the generation of ATP through oxidative phosphorylation [54]. Oxidative stress can cause mitochondrial dysfunction by disturbing cellular bioenergetics [55]. Herein, we found that PQ-induced oxidative stress reduced the activities of mitochondrial complexes I and V and lowered the mRNA expression level of *NDUFV2* (complex I) and *ATP5H* (complex V) in the liver. However, silybin supplementation alleviated the deleterious effects of PQ and normalized abnormal expression of the respiratory chain genes and subsequently recovered mitochondrial complex activity in the liver of the piglets, thereby enhancing the production of ATP. This is consistent with our previous research, which suggested that dietary silybin improved mitochondria function by increasing mitochondrial complex activity and ATP content in the intestine of weaned piglets [19]. In addition, another study also showed that silybin improved liver injury induced by NAFLD in mice by restoring hepatic mitochondrial respiratory chain activity (for all five complexes) [13]. The maintenance of mitochondrial function requires a proper balance between fission and fusion processes [56]. In this study, we found that PQ challenge increased the expression level of the mitochondrial fission-associated protein Drp1 and decreased the expression level of the mitochondrial fusion-associated protein Mfn2, implying that PQ treatment disrupted the balance of mitochondrial fission and fusion. Conversely, silybin supplementation markedly normalizes mitochondrial dynamic imbalance under PQ challenge. An in vitro study also suggested that silybin treatment protects human neuroblastoma SH-SY5Y cells from H_2O_2-induced mitochondrial damage by regulating OPA1 and Drp1 protein expression [57]. Collectively, silybin supplementation protected against PQ-induced liver injury in piglets by improving mitochondrial functions.

Previous research has demonstrated that mitochondrial dysfunction can exacerbate oxidative stress and trigger cell apoptosis, ultimately leading to liver injury and dysfunction [58]. Caspase 3 is an essential executioner in the process of apoptosis that can be heterologous activated by caspase 9 [59]. Herein, we found that dietary silybin inhibited PQ-triggered apoptosis by antagonizing an increase in caspase 3 and caspase 9 activity, which is in keeping with previous studies showing that pretreatment with silybin decreased the percentage of caspase-3-positive cells induced by H_2O_2 and silybin supplementation inhibited the increase in intestinal caspase 3 and caspase 9 activity induced by PQ challenge [19,60]. Interestingly, dietary silybin also decreased caspase 3 and 9 activity in the liver of non-challenged piglets. These results indicated that silybin has promising anti-apoptotic capacity. Among members of the Bcl-2 family, Bcl-2 and Bax proteins play an essential role in regulating cell apoptosis, and the former has anti-apoptotic effects, while the latter has the opposite [61]. In this study, silybin supplementation prevented PQ-induced cell apoptosis by enhancing the expression of anti-apoptotic proteins such as Bcl-2 and the ratio of Bcl-2 to Bax and inhibiting the expression of the proapoptotic protein Cleaved caspase 3, which may further clarify the underlying mechanism by which dietary silybin alleviated liver injury induced by PQ. Similar results were also found in both in vivo and in vitro models [62,63]. Overall, the results suggested that silybin supplementation suppressed PQ-induced hepatocyte apoptosis in piglets by regulating the Bcl-2/Bax signaling pathway.

5. Conclusions

Our study demonstrated that dietary silybin supplementation mitigated hepatic injury from oxidative stress induced by PQ in weaned piglets, which is closely associated with enhancing antioxidant capacity, mitigating inflammation, improving mitochondrial function, and inhibiting apoptosis. These findings provide a theoretical basis for applying silybin as an antioxidant in animal feed and improving liver health through dietary interventions.

Author Contributions: Conceptualization, L.C. and X.J.; investigation, L.C. and D.M.; data curation, L.C. and D.M.; software, L.C. and Y.Z.; visualization, X.J. and X.L.; formal analysis, L.C. and W.C.; writing—original draft preparation, L.C.; Writing—review and editing, Y.L., W.S., Y.P., X.J. and X.L.; supervision, X.J. and X.L.; project administration, X.L. All authors have read and agreed to the published version of the manuscript.

Funding: This study was financially supported by the National Natural Science Foundation of China (32260850), the Agricultural Science and Technology Innovation Program of the Chinese Academy of Agricultural Sciences (CAAS-IFR-ZDRW202302), and the Open Project of the Key Lab of Animal Production and Product Quality and Security of Ministry of Education of Jilin Agricultural University (202202).

Institutional Review Board Statement: The animal protocol in this study was approved by the Animal Care and Use Committee of the Institute of Feed Research of the Chinese Academy of Agricultural Sciences (IFR-CAAS20220428).

Informed Consent Statement: Not applicable.

Data Availability Statement: The datasets of the current study are available.

Conflicts of Interest: The authors declare no potential conflicts of interest.

References

1. Robinson, M.W.; Harmon, C.; O'Farrelly, C. Liver immunology and its role in inflammation and homeostasis. *Cell. Mol. Immunol.* **2016**, *13*, 267–276. [CrossRef] [PubMed]
2. Zhang, H.; Chen, Y.; Li, Y.; Jia, P.; Ji, S.; Chen, Y.; Wang, T. Protective effects of pterostilbene against hepatic damage, redox imbalance, mitochondrial dysfunction, and endoplasmic reticulum stress in weanling piglets. *J. Anim. Sci.* **2020**, *98*, skaa328. [CrossRef] [PubMed]
3. Machado, I.F.; Miranda, R.G.; Dorta, D.J.; Rolo, A.P.; Palmeira, C.M. Targeting Oxidative Stress with Polyphenols to Fight Liver Diseases. *Antioxidants* **2023**, *12*, 1212. [CrossRef] [PubMed]
4. Villanueva-Paz, M.; Morán, L.; López-Alcántara, N.; Freixo, C.; Andrade, R.J.; Lucena, M.I.; Cubero, F.J. Oxidative Stress in Drug-Induced Liver Injury (DILI): From Mechanisms to Biomarkers for Use in Clinical Practice. *Antioxidants* **2021**, *10*, 390. [CrossRef]
5. Wu, D.; Cederbaum, A.I. Oxidative stress and alcoholic liver disease. *Semin. Liver Dis.* **2009**, *29*, 141–154. [CrossRef]
6. Yang, J.; Fernández-Galilea, M.; Martínez-Fernández, L.; González-Muniesa, P.; Pérez-Chávez, A.; Martínez, J.A.; Moreno-Aliaga, M.J. Oxidative Stress and Non-Alcoholic Fatty Liver Disease: Effects of Omega-3 Fatty Acid Supplementation. *Nutrients* **2019**, *11*, 872. [CrossRef]
7. Guerbette, T.; Boudry, G.; Lan, A. Mitochondrial function in intestinal epithelium homeostasis and modulation in diet-induced obesity. *Mol. Metab.* **2022**, *63*, 101546. [CrossRef]
8. Dong, Y.; Zhao, J.; Zhu, Q.; Liu, H.; Wang, J.; Lu, W. Melatonin inhibits the apoptosis of rooster Leydig cells by suppressing oxidative stress via AKT-Nrf2 pathway activation. *Free Radic. Biol. Med.* **2020**, *160*, 1–12. [CrossRef]
9. Mooli, R.G.R.; Mukhi, D.; Ramakrishnan, S.K. Oxidative Stress and Redox Signaling in the Pathophysiology of Liver Diseases. *Compr. Physiol.* **2022**, *12*, 3167–3192.
10. Surai, P.F. Silymarin as a Natural Antioxidant: An Overview of the Current Evidence and Perspectives. *Antioxidants* **2015**, *4*, 204–247. [CrossRef]
11. Dobiasova, S.; Rehorova, K.; Kucerova, D.; Biedermann, D.; Kanova, K.; Petraskova, L.; Koucka, K.; Vaclavikova, R.; Valentova, K.; Ruml, T.; et al. Multidrug Resistance Modulation Activity of Silybin Derivatives and Their Anti-Inflammatory Potential. *Antioxidants* **2020**, *9*, 455. [CrossRef] [PubMed]
12. Liu, Y.; Xu, W.; Zhai, T.; You, J.; Chen, Y. Silibinin ameliorates hepatic lipid accumulation and oxidative stress in mice with non-alcoholic steatohepatitis by regulating CFLAR-JNK pathway. *Acta Pharm. Sin. B* **2019**, *9*, 745–757. [CrossRef] [PubMed]
13. Salamone, F.; Volti, G.L. S1830 Silybin Enhances Mitochondrial Function and Inhibits NFkB Activation in Experimental NAFLD. *Gastroenterology* **2010**, *138*, S-798. [CrossRef]
14. Serviddio, G.; Bellanti, F.; Stanca, E.; Lunetti, P.; Blonda, M.; Tamborra, R.; Siculella, L.; Vendemiale, G.; Capobianco, L.; Giudetti, A.M. Silybin exerts antioxidant effects and induces mitochondrial biogenesis in liver of rat with secondary biliary cirrhosis. *Free Radic. Biol. Med.* **2014**, *73*, 117–126. [CrossRef] [PubMed]
15. Jiang, X.; Lin, S.; Lin, Y.; Fang, Z.; Xu, S.; Feng, B.; Zhuo, Y.; Li, J.; Che, L.; Jiang, X.; et al. Effects of silymarin supplementation during transition and lactation on reproductive performance, milk composition and haematological parameters in sows. *J. Anim. Physiol. Anim. Nutr.* **2020**, *104*, 1896–1903. [CrossRef] [PubMed]
16. Hossain, M.M.; Hwang, H.S.; Jang, S.Y.; Yu, S.; Kim, I.H. Supplemental impact of silymarin in growing pig diet on the growth performance, total tract digestibility, faecal microflora, faecal noxious gas emission and absorption rate in blood. *J. Anim. Physiol. Anim. Nutr.* **2024**, *108*, 206–214. [CrossRef] [PubMed]

17. Shanmugam, S.; Park, J.H.; Cho, S.; Kim, I.H. Silymarin seed extract supplementation enhances the growth performance, meat quality, and nutrients digestibility, and reduces gas emission in broilers. *Anim. Biosci.* **2022**, *35*, 1215–1222. [CrossRef]
18. Zhang, Z.; Shi, B.; Lv, X.; Dong, Y.; Li, L.; Xia, Z. Effects of silybin supplementation on growth performance, serum indexes and liver transcriptome of Peking ducks. *Front. Vet. Sci.* **2024**, *10*, 1325115. [CrossRef]
19. Cai, L.; Gao, G.; Yin, C.; Bai, R.; Li, Y.; Sun, W.; Pi, Y.; Jiang, X.; Li, X. The Effects of Dietary Silybin Supplementation on the Growth Performance and Regulation of Intestinal Oxidative Injury and Microflora Dysbiosis in Weaned Piglets. *Antioxidants* **2023**, *12*, 1975. [CrossRef]
20. Chen, J.; Su, Y.; Lin, F.; Iqbal, M.; Mehmood, K.; Zhang, H.; Shi, D. Effect of paraquat on cytotoxicity involved in oxidative stress and inflammatory reaction: A review of mechanisms and ecological implications. *Ecotox. Environ. Safe* **2021**, *224*, 112711. [CrossRef]
21. Dilberger, B.; Baumanns, S.; Schmitt, F.; Schmiedl, T.; Hardt, M.; Wenzel, U.; Eckert, G.P. Mitochondrial Oxidative Stress Impairs Energy Metabolism and Reduces Stress Resistance and Longevity of *C. elegans*. *Oxid. Med. Cell. Longev.* **2019**, *2019*, 6840540. [CrossRef]
22. Zhang, R.; Zhao, C.; Gong, X.; Yang, J.; Zhang, G.; Zhang, W. Ginkgolide C Alleviates Acute Lung Injury Caused by Paraquat Poisoning via Regulating the Nrf2 and NF-κB Signaling Pathways. *Oxid. Med. Cell. Longev.* **2022**, *2022*, 7832983. [CrossRef]
23. Zheng, Q.; Zhang, Y.; Zhao, Z.; Shen, H.; Zhao, H.; Zhao, M. Isorhynchophylline ameliorates paraquat-induced acute kidney injury by attenuating oxidative stress and mitochondrial damage via regulating toll-interacting expression. *Toxicol. Appl. Pharmacol.* **2021**, *420*, 115521. [CrossRef]
24. Liu, X.; Yang, H.; Liu, Z. Signaling pathways involved in paraquat-induced pulmonary toxicity: Molecular mechanisms and potential therapeutic drugs. *Int. Immunopharmacol.* **2022**, *113*, 109301. [CrossRef]
25. Qi, M.; Wang, N.; Xiao, Y.; Deng, Y.; Zha, A.; Tan, B.; Wang, J.; Yin, Y.; Liao, P. Ellagic acid ameliorates paraquat-induced liver injury associated with improved gut microbial profile. *Environ. Pollut.* **2022**, *293*, 118572. [CrossRef] [PubMed]
26. Kilkenny, C.; Browne, W.J.; Cuthill, I.C.; Emerson, M.; Altman, D.G. Improving bioscience research reporting: The ARRIVE guidelines for reporting animal research. *Osteoarthr. Cartil.* **2012**, *20*, 256–260. [CrossRef] [PubMed]
27. Council, N.R. *Nutrient Requirements of Swine: Eleventh Revised Edition*; The National Academies Press: Washington, DC, USA, 2012; p. 420.
28. Dou, X.; Yan, D.; Liu, S.; Gao, L.; Shan, A. Thymol Alleviates LPS-Induced Liver Inflammation and Apoptosis by Inhibiting NLRP3 Inflammasome Activation and the AMPK-mTOR-Autophagy Pathway. *Nutrients* **2022**, *14*, 2809. [CrossRef] [PubMed]
29. Cai, L.; Wei, Z.; Zhao, X.; Li, Y.; Li, X.; Jiang, X. Gallic acid mitigates LPS-induced inflammatory response via suppressing NF-κB signalling pathway in IPEC-J2 cells. *J. Anim. Physiol. Anim. Nutr.* **2022**, *106*, 1000–1008. [CrossRef] [PubMed]
30. Costa, M.D.; de Freitas, M.L.; Dalmolin, L.; Oliveira, L.P.; Fleck, M.A.; Pagliarini, P.; Acker, C.; Roman, S.S.; Brandão, R. Diphenyl diselenide prevents hepatic alterations induced by paraquat in rats. *Environ. Toxicol. Pharmacol.* **2013**, *36*, 750–758. [CrossRef] [PubMed]
31. Han, J.; Zhang, Z.; Yang, S.; Wang, J.; Yang, X.; Tan, D. Betanin attenuates paraquat-induced liver toxicity through a mitochondrial pathway. *Food Chem. Toxicol.* **2014**, *70*, 100–106. [CrossRef]
32. Liu, Z.; Wang, X.; Li, L.; Wei, G.; Zhao, M. Hydrogen Sulfide Protects against Paraquat-Induced Acute Liver Injury in Rats by Regulating Oxidative Stress, Mitochondrial Function, and Inflammation. *Oxid. Med. Cell. Longev.* **2020**, *2020*, 6325378. [CrossRef] [PubMed]
33. Baradaran, A.; Samadi, F.; Ramezanpour, S.S.; Yousefdoust, S. Hepatoprotective effects of silymarin on CCl4-induced hepatic damage in broiler chickens model. *Toxicol. Rep.* **2019**, *6*, 788–794. [CrossRef]
34. Wan, J.-Y.; Gong, X.; Zhang, L.; Li, H.-Z.; Zhou, Y.-F.; Zhou, Q.-X. Protective effect of baicalin against Lipopolysaccharide/d-galactosamine-induced liver injury in mice by up-regulation of Heme oxygenase-1. *Eur. J. Pharmacol.* **2008**, *587*, 302–308. [CrossRef]
35. Zhang, Q.; Ahn, J.M.; Kim, I.H. Micelle silymarin supplementation to sows' diet from day 109 of gestation to entire lactation period enhances reproductive performance and affects serum hormones and metabolites. *J. Anim. Sci.* **2021**, *99*, skab354. [CrossRef] [PubMed]
36. Lin, Q.; Li, S.; Jiang, N.; Shao, X.; Zhang, M.; Jin, H.; Zhang, Z.; Shen, J.; Zhou, Y.; Zhou, W.; et al. PINK1-parkin pathway of mitophagy protects against contrast-induced acute kidney injury via decreasing mitochondrial ROS and NLRP3 inflammasome activation. *Redox Biol.* **2019**, *26*, 101254. [CrossRef]
37. Zhang, Q.H.; Yin, R.X.; Gao, H.; Huang, F.; Wu, J.Z.; Pan, S.L.; Lin, W.X.; Yang, D.Z. Association of the SPTLC3 rs364585 polymorphism and serum lipid profiles in two Chinese ethnic groups. *Lipids Health Dis.* **2017**, *16*, 1. [CrossRef] [PubMed]
38. Duan, G.; Huang, P.; Zheng, C.; Zheng, J.; Yu, J.; Zhang, P.; Wan, M.; Li, F.; Guo, Q.; Yin, Y.; et al. Development and Recovery of Liver Injury in Piglets by Incremental Injection of LPS. *Antioxidants* **2023**, *12*, 1143. [CrossRef]
39. Rolo, A.P.; Teodoro, J.S.; Palmeira, C.M. Role of oxidative stress in the pathogenesis of nonalcoholic steatohepatitis. *Free Radic. Biol. Med.* **2012**, *52*, 59–69. [CrossRef]
40. Farmer, C.; Lapointe, J.; Palin, M.-F. Effects of the plant extract silymarin on prolactin concentrations, mammary gland development, and oxidative stress in gestating gilts1. *J. Anim. Sci.* **2014**, *92*, 2922–2930. [CrossRef]
41. Kobayashi, M.; Yamamoto, M. Molecular Mechanisms Activating the Nrf2-Keap1 Pathway of Antioxidant Gene Regulation. *Antioxid. Redox Signal.* **2005**, *7*, 385–394. [CrossRef]

42. Ryan, S.W.; Quynh, T.T.; Samreen, F.; Carrie, H.S.; Thomas, W.K.; Thomas, R.S. Pharmacogenomics of Chemically Distinct Classes of Keap1-Nrf2 Activators Identify Common and Unique Gene, Protein, and Pathway Responses In Vivo. *Mol. Pharmacol.* **2018**, *93*, 297. [CrossRef]
43. Hussein, R.M.; Sawy, D.M.; Kandeil, M.A.; Farghaly, H.S. Chlorogenic acid, quercetin, coenzyme Q10 and silymarin modulate Keap1-Nrf2/heme oxygenase-1 signaling in thioacetamide-induced acute liver toxicity. *Life Sci.* **2021**, *277*, 119460. [CrossRef]
44. Guo, H.; Wang, Y.; Liu, D. Silibinin ameliorats H2O2-induced cell apoptosis and oxidative stress response by activating Nrf2 signaling in trophoblast cells. *Acta Histochem.* **2020**, *122*, 151620. [CrossRef]
45. Vezza, T.; Molina-Tijeras, J.A.; Rodríguez-Nogales, A.; Garrido-Mesa, J.; Cádiz-Gurrea, M.d.l.L.; Segura-Carretero, A.; González-Tejero, M.R.; Rodríguez-Cabezas, M.E.; Gálvez, J.; Algieri, F. The Antioxidant Properties of Salvia verbenaca Extract Contribute to Its Intestinal Antiinflammatory Effects in Experimental Colitis in Rats. *Antioxidants* **2023**, *12*, 2071. [CrossRef]
46. Cui, W.; Zhou, S.; Wang, Y.; Shi, X.; Liu, H. Cadmium exposure activates the PI3K/AKT signaling pathway through miRNA-21, induces an increase in M1 polarization of macrophages, and leads to fibrosis of pig liver tissue. *Ecotoxicol. Environ. Saf.* **2021**, *228*, 113015. [CrossRef] [PubMed]
47. Wang, L.; Huang, Q.-H.; Li, Y.-X.; Huang, Y.-F.; Xie, J.-H.; Xu, L.-Q.; Dou, Y.-X.; Su, Z.-R.; Zeng, H.-F.; Chen, J.-N. Protective effects of silymarin on triptolide-induced acute hepatotoxicity in rats. *Mol. Med. Rep.* **2018**, *17*, 789–800. [CrossRef] [PubMed]
48. Xu, S.; Jiang, X.; Jia, X.; Jiang, X.; Che, L.; Lin, Y.; Zhuo, Y.; Feng, B.; Fang, Z.; Li, J.; et al. Silymarin Modulates Microbiota in the Gut to Improve the Health of Sow from Late Gestation to Lactation. *Animals* **2022**, *12*, 2202. [CrossRef] [PubMed]
49. Marquardt, J.U.; Gomez-Quiroz, L.; Arreguin Camacho, L.O.; Pinna, F.; Lee, Y.-H.; Kitade, M.; Domínguez, M.P.; Castven, D.; Breuhahn, K.; Conner, E.A.; et al. Curcumin effectively inhibits oncogenic NF-κB signaling and restrains stemness features in liver cancer. *J. Hepatol.* **2015**, *63*, 661–669. [CrossRef]
50. Yoon, W.J.; Moon, J.Y.; Song, G.; Lee, Y.K.; Han, M.S.; Lee, J.S.; Ihm, B.S.; Lee, W.J.; Lee, N.H.; Hyun, C.G. Artemisia fukudo essential oil attenuates LPS-induced inflammation by suppressing NF-κB and MAPK activation in RAW 264.7 macrophages. *Food Chem. Toxicol.* **2010**, *48*, 1222–1229. [CrossRef] [PubMed]
51. Abhilash, P.A.; Harikrishnan, R.; Indira, M. Ascorbic acid is superior to silymarin in the recovery of ethanol-induced inflammatory reactions in hepatocytes of guinea pigs. *J. Physiol. Biochem.* **2013**, *69*, 785–798. [CrossRef]
52. Sharma, S.; Le Guillou, D.; Chen, J.Y. Cellular stress in the pathogenesis of nonalcoholic steatohepatitis and liver fibrosis. *Nat. Rev. Gastroenterol. Hepatol.* **2023**, *20*, 662–678. [CrossRef]
53. Nunnari, J.; Suomalainen, A. Mitochondria: In Sickness and in Health. *Cell* **2012**, *148*, 1145–1159. [CrossRef]
54. Bellance, N.; Lestienne, P.; Rossignol, R. Mitochondria: From bioenergetics to the metabolic regulation of carcinogenesis. *Front. Biosci.-Landmark* **2009**, *14*, 4015–4034. [CrossRef]
55. Chen, Y.; Zhang, H.; Ji, S.; Jia, P.; Chen, Y.; Li, Y.; Wang, T. Resveratrol and its derivative pterostilbene attenuate oxidative stress-induced intestinal injury by improving mitochondrial redox homeostasis and function via SIRT1 signaling. *Free Radic. Biol. Med.* **2021**, *177*, 1–14. [CrossRef]
56. Anand, R.; Wai, T.; Baker, M.J.; Kladt, N.; Schauss, A.C.; Rugarli, E.; Langer, T. The i-AAA protease YME1L and OMA1 cleave OPA1 to balance mitochondrial fusion and fission. *J. Cell Biol.* **2014**, *204*, 919–929. [CrossRef]
57. Tie, F.; Fu, Y.; Hu, N.; Wang, H. Silibinin Protects against H2O2-Induced Oxidative Damage in SH-SY5Y Cells by Improving Mitochondrial Function. *Antioxidants* **2022**, *11*, 1101. [CrossRef] [PubMed]
58. Sinha, K.; Das, J.; Pal, P.B.; Sil, P.C. Oxidative stress: The mitochondria-dependent and mitochondria-independent pathways of apoptosis. *Arch. Toxicol.* **2013**, *87*, 1157–1180. [CrossRef] [PubMed]
59. D'Arcy, M.S. Cell death: A review of the major forms of apoptosis, necrosis and autophagy. *Cell Biol. Int.* **2019**, *43*, 582–592. [CrossRef]
60. Wang, Y.-K.; Hong, Y.-J.; Huang, Z.-Q. Protective effects of silybin on human umbilical vein endothelial cell injury induced by H_2O_2 in vitro. *Vasc. Pharmacol.* **2005**, *43*, 198–206. [CrossRef] [PubMed]
61. Warren, C.F.A.; Wong-Brown, M.W.; Bowden, N.A. BCL-2 family isoforms in apoptosis and cancer. *Cell Death Dis.* **2019**, *10*, 177. [CrossRef] [PubMed]
62. Abd Eldaim, M.A.; Barakat, E.R.; Alkafafy, M.; Elaziz, S.A.A. Antioxidant and anti-apoptotic prophylactic effect of silymarin against lead-induced hepatorenal toxicity in rats. *Environ. Sci. Pollut. Res.* **2021**, *28*, 57997–58006. [CrossRef] [PubMed]
63. Liu, Z.; Sun, M.; Wang, Y.; Zhang, L.; Zhao, H.; Zhao, M. Silymarin attenuated paraquat-induced cytotoxicity in macrophage by regulating Trx/TXNIP complex, inhibiting NLRP3 inflammasome activation and apoptosis. *Toxicol. Vitr.* **2018**, *46*, 265–272. [CrossRef] [PubMed]

antioxidants

MDPI

Article

Effect of Dietary Supplementation of Chestnut and Quebracho Tannin Supplementation on Neonatal Diarrhoea in Preweaning Calves

Matteo Dell'Anno [1], Sara Frazzini [1], Irene Ferri [1], Susanna Tuberti [1], Elisa Bonaldo [1], Benedetta Botti [2], Silvia Grossi [1], Carlo Angelo Sgoifo Rossi [1] and Luciana Rossi [1,*]

[1] Department of Veterinary Medicine and Animal Sciences—DIVAS, Università degli Studi di Milano, 26900 Lodi, Italy; matteo.dellanno@unimi.it (M.D.); sara.frazzini@unimi.it (S.F.); irene.ferri@unimi.it (I.F.); susanna.tuberti@studenti.unimi.it (S.T.); elisa.bonaldo@studenti.unimi.it (E.B.); silvia.grossi@unimi.it (S.G.); carlo.sgoifo@unimi.it (C.A.S.R.)

[2] Freelance Veterinarian, Via Alessandrini, 4, Bogolese di Sorbolo, 43058 Parma, Italy; benedetta.botti@gmail.com

* Correspondence: luciana.rossi@unimi.it

Abstract: Neonatal calf diarrhoea (NCD) poses a significant health challenge in cattle herds, resulting in considerable economic losses and antimicrobial use. In response to the escalating threat of antimicrobial resistance, viable alternatives are imperative, aligning with European policies. This study evaluated the in-milk supplementation of the chestnut and quebracho tannin extract in preweaning calves on performance, diarrhoea occurrence, *Cryptosporidium* spp. shedding, protein digestibility, and intestinal health. Twenty newborn calves were divided, after colostrum administration, into two experimental groups for 30 days as follows: the control (CTRL) was fed with whole milk and solid feed, and tannins (TAN) were fed whole milk supplemented with 6/g day of tannin extract and solid feed. Faecal samples were collected on days 0, 3, 7, 14, and 30 for the evaluation of *Cryptosporidium* oocyst shedding and protein digestibility. Faecal consistency was evaluated during the sampling using the faecal score scale (0–3 scale, considering diarrhoea > 1). The results showed a significant reduction in diarrhoea frequency in the TAN compared to the CTRL group ($p < 0.05$) over 30 days of the trial. The prevalence of *Cryptosporidium* spp. was generally low (12%), considering all analysed samples. Protein digestibility revealed comparable values for the TAN and CTRL groups, suggesting that tannins did not negatively affect milk protein availability. In conclusion, the in-milk supplementation of 6/g day of the chestnut and quebracho tannin extract could be considered a valuable functional feed additive to decrease NCD occurrence, thus supporting animal health and decreasing antibiotic use in livestock.

Keywords: natural extracts; polyphenols; phytochemicals; antioxidants; antimicrobial; alternative to antibiotics; *Cryptosporidium parvum*; *Castanea sativa*; *Schinopsis* spp.

Citation: Dell'Anno, M.; Frazzini, S.; Ferri, I.; Tuberti, S.; Bonaldo, E.; Botti, B.; Grossi, S.; Sgoifo Rossi, C.A.; Rossi, L. Effect of Dietary Supplementation of Chestnut and Quebracho Tannin Supplementation on Neonatal Diarrhoea in Preweaning Calves. *Antioxidants* **2024**, *13*, 237. https://doi.org/10.3390/antiox13020237

Academic Editor: Serkos A. Haroutounian

Received: 19 January 2024
Revised: 9 February 2024
Accepted: 11 February 2024
Published: 15 February 2024

1. Introduction

Neonatal calf diarrhoea has an important impact on economic losses in dairy farms. It is estimated that diarrhoea is one of the primary causes of mortality in calves (53–57%) [1]. Moreover, the high morbidity, mortality, and antibiotic treatments associated with NCD occurrence significantly affect animal health and future dairy performance during the first lactation [2]. This disease is defined as multifactorial, where the interaction among microorganisms, the host, and the environment plays a pivotal role in its determinism. The main pathogens involved in NCD are *Rotavirus*, *Coronavirus*, *Escherichia coli* K99, *Salmonella* spp., *Clostridium perfrigens* and *Cryptosporidium parvum* [3,4]. In particular, *C. parvum* is a protozoan etiological agent of enteric infections and diarrhoea in various mammalian species, with studies estimating its prevalence ranging from 50 to 100% in dairy cattle worldwide [5,6], indicating a significant and widespread occurrence.

C. parvum is characterized by high resistance to different environmental conditions and high contagiousness [7]. The oocysts of *C. parvum*, after ingestion, release sporozoites, which colonises intestinal villi, establishing the clinical symptoms [8]. The infection often occurs after 3 days of life, reaching a peak of incidence 4–18 days after birth, depending on the nutritional and immune status of the calf [9]. The parasite causes diarrhoea with possible complications due to metabolic acidosis and an electrolyte unbalance, which can lead to severe dehydration and death in some cases [10]. Even if not always lethal, the *C. parvum* infection alters the gut barrier integrity and the absorption ability causing diarrhoea and facilitating invasion by other pathogens [11]. In the context of NCD, which significantly impacts economic losses in dairy farms, *C. parvum* is a noteworthy contributor.

Animals affected by NCD require treatment with antimicrobial drugs. Nevertheless, the growing issue of antimicrobial resistance in the livestock sector underlines the need for substitutes and to reduce the use of veterinary drugs in food-producing animals. Recently, this phenomenon prompted the European Union to introduce more limitations on veterinary drug use [12]. Nutrition plays a pivotal role in the health and welfare of humans and animals, and it is no longer intended only to satisfy nutritional requirements [13]. The supplementation of functional feed additives such as probiotics, prebiotics, phytochemicals, organic acids, and essential oils can sustain the health status and reduce the risk of pathologies in livestock [14]. Tannins are a polyphenolic class of secondary metabolites with well-known antioxidant, anti-inflammatory, and antimicrobial activities [15]. They can be classified as hydrolysable (pyrogallol) or condensed tannins (proanthocyanidins) [16]. Hydrolysable tannins can be hydrolysed in monomers by chemical or enzymatic treatments. These are composed of phenolic acids and polyols (commonly glucose) grouped into gallotannins and ellagitannins [17]. Condensed tannins (1–30 kDa) are more abundant in plants compared to hydrolysable ones (500–3000 Da) [18].

The efficacy of tannins as antioxidants stems from the presence of phenolic hydroxyl groups, which facilitate electron donation to quench reactive oxygen species and mitigate oxidative damage [19]. Extensive research has established the ability of tannins to scavenge free radicals, providing cellular protection against oxidative stress [20–22]. Moreover, their anti-inflammatory and anti-cancer properties make them subjects of interest in therapeutic applications [23]. The diverse dietary sources of tannins, such as tea, fruits, and nuts, emphasize their potential as natural antioxidants for human and animal health [24,25]. Tannins can inhibit the growth of several microorganisms, showing both a bactericidal and bacteriostatic effect depending on their concentration, class of tannins, and structural properties. An important antimicrobial property of tannins is their capacity to inhibit cell wall synthesis by directly binding to peptidoglycan, destroying the integrity of the bacterial wall and increasing the susceptibility to osmotic lysis [26]. In addition, tannins can affect the membrane potential, increasing the permeability of bacterial cell membranes and leading to cell death. Some classes of tannins can interact with the lipopolysaccharides contained in the membrane of Gram-negative bacteria. Proanthocyanidins have been shown to bind with lipopolysaccharides, impairing membrane integrity [27].

The use of moderate concentrations of tannins in ruminant nutrition (0.2–0.6% of dry matter in steers and 10 g/day for calves) was shown to improve animal performance and decrease gastrointestinal parasitism and nitrogen pollution [28–30]. However, the effect of tannin supplementation in the liquid feeds (whole milk and milk replacer) of calves was investigated by a limited number of studies with controversial findings. Krueger et al. [31] showed no effect of tannin supplementation on the growth performance and feed efficiency in calves. Demarco et al. [32] reported that the combination of tannins and probiotics increased the feed conversion without affecting growth performance and reduced the haptoglobin concentration after transportation in beef calves. The inclusion of 4 g/day of hydrolysable tannins in milk significantly raised the growth and antioxidant capacity of plasma and lowered diarrhoea occurrence and faecal shedding of pathogens in preweaning calves [33]. Soleiman et al. [34] highlighted that the supplementation of 4, 6, or 12 mg/L of tannic acid decreases the faecal score in Holstein calves.

However, tannins are reported to have the ability to bind dietary proteins, forming insoluble complexes, thus reducing their bioavailability for gut absorption and decreasing animal performance [35]. Therefore, it is clear that the negative impact of tannins on ruminants is not specific to the type of tannin but may depend on their inclusion rate [36]. Recent studies showed that chestnut and quebracho tannins could affect feed palatability, digestibility, and the utilization of dietary protein in swine [16,37]. The aim of the following study was to evaluate the effect of dietary supplementation of 6 g/day of the chestnut and quebracho tannin extract in milk on diarrhoea occurrence, *C. parvum* shedding, and protein digestibility in preweaning calves.

2. Materials and Methods

2.1. Animal Housing and Experimental Design

The experimental trial was approved by the Animal Welfare Organization of the University of Milan (OPBA_03_2021) and performed on a commercial dairy farm involved in the consortium of Parmigiano Reggiano (Emilia Romagna, Italy) in line with European regulations [38,39].

Within three hours of birth, twenty Holstein calves (7 males and 13 females; 44.25 ± 3.97 kg) were fed with 4 L of high-quality colostrum (checked with a refractometer ≥22 °Brix, ≥50 g/L of IgG) by bottle feeding with two portions. The day after, animals were allotted into two groups, balanced per weight: the control group (CTRL, n = 10) fed whole milk, and the tannins group (TAN, n = 10) fed whole milk supplemented with 6 g/day per calves of chestnut (*Castanea sativa*) and the quebracho (*Schinopsis* spp.) tannin extract (Silvafeed ByPro, Silvateam S.p.A., Cuneo, Italy; with a total phenol concentration of 0.70 g/g of tannic acid equivalent [40]; the feed additive was approved by the Reg. EC 1831/2003, identification number following Reg. EC 767/2009: IT000431CN) solubilized in a premixture of 50 mL of polypropylene glycol for 30 days. Dietary treatments were provided after colostrum administration, animals were fed whole milk twice per day, and ad libitum access to solid feed and water was guaranteed. In particular, calves were fed 10% of their body weight in milk for the entire trial [41]. Animals were housed in individual straw-bedded pens (0.90 × 1.80 m) with free access to water under homogeneous environmental conditions for 30 days.

2.2. Zootechnical Performance, Diarrhoea Frequency and Sample Collection

Body weight (BW) was individually recorded on days 0 (d0), 14 (d14), and 30 (d30). The average daily gain (ADG) was calculated by dividing the weight gain for the considered time period. Faecal consistency was scored on a daily basis for each animal using a four-point scale (0 = dry, 1 = normal, 2 = runny, 3 = watery), considering > 1 as diarrhoea [42,43]. The frequency of diarrhoea was evaluated as the percentage of observations of animals with clinical signs of diarrhoea divided by the total observation performed for the considered period (diarrhoea frequency = n° of faeces with score 2 and 3/total number of observations; moderate diarrhoea frequency = n° of faeces with score 2/total number of observations; severe diarrhoea frequency = n° of faeces with score 3/total number of observations).

Faecal samples were collected from the rectal ampulla on days 0 (d0), 3 (d3), 7 (d7), 14 (d14), and 30 (d30) for the evaluation of protein digestibility and *Cryptosporidium* spp. oocyst count via optical microscopy after floatation in 10 randomly selected observation fields at 400× magnification.

At d0 and d30, blood samples were collected from the jugular vein using vacuum tubes without anticoagulants for immunoenzymatic analyses.

2.3. Tannin Solubilization and Evaluation of Antioxidant Activity of Tannins Premixture

Every 7 days, a tannin-soluble premixture was prepared in order to facilitate the in-milk dispersion of the tannin extract during the individual milk-feeding procedure. Chestnut and quebracho tannin extract was diluted in a 20% propylene glycol solution in order to obtain a final concentration of 120 mg/mL, and 50 mL aliquots of premixture were prepared for daily administration in milk. Aliquot samples from the tannin premixture were collected at 0, 1, 2, 3, 5, 7, and 10 days from preparation for the evaluation of the antioxidant capacity stability over time.

The antioxidant capacity was assessed using the Trolox Equivalent Antioxidant Capacity assay according to Frazzini et al. [44]. Briefly, the reaction mixture with 2,2′-azino-bis (3-ethylbenzothiazoline-6-sulfonic acid) (ABTS) was generated by combining 5 mL of 7 mM ABTS with 88 μ of 140 mM of potassium persulfate ($K_2S_2O_8$). After 16 h of incubation of the reaction mixture in the dark at room temperature, the working solution of $ABTS^{\bullet+}$ radical cation was obtained diluting $ABTS^{\bullet+}$ with deionized water until reaching 0.700 ± 0.02 of absorbance at 734 nm, room temperature, using a spectrophotometer (Jasco V-630 UV-Vis, Jasco Deutschland GmbH, Pfungstadt, Germany). Trolox (6-hydroxy-2,5,7,8-tetramethychroman-2-carboxylic acid) was used as a standard for the generation of a calibration curve from 2000 μM to 125 μM via serial dilution. The test was performed in triplicate by adding 10 μL of the diluted sample to 1 mL of the $ABTS^{\bullet+}$ working solution. Absorbances were measured after 6 min of incubation at room temperature, and the total antioxidant capacity was expressed as μmol Trolox Equivalent/g of tannin extract powder (μM TroloxEq/g).

2.4. Nutrient Composition of the Whole Milk

The milk used in this study for animal feeding was analysed in terms of the principal nutrients according to the Official Methods of Analysis [45]. In particular, the dry matter (DM), crude protein (CP), ether extract (EE), and ash content were determined. DM was obtained by drying whole milk in a forced air oven until reaching weight stability after two subsequent attempts at weighing (AOAC method 930.15). CP was determined by measuring the total nitrogen concentration in milk samples multiplied by 6.25 as the conversion factor according to the Kjeldahl method (AOAC method 2001.11). EE was determined using a butyrometer according to the Gerber method (AOAC method 2000.18). Ash content was obtained by incinerating dried milk samples in a muffle furnace at 550 °C (AOAC method 942.05). Each determination was performed in triplicate.

2.5. Faecal Parameters and Apparent Total Tract Digestibility of Dietary Protein

DM and nitrogen concentrations in faeces were measured according to the Official Methods of Analysis as previously described [45]. The apparent total tract digestibility of the dietary (ATTD) protein was evaluated using acid-insoluble ash (AIA) as a marker on milk and faecal samples. Pre-dried samples of milk and faeces were incinerated in a muffle furnace at 550 °C for 3 h (AOAC method 942.05). Ashes were subsequently diluted in HCl (3N) and boiled for 15 min. Residual ashes were filtered on paper filters (Whatman 41, Cytiva, Pall Corporation, New York, NY, USA) and washed with hot deionized water until a neutral pH was obtained using a litmus test. Filters were incinerated according to the previously described methodology for obtaining the weight of acid-insoluble ashes. The apparent total protein digestibility was calculated according to the following equation:

$$\text{Apparent protein digestibility } (\%) = 100 \times \left(1 - \frac{acid\ insoluble\ ashes\ in\ feed}{acid\ insouble\ ashes\ in\ faeces} \times \frac{nitrogen\ content\ in\ faeces}{nitrogen\ content\ in\ feed}\right)$$

2.6. Enzyme-Linked Immunosorbent Assay of Glucagon-like Peptide 2 and Diamine Oxidase in Serum Samples

Blood samples collected on days 0 and 30 were centrifuged (3000 rpm, 15 min, RT) to obtain the serum. Glucagon-Like Peptide 2 (GLP2) and Diamine Oxidase (DAO) levels were quantified in duplicate using enzyme-linked immunosorbent assay (ELISA) kits specific for bovine species, following the manufacturer's instructions (Bioassay Technology Laboratory, Shanghai, China). After the addition of 50 μL of the stop solution, absorbances were measured with a microplate reader (Bio-Rad 680 microplate reader, Bio-Rad Laboratories, Inc., Hercules, CA, USA) at 450 nm, and concentrations were calculated according to the respective standard curves using CurveExpert 1.4 software.

2.7. Statistical Analysis

Obtained data were statistically analysed through JMP Pro 15® (SAS Inst. Inc., Cary, NC, USA) software. The antioxidant activity of the tannin premixture was evaluated with an analysis of variances (one-way ANOVA). The results of growth performance, faecal parameters, protein digestibility, and serum levels of GLP2 and DAO were analysed using a linear model, including the fixed effect of treatment (Trt), time (Time), and the interaction between treatment and time (Trt × Time), while each animal was included as a random factor. Pairwise comparisons were evaluated using Tukey's Honestly Significant Difference test (Tukey's HSD). Total ADG (0–30 days) was analysed using Student's unpaired t-test. The frequencies of the faecal score were converted into a dichotomous variable (normal/pathological) considering diarrhoea for a registered faecal score > 1, moderate diarrhoea frequency for each faecal score = 2, and severe for a faecal score = 3. The observed frequencies were assessed using the chi-squared test, allowing us to evaluate whether differences between observed and expected frequencies were caused by treatments or only by casualty. The results are presented as means ± standard error. Means or frequencies were considered statistically different when $p \leq 0.05$.

3. Results

3.1. Antioxidant Activity of Tannin Extract Premixture and Whole Milk Composition

The tannin premixture showed a stable antioxidant activity for 7 days. After 10 days from solubilization, the total antioxidant activity was significantly lowered to about 15.5% compared to 7 days after the preparation of the tannins premixture (Figure 1) ($p < 0.01$). The milk used for calf feeding showed a DM content of 12.98 ± 0.30%, CP 3.61 ± 0.04%, EE 4.28 ± 0.15% and 0.43 ± 0.14% of ashes.

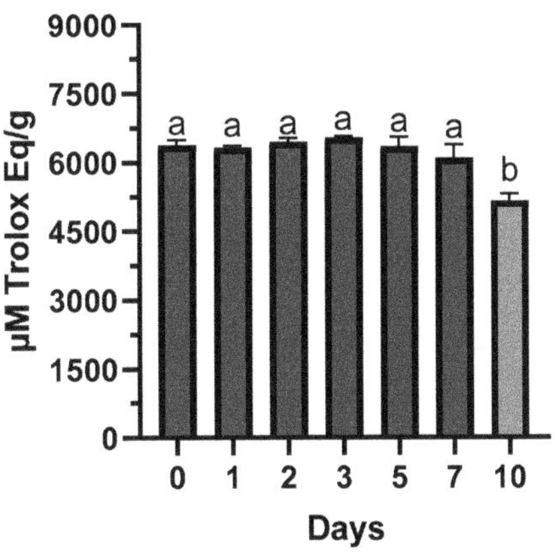

Figure 1. Antioxidant activity of tannins premixture (120 mg/mL in 20% propylene glycol solution) measured over 10 days of storage at room temperature. Data are presented as means ± standard error. [a,b] Different lowercase letters indicate statistically significant differences between days ($p < 0.01$).

3.2. Zootechnical Performance and Diarrhoea Occurrence

The calves' growth showed comparable values for both groups over 30 days of the trial without registering significant differences in terms of growth performance (Table 1).

Diarrhoea occurrence showed a significant reduction in the TAN group compared to the CTRL during the 30 days of the trial ($p < 0.05$) (Table 2). A significantly lower frequency of moderate diarrhoea (faecal score = 2) was observed in the TAN compared to the CTRL group from 0 to 30 days ($p < 0.05$). The observed decrease was registered after 3 days of age, underlining −25% of moderate diarrhoea in the TAN-supplemented group ($p < 0.05$).

Table 1. Growth performance of calves in control (CTRL) and tannin-supplemented group (TAN) during 30 days of trial.

	Group		*p*-Values		
	CTRL	TAN	Treatment	Time	Time × Treatment
BW (kg)			0.0417	<0.0001	0.9115
d0	45.30 ± 1.39	43.20 ± 1.39			
d14	45.80 ± 1.39	42.20 ± 1.39			
d30	60.80 ± 1.39	58.00 ± 1.39			
ADG (g/day)			0.7979	<0.0001	0.9423
d0–14	35.72 ± 79.90	35.71 ± 79.90			
d15–30	937.50 ± 79.90	893.75 ± 79.90			
d0–30	516.67 ± 63.25	493.33 ± 63.25	0.7971		

BW: body weight; ADG: average daily gain; CTRL: control group fed with whole milk; TAN: treatment group fed with whole milk supplemented with 6 g/day of tannin extract. Data are presented as means ± standard error.

Table 2. Diarrhoea frequency divided as presence (faecal score > 1), moderate (faecal score = 2), and severe (faecal score = 3) in calves of the control (CTRL) and tannin (TAN) groups registered during the 30 days of trial.

		CTRL	TAN	*p*-Value
Frequency of diarrhoea (%)	d0	10.00	0.00	0.1360
	d3	30.00 [a]	5.00 [b]	0.0191
	d7	36.00	25.00	0.3613
	d14	40.00	30.00	0.3291
	d30	20.00	15.00	0.6392
	d0–30	27.00 [a]	15.00 [b]	0.0150
Moderate diarrhoea (%)	d0	5.00	0.00	0.3049
	d3	30.00 [a]	5.00 [b]	0.0191
	d7	25.00	30.00	0.3613
	d14	35.00	25.00	0.3613
	d30	15.00	10.00	0.6056
	d0–30	22.00 [a]	11.00 [b]	0.0193
Severe diarrhoea (%)	d0	5.00	0.00	0.3049
	d3	0.00	0.00	-
	d7	10.00	10.00	1.0000
	d14	5.00	5.00	1.0000
	d30	5.00	5.00	1.0000
	d0–30	5.00	4.00	0.7268

Frequencies are presented as the percentage of observations of animals with clinical signs of diarrhoea for the considered period. [a,b] Different lowercase letters indicate statistically significant differences between groups ($p < 0.05$).

Cryptosporidium spp. showed that 50% of animals were positive for the presence of oocysts (4 from the CTRL and 6 from the TAN group); however, the prevalence of *Cryptosporidium* spp. was low (median value for both groups: 0) during the entire trial. The total number of measured oocysts was 828 for the CTRL and 302 for the TAN group, respectively. The presence of oocysts characterized the period from 3 to 14 days without registering any positivity on day 30 for both groups. The highest shedding was observed in the CTRL group after 7 days which registered 765 oocysts, while the TAN group registered a delayed peak at 14 days with 277 oocysts in faecal samples (Figure 2).

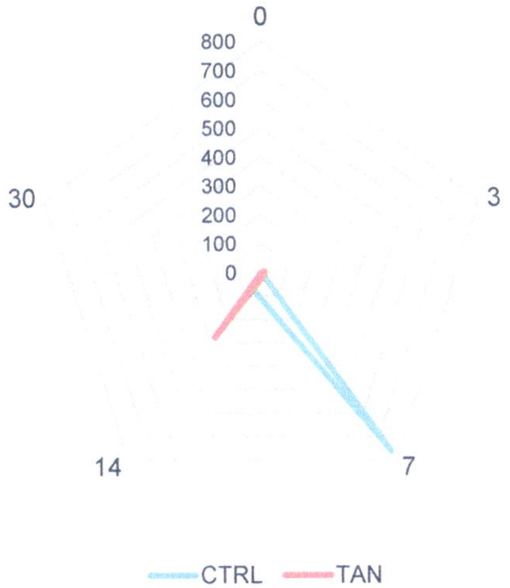

Figure 2. Spider plot of number of total faecal *Cryptosporidium* spp. oocysts measured in control (CTRL) and tannin (TAN) groups during the entire experimental trial of 30 days. Data are presented as the sum of total oocysts for each sampling timepoint (0, 3, 7, 14, and 30 days).

3.3. Faecal Parameters and Apparent Total Tract Digestibility of Dietary Protein

Concentrations of dry matter and total nitrogen showed a comparable trend between the two groups during the experimental trial without underlining significant differences (Figure 3A,B). The ATTD of nitrogen displayed a high individual variability after one week of age without observing significant differences over the entire period (Figure 3C).

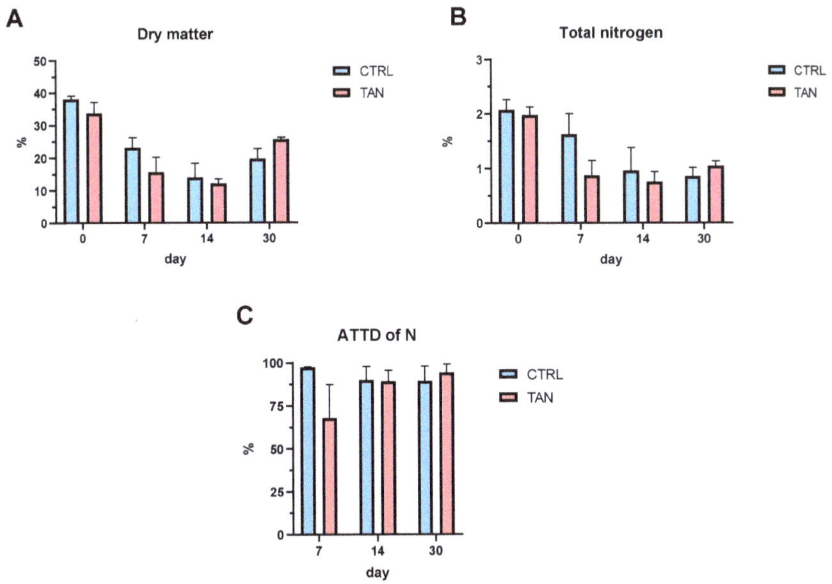

Figure 3. Faecal parameters and apparent total tract digestibility (ATTD) of nitrogen divided into control (CTRL) and tannin (TAN) groups at different timepoints of the experimental trial. (**A**) Dry matter of faecal samples at 0, 7, 14 and 30 days of trial; (**B**) Total nitrogen concentration of faecal samples at 0, 7, 14 and 30 days of trial on fresh matter basis; (**C**) Apparent total tract digestibility of nitrogen at 7, 14 and 30 days of trial. Data are presented as the means ± standard error.

3.4. Serum Concentration of GLP2 and DAO

Blood levels of GLP2 and DAO showed comparable values between the CTRL and TAN groups on days 0 and 30 of the experimental trial (Figure 4).

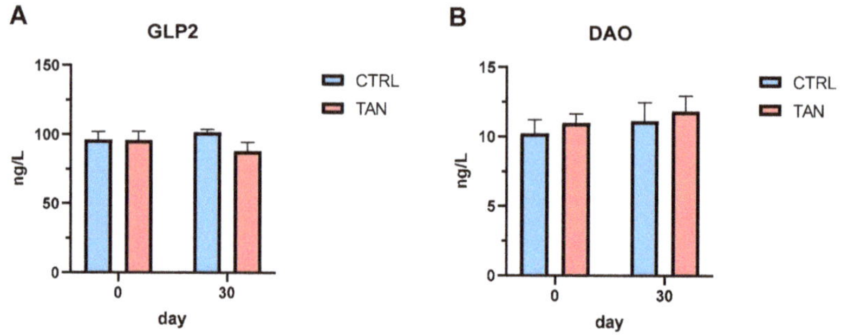

Figure 4. Serum concentration of blood metabolites titrated with enzyme-linked immunoassays in control (CTRL) and tannin (TAN) groups on days 0 and 30 of experimental trial. (**A**) Glucagon-like peptide 2 (GLP2) levels. (**B**) Diamine Oxidase (DAO) levels. Data are presented as means ± standard error.

4. Discussion

Due to global concerns regarding antimicrobial resistance, there is a current demand for alternative treatments. Tannins extracted from different plants have been suggested for their potential to modulate rumen fermentation and reduce methane emissions in ruminants [46–48]. However, limited information is available on the impact of tannins on NCD and protein digestibility. The present study evaluated the effect of tannin extract supplementation in whole milk on calves' performance, diarrhoea frequency, and *Cryptosporidium* spp. shedding, protein digestibility, and intestinal barrier status through indirect markers in serum. The tannin premixture showed high stability for 7 days after the preparation and registered a drop after 10 days. The registered antioxidant capacity was in line with the scavenging activity evaluated in our previous study on the tannin extract from chestnut and quebracho trees [49]. The liquid premixture optimized the procedure for the administration of 6 g/day of tannin extract within the individual milk-feeding procedure of the farm. This protocol ensured the preservation of antioxidant capacity for a week of storage, indicating this method is a practical and efficient approach for tannins supplementation in the diet of animals.

The nutritional content of the milk used in the present study exhibited nutrient levels similar to the Italian average for Holstein cow milk quality, especially in the context of Parmigiano Reggiano production [50–52]. The inclusion of tannins in milk in preweaning calves did not affect the growth curve over 30 days of the trial. A slight numerical difference was observed in the body weight of the TAN compared to the CTRL group after 30 days, even if not statistically significant. The lower average BW could be due to the presence of seven male calves in the CTRL group, which typically achieve more rapid growth compared to females. In line with our results, different studies have shown that tannin supplementation does not influence growth performance during the first month of life in calves [34,53,54]. Serri et al. [33] observed that the supplementation of 4/g day of the hydrolysable tannin extract in milk significantly improved ADG and final body weight after 76 days compared to the 2 g/day, 6 g/day, and control groups. Oliveira et al. [54] revealed that the tannin-rich pomegranate extract supplemented in feed significantly increased calves' performance and feed intake from 30 to 70 days of age. These data suggest that regular tannin supplementation may have a positive impact on performance during the second month of life. Although it is challenging to directly compare natural extracts due to the significant variability in doses, batches, and the wide difference of commercial products on the market, the positive effects of tannins on calves' performance could become more evident from the second month of life.

The average age at which diarrhoea onset occurred was consistent with data reported in the existing literature [55,56]. The frequency of diarrhoea was significantly decreased in the TAN group compared to the CTRL during 30 days of the trial. In particular, moderate diarrhoea occurrence was reduced by 25% after 3 days from birth in the TAN compared to the CTRL group. Faecal *Cryptosporidium* spp. oocysts registered a low positivity (12% of total analysed samples) [57,58], suggesting that their presence was not the main cause driving diarrhoeic episodes. Only three animals (2 CTRL and 1 TAN) showed high parasitic shedding (>30 oocysts) according to the scale proposed by Delafosse et al. [59] on days 7 and 14 of the trial. Neonatal calves typically face an elevated susceptibility to gastrointestinal disorders, notably diarrhoea. Dietary strategies that reduce the probability of NCD are preferred, obviating the necessity for antimicrobial treatments in line with One Health principles. The intimate connection between oxidative stress and the immune system underscores the significance of maintaining a delicate equilibrium between oxidants and antioxidants within immune cells. This balance is crucial, as immune cells require the production of reactive oxygen species to complete their functions [60]. Plant extracts rich in polyphenols are extensively employed as feed additives, primarily due to their claimed beneficial effects associated with antioxidant and antimicrobial properties. Additionally, plant polyphenols have been reported for their ability to deactivate bacterial enterotoxin in vitro [61]. In this context, tannins are strong antioxidants that have been proposed as a valid substitute for substituting artificial antioxidants for different applications [62]. Numerous studies have suggested that oxidative stress characterizes pathological conditions such as NCD [63–65]; thus, dietary antioxidants could provide important support to restore oxidative balance. Consequently, raising the antioxidant capacity in newborn calves has the potential to promote immune system development, enhance health status, and consequently decrease calf mortality. The supplementation of tannins in animal nutrition is proposed as particularly advantageous for animal welfare and health and is primarily attributed to the enhancement of antioxidant status [33].

Tannins are also effective for their ability to inhibit bacterial growth and protease activity by damaging the cell wall and cytoplasm, causing rapid structural destruction. Previous studies have suggested that the antimicrobial effect of tannins may be related to their ability to impair microbial adhesions and inhibit hydrolytic enzymes such as proteases, carbohydrolases, and cell envelope transport proteins [62,66–68]. Consistent with our findings, several studies have reported enhanced faecal consistency following tannins supplementation, contributing to the reduction in the duration of diarrhoea, delaying its onset, and serving as a preventive measure in calves [29,33,53,69]. On the contrary, some studies did not register a preventive effect on diarrhoea [54,70]; however, a notable improvement in faecal consistency was recorded [54]. This difference could be linked to the broad range of tested doses of tannins and the different nature and composition of tannin extracts used in the literature. The extensive variation in dosages and diverse tannin extract profiles may influence the effectiveness of diarrhoea prevention, underscoring the need for a correct consideration of both the dosage and extract composition in feed applications.

The antiprotozoal ability of polyphenols against coccidia has been reported in previous studies [71–73], suggesting that tannins possess the ability to directly reduce the viability of the larval stage and disrupt egg hatching. Bhatta et al. [74] observed a reduction in protozoa when applying hydrolysable tannins from different plant sources. Benchaar et al. [75] reported that condensed tannins from quebracho can decrease protozoa abundance. In contrast, Vasta et al. [76] found that quebracho tannins were capable of increasing protozoa in rumen liquor. Discrepancies in the tannin concentrations, plant sources, protozoal species, and environmental factors considered in these studies may elucidate the conflicting results obtained, as these parameters significantly influence the antiprotozoal activity of polyphenols. Even if the recorded *Cryptosporidium* shedding did not reveal significant differences, a lower peak of oocysts in faecal samples was observed in the TAN compared to the CTRL group delaying its onset. The average number of days of *C. parvum* oocysts shedding was registered from 6 to 9 days of age in calves after experimental infection [57]. Zambriski et al. [58] observed a comparable faecal shedding pattern in calves regardless of the administered dose of *C. parvum* in experimentally challenged calves. Animals subjected to a lower oral dose delayed the shedding with a lower number of oocysts. We hypothesize that the antiparasitic properties of tannins could potentially limit the shedding of oocysts in calves, also causing a delay in faecal elimination. However, to comprehensively understand the impact of the combination of chestnut and quebracho tannins on calves affected by *Cryptosporidium* spp., further studies, including experimental challenges, will be necessary to provide a detailed examination of the interplay between these tannins and the parasitic infection.

Faecal DM, N concentrations, and the ATTD of dietary protein showed comparable levels in CTRL and TAN groups during the entire trial. Tannins are recognized for their pronounced affinity with proteins, giving rise to their astringent effect. By forming non-specific bonds with dietary proteins, tannins can create complexes resistant to gastrointestinal proteases. Unlike ruminants, tannins have traditionally been considered antinutritional factors in animal nutrition, leading to adverse effects on feed intake, nutrient digestibility, and growth performance [77]. Consequently, the feed industry aimed to minimize the inclusion of tannin-rich feed ingredients in the diets of pigs and poultry or to adopt measures for reducing their concentrations in the diet. However, recent studies have indicated that low concentrations of various tannin sources enhance the health status and performance of monogastric animals [16,78]. In the literature, only a few studies have assessed the effect of tannins on protein digestibility in calves. Soleiman and Kheiri [34] found that the supplementation of 4, 6, and 12 mg/L of tannic acid in milk did not influence the digestibility of protein and dietary lipids. The observed results align with registered growth performance, assuming similar milk digestibility. These findings present promising outcomes for supplementing the tannin extract into calf milk without compromising either milk palatability or the digestibility of the protein component.

No differences were observed for GLP2 and DAO serum concentrations after 30 days of tannin supplementation. GLP-2 is a 33-amino acid peptide primarily produced by intestinal L cells after food ingestion, derived from proglucagon (a hormonal peptide precursor). This hormone plays a critical role in the trophism of intestinal crypts, stimulating their proliferation and inhibiting apoptosis processes [79]. The GLP-2 increase in serum has been associated with improved morphology and intestinal functionality, positively impacting nutrient metabolism [80]. The observed values of GLP2 suggest that the supplementation of tannins did not negatively influence the production of this peptide by intestinal L cells; consequently, its role related to the trophism of intestinal crypts as a stimulator of proliferation and inhibitor of apoptosis remained unaffected after 30 days of the trial.

DAO is an antihistaminic enzyme produced by the intestinal epithelium of mammals. This enzyme is minimally present in serum under normal conditions. High serum levels of DAO may occur following damage to the intestinal mucosa; hence, the DAO concentration can be used as an indirect marker of intestinal integrity [81]. A previous study suggested that serum DAO levels may be modulated by feed additive supplementation in preweaning calves [82]. Serum DAO titres suggest that the animal welfare levels and health status were generally correct, in line with the absence of diarrhoea occurrence at the end of the trial.

The limitation of this study is that it relies on the short duration of the trial, which did not allow the evaluation of the long-term effects of tannins on growth and production performance. Furthermore, previous studies indicated the possible inhibitory activity of tannins on *Cryptosporidium* spp., but the low prevalence of this pathogen in our study did not allow us to emphasize this activity. However, our data provide interesting findings for practical applications. The use of 6 g/day of the investigated tannin extract could be considered a straightforward strategy to reduce the occurrence of diarrhoea, thereby enhancing both farm profitability and animal welfare. It is also important to underline the variability in feed additives available on the market, including differences in dosages and natural extract profiles. This variability makes it challenging to compare previous studies and establish a globally applicable protocol for alleviating NCD based on phytochemicals.

Considering the widespread issue of fake news in livestock farming, the need to share scientific findings through contemporary channels such as social media becomes fundamental. Recent studies conducted on Instagram as a tool for study engagement and information on the livestock system found that the viewing of the posts had a favourable impact on consumers' opinions, even showing little change in participants' attitudes [83,84]. These findings emphasize the importance of utilizing social media platforms for effective communication and outreach in animal science and veterinary education [85]. Furthermore, our study offers valuable suggestions and insights for creating social media content, contributing to the broader field of innovative communication strategies in animal and veterinary sciences.

5. Conclusions

The supplementation of 6 g/day of chestnut and quebracho tannin extract in milk did not influence the growth performance of calves over the course of 30 days. Tannins showed a positive impact on animal health by reducing diarrhoea frequency and improving faecal consistency. Throughout the trial, both the control and tannin groups did not show differences in the presence of faecal *Cryptosporidium* spp. oocysts. The tannin extract did not impair protein digestibility and milk palatability over the 30-day supplementation period. Further studies will be useful to comprehensively understand the specific effects of the chestnut and quebracho tannin extract on cryptosporidiosis. In conclusion, this study provides valuable insights, supporting the potential use of the combination of chestnut and quebracho tannins as functional feed additive to enhance health, reduce neonatal diarrhoea, and minimize antibiotic treatments in preweaning calves.

Author Contributions: Conceptualisation: L.R.; methodology: M.D., L.R., B.B., S.T. and E.B.; investigation: M.D., S.T., E.B., S.F., I.F., B.B., S.G. and C.A.S.R.; formal analysis: M.D., S.T., E.B., S.F. and I.F.; data curation: M.D.; writing original draft preparation: M.D.; writing review and editing: M.D., S.F. and I.F.; visualisation: M.D.; supervision: L.R. and C.A.S.R.; funding acquisition: L.R. All authors have read and agreed to the published version of the manuscript.

Funding: The research received no external funding.

Institutional Review Board Statement: This study was conducted following the guidelines of the Declaration of Helsinki. The experimental trial was approved by the Animal Welfare Organisation of University of Milan (OPBA_03_2021).

Informed Consent Statement: Not applicable.

Data Availability Statement: Data are available within the article and from the corresponding author upon reasonable request.

Acknowledgments: The authors are grateful to Carra Mangimi S.p.A. (43058 Bolognese di Sorbolo, Parma, Italy) and Andrea Luppi from the Istituto Zooprofilattico Sperimentale della Lombardia e dell'Emilia Romagna (IZSLER, 43126 Parma, Italy) for unwavering support during the research study.

Conflicts of Interest: The authors declare no conflicts of interests.

References

1. Schinwald, M.; Creutzinger, K.; Keunen, A.; Winder, C.; Haley, D.; Renaud, D. Predictors of diarrhea, mortality, and weight gain in male dairy calves. *J. Dairy Sci.* **2022**, *105*, 5296–5309. [CrossRef]
2. Heinrichs, A.J.; Heinrichs, B. A prospective study of calf factors affecting first-lactation and lifetime milk production and age of cows when removed from the herd. *J. Dairy Sci.* **2011**, *94*, 336–341. [CrossRef] [PubMed]
3. Cho, Y.-i.; Yoon, K.-J. An overview of calf diarrhea-infectious etiology, diagnosis, and intervention. *J. Vet. Sci.* **2014**, *15*, 1–17. [CrossRef]
4. Brunauer, M.; Roch, F.-F.; Conrady, B. Prevalence of worldwide neonatal calf diarrhoea caused by bovine rotavirus in combination with Bovine Coronavirus, *Escherichia coli* K99 and *Cryptosporidium* spp.: A meta-analysis. *Animals* **2021**, *11*, 1014. [CrossRef] [PubMed]

5. Santín, M.; Trout, J.M.; Fayer, R. A longitudinal study of cryptosporidiosis in dairy cattle from birth to 2 years of age. *Vet. Parasitol.* **2008**, *155*, 15–23. [CrossRef]
6. Lassen, B.; Viltrop, A.; Raaperi, K.; Järvis, T. Eimeria and Cryptosporidium in Estonian dairy farms in regard to age, species, and diarrhoea. *Vet. Parasitol.* **2009**, *166*, 212–219. [CrossRef] [PubMed]
7. Zahedi, A.; Ryan, U. Cryptosporidium–an update with an emphasis on foodborne and waterborne transmission. *Res. Vet. Sci.* **2020**, *132*, 500–512. [CrossRef]
8. Garcia, L.S.; Arrowood, M.; Kokoskin, E.; Paltridge, G.P.; Pillai, D.R.; Procop, G.W.; Ryan, N.; Shimizu, R.Y.; Visvesvara, G. Practical guidance for clinical microbiology laboratories: Laboratory diagnosis of parasites from the gastrointestinal tract. *Clin. Microbiol. Rev.* **2018**, *31*, e00025-17. [CrossRef]
9. Quilez, J.; Sanchez-Acedo, C.; Del Cacho, E.; Clavel, A.; Causape, A. Prevalence of Cryptosporidium and Giardia infections in cattle in Aragon (northeastern Spain). *Vet. Parasitol.* **1996**, *66*, 139–146. [CrossRef]
10. Weyl-Feinstein, S.; Markovics, A.; Eitam, H.; Orlov, A.; Yishay, M.; Agmon, R.; Miron, J.; Izhaki, I.; Shabtay, A. effect of pomegranate-residue supplement on Cryptosporidium parvum oocyst shedding in neonatal calves. *J. Dairy Sci.* **2014**, *97*, 5800–5805. [CrossRef]
11. Al Mawly, J. Epidemiological Studies of Enteropathogens of Newborn Calves in New Zealand Dairy Farms. Ph.D. Thesis, Massey University, Palmerston North, New Zealand.
12. *Regulation (EU) 2019/6 of the European Parliament and of the Council of 11 December 2018 on Veterinary Medicinal Products and Repealing Directive 2001/82/EC (Text with EEA Relevance)*; European Parliament: Strasbourg, France, 2018.
13. Bobeck, E.A. Nutrition and health: Companion animal applications: Functional nutrition in Livestock and Companion Animals to Modulate the Immune Response. *J. Anim. Sci.* **2020**, *98*, skaa035. [CrossRef]
14. Jouany, J.-P.; Morgavi, D. Use of 'natural' products as alternatives to antibiotic feed additives in ruminant production. *Animal* **2007**, *1*, 1443–1466. [CrossRef]
15. Silva, V.; Falco, V.; Dias, M.I.; Barros, L.; Silva, A.; Capita, R.; Alonso-Calleja, C.; Amaral, J.S.; Igrejas, G.; CFR Ferreira, I. Evaluation of the phenolic profile of *Castanea sativa* Mill. by-products and their antioxidant and antimicrobial activity against multiresistant bacteria. *Antioxidants* **2020**, *9*, 87. [CrossRef]
16. Caprarulo, V.; Giromini, C.; Rossi, L. Chestnut and quebracho tannins in pig nutrition: The effects on performance and intestinal health. *Animal* **2020**, *15*, 100064. [CrossRef]
17. Khanbabaee, K.; Van Ree, T. Tannins: Classification and definition. *Nat. Prod. Rep.* **2001**, *18*, 641–649.
18. Ebrahimnejad, H.; Burkholz, T.; Jacob, C. Flavanols and proanthocyanidins. In *Recent Advances in Redox Active Plant and Microbial Products*; Springer: Berlin/Heidelberg, Germany, 2014; pp. 211–232.
19. Wu, M.; Hou, J.; Yang, Z.; Wang, A.; Wang, Q.; Wu, J.; Miao, L. Enhanced degradation of methionine by eco-friendly and low-cost tannin extract modified Fenton oxidation: Promotion of the Fe (III)/Fe (II) cycle. *Chem. Eng. J.* **2023**, *476*, 146656. [CrossRef]
20. Peng, K.; Lv, X.; Zhao, H.; Chen, B.; Chen, X.; Huang, W. Antioxidant and intestinal recovery function of condensed tannins in Lateolabrax maculatus responded to in vivo and in vitro oxidative stress. *Aquaculture* **2022**, *547*, 737399. [CrossRef]
21. Gourlay, G.; Constabel, C.P. Condensed tannins are inducible antioxidants and protect hybrid poplar against oxidative stress. *Tree Physiol.* **2019**, *39*, 345–355. [CrossRef] [PubMed]
22. de Veras, B.O.; da Silva, M.V.; Ribeiro, P.P.C. Tannic acid is a gastroprotective that regulates inflammation and oxidative stress. *Food Chem. Toxicol.* **2021**, *156*, 112482. [CrossRef] [PubMed]
23. Mhlanga, P.; Perumal, P.O.; Somboro, A.M.; Amoako, D.G.; Khumalo, H.M.; Khan, R.B. Mechanistic insights into oxidative stress and apoptosis mediated by tannic acid in human liver hepatocellular carcinoma cells. *Int. J. Mol. Sci.* **2019**, *20*, 6145. [CrossRef]
24. Lourenço, S.C.; Moldão-Martins, M.; Alves, V.D. Antioxidants of natural plant origins: From sources to food industry applications. *Molecules* **2019**, *24*, 4132. [CrossRef]
25. Rahaman, M.M.; Hossain, R.; Herrera--Bravo, J.; Islam, M.T.; Atolani, O.; Adeyemi, O.S.; Owolodun, O.A.; Kambizi, L.; Daştan, S.D.; Calina, D. Natural antioxidants from some fruits, seeds, foods, natural products, and associated health benefits: An update. *Food Sci. Nutr.* **2023**, *11*, 1657–1670. [CrossRef] [PubMed]
26. Dong, G.; Liu, H.; Yu, X.; Zhang, X.; Lu, H.; Zhou, T.; Cao, J. Antimicrobial and anti-biofilm activity of tannic acid against Staphylococcus aureus. *Nat. Prod. Res.* **2018**, *32*, 2225–2228. [CrossRef]
27. Alshaibani, D.; Zhang, R.; Wu, V.C. Antibacterial characteristics and activity of Vaccinium macrocarpon proanthocyanidins against diarrheagenic Escherichia coli. *J. Funct. Foods* **2017**, *39*, 133–138. [CrossRef]
28. Fraga-Corral, M.; Otero, P.; Echave, J.; Garcia-Oliveira, P.; Carpena, M.; Jarboui, A.; Nuñez-Estevez, B.; Simal-Gandara, J.; Prieto, M.A. By-Products of Agri-Food Industry as Tannin-Rich Sources: A Review of Tannins' Biological Activities and Their Potential for Valorization. *Foods* **2021**, *10*, 137. [CrossRef]
29. Bonelli, F.; Turini, L.; Sarri, G.; Serra, A.; Buccioni, A.; Mele, M. Oral administration of chestnut tannins to reduce the duration of neonatal calf diarrhea. *BMC Vet. Res.* **2018**, *14*, 227. [CrossRef] [PubMed]
30. Rivera-Méndez, C.; Plascencia, A.; Torrentera, N.; Zinn, R. Effect of level and source of supplemental tannin on growth performance of steers during the late finishing phase. *J. Appl. Anim. Res.* **2017**, *45*, 199–203. [CrossRef]
31. Krueger, W.; Gutierrez-Bañuelos, H.; Carstens, G.E.; Min, B.; Pinchak, W.; Gomez, R.; Anderson, R.; Krueger, N.; Forbes, T. Effects of dietary tannin source on performance, feed efficiency, ruminal fermentation, and carcass and non-carcass traits in steers fed a high-grain diet. *Anim. Feed Sci. Technol.* **2010**, *159*, 1–9. [CrossRef]

32. Demarco, C.F.; Paisley, S.; Goodall, R.; Brauner, C.C.; Lake, S. Effects of bacterial DFM and tannins on measures of immunity and growth performance of newly weaned beef calves. *Livest. Sci.* **2021**, *250*, 104571. [CrossRef]
33. Serri, M.; Mahdavi, A.; Riasi, A.; Hashemzadeh, F.; Kowsar, R. The addition of hydrolyzable tannin extract to milk affects calves' performance, health, blood metabolites, and pathogen shedding. *Anim. Feed Sci. Technol.* **2022**, *292*, 115451. [CrossRef]
34. Soleiman, P.; Kheiri, F. The effect of different levels of tannic acid on some performance traits in holstein dairy calves. *Iran. J. Appl. Anim. Sci.* **2018**, *8*, 19–23.
35. Hassan, Z.M.; Manyelo, T.G.; Selaledi, L.; Mabelebele, M. The effects of tannins in monogastric animals with special reference to alternative feed ingredients. *Molecules* **2020**, *25*, 4680. [CrossRef]
36. Aboagye, I.A.; Beauchemin, K.A. Potential of molecular weight and structure of tannins to reduce methane emissions from ruminants: A review. *Animals* **2019**, *9*, 856. [CrossRef] [PubMed]
37. Caprarulo, V.; Hejna, M.; Giromini, C.; Liu, Y.; Dell'Anno, M.; Sotira, S.; Reggi, S.; Sgoifo-Rossi, C.A.; Callegari, M.L.; Rossi, L. Evaluation of Dietary Administration of Chestnut and Quebracho Tannins on Growth, Serum Metabolites and Fecal Parameters of Weaned Piglets. *Animals* **2020**, *10*, 1945. [CrossRef] [PubMed]
38. *Directive 2010/63/EU of the European Parliament and of the Council of 22 September 2010 on the Protection of Animals Used for Scientific Purposes. Text with EEA Relevance*; European Parliament: Strasbourg, France, 2010.
39. *Directive 2008/119/EC of 18 December 2008 Laying Down Minimum Standards for the Protection of Calves*; European Parliament: Strasbourg, France, 2008.
40. Adejoro, F.A.; Hassen, A.; Akanmu, A.M. Effect of lipid-encapsulated acacia tannin extract on feed intake, nutrient digestibility and methane emission in sheep. *Animals* **2019**, *9*, 863. [CrossRef]
41. Amaral-Phillips, D.M.; Scharko, P.B.; Johns, J.T.; Franklin, S. Feeding and Managing Baby Calves from Birth to 3 Months of Age. UK Cooperative Extension Service, University of Kentucky, ASC-161. 2006. Available online: https://www2.ca.uky.edu/agcomm/pubs/asc/asc161/ASC161.PDF (accessed on 18 January 2024).
42. Santos, F.; De Paula, M.; Lezier, D.; Silva, J.; Santos, G.; Bittar, C. Essential oils for dairy calves: Effects on performance, scours, rumen fermentation and intestinal fauna. *Animal* **2015**, *9*, 958–965. [CrossRef]
43. Gomez, D.; Arroyo, L.; Costa, M.; Viel, L.; Weese, J. Characterization of the fecal bacterial microbiota of healthy and diarrheic dairy calves. *J. Vet. Intern. Med.* **2017**, *31*, 928–939. [CrossRef]
44. Frazzini, S.; Scaglia, E.; Dell'Anno, M.; Reggi, S.; Panseri, S.; Giromini, C.; Lanzoni, D.; Sgoifo Rossi, C.A.; Rossi, L. Antioxidant and Antimicrobial Activity of Algal and Cyanobacterial Extracts: An In Vitro Study. *Antioxidants* **2022**, *11*, 992. [CrossRef]
45. AOAC. *Official Methods of Analysis*, 21st ed.; Association of Official Analytical Chemists: Washington, DC, USA, 2019.
46. Kelln, B.; Penner, G.B.; Acharya, S.N.; McAllister, T.A.; Lardner, H.A. Impact of condensed tannin-containing legumes on ruminal fermentation, nutrition, and performance in ruminants: A review. *Can. J. Anim. Sci.* **2020**, *101*, 210–223. [CrossRef]
47. Addisu, S. Effect of dietary tannin source feeds on ruminal fermentation and production of cattle; a review. *Online J. Anim. Feed Res.* **2016**, *6*, 45–56.
48. Orzuna-Orzuna, J.F.; Dorantes-Iturbide, G.; Lara-Bueno, A.; Mendoza-Martínez, G.D.; Miranda-Romero, L.A.; Hernández-García, P.A. Effects of dietary tannins' supplementation on growth performance, rumen fermentation, and enteric methane emissions in beef cattle: A meta-analysis. *Sustainability* **2021**, *13*, 7410. [CrossRef]
49. Reggi, S.; Giromini, C.; Dell'Anno, M.; Baldi, A.; Rebucci, R.; Rossi, L. In Vitro Digestion of Chestnut and Quebracho Tannin Extracts: Antimicrobial Effect, Antioxidant Capacity and Cytomodulatory Activity in Swine Intestinal IPEC-J2 Cells. *Animals* **2020**, *10*, 195. [CrossRef] [PubMed]
50. Hoffmann, A.; Görlich, S.; Steingass, H.; Terry, H.; Schollenberger, M.; Hartung, K.; Mosenthin, R. Milk production and milk fatty acids in dairy cows fed crushed rapeseed or rapeseed oil. *Livest. Sci.* **2016**, *190*, 31–34. [CrossRef]
51. Franceschi, P.; Malacarne, M.; Bortolazzo, E.; Coloretti, F.; Formaggioni, P.; Garavaldi, A.; Musi, V.; Summer, A. Automatic milking systems in the production of Parmigiano Reggiano cheese: Effects on the milk quality and on cheese characteristics. *Agriculture* **2022**, *12*, 104. [CrossRef]
52. Malacarne, M.; Summer, A.; Fossa, E.; Formaggioni, P.; Franceschi, P.; Pecorari, M.; Mariani, P. Composition, coagulation properties and Parmigiano-Reggiano cheese yield of Italian Brown and Italian Friesian herd milks. *J. Dairy Res.* **2006**, *73*, 171–177. [CrossRef]
53. Bonelli, F.; Turini, L.; Paganelli, A.; Conte, G.; Meucci, V.; Sgorbini, M.; Buccioni, A.; Mele, M. Evaluation of oral administration of chestnut tannins in preventing calf diarrhoea. *Ital. J. Anim. Sci.* **2021**, *20*, 640–649. [CrossRef]
54. Oliveira, R.; Narciso, C.; Bisinotto, R.; Perdomo, M.; Ballou, M.; Dreher, M.; Santos, J. Effects of feeding polyphenols from pomegranate extract on health, growth, nutrient digestion, and immunocompetence of calves. *J. Dairy Sci.* **2010**, *93*, 4280–4291. [CrossRef]
55. Smith, D.R. Field disease diagnostic investigation of neonatal calf diarrhea. *Vet. Clin. Food Anim. Pract.* **2012**, *28*, 465–481. [CrossRef]
56. Muktar, Y.; Mamo, G.; Tesfaye, B.; Belina, D. A review on major bacterial causes of calf diarrhea and its diagnostic method. *J. Vet. Med. Anim. Health* **2015**, *7*, 173–185.
57. Nydam, D.V.; Wade, S.E.; Schaaf, S.L.; Mohammed, H.O. Number of Cryptosporidium parvum oocysts or Giardia spp cysts shed by dairy calves after natural infection. *Am. J. Vet. Res.* **2001**, *62*, 1612–1615. [CrossRef] [PubMed]

58. Zambriski, J.A.; Nydam, D.V.; Bowman, D.D.; Bellosa, M.L.; Burton, A.J.; Linden, T.C.; Liotta, J.L.; Ollivett, T.L.; Tondello-Martins, L.; Mohammed, H.O. Description of fecal shedding of Cryptosporidium parvum oocysts in experimentally challenged dairy calves. *Parasitol. Res.* **2013**, *112*, 1247–1254. [CrossRef]
59. Delafosse, A.; Chartier, C.; Dupuy, M.-C.; Dumoulin, M.; Pors, I.; Paraud, C. Cryptosporidium parvum infection and associated risk factors in dairy calves in western France. *Prev. Vet. Med.* **2015**, *118*, 406–412. [CrossRef]
60. Álvarez, P.; Alvarado, C.; Puerto, M.; Schlumberger, A.; Jiménez, L.; De la Fuente, M. Improvement of leukocyte functions in prematurely aging mice after five weeks of diet supplementation with polyphenol-rich cereals. *Nutrition* **2006**, *22*, 913–921. [CrossRef]
61. Verhelst, R.; Schroyen, M.; Buys, N.; Niewold, T. Dietary polyphenols reduce diarrhea in enterotoxigenic Escherichia coli (ETEC) infected post-weaning piglets. *Livest. Sci.* **2014**, *160*, 138–140. [CrossRef]
62. Sung, S.H.; Kim, K.H.; Jeon, B.T.; Cheong, S.H.; Park, J.H.; Kim, D.H.; Kweon, H.J.; Moon, S.H. Antibacterial and antioxidant activities of tannins extracted from agricultural by-products. *J. Med. Plants Res.* **2012**, *6*, 3072–3079. [CrossRef]
63. Fu, Z.; Yang, Y.; Ma, L.; Malmuthuge, N.; Guan, L.; Bu, D. Dynamics of oxidative stress and immune responses in neonatal calves during diarrhea. *J. Dairy Sci.* **2023**, *107*, 1286–1298. [CrossRef]
64. Yanar, K.E.; Eren, E.; Aktaş, M.S.; Eroğlu, M.S.; Kandemir, Ö.; Aydın, G. Prognostic potential of inflammatory markers, oxidative status, thrombocyte indices, and renal biochemical markers in neonatal calf diarrhoea-induced systemic inflammatory response syndrome. *Vet. Immunol. Immunopathol.* **2023**, *265*, 110680. [CrossRef]
65. Kumar, S.; Jakhar, K. Haematological, biochemical and oxidative stress parameter as prognostic indicators in calf diarrhoea. *Pharma Innov. J.* **2020**, *9*, 10–13.
66. Farha, A.K.; Yang, Q.-Q.; Kim, G.; Li, H.-B.; Zhu, F.; Liu, H.-Y.; Gan, R.-Y.; Corke, H. Tannins as an alternative to antibiotics. *Food Biosci.* **2020**, *38*, 100751. [CrossRef]
67. Kaczmarek, B. Tannic acid with antiviral and antibacterial activity as a promising component of biomaterials—A minireview. *Materials* **2020**, *13*, 3224. [CrossRef] [PubMed]
68. Ekambaram, S.P.; Perumal, S.S.; Balakrishnan, A. Scope of hydrolysable tannins as possible antimicrobial agent. *Phytother. Res.* **2016**, *30*, 1035–1045. [CrossRef] [PubMed]
69. Turini, L.; Mantino, A.; Tozzi, B.; Bonelli, F.; Silvi, A.; Mele, M.; Sgorbini, M.; Meucci, V.; Minieri, S. Effect of a Phytogenic Feed Additive in Preventing Calves' Diarrhea. *Front. Vet. Sci.* **2022**, *9*, 873194. [CrossRef] [PubMed]
70. Gupta, N.; Rajora, V.; Gupta, K.K.; Arora, N. Evaluation of novel polyherbal formulation for the clinico-therapeutic management of diarrhea in calves. *J. Entomol. Zool. Stud.* **2020**, *8*, 193–199.
71. Landau, S.; Azaizeh, H.; Muklada, H.; Glasser, T.; Ungar, E.; Baram, H.; Abbas, N.; Markovics, A. Anthelmintic activity of *Pistacia lentiscus* foliage in two Middle Eastern breeds of goats differing in their propensity to consume tannin-rich browse. *Vet. Parasitol.* **2010**, *173*, 280–286. [CrossRef]
72. Markovics, A.; Cohen, I.; Muklada, H.; Glasser, T.; Dvash, L.; Ungar, E.; Azaizeh, H.; Landau, S. Consumption of *Pistacia lentiscus* foliage alleviates coccidiosis in young goats. *Vet. Parasitol.* **2012**, *186*, 165–169. [CrossRef]
73. Burke, J.; Miller, J.; Terrill, T.; Orlik, S.; Acharya, M.; Garza, J.; Mosjidis, J. Sericea lespdeza as an aid in the control of *Emeria* spp. in lambs. *Vet. Parasitol.* **2013**, *193*, 39–46. [CrossRef]
74. Bhatta, R.; Uyeno, Y.; Tajima, K.; Takenaka, A.; Yabumoto, Y.; Nonaka, I.; Enishi, O.; Kurihara, M. Difference in the nature of tannins on in vitro ruminal methane and volatile fatty acid production and on methanogenic archaea and protozoal populations. *J. Dairy Sci.* **2009**, *92*, 5512–5522. [CrossRef]
75. Benchaar, C.; McAllister, T.; Chouinard, P. Digestion, ruminal fermentation, ciliate protozoal populations, and milk production from dairy cows fed cinnamaldehyde, quebracho condensed tannin, or *Yucca schidigera* saponin extracts. *J. Dairy Sci.* **2008**, *91*, 4765–4777. [CrossRef]
76. Vasta, V.; Yáñez-Ruiz, D.R.; Mele, M.; Serra, A.; Luciano, G.; Lanza, M.; Biondi, L.; Priolo, A. Bacterial and protozoal communities and fatty acid profile in the rumen of sheep fed a diet containing added tannins. *Appl. Environ. Microbiol.* **2010**, *76*, 2549–2555. [CrossRef]
77. Redondo, L.M.; Chacana, P.A.; Dominguez, J.E.; Fernandez Miyakawa, M.E. Perspectives in the use of tannins as alternative to antimicrobial growth promoter factors in poultry. *Front. Microbiol.* **2014**, *5*, 118. [CrossRef]
78. Dell'Anno, M.; Reggi, S.; Caprarulo, V.; Hejna, M.; Sgoifo Rossi, C.A.; Callegari, M.L.; Baldi, A.; Rossi, L. Evaluation of Tannin Extracts, Leonardite and Tributyrin Supplementation on Diarrhoea Incidence and Gut Microbiota of Weaned Piglets. *Animals* **2021**, *11*, 1693. [CrossRef]
79. Khan, D.; Vasu, S.; Moffett, R.C.; Irwin, N.; Flatt, P.R. Differential expression of glucagon-like peptide-2 (GLP-2) is involved in pancreatic islet cell adaptations to stress and beta-cell survival. *Peptides* **2017**, *95*, 68–75. [CrossRef] [PubMed]
80. Hatew, B.; Inabu, Y.; Sugino, T.; Steele, M. Effects of pulse-dose ruminal infusion of butyrate on plasma glucagon-like peptide 1 and 2 concentrations in dairy calves. *J. Dairy Sci.* **2019**, *102*, 2254–2265. [CrossRef]
81. García-Martín, E.; Ayuso, P.; Martínez, C.; Agúndez, J.A. Improved analytical sensitivity reveals the occurrence of gender-related variability in diamine oxidase enzyme activity in healthy individuals. *Clin. Biochem.* **2007**, *40*, 1339–1341. [CrossRef] [PubMed]
82. Dell'Anno, M.; Scaglia, E.; Reggi, S.; Grossi, S.; Rossi, C.A.S.; Frazzini, S.; Caprarulo, V.; Rossi, L. Evaluation of tributyrin supplementation in milk replacer on diarrhoea occurrence in preweaning Holstein calves. *Animal* **2023**, *17*, 100791. [CrossRef] [PubMed]

83. Thomas, V.L.; Chavez, M.; Browne, E.N.; Minnis, A.M. Instagram as a tool for study engagement and community building among adolescents: A social media pilot study. *Digit. Health* **2020**, *6*, 2055207620904548. [CrossRef]
84. Locke, S. Instagram as a Tool of Diffusion for the Livestock Industry. *J. Appl. Commun.* **2023**, *107*, 1. [CrossRef]
85. Muca, E.; Buonaiuto, G.; Lamanna, M.; Silvestrelli, S.; Ghiaccio, F.; Federiconi, A.; De Matos Vettori, J.; Colleluori, R.; Fusaro, I.; Raspa, F. Reaching a wider audience: Instagram's role in dairy cow nutrition education and engagement. *Animals* **2023**, *13*, 3503. [CrossRef]

antioxidants

Article

Impact of Chlorogenic Acid on Peripheral Blood Mononuclear Cell Proliferation, Oxidative Stress, and Inflammatory Responses in Racehorses during Exercise

Izabela Dąbrowska, Jowita Grzędzicka, Adrianna Niedzielska and Olga Witkowska-Piłaszewicz *

Department of Large Animal Diseases and Clinic, Institute of Veterinary Medicine, Warsaw University of Life Sciences, 02-787 Warsaw, Poland
* Correspondence: olga_witkowska_pilaszewicz@sggw.edu.pl

Abstract: Green coffee extract is currently of great interest to researchers due to its high concentration of chlorogenic acid (CGA) and its potential health benefits. CGA constitutes 6 to 10% of the dry weight of the extract and, due to its anti-inflammatory properties, is a promising natural supplement and agent with therapeutic applications. The purpose of our study was to discover the effects of CGA on peripheral blood mononuclear cell proliferation, and the production of pro- and anti-inflammatory cytokines as well as reactive oxidative species (ROS) in horses during exercise. According to the findings, CGA can affect the proliferation of T helper cells. In addition, at a dose of 50 g/mL, CGA increased the activation of CD4+FoxP3+ and CD8+FoxP3+ regulatory cells. Physical activity decreases ROS production in CD5+ monocytes, but this effect depends on the concentration of CGA, and the effect of exercise on oxidative stress was lower in CD14+ than in CD5+ cells. Regardless of CGA content, CGA significantly increased the release of the anti-inflammatory cytokine IL-10. Moreover, the production of IL-17 was greater in cells treated with 50 g/mL of CGA from beginners compared to the control and advanced groups of horses. Our findings suggest that CGA may have immune-enhancing properties. This opens new avenues of research into the mechanisms of action of CGA and possible applications in prevention and health promotion in sport animals.

Keywords: green coffee extract; chlorogenic acid; lymphocytes; anti-inflammatory cytokines; racing; natural supplements; exercise; monocytes; sport; oxidative stress

Citation: Dąbrowska, I.; Grzędzicka, J.; Niedzielska, A.; Witkowska-Piłaszewicz, O. Impact of Chlorogenic Acid on Peripheral Blood Mononuclear Cell Proliferation, Oxidative Stress, and Inflammatory Responses in Racehorses during Exercise. *Antioxidants* **2023**, *12*, 1924. https://doi.org/10.3390/antiox12111924

Academic Editors: Matteo Dell'Anno and Luciana Rossi

Received: 9 August 2023
Revised: 11 October 2023
Accepted: 26 October 2023
Published: 28 October 2023

1. Introduction

There has been a recent increase in the social awareness and understanding of the health-promoting properties of food. Eating less processed and/or more additive-free natural foods seems to be particularly common. Among unprocessed beverages, apart from water, coffee is the most popular drink in the world, with consumption increasing by as much as 3.3% and global sales of 170.3 million 60 kg bags in 2021/2022 compared to 2020/2021 [1]. The most common commercially available coffee is *Coffea arabica* L., *C. canephora Pierre ex A. Froehner* (also known as Robusta coffee), and *C. liberica Bull.* ex Hiern [2]. The chemical composition of roasted coffee varies depending on the kind. Coffee acids have a considerable impact on taste and aroma as well as serving as flavor precursors [3,4]. Caffeic acids are classified as organic acids (OA), chlorogenic acids (CGA), and inorganic acids like phosphoric acid [5], with CGA receiving the most interest in recent years. CGA is not only a natural diet supplement [6,7], studies indicate that it is also clinically safe regarding side effects, and its long-term administration has not caused changes in serum biochemical variables, including no significant changes in serum iron, magnesium, copper, zinc or vitamin B1 [8].

Chlorogenic acid (CGA) belongs to the group of hydroxycinnamates, esters of cinnamic acid derivatives with quinic acid, which are widely distributed in plant materials such as apples [9], pineapple [10], blueberries [11] and coffee seeds, in which the concentration of CGA is as much as 6–10% of dry matter, leading such plants to contain its highest concentration [12]. 5-O-caffeoylquinic acid (5-CQA) is the most prevalent type and accounts for 76–84% of all CGAs [13], but monoesters from caffeic acids (caffeolic acids, CQA), p-coumaric acid (p-coumaroylquinic acids, p-CoQA), and ferulic acid (feruloylquinic acids, FQA) are also prevalent [14]. CGA is proven to possess antioxidant [9,15], anti-inflammatory [16–20], antibacterial [21–25] and anticancer [26–30] effects both in humans and animals. It was proven that CGA at various doses (0, 2, 5, and 20 lM) reduces the production of pro-inflammatory cytokines like IL-1, TNF-α, and IL-6 in a dose-dependent manner in mouse macrophage cell line RAW 264.7 and BV2 microglial cells stimulated by lipopolysaccharide (LPS) [19]. Another study conducted on human peripheral blood mononuclear cells (PBMSs) proved that CGA can reduce the production of IL-1b, IL-6, TNFa, and IFNg at 86%, 90%, 84%, and 95%, respectively, at 200 mg/mL dose while levels of MCP-1, MIP-1a, and MIP-1b were decreased by 99%, 77%, and 91%, respectively [16]. Worth mentioning is that CGA is not only safe for the liver and plasma [6] but systematic consumption also leads to lower fasting plasma glucose by reducing its absorption of glycosylated hemoglobin and insulin levels by up to 6.9% [31–35].

The promotion of the anti-inflammatory process shows promise as an effective therapeutic target for treating traumatic muscle injuries [36]. Studies reveal that during physical activity in both race and endurance horses, muscle cramps accompanying muscle fiber damage initiate an exercise-induced inflammatory response [37–39]. It has been found that physical activity such as an ultramarathon stimulates the release of IL-6, IL-10, and IL-1ra [40]. It is crucial since they can suppress the production of IL-1-type cytokines and regulate the early inflammatory response to exercise in both people and horses. IL-6 release regulates metabolism and is vital in boosting satellite cell renewal and proliferation. Satellite cells, in turn, can influence muscle regeneration by secreting platelet-derived growth factor (PDGF) and IL-6, both of which encourage cell proliferation and differentiation in the healing area. The following findings suggest that extensive endurance training in horses leads to the development of a reduced inflammatory capacity, as seen by a drop in type 1 pro-inflammatory cytokine concentrations over time, and maybe the development of an anti-inflammatory condition [37]. The horse's adaptability to training, particularly the first three months of regular development, appears to be crucial to the adaptation process, with the main anti-inflammatory changes occurring in the second and third months, and the cumulative effect of the training process may be a consideration. A similar finding was confirmed in racehorses. It has been demonstrated that IL-6, IL-13, IL-10, or IL-1ra can mediate the protective long-term anti-inflammatory benefits of exercise by increasing the anti-inflammatory response [38].

Racehorses, due to their intensive training regimen are constantly exposed to physical stress stimuli. Intensive physical exercise is an acute mechanical and metabolic load that induces a wide systematic response to preserve homeostasis. It has been widely described that one of the consequences of physical exercise is changes in immunological parameters [41–43]. The inflammatory signaling cascade triggered by working muscle involves changes in peripheral blood cell numbers, granulocyte activity, NK cell cytotoxic activity, lymphocyte proliferation, and cytokine levels in plasma among others [44–47]. The inflammatory process is crucial for musculoskeletal system remodeling; however, an unbalanced inflammatory response may lead to tissue destruction [48]. The hunt for alternative remedies that can balance this inflammatory response may be critical to improving horse health and sports performance. Supplements that may have a positive impact on horse athletes' performance are becoming more and more popular nowadays, especially those of natural origin. Plant-origin immunomodulators such as CGA are used in animal nutrition. However, there is a lack of studies concerning the CGA influence on horses. In addition, it is possible to use decaffeinated green coffee seeds containing CGA which is very important especially for athletes to avoid doping accusations. Thus, the goal of this study is to evaluate how natural and safe components such as CGA impact the inflammatory reaction and oxidative stress triggered by physical exertion in horses.

2. Materials and Methods

2.1. Animals and Blood Sampling

The investigation encompassed the impact of CGA on immune cells in racehorses at different fitness levels. A total of 29 healthy racehorses, aged 2–7 years, were included in the study, with 15 males and 14 females. The horses were divided into two groups: the experienced group (advanced group), consisting of 7 Thoroughbreds (average age: 3 ± 0.37 years) and 7 Arabians (average age: 6 ± 1.29) with a history of good performance in the previous training seasons, and the inexperienced group (beginners), comprising 9 Thoroughbreds (average age: 2 ± 0) and 6 Arabians (average age: 4 ± 0.82) at the beginning of their race training career. The good performance was determined based on the history of the previous training season. The inexperienced group consisted of horses during their first training season.

Both groups were subjected to the same environmental conditions and training regimen, considering their respective fitness levels. All horses were stabled and trained by a single trainer. Prior to the study, a veterinary practitioner conducted clinical examinations, including assessment of heart rate, mucous membrane color and moisture, capillary refill time, and dehydration (measured by the time it takes for a pinched skin to fold over the point of the shoulder and flatten). Additionally, basic blood hematological and biochemical tests were performed. No clinical symptoms of diseases were observed in any of the horses.

The exercise session took place on an 800 m track, with all horses running at a speed of approximately 800 m/min. Blood samples were collected before, immediately after, and thirty minutes after the exercise session according to the routine protocol for fitness monitoring from all horses through jugular venipuncture, using K2EDTA tubes within a BD vacutainer system (BD Vacutainer®, Franklin Lakes, NJ, USA), for the isolation of peripheral blood mononuclear cells (PBMCs). For the purposes of this study, only the samples taken before and immediately after the exercise were analyzed. Only excess peripheral blood collected for routine diagnostic tests was used for this study. It is important to note that all sampling procedures adhered to the standard veterinary diagnostic protocol and were performed in compliance with Polish legal regulations and the European directive EU/2010/63. Ethical approval from the Local Commission for Ethics in Animal Experiments was not required for this study.

2.2. Chlorogenic Acid

Chlorogenic acid (CGA) obtained from Sigma-Aldrich (Sigma-Aldrich, Saint Louis, MO, USA) was solubilized in phosphate-buffered saline (Life Technologies, Bleiswijk, The Netherlands). Cells were cultured with 15 μg/mL and 50 μg/mL CGA which was used based on previous human, laboratory animal, and cell culture publications [8–32]. For the control treatments, the same procedures were followed, but without the addition of CGA. Samples from corresponding horses were used for the control treatments.

2.3. Cell Isolation and Culture

PBMCs were obtained from the K2EDTA tube blood of all horses using density gradient centrifugation (SepMate™-Lymphoprep™ System, Cologne, Germany). The cells were centrifuged at $1200\times g$ for 10 min following the manufacturer's instructions. The cells were washed twice in 2% BSA and frozen in 10% DMSO in heat-inactivated horse serum (IHS) at −80 °C for further analysis.

Subsequently, the cells were refrozen, washed twice in 2% BSA and cell cultures were established using RPMI 1640 Medium with GlutaMAX™ (Gibco, Life Technologies, Bleiswijk, The Netherlands), supplemented with 10% IHS, penicillin (100 IU/mL), streptomycin (100 μg/mL), nonessential amino acids (1%), MEM vitamins (100 μM), sodium pyruvate (1 mM), and amphotericin B (1 μg/mL) (Gibco™, Life Technologies, Bleiswijk, The Netherlands).

PBMCs were placed in a 96-well flat-bottom plate (353072, Falcon, BD, Franklin Lakes, NJ, USA) at a concentration of 2×10^5 cells in 200 μL of the culture medium per well. PBMCs were cultured in the absence or presence of phytohemagglutinin (PHA) (Sigma-Aldrich, St. Louis, MO, USA; 5 μg/mL). After 24 h, the cells were washed, and recombinant equine IL-2 (R&D Systems, Abingdon, UK; 100 U/mL) was added. The cells were then incubated for an additional 3 days at 37 °C with 5% CO_2.

2.4. Cell Staining

Samples stained with CellTrace™ Violet Cell Proliferation Kit (Life Technologies, Bleiswijk, The Netherlands) prior to culturing were used in a cell proliferation assay. The procedure was conducted using the manufacturer's instructions.

After 4 days of culture, the production of reactive oxygen species (ROS) by isolated PBMCs was assessed using the CellRox (CR) Deep Red Assay Kit (Life Technologies, Paisley, Scotland), following the manufacturer's protocol.

To analyze lymphocytes, non-adherent cells were specifically collected (these were also used for cell proliferation assessment). Equine-specific antibodies or antibodies with documented cross-reactivity (as indicated in Table 1) were utilized to evaluate the surface marker expression on PBMCs. To minimize nonspecific antibody binding, 10% BSA was used for blocking (15 min at 4 °C) prior to staining with antibodies. The cells were incubated with antibodies in eBioscience™ Flow Cytometry Staining Buffer (Life Technologies, Bleiswijk, The Netherlands) in the dark for 20 min at 4 °C. Subsequently, the cells were washed twice with 2% BSA, resuspended in 200 μL of flow cytometry staining buffer, and immediately analyzed using a cytometer.

Table 1. Monoclonal antibodies employed to tag peripheral blood mononuclear cells (PBMCs) for flow cytometry labeling.

Antibody	Clone; Dilution	Source	Target Cell
CD4:PE	CVS4; 1:10	BioRad, Berkeley, CA, USA	Lymphocytes
CD8:FITC	CVS21; 1:10	BioRad, Berkeley, CA, USA	Lymphocytes
CD5:PE	CVS5; 1:10	BioRad, Berkeley, CA, USA	Lymphocytes
CD14:AF405	433423; 1:10	R&D Systems, Minneapolis, MI, USA	Monocytes
MHCII:FITC	CVS20; 1:20	BioRad, Berkeley, CA, USA	Monocytes
FoxP3:APC	FJK-16s; 1:10	Life Technologies, Bleiswijk, The Netherland	Lymphocytes

For FoxP3 staining, the eBioscience™ FoxP3/Transcription Factor Staining Buffer Set (Life Technologies, Bleiswijk, The Netherlands) was utilized following the manufacturer's protocol.

2.5. Flow Cytometry Analysis

A similar gating strategy utilized in this study was previously described in a previous publication [47]. To ensure the analysis of single cells, doublets were excluded by setting a gate based on the FSC-area (FSC-A) vs. FSC-high (FSC-H) dot plot. Cell proliferation calculations were performed exclusively on singlet cells. The gate specifically included lymphocytes, and further analyses were conducted on CD4+, CD8+, and FoxP3+ cell populations. In the case of the second sample, the gate encompassed CD5+, CD14+, MHCII+ cells, and the median fluorescence intensity (MFI) of reactive oxygen species (ROS) was determined within that specific cell population.

Flow cytometric analysis was carried out using a FACSCanto II flow cytometer and FlowJo™ version 10.9 software (Becton, Dickinson, NJ, USA). A total of 10.000 cells from each sample were acquired for analysis.

2.6. ELISA

Cytokine concentrations (IL-1β, IL-4, IL-8, IL-10, IL-17, INF-γ, and TNF-α) were assessed in post-exercise samples treated by CGA in different doses using immunoenzymatic commercial assays specifically designed for equine species (ELK Biotechnology, Wuhan, China). Absorbance readings were obtained with a Multiscan Reader (Labsystem, Helsinki, Finland) and analyzed using Genesis V 3.00 software.

2.7. Statistical Analysis

Statistical analysis was conducted using the OriginPro 2022 statistics package (OriginLab Corporation, Northampton, MA, USA). A one-way repeated measures ANOVA was utilized to assess the impact of CGA on horses' PBMCs across three different concentrations. To compare individual groups, a post-hoc pairwise comparison was performed with the Bonferroni test. Significance was determined at a p-value of 0.05.

In cases where the data violated the normality assumption, a non-parametric Friedman ANOVA test was employed. A post-hoc analysis for the aforementioned test involved the Wilcoxon–Nemenyi–McDonald–Thompson test. Significance was evaluated at a 0.05 level.

Prior to the analysis, outlier detection was conducted using Grubbs' test or based on linear regression to ensure data integrity.

3. Results

3.1. Chlorogenic Acid Effect on Lymphocyte Phenotype and Proliferation

The lack of statistical significance indicated CGA had no effect on percentage of CD4+ cells (Figure 1A). Although CGA left the CD4+ percentage unaffected, the compound significantly influenced the proliferation of helper T cells. The highest concentration used (50 μg/mL) enhanced the activity of CD4+ cells in comparison to the control treatment ($p = 6.6 \times 10^{-6}$) (Figure 1B). Moreover, there was a significant difference when comparing it to the 15 μg/mL CGA treatment ($p = 5.6 \times 10^{-7}$). Regardless of CGA treatment, the CD4+ % remained unchanged in response to exercise. Furthermore, considering the horses' level of advancement, the percentage of CD4+ cells showed no variation following exercise. (Figure S1).

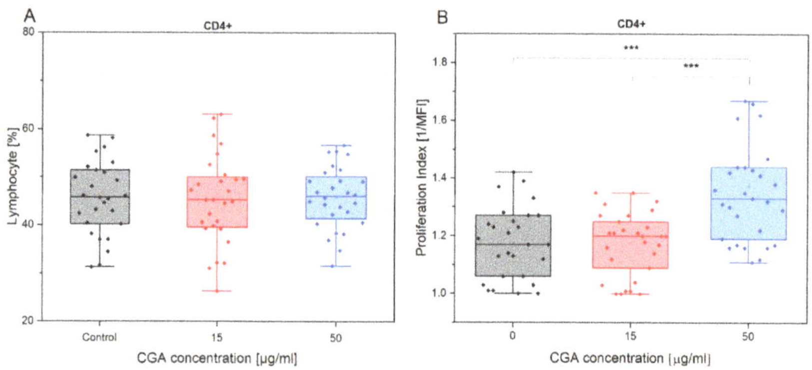

Figure 1. Presentation of CD4+ % (**A**) and its proliferation index (**B**) variations in the absence and presence of CGA at 15 μg/mL and 50 μg/mL concentrations. Each dot represents one horse sample in a particular treatment condition, and means ± SEM (standard error of the mean) are presented. Significance levels are: *** $p < 0.001$.

Statistically significant findings were only observed between two of the CGA concentrations used ($p = 0.017$). Specifically, CGA showed no significant effect on the % of CD8+ cells compared to the control treatment (Figure 2A). Similar results were also confirmed for the measured proliferation (Figure 2B).

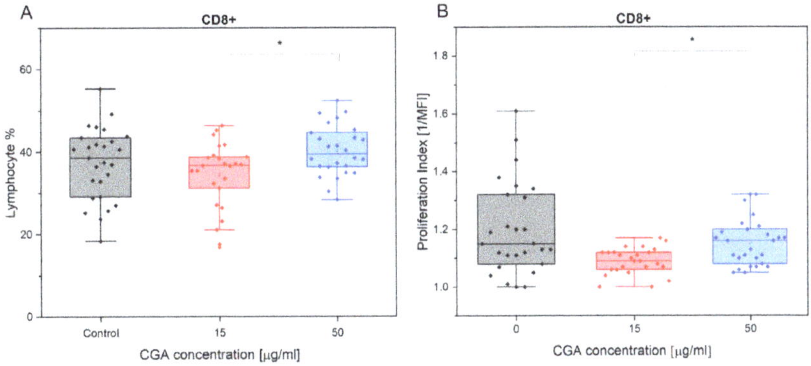

Figure 2. CD8+ % (**A**) and proliferation index (**B**) variations in the absence and presence of CGA at 15 μg/mL and 50 μg/mL concentrations. Each dot represents one horse sample in a particular treatment condition, and means ± SEM (standard error of the mean) are presented. Significance levels are: * $p < 0.05$.

Similar to the results in the case of CD4+ lymphocytes, exercise established no effect on CD8+ lymphocyte count regardless of CGA treatment (Figure S2).

3.2. *Chlorogenic Acid Effect on T Regulatory Cells*

The analysis of CD4+FoxP3+ cell % variation between the absence of CGA treatment and 50 μg/mL CGA treatment revealed a statistically significant increase in regulatory lymphocytes after CGA treatment ($p = 0.02558$) (Figure 3). The observed difference could imply that CGA at 50 μg/mL plays a role in modulating the number of regulatory CD4+FoxP3+ T cells.

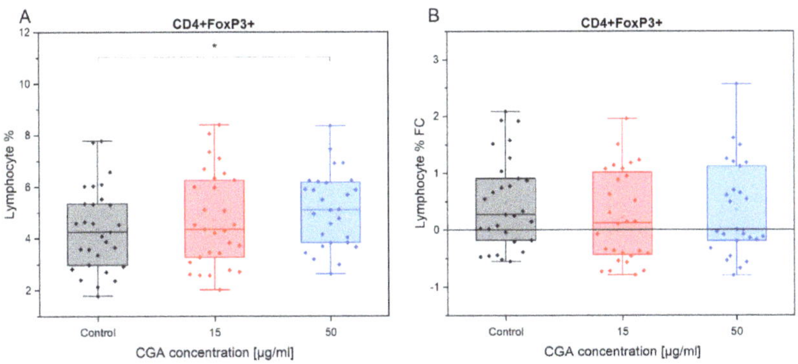

Figure 3. Presentation of CD4+FoxP3+ count variations in the absence and presence of CGA at 15 μg/mL and 50 μg/mL concentrations (**A**), and its fold change in response to exercise (**B**). Each dot represents one horse sample in a particular treatment condition, and means ± SEM (standard error of the mean) are presented. Significance levels are: * $p < 0.05$.

A statistically significant increase in CD8+FoxP3+ cell production was observed in response to CGA treatment at a concentration of 50 μg/mL compared to the control treatment ($p = 0.00015$) (Figure 4). However, treatment with CGA at a dose of 15 μg/mL did not significantly stimulate regulatory lymphocytes compared to the control treatment. Notably, a significant difference was observed when comparing the highest concentration (50 μg/mL) to the 15 μg/mL treatment, confirming a strong stimulation of CD8+FoxP3+ ($p = 0.047$). These findings suggest that CGA may stimulate the polarization of CD8+FoxP3+ regulatory lymphocytes specifically at the concentration of 50 μg/mL.

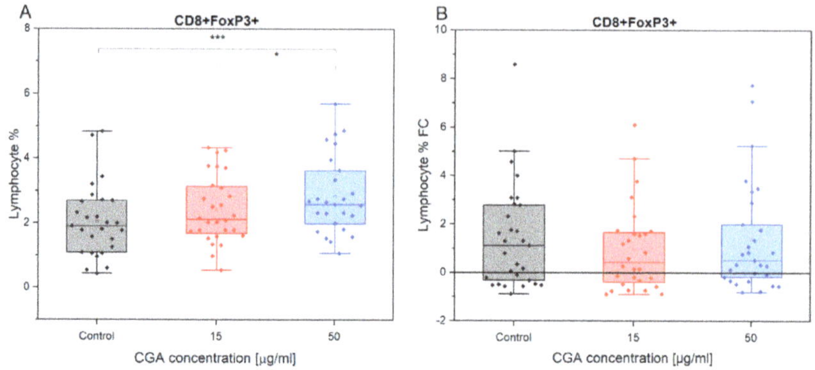

Figure 4. Presentation of CD4+FoxP3+ count variations (**A**), and its fold change in response to exercise (**B**) in the absence and presence of CGA at 15 µg/mL and 50 µg/mL concentrations. Each dot represents one horse sample in a particular treatment condition, and means ± SEM (standard error of the mean) are presented. Significance levels are: * $p < 0.05$ and *** $p < 0.001$.

The administration of CGA treatment led to an increase in the T reg population in response to exercise. While the control treatment (FC = 1.19) showed the most significant change. However, there were no significant differences observed in the magnitude of this change between different treatment conditions in response to exercise.

3.3. Chlorogenic Acid Effect on Monocyte Phenotype

According to the established criteria [47], changes in the proportion of CD14−MHCII+ non-classical, CD14+MHCII+ intermediate, and CD14+MHCII− classical monocyte cells observed in both inexperienced and experienced equine subjects following treatment with varying concentrations of CGA, as well as the FC of pre- to post-exercise did not show statistically significant findings (Figures S4–S7).

However, it can be noted that overall, the proportion of CD14+MHCII− classical monocytes prevails in the investigated horse PBMCs, while CD14−MHCII+ non-classical monocytes are the least abundant (Figure S3).

3.4. Chlorogenic Acid Effect on ROS Production in Monocytes and Lymphocytes

CGA exhibited a notable reduction in percentage of the CD5+ T lymphocytes and CD14+ monocytes that produced ROS, as assessed by the CellRox assay. This effect was most pronounced at the highest concentration of CGA tested, compared to the control treatment (CD5+: $p = 3.1 \times 10^{-5}$, CD14+: $p = 1.34 \times 10^{-6}$). However, the observed effect at 50 µg/mL did not significantly differ from the cells treated with 15 µg/mL (Figure 5A,B). Furthermore, CGA treatment significantly reduced the overall ROS accumulation (MFI) in the CD5+ and CD14+ cell subtypes in comparison to the control treatment. The extent of this effect depended on the concentration of the compound, and it is the highest for 50 µg/mL (CD5+: $p = 7.88 \times 10^{-5}$; CD14+: $p = 1.34 \times 10^{-6}$) (Figure 5C,D).

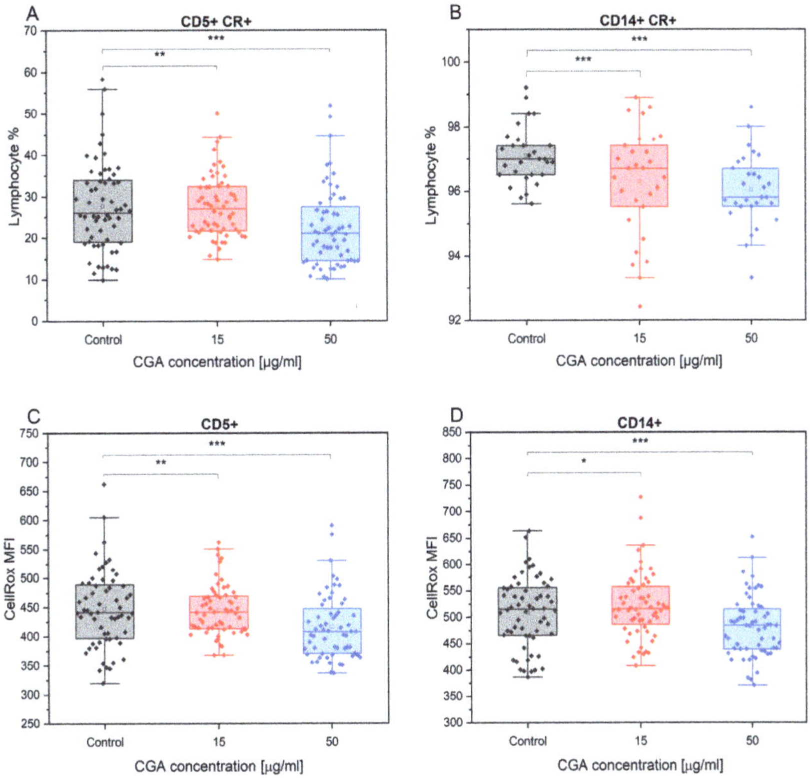

Figure 5. The CD5+ (**A**) and CD14+ (**B**) percentage gated from total lymphocytes/monocytes positive for CellRox fluorescence and the mean fluorescence intensity (MFI) of CellRox (ROS accumulation quantity) in each subtype (**C**,**D**) in response to the absence and presence of CGA at 15 µg/mL and 50 µg/mL concentrations. Each dot represents one individual horse's sample in a particular treatment condition, and means ± SEM (standard error of the mean) are presented. Significance levels are * $p < 0.05$, ** $p < 0.01$ and *** $p < 0.001$.

Notably, exercise diminished the CD5+ lymphocyte proportion identified as ROS source (FC = −0.4). This effect was clearly observed under CGA treatment conditions. However, at 15 µg/mL CGA, CD5+ lymphocyte proportion that produced ROS seemed to be reduced to a significantly lower extent in comparison to the control ($p = 4.8 \times 10^{-6}$) and 50 µg/mL treatment ($p = 6.1 \times 10^{-7}$).

The tendency of these changes was observed in both horses' advancement level groups. The average fold change (FC) of ROS production in the absence of CGA was comparable between the inexperienced (FC = −0.42) and experienced (FC = 0.41) groups (Figure 6A). However, following CGA treatment, the oxidative stress response was different. Inexperienced horses' samples demonstrated a greater reduction in oxidative stress after training when treated with 50 µg/mL of CGA, compared to the experienced group ($p = 0.003$) (Figure 6B).

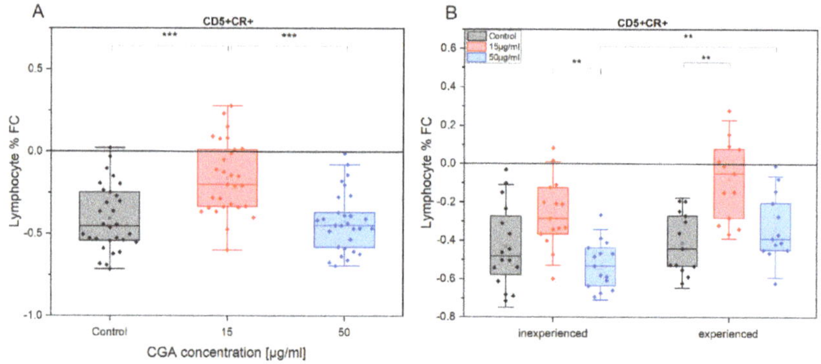

Figure 6. The fold change (FC) of CR (CellRox) intensity of CD5+ percentage gated from total lymphocytes/monocytes in response to exercise (**A**) in the absence and presence of CGA at 15 µg/mL and 50 µg/mL concentrations and at different horses' advancement level (inexperienced, experienced) (**B**). Each dot represents one individual horse sample in a particular treatment condition, and means ± SEM (standard error of the mean) are presented. Significance levels are: ** $p < 0.01$, and *** $p < 0.001$.

The impact of training had a lesser effect on oxidative stress in CD14+ monocytes compared to CD5+ lymphocytes (FC = −0.02). Moreover, treatment with CGA at 15 µg/mL did not appear to elicit changes in ROS production in CD14+ cells. However, this significantly differed from the control treatment and the 50 µg/mL CGA treatment (control: $p = 7.2 \times 10^{-6}$, 50 µg/mL: $p = 3.5 \times 10^{-5}$) (Figure 7). These trends have also been observed in both groups of horses with varying levels of advancement. It is worth noting, that 15 µg/mL CGA treatment led to a slight increase in ROS production after training (FC = 0.01).

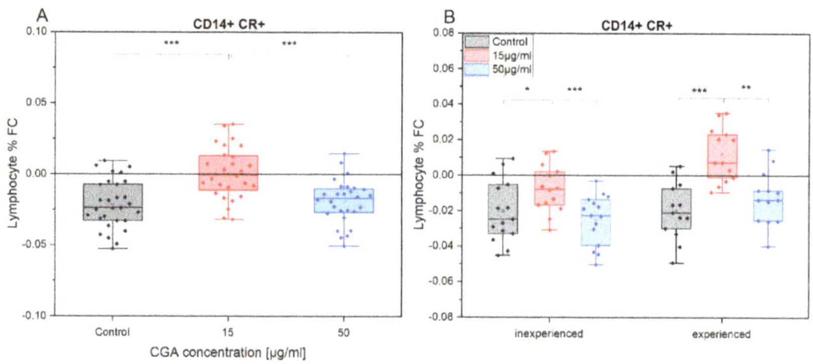

Figure 7. The fold change (FC) in CR (CellRox) intensity of CD14+ percentage gated from total lymphocytes/monocytes in response to exercise (**A**) in the absence and presence of CGA at 15 µg/mL and 50 µg/mL concentrations and at different horses' advancement level (inexperienced, experienced) (**B**). Each dot represents one horse sample in a particular treatment condition, and means ± SEM (standard error of the mean) are presented. Significance levels are: * $p < 0.05$, ** $p < 0.01$, and *** $p < 0.001$.

After exercise, a noticeable decrease in ROS accumulation was observed in both CD5+ lymphocytes and CD14+ monocytes, regardless of CGA treatment (Figure 8A,B). However, among the various CGA-treated cell groups, those exposed to a concentration of 15 µg/mL displayed the least pronounced changes.

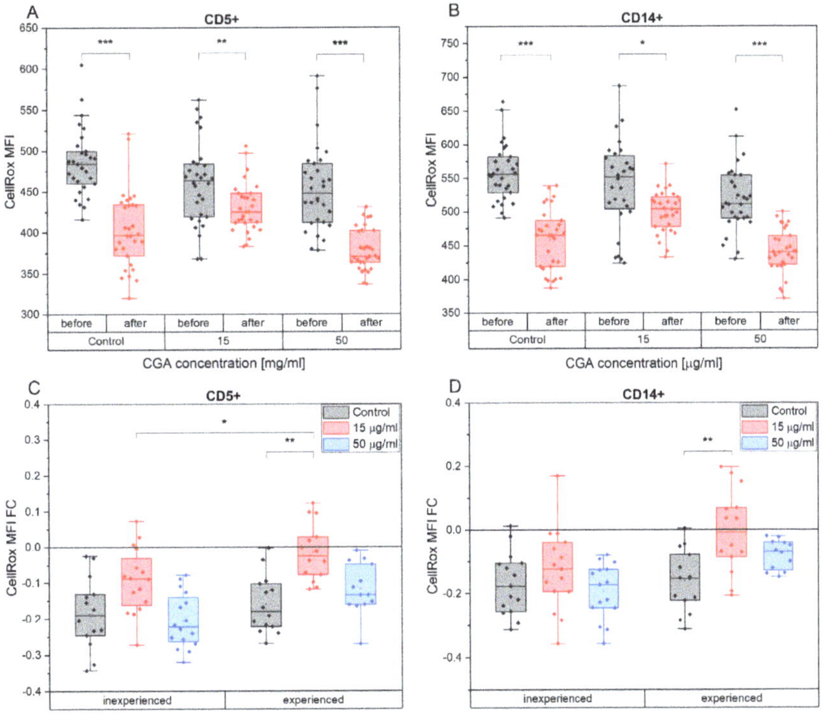

Figure 8. The ROS accumulation quantity expressed as CellRox mean fluorescence intensity (MFI) of CD5+ (**A**) and CD14+ (**B**) cell subtypes and the fold change (FC) in mean fluorescence intensity (MFI) of CD5+ (**C**) and CD14+ (**D**) in the two horses' advancement groups, inexperienced and experienced, in response to the absence and presence of CGA at 15 μg/mL and 50 μg/mL concentrations. Each dot represents one individual horse's sample in a particular treatment condition, and means ± SEM (standard error of the mean) are presented. Significance levels are * $p < 0.05$, ** $p < 0.01$, and *** $p < 0.001$.

The extent of ROS accumulation changed in CD5+ lymphocytes, expressed as fold change (FC) in response to exercise, exhibited a lesser reduction in cells derived from the more experienced group of horses compared to the inexperienced group (Figure 8C). Notably, this difference reached statistical significance in cells treated with 15 μg/mL CGA ($p = 0.049$) (Figure 8C).

In the group of inexperienced horses' cells, CGA treatment did not influence the post-exercise alterations in ROS accumulation for both CD5+ and CD14+ cell subtypes. However, within the experienced group, cells treated with 15 μg/mL CGA showed a subtle reduction in ROS accumulation after exercise (mean FC: CD5+ = −0.015 and CD14+ = −0.0056) in contrast to a more pronounced reduction in the control group (mean FC: CD5+ = −0.16 and CD14+ = −0.16) (CD5+: $p = 0.0036$; CD14+: $p = 0.0088$) (Figure 8D).

3.5. Chlorogenic Acid Effect on PBMCs Cytokine Production after Training

The production of pro-inflammatory cytokines, including IL-1β, IL-8, INF-γ, TNF-α, and IL-17 as well as anti-inflammatory cytokine IL-10, was investigated in PBMCs collected after the training session (Figure 9). The supernatant was analyzed for cytokine levels. No significant changes were observed in the production of pro-inflammatory cytokines following CGA treatment. However, CGA treatment had a significant impact on the production of anti-inflammatory cytokines. Our study revealed an increase in IL-10 secretion compared to the control treatment, regardless of the concentration of CGA used.

Additionally, IL-17 secretion was significantly higher in cells treated with 50 μg/mL CGA compared to the control treatment.

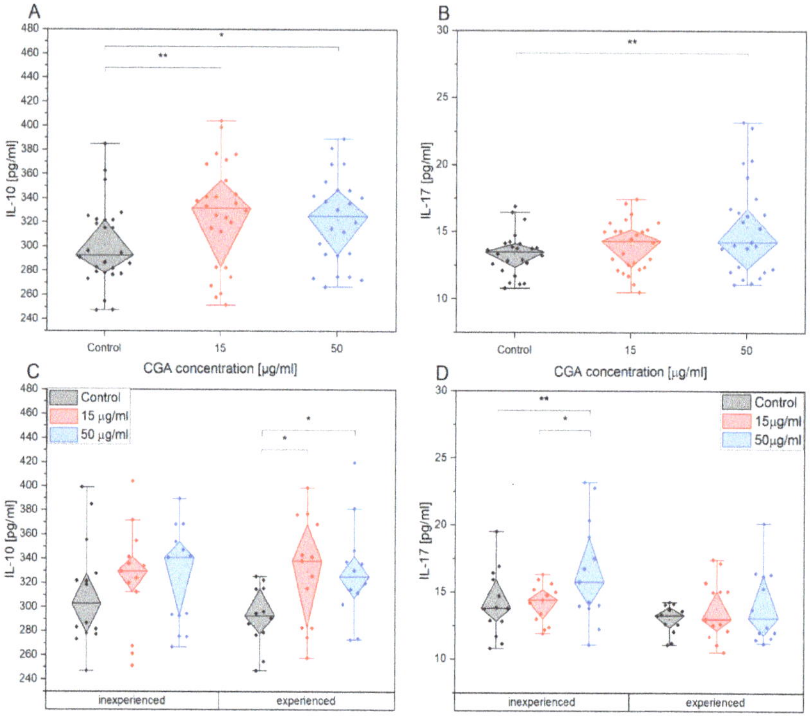

Figure 9. The change in IL-10 (**A**) and IL-17 (**B**) secretion in the absence and presence of CGA at 15 μg/mL and 50 μg/mL concentrations and its comparison between groups of experienced and inexperienced horses during exercise (**C**,**D**). Each dot represents one horse sample in a particular treatment condition, and means ± SEM (standard error of the mean) are presented. Significance levels are: * $p < 0.05$, ** $p < 0.01$.

IL-10 is upregulated significantly after CGA stimulation in PBMCs of experienced horses, while IL-17 is upregulated in inexperienced horses.

4. Discussion

Our findings provide the first confirmation of the immunomodulatory function of CGA on PBMCs in racehorses. Previous studies conducted on humans have established the anti-inflammatory and anti-oxidative effects of CGA supplementation, positioning it as a therapeutic agent with diverse applications [7,49,50]. Given the significant physiological similarity between horses and humans in the field of exercise immunology [37–39,47], the addition of CGA to the equine diet holds the potential to offer various benefits to equine athletes.

4.1. Lymphocyte Phenotype and Proliferation

T lymphocytes play a critical role in the cellular immune system. The CD4+/CD8+ ratio is defined as the ratio of helper T cells (surface marker CD4) to cytotoxic T cells (surface marker CD8), where CD8+ cells are pro-inflammatory and CD4+ cells are anti-inflammatory. Recent studies have found that CGA may affect both the population of T cells and their proliferation. In human studies, it was confirmed that CGA at a concentration of 400 mg/kg improves cellular immunity by enhancing CD4+/CD8+ proliferation by decreasing the number of CD8+ T cells while at the same time increasing CD4+ T cell proliferation in a dose-dependent manner [51]. Similar findings regarding CD4+ lymphocytes were confirmed in our study.

In addition, several studies confirmed that CGA treatment reveals a significant role in regulating the CD4+/CD8+ ratio through the activation of CD4+ T cells and in stimulating T helper cells, which may directly contribute to the regulation of inflammation [52,53]. Reaching out to another publication, it was shown that CGA exhibits its anti-cancer benefits in breast cancer treatment by increasing the percentage of CD4+ and CD8+ T cells in the spleens of experimental mice [54]. It has been clarified that CGA can activate CD4 T cells by suppressing Toll-like receptor (TLR) 4 signaling molecules, including TLR4, p-IRAK1, p-IκB, and p-p38 [55]. Our results show that the presence of CGA does not affect the percentage of CD4+ and CD8+ cells, although it influences CD4+ proliferation in a dose-dependent manner. However, the lack of significant differences in the percent of T lymphocytes after CGA and exercise may be related to the fact that intense physical activity may inhibit the proliferation of CD4+ and CD8+ cells in humans as well as equine athletes [47,55]. It was concluded that those T cells tend to show a reduced response to mitogens and antigen-specific stimulation during intense exercise.

An additional objective of our research was to show how CGA affected CD4+FoxP3+ and CD8+FoxP3+ T-cell populations. FOXP3, which encodes a transcriptional repressor protein, is primarily expressed by CD4+ or CD8+ innate regulatory cells, ensuring immunological homeostasis and self-tolerance [55,56]. CGA administration at a dose of 50 μg/mL generates a significant increase in the percent of CD4 + FOXP3 + and CD8 + FOXP3 + T cells, according to our findings. It is worth noting that a significant difference was seen when comparing the maximum concentration (50 μg/mL) with the 15 μg/mL treatment, indicating substantial activation of CD8+FoxP3+ ($p = 0.047$). As CD4+FoxP3+ and CD8+FoxP3+ have strong immunoregulating activity [57,58], we confirmed that CGA may regulate the immune system response during exercise. Also, in response to exercise, the administration of CGA therapy increases the T reg population. The available data obtained on humans are consistent with our results [51,53].

4.2. ROS Production by T Lymphocytes and Monocytes

ROS develop naturally as byproducts of normal oxygen metabolism and are crucial intracellular signaling molecules [59]. Mitochondrial ROS tends to activate and regulate the development of Th17 and Th1 lymphocytes, and their low levels trigger the immunoregulatory enzyme indoleamine-2,3-dioxygenase and increase the activity of T reg lymphocytes. It has been proven that ROS plays a pivotal role as a T cell receptor (TCR) signaling molecule in various aspects of T lymphocyte-mediated immunity encompassing T cell proliferation, effector function, and apoptosis [60]. Moreover, ROS production by monocytes serves as a key machinery in innate immunity and inflammation [61]. Despite its critical immunostimulatory role, excess ROS can react with and damage biomolecules such as proteins, lipids, and DNA, which can disrupt redox equilibrium. Increased exposure to ROS leads to reduced phosphorylation and activation of NF-B, resulting in hyporeactivity of T lymphocytes. It is well-studied that excessive ROS production is triggered within the contracting muscle during physical exercise [62]. Therefore, it is essential to consider the importance of maintaining a balance between ROS and antioxidant proteins within the cell, especially during an intensive training regimen.

This balance plays a critical role in preserving the integrity of cell-mediated immunity. Several studies have shown that CGA has the biological ability to reduce ROS levels in a variety of concentrations, including 250 M CGA [63], 125 and 250 g/mL [64], 64 g/mL [65], 50 mg/mL [66], and 10 M [67]. This aligns with the results of our work. We confirmed that CGA significantly reduces the percentage of ROS-positive T lymphocytes CD5+ and CD14+ monocytes in racehorses. This effect is most pronounced at the highest CGA concentration tested compared to the control treatment (CD5+: $p = 3.1 \times 10^{-5}$, CD14+: $p = 1.34 \times 10^{-6}$).

In addition, in this study, it has been noticed that the important balance in ROS production, in response to exercise, can be preserved by the CGA treatment. Studies report that physical exercise induces upregulation of antioxidant enzyme pathways and thus reduces the oxidative state in skeletal muscles, liver, and heart [68,69]. We have observed a lowered percentage of CD5+ and CD14+ cells positive for ROS production after exercise. A uniform decrease in ROS generation is observed in both horse performance level groups, which is in contrast to our previous study [47]. However, in a previous study, tert-butyl hydroperoxide solution (TBHP) was used as an inducer of reactive oxygen species (ROS) production in all samples. In this study, we studied ROS production without additional stimulation besides exercise.

ROS generation plays an important role in muscle remodeling, and hypertrophy, through stimulating molecular pathways via proteins, including peroxisome proliferator-activated receptor-c coactivator (PGC1-α) and mitogen-activated protein kinases (MAPK) [70]. However, when cells were treated with 15 μg/mL of CGA, ROS production appeared to remain unchanged in post-exercise cells compared to pre-exercise levels. Notably, this observation is limited to the experienced group of horses, suggesting that CGA may enhance the positive adaptation specifically in the experienced individuals. ROS contributes to the maintenance of cartilage homeostasis and regulates apoptosis, extracellular matrix synthesis and breakdown [71–73], and cytokine production in chondrocytes [63], which is stronger in horses starting their training. Comparing CGA effects over PBMCs, we show that training has less effect on oxidative stress in CD14+ monocytes compared to CD5+ T lymphocytes. In our previous study, exercise had only an effect on ROS accumulation in CD14+ cells [47]. The ability of CGA to modulate ROS production is particularly important since low levels of ROS play an important role in controlling cellular functions by maintaining cellular antioxidant systems against oxidative stress as well as cytokine production. In summary, under physiological conditions, the equilibrium between ROS and antioxidant systems guarantees the proper functioning of T cells and the mounting of a controlled immune response.

4.3. Cytokines

CGA is known to promote an anti-inflammatory state by decreasing the level of the pro-inflammatory cytokine, as has been described in in vitro and in vivo animal and human studies [20,74–76]. The IL-8 diminished production has been observed in Caco-2 cells treated with 0.5–2 mM CGA [20]. Other studies confirmed that CGA, in a dose-dependent manner (2 to 20 μM), decreases the production of pro-inflammatory cytokines, mainly TNF-α, IL-8, IL-6, and IL-1β in vivo [20,74–76]. CGA's ability to modulate cytokine secretion is believed to be associated with its capacity to reduce oxidative stress as it was mentioned earlier. This relationship has also been observed in our study, where CGA was found to decrease ROS production in immune cells. It has been suggested that by reducing oxidative stress, CGA can inhibit the activation of the NF-κB signaling pathway, leading to a decrease in the production of pro-inflammatory cytokines and other cellular mediators.

Indirect confirmation for that mechanism, despite the evident effect of reduced ROS production, is the increased production of IL-10 after CGA stimulation by PBMCs collected post-exercise. IL-10 exhibits diverse biological effects across different cell types [77], and it is the strongest anti-inflammatory cytokine. Its action promotes the downregulation of pro-inflammatory cytokines and Th1 lymphocyte functioning [78]. The observed elevation in CGA-induced IL-10 secretion after exercise could explain the subdued inflammatory response noted in our assessment of lymphocyte and monocyte activity changes from pre- to post-exercise cells. Remarkably, our study demonstrates that the anti-inflammatory effects of CGA are particularly pronounced in PBMCs from the well-trained group of horses, which is consistent with our previous findings. It is worth noting that the elevated IL-10 level is characteristic of progressing adaptation to exercise [47]. CGA stimulation might have been responsible for the further promotion of positive adaptation to training.

IL-17 is the only investigated pro-inflammatory cytokine that responds to CGA treatment. In fact, it has been confirmed that IL-17 is one of the highly enriched molecular pathways by CGA [79]. The CGA stimulation increases IL-17 secretion but only in post-exercise PBMCs of inexperienced horses. IL-17 is recognized for its ability to induce the production of various other pro-inflammatory factors. Through its action on endothelial cells, this cytokine facilitates the migration of neutrophils into inflamed tissues. The acute bout of exercise, as untrained horses experienced, might provoke the muscle damage inflammatory response, and it is possible that CGA treatment further promotes the IL-17 secretion by stimulating the CD4+ activity as was also confirmed in our study [80]. This thesis may be forced by the stimulation of IL-17 synthesis by PBMCs after exercise in untrained racehorses, which was confirmed in our previous study [47]. As was mentioned in our previous study, the inflammatory process and tendon remodeling are stimulated by IL-17 production not only by local tissues but also by PBMCs and it is balanced mainly by IL-10 upregulation [81]. IL-17 production is needed; however, its upregulation may not always be beneficial, especially in inexperienced horses. Thus, it confirms that CGA has a strong immunomodulatory effect and can be beneficial to horses at different fitness levels.

5. Limitations

The limitations of this study are the relatively low sample size (n = 29) and diversity of the horses as only two breeds were evaluated. On the other hand, Thoroughbreds and Arabians are the most popular racing breeds. In addition, the study was not performed in vivo; however, this was decided based on good ethical practices in performing experiments in animal models. The long-term effects and safety of CGA supplementation remain ambiguous, and its interactions with other common supplements or medications given to racehorses are unexplored. It should be noted that the effects of varying exercise intensities, durations, and other environmental factors influencing immune responses and ROS production could also impact the results.

6. Conclusions

The research presented here holds significant value for harnessing CGA's potential as a nutritional supplementation to enhance the training performance of racehorses. The demonstrated immunomodulatory and oxidative stress reduction properties of this phenolic acid on PBMCs may greatly benefit athletes' performance and recovery. The CGA anti-inflammatory effect is based on the stimulation of CD4+ proliferation and the immunophenotype of CD4+FoxP3+ and CD8+FoxP3+ regulatory T cells as well as IL-10 production. By understanding these effects, CGA could prove to be an asset in optimizing athletic performance and aiding in post-injury recovery. Thus, our findings have great meaning for future research or may have some practical applications for producing supplements for equine athletes. To fortify the study, it might be beneficial to delve into a long-term analysis of CGA's effects, understand its interactions with other supplements, and assess its performance under diverse exercise regimens. Investigating the molecular mechanisms through which CGA operates, its impact on other equine cells beyond PBMCs, and linking the cellular benefits to behavioral or performance outcomes in horses could provide a more holistic view of CGA's potential benefits in equine health.

Supplementary Materials: The following supporting information can be downloaded at: https://www.mdpi.com/article/10.3390/antiox12111924/s1, Figure S1: CD4+ count fold change (FC) in response to exercise (A) at different advancement levels (B) in the absence and presence of CGA at 15 μg/mL and 50 μg/mL concentrations. Each dot represents one horse sample in a particular treatment condition, and means ± SEM (standard error of the mean) are presented. Significance levels are: * $p < 0.05$, ** $p < 0.01$, and *** $p < 0.001$; Figure S2: Presentation of CD8+ count fold change (FC) in response to exercise in general (A) and at different advancement levels (B) in the absence and presence of CGA at 15 μg/mL and 50 μg/mL concentrations. Each dot represents one horse sample in a particular treatment condition, and means ± SEM (standard error of the mean) are presented. Significance levels are: * $p < 0.05$, ** $p < 0.01$, and *** $p < 0.001$; Figure S3: Percentages of positive cells: CD14+MHCII−, CD14−MHCII+, CD14−MHCII−, and CD14+MHCII+ gated from total monocytes (A), and its FC in response to exercise (B) in the absence and presence of CGA at 15 μg/mL and 50 μg/mL concentrations. Each dot represents one horse sample in a particular treatment condition, and means ± SEM (standard error of the mean) are presented. Significance levels are: * $p < 0.05$, ** $p < 0.01$, and *** $p < 0.001$; Figure S4: Presentation of CD14+MHCII+ count fold change (FC) in response to exercise in general (A) and at different advancement levels (B) in the absence and presence of CGA at 15 μg/mL and 50 μg/mL concentrations. Each dot represents one horse sample in a particular treatment condition, and means ± SEM (standard error of the mean) are presented. Significance levels are: * $p < 0.05$, ** $p < 0.01$, and *** $p < 0.001$; Figure S5: Presentation of CD14+MHCII− count fold change (FC) in response to exercise in general (A) and at different advancement levels (B) in the absence and presence of CGA at 15 μg/mL and 50 μg/mL concentrations. Each dot represents one horse sample in a particular treatment condition, and means ± SEM (standard error of the mean) are presented. Significance levels are: * $p < 0.05$, ** $p < 0.01$, and *** $p < 0.001$; Figure S6: Presentation of CD14−MHCII+ count fold change (FC) in response to exercise in general (A) and at different advancement levels (B) in the absence and presence of CGA at 15 μg/mL and 50 μg/mL concentrations. Each dot represents one horse sample in a particular treatment condition, and means ± SEM (standard error of the mean) are presented. Significance levels are: * $p < 0.05$, ** $p < 0.01$, and *** $p < 0.001$; Figure S7: Presentation of CD14−MHCII− count fold change (FC) in response to exercise in general (A) and at different advancement levels (B) in the absence and presence of CGA at 15 μg/mL and 50 μg/mL concentrations. Each dot represents one horse sample in a particular treatment condition, and means ± SEM (standard error of the mean) are presented. Significance levels are: * $p < 0.05$, ** $p < 0.01$, and *** $p < 0.001$.

Author Contributions: Conceptualization, O.W.-P.; methodology, O.W.-P.; formal analysis, O.W.-P., I.D. and J.G.; investigation, O.W.-P., I.D., J.G. and A.N.; resources, O.W.-P.; data curation, I.D. and J.G.; writing—original draft preparation, O.W.-P., I.D. and J.G.; writing—review and editing, O.W.-P., I.D. and J.G.; visualization, I.D. and J.G.; supervision, O.W.-P.; project administration, O.W.-P.; funding acquisition, O.W.-P. All authors have read and agreed to the published version of the manuscript.

Funding: This research was partly funded by the National Science Centre, Poland, grant number 2021/41/B/NZ7/03548 (O.W.P.).

Institutional Review Board Statement: Ethical review and approval was not required for the animal study because blood samples were collected as part of standard veterinary diagnostic procedures. Thus, no approval of the Local Commission for Ethics in Animal Experiments was required, according to the Polish legal regulations and the European directive EU/2010/6.

Informed Consent Statement: Informed consent was obtained from the owners for the participation of their animals in this study.

Data Availability Statement: All datasets generated and/or analyzed during the current study are presented in the article, the accompanying SourceData (https://doi.org/10.18150/LAVBG5, RepOD, V1) or Supplementary Information files, are available from the corresponding author upon reasonable request.

Conflicts of Interest: The authors declare no conflict of interest.

References

1. International Coffee Organization—Monthly Coffee Market Report (2022/23). 2022. Available online: https://www.ico.org/Market-Report-22-23-e.asp (accessed on 13 July 2023).
2. Davis, A.P.; Govaerts, R.; Bridson, D.M.; Stoffelen, P. An annotated taxonomic conspectus of the genus Coffea (Rubiaceae). *Bot. J. Linn. Soc.* **2006**, *152*, 465–512. [CrossRef]
3. Ginz, M.; Balzer, H.H.; Bradbury, A.G.W.; Maier, H.G. Formation of aliphatic acids by carbohydrate degradation during roasting of coffee. *Eur. Food Res. Technol.* **2000**, *211*, 404–410. [CrossRef]
4. Vio, M.B.E.M.F.A.; Luisa, P.F.; Fabiana, C.R.; Eacute, H.S.T.J.; Gerson, S.G.; Terezinha, J.G.S. The relationship between organic acids, sucrose and the quality of specialty coffees. *Afr. J. Agric. Res.* **2016**, *11*, 709–717. [CrossRef]
5. Rune, C.J.; Giacalone, D.; Steen, I.; Duelund, L.; Münchow, M.; Clausen, M.P. Acids in brewed coffees: Chemical composition and sensory threshold. *Curr. Res. Food Sci.* **2023**, *6*, 100485. [CrossRef]
6. Almoosawi, S.; Dickinson, A.; Fyfe, L.; Kenyon, C.J.; Al-Dujaili, E.A.S. Effect of green coffee bean extract and chlorogenic acid consumption on 11[beta]HSD activity in humans and mice. *Endocr. Abstr.* **2023**, *19*, 122.
7. Naveed, M.; Hejazi, V.; Abbas, M.; Kamboh, A.A.; Khan, G.J.; Shumzaid, M.; Ahmad, F.; Babazadeh, D.; FangFang, X.; Modarresi-Ghazani, F.; et al. Chlorogenic acid (CGA): A pharmacological review and call for further research. *Biomed. Pharmacother.* **2019**, *97*, 67–74. [CrossRef]
8. Watanabe, T.; Arai, Y.; Mitsui, Y.; Kusaura, T.; Okawa, W.; Kajihara, Y.; Saito, I. The blood pressure-lowering effect and safety of chlorogenic acid from green coffee bean extract in essential hypertension. *Clin. Exp. Hypertens.* **2006**, *28*, 439–449. [CrossRef]

9. Upadhyay, R.; Mohan Rao, L.J. An outlook on chlorogenic acids-occurrence, chemistry, technology, and biological activities. *Crit. Rev. Food Sci. Nutr.* **2013**, *53*, 968–984. [CrossRef]
10. Shafaghatlonbar, M.; Bagherzade, G. Dual role of chlorogenic acid as an influential precursor in synthesizing nano-sized Cu(II) complexes and investigating its catalytic role in the oxidation of alcohols and its antibacterial activity. *J. Organomet. Chem.* **2023**, *996*, 122758. [CrossRef]
11. Rodriguez-Mateos, A.; Cifuentes-Gomez, T.; Tabatabaee, S.; Lecras, C.; Spencer, J.P.E. Procyanidin, Anthocyanin, and Chlorogenic Acid Contents of Highbush and Lowbush Blueberries. *J. Agric. Food Chem.* **2012**, *60*, 5772–5778. [CrossRef]
12. Clifford, M.N. Chlorogenic acids and other cinnamates—Nature, occurrence and dietary burden. *J. Sci. Food Agric.* **1999**, *79*, 362–372. [CrossRef]
13. Liang, N.; Kitts, D.D. Role of Chlorogenic Acids in Controlling Oxidative and Inflammatory Stress Conditions. *Nutrients* **2015**, *8*, 16. [CrossRef] [PubMed]
14. Clifford, M.N. Chlorogenic acids and other cinnamates—Nature, occurrence, dietary burden, absorption and metabolism. *J. Sci. Food Agric.* **2000**, *80*, 1033–1043. [CrossRef]
15. Chiang, H.-M.; Chen, C.-W.; Chen, C.-C.; Wang, H.-W.; Jhang, J.-H.; Huang, Y.-H.; Wen, K.-C. Chapter 58—Role of Coffea arabica Extract and Related Compounds in Preventing Photoaging and Photodamage of the Skin. In *Coffee in Health and Disease Prevention*; Preedy, V.R., Ed.; Academic Press: San Diego, CA, USA, 2015; pp. 523–530. [CrossRef]
16. Krakauer, T. The polyphenol chlorogenic acid inhibits staphylococcal exotoxin-induced inflammatory cytokines and chemokines. *Immunopharmacol. Immunotoxicol.* **2002**, *24*, 113–119. [CrossRef] [PubMed]
17. de Magalhães, P.M.; Dupont, I.; Hendrickx, A.; Joly, A.; Raas, T.; Dessy, S.; Sergent, T.; Schneider, Y.-J. Anti-inflammatory effect and modulation of cytochrome P450 activities by Artemisia annua tea infusions in human intestinal Caco-2 cells. *Food Chem.* **2012**, *134*, 864–871. [CrossRef]
18. Yun, N.; Kang, J.-W.; Lee, S.-M. Protective effects of chlorogenic acid against ischemia/reperfusion injury in rat liver: Molecular evidence of its antioxidant and anti-inflammatory properties. *J. Nutr. Biochem.* **2012**, *23*, 1249–1255. [CrossRef]
19. Hwang, S.J.; Kim, Y.-W.; Park, Y.; Lee, H.-J.; Kim, K.-W. Anti-inflammatory effects of chlorogenic acid in lipopolysaccharide-stimulated RAW 264.7 cells. *Inflamm. Res.* **2014**, *63*, 81–90. [CrossRef]
20. Shin, H.S.; Satsu, H.; Bae, M.-J.; Zhao, Z.; Ogiwara, H.; Totsuka, M.; Shimizu, M. Anti-inflammatory effect of chlorogenic acid on the IL-8 production in Caco-2 cells and the dextran sulphate sodium-induced colitis symptoms in C57BL/6 mice. *Food Chem.* **2015**, *168*, 167–175. [CrossRef]
21. Lou, Z.; Wang, H.; Zhu, S.; Ma, C.; Wang, Z. Antibacterial activity and mechanism of action of chlorogenic acid. *J Food Sci* **2011**, *76*, M398–M403. [CrossRef]
22. Sun, Z.; Zhang, X.; Wu, H.; Wang, H.; Bian, H.; Zhu, Y.; Xu, W.; Liu, F.; Wang, D.; Fu, L. Antibacterial activity and action mode of chlorogenic acid against Salmonella Enteritidis, a foodborne pathogen in chilled fresh chicken. *World J. Microbiol. Biotechnol.* **2020**, *36*, 24. [CrossRef]
23. Wu, Y.; Liang, S.; Zhang, M.; Wang, Z.; Wang, Z.; Ren, X. The Effect of Chlorogenic Acid on Bacillus subtilis Based on Metabolomics. *Molecules* **2020**, *25*, 4038. [CrossRef] [PubMed]
24. Chen, K.; Peng, C.; Chi, F.; Yu, C.; Yang, Q.; Li, Z. Antibacterial and Antibiofilm Activities of Chlorogenic Acid Against Yersinia enterocolitica. *Front. Microbiol.* **2022**, *13*, 885092. [CrossRef] [PubMed]
25. Yang, X.; Lan, W.; Xie, J. Antimicrobial and anti-biofilm activities of chlorogenic acid grafted chitosan against Staphylococcus aureus. *Microb. Pathog.* **2022**, *173*, 105748. [CrossRef] [PubMed]
26. Kang, T.Y.; Yang, H.R.; Zhang, J.; Li, D.; Lin, J.; Wang, L.; Xu, X. The Studies of Chlorogenic Acid Antitumor Mechanism by Gene Chip Detection: The Immune Pathway Gene Expression. *J. Anal. Methods Chem.* **2013**, *2013*, 617243. [CrossRef]
27. Yan, Y.; Li, J.; Han, J.; Hou, N.; Song, Y.; Dong, L. Chlorogenic acid enhances the effects of 5-fluorouracil in human hepatocellular carcinoma cells through the inhibition of extracellular signal-regulated kinases. *Anticancer Drugs* **2015**, *26*, 540–546. [CrossRef]
28. Zhang, F.; Yin, G.; Han, X.; Jiang, X.; Bao, Z. Chlorogenic acid inhibits osteosarcoma carcinogenesis via suppressing the STAT3/Snail pathway. *J. Cell. Biochem.* **2019**, *120*, 10342–10350. [CrossRef]
29. Sapio, L.; Salzillo, A.; Illiano, M.; Ragone, A.; Spina, A.; Chiosi, E.; Pacifico, S.; Catauro, M.; Naviglio, S. Chlorogenic acid activates ERK1/2 and inhibits proliferation of osteosarcoma cells. *J. Cell. Physiol.* **2020**, *235*, 3741–3752. [CrossRef]
30. Gupta, A.; Atanasov, A.G.; Li, Y.; Kumar, N.; Bishayee, A. Chlorogenic acid for cancer prevention and therapy: Current status on efficacy and mechanisms of action. *Pharmacol. Res.* **2022**, *186*, 106505. [CrossRef]
31. Thom, E. The effect of chlorogenic acid enriched coffee on glucose absorption in healthy volunteers and its effect on body mass when used long-term in overweight and obese people. *J. Int. Med. Res.* **2007**, *35*, 900–908. [CrossRef]
32. Jin, S.; Chang, C.; Zhang, L.; Liu, Y.; Huang, X.; Chen, Z. Huang, and Z. Chen. Chlorogenic Acid Improves Late Diabetes through Adiponectin Receptor Signaling Pathways in db/db Mice. *PLoS ONE* **2015**, *10*, e0120842. [CrossRef]
33. Roshan, H.; Nikpayam, O.; Sedaghat, M.; Sohrab, G. Sedaghat, and G. Sohrab. Effects of green coffee extract supplementation on anthropometric indices, glycaemic control, blood pressure, lipid profile, insulin resistance and appetite in patients with the metabolic syndrome: A andomized clinical trial. *Br. J. Nutr.* **2018**, *119*, 250–258. [CrossRef] [PubMed]
34. Asbaghi, O.; Sadeghian, M.; Nasiri, M.; Khodadost, M.; Shokri, A.; Panahande, B.; Pirouzi, A.; Sadeghi, O. The effects of green coffee extract supplementation on glycemic indices and lipid profile in adults: A systematic review and dose-response meta-analysis of clinical trials. *Nutr. J.* **2020**, *19*, 71. [CrossRef] [PubMed]

35. Pourmasoumi, M.; Hadi, A.; Marx, W.; Najafgholizadeh, A.; Kaur, S.; Sahebkar, A. The Effect of Green Coffee Bean Extract on Cardiovascular Risk Factors: A Systematic Review and Meta-analysis. *Adv. Exp. Med. Biol.* **2021**, *1328*, 323–345. [CrossRef] [PubMed]
36. Urso, M.L.; Sawka, M.N. Inflammation: Sustaining the balance to optimize recovery of skeletal muscle, connective tissue, and exertional injuries. *J. Appl. Physiol.* **1985**, *115*, 877–878. [CrossRef]
37. Witkowska-Piłaszewicz, O.; Bąska, P.; Czopowicz, M.; Żmigrodzka, M.; Szarska, E.; Szczepaniak, J.; Nowak, Z.; Winnicka, A.; Cywińska, A. Anti-Inflammatory State in Arabian Horses Introduced to the Endurance Training. *Animals* **2019**, *9*, 616. [CrossRef]
38. Plisak, U.; Szczepaniak, J.; Żmigrodzka, M.; Giercuszkiewicz-Hecold, B.; Witkowska-Piłaszewicz, O. Changes in novel anti-infalmmatory cytokine concetration in the bood of endurance and race horses at different levels of training. *Comput. Struct. Biotechnol. J.* **2023**, *21*, 418–424. [CrossRef]
39. Horohov, D.W.; Sinatra, S.T.; Chopra, R.K.; Jankowitz, S.; Betancourt, A.; Bloomer, R.J. The Effect of Exercise and Nutritional Supplementation on Proinflammatory Cytokine Expression in Young Racehorses During Training. *J. Equine Vet. Sci.* **2012**, *32*, 805–815. [CrossRef]
40. Fallon, K.E. The acute phase response and exercise: The ultramarathon as prototype exercise. *Clin. J. Sport. Med.* **2001**, *11*, 38–43. [CrossRef]
41. Pedersen, B.K.; Toft, A.D. Effects of exercise on lymphocytes and cytokines. *Br. J. Sports Med.* **2000**, *34*, 246–251. [CrossRef]
42. Capomaccio, S.; Cappelli, K.; Spinsanti, G.; Mencarelli, M.; Muscettola, M.; Felicetti, M.; Supplizi, A.V.; Bonifazi, M. Athletic humans and horses: Comparative analysis of interleukin-6 (IL-6) and IL-6 receptor (IL-6R) expression in peripheral blood mononuclear cells in trained and untrained subjects at rest. *BMC Physiol.* **2011**, *11*, 3. [CrossRef]
43. Cappelli, K.; Felicetti, M.; Capomaccio, S.; Nocelli, C.; Silvestrelli, M.; Verini-Supplizi, A. Effect of training status on immune defence related gene expression in Thoroughbred: Are genes ready for the sprint? *Vet. J.* **2013**, *195*, 373–376. [CrossRef] [PubMed]
44. Moldoveanu, A.I.; Shephard, R.J.; Shek, P.N. The cytokine response to physical activity and training. *Sports Med.* **2001**, *31*, 115–144. [CrossRef]
45. Walsh, N.P.; Gleeson, M.; Shephard, R.J.; Gleeson, M.; A Woods, J.; Bishop, N.C.; Fleshner, M.; Green, C.; Pedersen, B.K.; Hoffman-Goetz, L.; et al. Position statement. Part one: Immune function and exercise. *Exerc. Immunol. Rev.* **2011**, *17*, 6–63. [PubMed]
46. Peake, J.M.; Neubauer, O.; Walsh, N.P.; Simpson, R.J. Recovery of the immune system after exercise. *J. Appl. Physiol.* **2017**, *122*, 1077–1087. [CrossRef] [PubMed]
47. Witkowska-Piłaszewicz, O.; Pingwara, R.; Winnicka, A. The Effect of Physical Training on Peripheral Blood Mononuclear Cell Ex Vivo Proliferation, Differentiation, Activity, and Reactive Oxygen Species Production in Racehorses. *Antioxidants* **2020**, *9*, 1155. [CrossRef]
48. Beiter, T.; Hoene, M.; Prenzler, F.; Mooren, F.C.; Steinacker, J.M.; Weigert, C.; Nieß, A.M.; Munz, B. Exercise, skeletal muscle and inflammation: ARE-binding proteins as key regulators in inflammatory and adaptive networks. *Exerc. Immunol. Rev.* **2015**, *21*, 42–57.
49. Şimşek, A.; Çiçek, B.; Turan, E. The effect of chlorogenic acid from green coffee as a natural antioxidant on the shelf life and composition of hazelnut paste. *Eur. Food Res. Technol.* **2023**, *8*, 2077–2086. [CrossRef]
50. Abaidullah, M.; Peng, S.; Song, X.; Zou, Y.; Li, L.; Jia, R.; Yin, Z. Chlorogenic acid is a positive regulator of MDA5, TLR7 and NF-κB signaling pathways mediated antiviral responses against Gammacoronavirus infection. *Int. Immunopharmacol.* **2021**, *96*, 107671. [CrossRef]
51. Wang, L.; Pan, X.; Jiang, L.; Chu, Y.; Gao, S.; Jiang, X.; Zhang, Y.; Chen, Y.; Luo, S.; Peng, C. The Biological Activity Mechanism of Chlorogenic Acid and Its Applications in Food Industry: A Review. *Front. Nutr.* **2022**, *9*, 943911. [CrossRef]
52. Zeng, L.; Xiang, R.; Fu, C.; Qu, Z.; Liu, C. The Regulatory effect of chlorogenic acid on gut-brain function and its mechanism: A systematic review. *Biomed. Pharmacother.* **2022**, *149*, 112831. [CrossRef]
53. Zeng, A.; Liang, X.; Zhu, S.; Liu, C.; Wang, S.; Zhang, Q.; Zhao, J.; Song, L. Chlorogenic acid induces apoptosis, inhibits metastasis and improves antitumor immunity in breast cancer via the NF-κB signaling pathway. *Oncol. Rep.* **2021**, *45*, 717–727. [CrossRef]
54. Yuan, Y.; Gong, X.; Zhang, L.; Jiang, R.; Yang, J.; Wang, B.; Wan, J. Chlorogenic acid ameliorated concanavalin A-induced hepatitis by suppression of Toll-like receptor 4 signaling in mice. *Int. Immunopharmacol.* **2017**, *44*, 97–104. [CrossRef] [PubMed]
55. Folsom, R.W.; Littlefield-Chabaud, M.A.; French, D.D.; Pourciau, S.S.; Mistric, L.; Horohov, D.W. Exercise alters the immune response to equine influenza virus and increases susceptibility to infection. *Equine Vet. J.* **2012**, *33*, 664–669. [CrossRef] [PubMed]
56. Sakaguchi, S. The origin of FOXP3-expressing CD4+ regulatory T cells: Thymus or periphery. *J. Clin. Investig.* **2003**, *112*, 1310–1312. [CrossRef] [PubMed]
57. Wang, J.; Gong, R.; Zhao, C.; Lei, K.; Sun, X.; Ren, H. Human FOXP3 and tumour microenvironment. *Immunology* **2023**, *168*, 248–255. [CrossRef] [PubMed]
58. Mayer, C.T.; Floess, S.; Baru, A.M.; Lahl, K.; Huehn, J.; Sparwasser, T. CD8+ Foxp3+ T cells share developmental and phenotypic features with classical CD4+ Foxp3+ regulatory T cells but lack potent suppressive activity. *Eur. J. Immunol.* **2011**, *41*, 716–725. [CrossRef]
59. Thannickal, V.J.; Fanburg, B.L.; Yasuoka, H.; Garrett, S.M.; Nguyen, X.-X.; Artlett, C.M.; Feghali-Bostwick, C.A.; Zimmer, A.; Bagchi, A.K.; Vinayak, K.; et al. Reactive oxygen species in cell signaling. *Am. J. Physiol. Lung Cell Mol. Physiol.* **2000**, *279*, L1005–L1028. [CrossRef]

60. Yarosz, E.L.; Chang, C.-H. The Role of Reactive Oxygen Species in Regulating T Cell-mediated Immunity and Disease. *Immune Netw.* **2018**, *18*, e14. [CrossRef]
61. Jacinto, T.A.; Meireles, G.S.; Dias, A.T.; Aires, R.; Porto, M.L.; Gava, A.L.; Vasquez, E.C.; Pereira, T.M.C.; Campagnaro, B.P.; Meyrelles, S.S. Increased ROS production and DNA damage in monocytes are biomarkers of aging and atherosclerosis. *Biol. Res.* **2018**, *51*, 33. [CrossRef]
62. Powers, S.K.; Deminice, R.; Ozdemir, M.; Yoshihara, T.; Bomkamp, M.P.; Hyatt, H. Exercise-induced oxidative stress: Friend or foe? *J. Sport. Health Sci.* **2020**, *9*, 415–425. [CrossRef]
63. Zada, S.; Pham, T.M.; Hwang, J.S.; Ahmed, M.; Lai, T.H.; Elashkar, O.; Kim, J.-H.; Kim, D.H.; Kim, D.R. Chlorogenic acid protects human chondrocyte C28/I2 cells from oxidative stress-induced cell death through activation of autophagy. *Life Sci.* **2021**, *285*, 119968. [CrossRef] [PubMed]
64. Le, Y.-J.; He, L.-Y.; Li, S.; Xiong, C.-J.; Lu, C.-H.; Yang, X.-Y. Chlorogenic acid exerts antibacterial effects by affecting lipid metabolism and scavenging ROS in Streptococcus pyogenes. *FEMS Microbiol. Lett.* **2022**, *369*, fnac061. [CrossRef] [PubMed]
65. Lee, B.; Lee, D.G. Depletion of reactive oxygen species induced by chlorogenic acid triggers apoptosis-like death in Escherichia coli. *Free Radic. Res.* **2018**, *52*, 605–615. [CrossRef] [PubMed]
66. Shi, H.; Shi, A.; Dong, L.; Lu, X.; Wang, Y.; Zhao, J.; Dai, F.; Guo, X. Chlorogenic acid protects against liver fibrosis in vivo and in vitro through inhibition of oxidative stress. *Clin. Nutr.* **2016**, *35*, 1366–1373. [CrossRef] [PubMed]
67. Yin, X.; He, X.; Wu, L.; Yan, D.; Yan, S. Chlorogenic Acid, the Main Antioxidant in Coffee, Reduces Radiation-Induced Apoptosis and DNA Damage via NF-E2-Related Factor 2 (Nrf2) Activation in Hepatocellular Carcinoma. *Oxid. Med. Cell Longev.* **2022**, *2022*, 4566949. [CrossRef]
68. Lawler, J.M.; Powers, S.K. Oxidative Stress, Antioxidant Status, and the Contracting Diaphragm. *Can. J. Appl. Physiol.* **1998**, *23*, 23–55. [CrossRef]
69. Leeuwenburgh, C.; Heinecke, J.W. Oxidative Stress and Antioxidants in Exercise. *CMC* **2001**, *8*, 829–838. [CrossRef]
70. Morrison, D.; Hughes, J.; Della Gatta, P.A.; Mason, S.; Lamon, S.; Russell, A.P.; Wadley, G.D. Vitamin C and E supplementation prevents some of the cellular adaptations to endurance-training in humans. *Free. Radic. Biol. Med.* **2015**, *89*, 852–862. [CrossRef]
71. Chen, X.; Song, M.; Zhang, B.; Zhang, Y. Reactive Oxygen Species Regulate T Cell Immune Response in the Tumor Microenvironment. *Oxidative Med. Cell. Longev.* **2016**, *2016*, e1580967. [CrossRef]
72. Bassoy, E.Y.; Walch, M.; Martinvalet, D. Reactive Oxygen Species: Do They Play a Role in Adaptive Immunity? *Front. Immunol.* **2021**, *12*, 755856.
73. Chauhan, P.S.; Satti, N.K.; Sharma, P.; Sharma, V.K.; Suri, K.A.; Bani, S. Differential effects of chlorogenic acid on various immunological parameters relevant to rheumatoid arthritis. *Phytother. Res.* **2012**, *26*, 1156–1165. [CrossRef] [PubMed]
74. Feng, Y.; Yu, Y.-H.; Wang, S.-T.; Ren, J.; Camer, D.; Hua, Y.-Z.; Zhang, Q.; Huang, J.; Xue, D.-L.; Zhang, X.-F.; et al. Chlorogenic acid protects d-galactose-induced liver and kidney injury via antioxidation and anti-inflammation effects in mice. *Pharm. Biol.* **2016**, *54*, 1027–1034. [CrossRef] [PubMed]
75. Ohkawara, T.; Takeda, H.; Nishihira, J. Protective effect of chlorogenic acid on the inflammatory damage of pancreas and lung in mice with l-arginine-induced pancreatitis. *Life Sci.* **2017**, *190*, 91–96. [CrossRef] [PubMed]
76. Saraiva, M.; Vieira, P.; O'garra, A. Biology and therapeutic potential of interleukin-10. *J. Exp. Med.* **2020**, *217*, e20190418. [CrossRef] [PubMed]
77. Del Prete, G.; De Carli, M.; Almerigogna, F.; Giudizi, M.G.; Biagiotti, R.; Romagnani, S. Human IL-10 is produced by both type 1 helper (Th1) and type 2 helper (Th2) T cell clones and inhibits their antigen-specific proliferation and cytokine production. *J. Immunol.* **1993**, *150*, 353–360. [CrossRef]
78. Patil, R. Chlorogenic acid may be a potent inhibitor of dimeric SARS-CoV-2 main protease 3CLpro: An in silico study. *Tradit. Med. Res.* **2021**, *2*, 135–148.
79. Cua, D.J.; Tato, C.M. Innate IL-17-producing cells: The sentinels of the immune system. *Nat. Rev. Immunol.* **2010**, *10*, 479–489. [CrossRef]
80. Millar, N.L.; Akbar, M.; Campbell, A.L.; Reilly, J.H.; Kerr, S.C.; McLean, M.; Frleta-Gilchrist, M.; Fazzi, U.G.; Leach, W.J.; Rooney, B.P.; et al. IL-17A mediates inflammatory and tissue remodelling events in early human tendinopathy. *Sci. Rep.* **2016**, *6*, 27149. [CrossRef]
81. Ackermann, P.W.; Domeij-Arverud, E.; Leclerc, P.; Amoudrouz, P.; Nader, G.A. Anti-inflammatory cytokine profile in early human tendon repair. *Knee Surg. Sports Traumatol. Arthrosc.* **2013**, *21*, 1801–1806. [CrossRef]

antioxidants

MDPI

Review

Microalgae as a Sustainable Source of Antioxidants in Animal Nutrition, Health and Livestock Development

Alexandros Mavrommatis [1], Eleni Tsiplakou [1,*], Anastasia Zerva [2], Panagiota D. Pantiora [2], Nikolaos D. Georgakis [2], Georgia P. Tsintzou [3], Panagiotis Madesis [3,4] and Nikolaos E. Labrou [2,*]

1 Laboratory of Nutritional Physiology and Feeding, Department of Animal Science, School of Animal Biosciences, Agricultural University of Athens, 75 Iera Odos Str., GR-11855 Athens, Greece; mavrommatis@aua.gr

2 Laboratory of Enzyme Technology, Department of Biotechnology, School of Applied Biology and Biotechnology, Agricultural University of Athens, 75 Iera Odos Str., GR-11855 Athens, Greece; anazer@aua.gr (A.Z.); pantiora@aua.gr (P.D.P.); n.georgakis@aua.gr (N.D.G.)

3 Laboratory of Molecular Biology of Plants, School of Agricultural Sciences, University of Thessaly, GR-38221 Volos, Greece; gtsintzou@uth.gr (G.P.T.); pmadesis@certh.gr (P.M.)

4 Institute of Applied Biosciences, CERTH, 6th km Charilaou-Thermis Road, P.O. Box 361, Thermi, GR-57001 Thessaloniki, Greece

* Correspondence: eltsiplakou@aua.gr (E.T.); lambrou@aua.gr (N.E.L.); Tel.: +30-210-529-4435 (E.T.); +30-210-529-4308 (N.E.L.)

Abstract: Microalgae are a renewable and sustainable source of bioactive compounds, such as essential amino acids, polyunsaturated fatty acids, and antioxidant compounds, that have been documented to have beneficial effects on nutrition and health. Among these natural products, the demand for natural antioxidants, as an alternative to synthetic antioxidants, has increased. The antioxidant activity of microalgae significantly varies between species and depends on growth conditions. In the last decade, microalgae have been explored in livestock animals as feed additives with the aim of improving both animals' health and performance as well as product quality and the environmental impact of livestock. These findings are highly dependent on the composition of microalgae strain and their amount in the diet. The use of carbohydrate-active enzymes can increase nutrient bioavailability as a consequence of recalcitrant microalgae cell wall degradation, making it a promising strategy for monogastric nutrition for improving livestock productivity. The use of microalgae as an alternative to conventional feedstuffs is becoming increasingly important due to food–feed competition, land degradation, water deprivation, and climate change. However, the cost-effective production and use of microalgae is a major challenge in the near future, and their cultivation technology should be improved by reducing production costs, thus increasing profitability.

Keywords: animal nutrition; antioxidants; livestock; microalgae; monogastric diet; sustainability

Citation: Mavrommatis, A.; Tsiplakou, E.; Zerva, A.; Pantiora, P.D.; Georgakis, N.D.; Tsintzou, G.P.; Madesis, P.; Labrou, N.E. Microalgae as a Sustainable Source of Antioxidants in Animal Nutrition, Health and Livestock Development. *Antioxidants* **2023**, *12*, 1882. https://doi.org/10.3390/antiox12101882

Academic Editors: Luciana Rossi and Matteo Dell'Anno

Received: 18 September 2023
Revised: 10 October 2023
Accepted: 13 October 2023
Published: 19 October 2023

1. Introduction

Over the past ten years, the feed, food, cosmetic, and nutraceutical industries have all become interested in the natural compounds from microalgae, due to the rising demand for natural antioxidants as an alternative to synthetic antioxidants [1,2]. Microalgae are an underutilized resource with the potential to produce antioxidants and novel bioactive molecules. They are attractive for developing biotechnological applications due to their high growth rate, simplicity in cultivation, production scalability, potential for genetic modification, low maintenance costs, and metabolic plasticity that can be directed and manipulated to produce target compounds by altering culture conditions [3,4]. Due to all these factors, the demand for algal biomass is expected to increase significantly in the coming years. It has been estimated that the global algae market will reach a value of USD 6.3 billion by 2028, up from USD 4.5 billion in 2021 [5].

From Ancient Greece to Icelandic sagas and all over the world, ample evidence suggests that both wild and domesticated animals approached coastal areas to feed on seaweed. After the intensification of livestock production systems and the consideration of current emerging issues such as food–feed competition and the hard-to-find arable lands for feedstuff production, the usage of algae was redefined. Corn and soybean, the two main conventional feedstuffs for animal feeding, are unsustainable, and therefore alternatives to these ingredients are required to maintain livestock performance, especially considering the latest exponential demands for animal protein [6,7]. Land degradation, water deprivation, and drastic changes in climate are also significant challenges for the future of the livestock sector signifying the need to explore highly sustainable alternatives to conventional feedstuffs, which are not affected by environmental conditions. The cultivation of microalgae does not require rainfall, which would decrease competition with areas more suited for agricultural production or biodiversity conservation [8]. Although until recently, the use of algae as an alternative to conventional feedstuffs was advocated as a panacea for the aforementioned challenges, when applied in animal diets, their assessment unveiled two significant drawbacks. The high production cost of algae limits their inclusion levels in animal diets since this cost cannot be remunerated by the price of animal products [7]. Additionally, their cell walls inhibit nutrient release in the duodenum of monogastric animals and decrease their digestibility. Considering these issues, in the last decade, seaweeds and microalgae have been explored in animal nutrition as feed additives (in low inclusion levels) rich in bioactive compounds for livestock animals and aquaculture with the aim of improving both animals' health and performance as well as product quality and livestock environmental impact.

This review article aims to cover recent research progress on the antioxidant molecules in microalgae and their role as feed additives for improving both animals' health and performance as well as product quality and livestock environmental impact.

2. Microalgal Diversity in Industrial Setting

Microalgae are a large group of organisms that are extremely diverse and heterogeneous from evolutionary and ecological viewpoints. Microalgal biomass is an excellent source of diverse bioactive compounds such as lipids, polysaccharides, carotenoids, vitamins, phenolics, and phycobiliproteins [1,2]. The bioactive properties of different industrially produced microalgae vary as a consequence of their physiology and biochemistry. The diversity of microalgal species gives rise to various antioxidant molecules, which makes microalgae the richest natural resource for nutritional and bioactive components [9–11]. The list of industrially produced microalgae includes genera spanning different classes, namely Chlorodendrophyceae (*Tetraselmis chui*, *Tetraselmis striata* CTP4); Chlorophyceae (*Haematococcus lacustris*, formerly *Haematococcus pluvialis*); Coccolithophyceae (*Tisochrysis lutea*); Bacillariophyceae (*Phaeodactylum tricornutum*, *Skeletonema* sp.); Eustigmatophyceae (*Nannochloropsis* sp.); Porphyridiophyceae (*Porphyridium* sp.); and Cyanophyceae (*Spirulina*). Among them, there are only a few microalgae that have "Generally Recognized as Safe" (GRAS) status as recognized by the FDA. These microalgae include *Arthrospira platensis* (*Spirulina, Cyanophyceae*), *Chlamydomonas reinhardtii*, *Auxenochlorella protothecoides* (Trebouxiophyceae), *Chlorella vulgaris*, *Dunaliella salina* (formerly *Dunaliella bardawil*) (Chlorophyceae), and *Euglena gracilis* (Euglenophyceae).

3. Antioxidant Compounds in Microalgae

Microalgae live in habitats under high solar irradiation and as a consequence have a wide range of antioxidant compounds, which protect them from radiation and oxidation damage. High-value antioxidant compounds produced by microalgae include polyunsaturated fatty acids (PUFAs); carotenoids, including astaxanthin and lutein; chlorophylls; phycobiliproteins; and phenolic compounds [1,2,9–13] (Figure 1a). These compounds have immunomodulation, anti-inflammatory, neuroprotective, antimicrobial, antiviral, and anthelmintic properties, beneficial to animal health [14] (Figure 1b). Antioxidants help to prevent oxidative damage, which mostly occurs from oxygen's reduced states. There are many types of reactive oxygen species [15]. The antioxidant function can be derived from two sources: the activity of antioxidant enzymes [12,13] or the production of molecules that serve as sacrificial scavengers of reactive oxygen species [2]. Additionally, antioxidant activity can be divided into two major mechanisms of action: limiting reactive oxygen species in the digestive tract to lessen oxidative stress on the gut microbiome and epithelial cells or transporting antioxidants into the circulation for distribution throughout the body. There are still knowledge gaps about the effectiveness of the antioxidant qualities of microalgal meals at all levels, from species differentiation to effects on the gut microbiota and movement through the gut lumen to their effects on animal physiology. This will be useful for further research in the next decade.

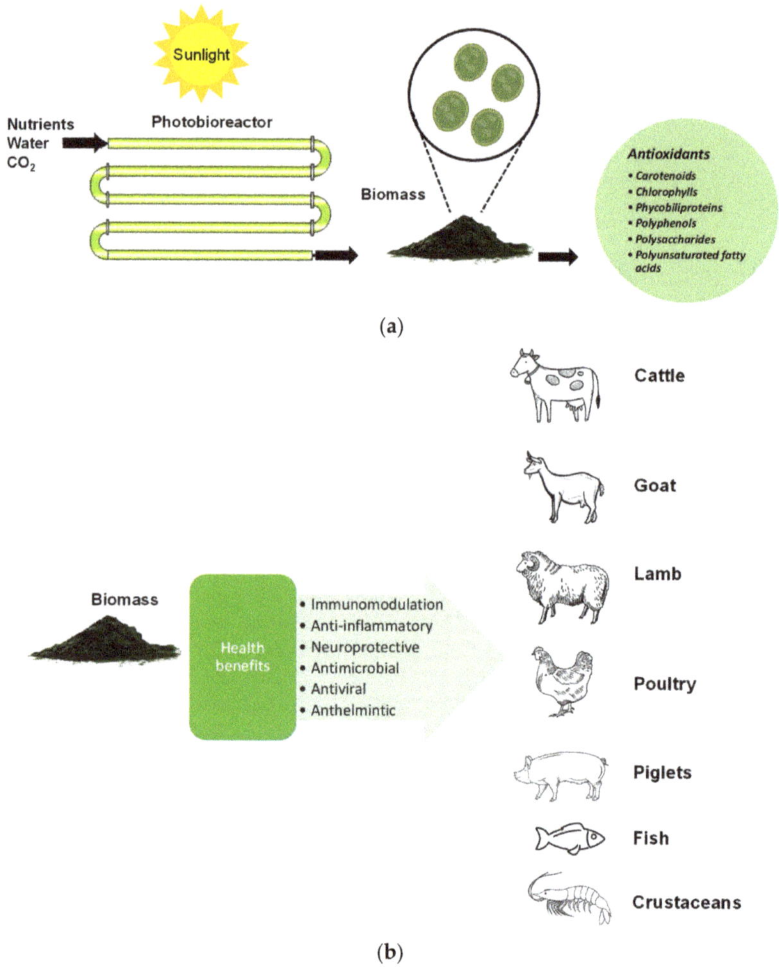

Figure 1. (**a**) Main antioxidant compounds produced by microalgae; (**b**) the beneficial properties of microalgae in animal health.

3.1. Polyunsaturated Fatty Acids (PUFAs)

Polyunsaturated fatty acids (PUFAs) include docosahexaenoic acid (DHA), eicosapentaenoic acid (EPA), arachidonic acid (AA), and γ-linolenic acid (GLA), and are well known for being advantageous to both animal and human health [16–18]. Several microalgal species have been recorded as sources of PUFAs, such as *Phaeodactylum tricornutum*; *Monodopsis subterranea* (formerly *Monodus subterranea*) (Xanthophyceae); *Porphyridium purpureum* (formerly *Porphyridium cruentum*) (Porphyridiophyceae); *Chaetoceros calcitrans* (Mediophyceae); *Nannochloropsis* sp. (Eustigmatophyceae); *Crypthecodinium cohnii* (Dinophyceae); *Isochrysis galbana* (Coccolithophyceae); and *Rebecca salina* (formerly *Pavlova salina*) (Pavlovophyceae) [17–20]. Among the physiologically important fatty acids, docosahexaenoic acid is the most sensitive to oxidation, whereas palmitic acid is the least oxidizable. The oxidation stability of fatty acids is directly connected with the degree of unsaturation.

3.2. Carotenoids

Carotenoids are abundant in microalgae [21,22]. These colorful compounds are well known for their antioxidant properties [14,23–25]. They are used as natural colorants [26–28] and as food and feed additives or health supplements. They are divided into two groups: carotenes and xanthophylls. The former group of compounds are oxygen-free hydrocarbons, such as α-carotene and β-carotene, while the latter compounds are oxygenated derivatives of carotenes (including lutein, violaxanthin, zeaxanthin, fucoxanthin, and astaxanthin) [25,29]. The main sources of carotenoids are microalgae belonging to the Chlorophyceae. Many microalgae accumulate carotenoids but the dominant species extensively studied are *Dunaliella salina*; *Haematococcus lacustris* (Chlorophyceae); *Chromochloris zofingiensis* (formerly *Chlorella zofingiensis*) (Trebouxiophyceae); and *Chlorella vulgaris* (Chlrophyceae), especially due to the capability of commercial production in large-scale cultures [30,31]. These microalgae may generate a variety of pigments, including carotenes (β-carotene and lycopene) and xanthophylls (astaxanthin, violaxanthin, antheraxanthin, zeaxanthin, neoxanthin, and lutein). Other microalgae phyla generate other compounds such as fucoxanthin, diatoxanthin, and diadinoxanthin [23]. Currently, the two pigments with the highest global market are β-carotene and astaxanthin from the genera *Dunaliella* and *Haematococcus*, respectively [23,32,33]. More specifically, β-carotene is commonly produced by *Tetradesmus almeriensis* (formerly *Scenedesmus almeriensis*), *Dunaliella salina* (formerly *Dunaliella bardawil*), and *Dunaliella tertiolecta* (Chlorophyceae) [14,19,20,23,34–36]; however, the best source for its production is *Dunaliella salina* [33,37].

Astaxanthin, a red xanthophyll pigment, is the second most widely used carotenoid [38]. It displays an efficacious antioxidant activity [39–41] and shows about ten times higher antioxidant activity than other carotenoids [19,23,34,35]. It is produced by several microalgae such as *Chlorella zofingiensis*, *Chlorococcum* sp., and *Scenedesmus* sp., as well as the yeast *Xanthophyllomyces dendrorhous* [14,20,24]. Interestingly, the microalga *Haematococcus lacustris* [35,42], under certain cultivation conditions, can accumulate up to 7% astaxanthin on a dry weight basis [37]. Therefore, *Haematococcus lacustris* is seen as the most favorable species for industrial scale production of natural astaxanthin [33,43]. Astaxanthin is widely used in aquaculture feed as a dye agent for fish and shellfish flesh due to its red color [14,35,42]. It is also exploited as an antioxidant supplement to improve the health and production performance of broiler chicken [44].

Lutein is another important carotenoid [45]. It is used for the pigmentation of animal tissues and products [46]. It also shows bioactive beneficial properties in chronic diseases, such as cataracts, atherosclerosis, blindness, or decreased vision [47–49]. Lutein is produced by several microalgae such as *Chlorella* sp. [50,51], *Muriellopsis* sp. [52], *Scenedesmus* sp. [53], and *Chlamydomonas* sp. [54]. It is widely used for the natural coloration of foods, drugs, and cosmetics [55]. Lutein-producing strains include *Muriellopsis* sp., *Auxenochlorella protothecoides* (formerly *Chlorella protothecoides*) (Trebouxiophyceae), *Chromochloris zofingiensis* (formerly *Chlorella zofingiensis*), *Pleurastrum insigne* (formerly *Chlorococcum citriforme*), *Neospongiococcum gelatinosum*, and *Tetradesmus almeriensis* (Chlorophyceae) [14,20,56].

Other important carotenoid pigments with high commercial value include lycopene, violaxanthin, and zeaxanthin [57]. Lycopene is extensively used in cosmetic formulations as a sunscreen and antiaging compound [20,34]. It also displays anticarcinogenic and antiatherogenic properties [20]. Violaxanthin, an orange carotenoid pigment, is well known for its anti-inflammatory and anticancer properties. It is produced by *Chloroidium ellipsoideum* (formerly *Chlorella ellipsoidea*) (Trebouxiophyceae) and *Dunaliella tertiolecta* [19,23] strains. Zeaxanthin is a yellow carotenoid that has found successful applications in the pharmaceutical, cosmetic, and food industries. The industrial production of zeaxanthin is mainly achieved by *Tetradesmus almeriensis* and *Nannochloropsis oculata* [20,35]. Other noteworthy carotenoids include canthaxanthin, β-cryptoxanthin, and fucoxanthin, which have shown significant tanning, anti-inflammatory, and anticancer properties, respectively [21,22,34,58,59], and therefore they have been used in the pharmaceutical or cosmetic industries.

3.3. Chlorophylls

Chlorophylls are found in all photosynthetic microalgae [60,61]. Due to their green pigmentation, they are becoming increasingly important as colorants in the food industry as well as in the pharmaceutical and cosmetic industries [14,21,24,57]. Chlorophyll a and chlorophyll b also occur in the form of sodium and copper derivatives. The latter types of derivatives are mainly used as food additives or in drinks [14]. Microalgae that belong to the genus *Chlorella* contain chlorophyll to about 7% of their biomass, five times more than that of *Arthrospira* [36,61].

3.4. Phycobiliproteins

Phycobiliproteins are only found in cyanobacteria and some red algae [62–64]. Phycobiliprotein, a high-potential molecule, has been utilized commercially as a natural dye and has a variety of applications in the pharmaceutical industry [63–65]. Due to their powerful and highly sensitive fluorescent properties, they are used as markers for certain immunological methods, such as flow cytometry, microscopy, and DNA tests [14,24,32,35,57,63,64,66]. On an industrial scale, these pigments are produced from the species *Porphyridium* sp., *Arthrospira* sp., and *Aphanizomenon flosaquae* (Cyanophyceae) [14,19,20,24,34,35].

The predominant pigment in the phycobiliprotein family is phycocyanin [63,64]. Phycocyanin is a blue protein in cyanobacteria, rhodophytes, and cryptophytes that possesses the blue tetrapyrrole chromophore, phycocyanobilin, with fluorescent and bioactive properties (Figure 2). Phycocyanin is a water-soluble, non-toxic, and blue-colored photosynthetic pigment that have been used in food, cosmetic, and pharmaceutical industries. Over the years, the biological function of phycocyanin has been extensively studied. For example, numerous studies have investigated its antioxidative, anti-inflammatory, anticancer, and antimicrobial activity, as well as its effects on neurodegeneration, diabetes, wound healing, and hyperpigmentation [67–69]. Extensive studies during the last two decades concerning purification, biochemical, and structural properties has resulted in a detailed description of phycocyanin's architecture, structure, and bioactivity [67,70]. The primary biotechnological potential of phycocyanin seems to be its application as a natural pigment, replacing toxic synthetic dyes. However, a growing number of evidences have demonstrated that phycocyanin also exhibits bioactive properties related to health benefits such as antiaging, antioxidant, anticancer, neuroprotective, and anti-inflammatory activities [70].

3.5. Polysaccharides

Polysaccharides are polymers consisting of saccharide units linked with glycosidic bonds attached to the cell wall or released into the medium (exopolysaccharides) [71]. Polysaccharides isolated from a range of microalgal species (e.g., *Arthrospira platensis*, *Porphyridium purpureum*, *Dunaliella salina*, *Dixoniella grisea* [formerly *Rhodella reticulata*) (Rhodellophyceae), and *Schizochytrium* sp.] exhibit in vitro antioxidant properties and ability to effectively scavenge superoxide radicals, hydroxyl radicals, and hydroxyl peroxides [2,43,71,72]. The diverse biological activities of polysaccharides in microalgae are due to their complex structural features, including molecular weight, composition of sugar residues, types of glycosidic linkages, nature of monosaccharides, and the presence of some sugar-free units (sulfate, methyl, organic acids, amino acids, or amines) in the main skeleton of polysaccharides [2,71,72].

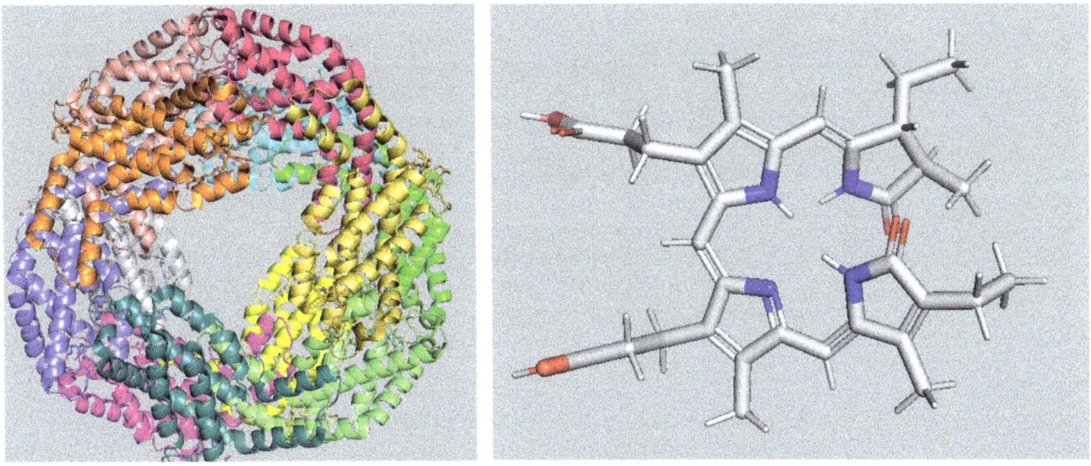

(a) (b)

Figure 2. (**a**) The structure of C-phycocyanin from *Arthrospira platensis* (formerly *Spirulina platensis*) at 2.2 Å resolution (PDB accession number 1GH0); (**b**) the structure of phycocyanobilin (PCB), a linear tetrapyrrole chromophore covalently attached to protein subunits of C-phycocyanin.

3.6. Polyphenols

Polyphenols are a large group of secondary metabolites comprising phenolic acids, flavonoids, isoflavonoids, stilbenes, lignans, and phenolic polymers [71–74]. Phenolic compounds, especially flavonoids and phenolic acids, exhibit high antioxidant function. The extraction and purification of polyphenols from microalgae is challenging, and improvements in analytical methodology are needed to facilitate more detailed characterization of their structure and function. Recently, a study was carried out on polyphenols extracted from two different microalgal species: *Nannochloropsis* sp. and *Arthrospira/Spirulina* sp. [75].

4. Antioxidants and Gut Microbiota

Several studies have investigated the relationship between antioxidants and the gut microbiota in animals [76–81]. Some studies have suggested that the antioxidants present in microalgae can influence the composition of the gut microbiota [82]. For example, dietary antioxidants like polyphenols and flavonoids found in fruits and vegetables have been shown to promote the growth of beneficial bacteria like *Bifidobacterium* and *Lactobacillus* while reducing the abundance of potentially harmful bacteria [79,82–85]. Some antioxidants, especially dietary fibers and polyphenols, can serve as substrates for bacterial fermentation in the gut. This fermentation process produces short-chain fatty acids (SCFAs), which have beneficial effects on gut health. SCFAs serve as an energy source for gut cells, regulate inflammation, and support the growth of beneficial bacteria [84,85]. Furthermore, antioxidants possess anti-inflammatory properties. Inflammation in the gut is associated with various gastrointestinal disorders. Therefore, by reducing inflammation, microalgae may help maintain a healthy gut environment and support the growth of beneficial gut bacteria [86–88].

Antioxidants may also contribute to the integrity and function of the gut barrier, which acts as a protective barrier between the gut microbiota and the rest of the body [89]. A healthy gut barrier prevents the translocation of harmful bacteria or their by-products into the bloodstream.

5. The Effect of Microalgae on Animal Performance

As mentioned before, the high production cost of microalgae due to current cultivation technology, together with their low digestibility in monogastric species, are the limiting factors for their inclusion in animals' diets substituting conventional feedstuffs. For this reason, research has focused on lower dietary supplementation levels with the aim of improving animal performance by enhancing nutritional physiology pathways. Moreover, the chemical composition of microalgae (rich in protein or fat content) affects animal performance differently. It has been reported that the dietary inclusion of high-protein species improved the body condition and average daily gain (ADG) of dairy cows and lambs, respectively [90,91], while diets containing high-fat microalgae namely *Schizochytrium* spp. negatively affected lambs' performance through a reduction in their dry matter intake (DMI) [92,93]. Mavrommatis and Tsiplakou (2020) [94] also observed a 30% reduction in DMI when *Schizochytrium* spp. included at 3% in goats' diet. This reduction in DMI is usually attributed to the fish-like odor of microalgae or to the high-fat content of *Schizochytrium* spp., which might impair the hypophagic effect on the brain's satiety center.

In broilers, the dietary inclusion of *Arthrospira* sp. in different levels (ranging from 0.5% to 21%) had no effects on performance parameters [95–97]. On the other hand, the low dietary inclusion of *Chlorella* sp. (0.00003% to 1%) consistently increased ADG and overall growth performance in broilers, possibly through beneficial cellular remodeling owing to microalgal secondary metabolites and bioactive compounds [98,99]. Thus, both nutrients' complementarity and the digestibility of microalgae, especially in the higher inclusion levels, might explain these results. Moreover, the dietary inclusion of fat-rich microalgae *Schizochytrium* spp. (3.7–7.4%) increased broilers' DMI and consequently their productive performance from 21 to 35 days old [100,101]. These findings indicate that not only the dietary inclusion level but also the chemical composition of microalgae needs to be taken into consideration in diet formulation.

In pigs, negligible improvement in growth performance was observed when fed with diets containing *Arthrospira platensis* (0.2–2%) [102]. In fattening pigs, the inclusion of 0.2% *Arthrospira platensis* (*Spirulina*) significantly increased ADG without affecting back fat thickness [103]. However, the combined dietary inclusion of Spirulina with *Chlorella vulgaris* at 1% on weaned piglets for only 14 days did not affect ADG even though a potential effect on intestinal development through the regulation of a mild digestive disorder was reported [104]. Thus, it could be hypothesized that the trial interval of the previous study was limited in order for substantial changes in animal performances to be unveiled. No effect on ADG, final body weight, and carcass traits of female pigs was observed either when 0.1% *Chlorella* spp. was included in their diet [105]. The low supplementation level of microalgae might be the reason for the absence of any significant effect. Indeed, variable dietary inclusion levels of *Schizochytrium* spp. (1.10% to 5.51% from day 79 to 106 and 0.39% to 1.94% from day 107 to 120) increased ADG and FCR without affecting DMI [106], while lower inclusion levels of the same microalgae (0.25–0.50%) did not change the growth performance of finishing pigs [107]. The former experimental trials in pigs signify once again the importance of supplementation levels and the duration of administration.

6. The Effect of Microalgae on Animal Health

The biochemical profiles of microalgal species (*Spirulina*, *Chlorella* sp., *Nannochloropsis granulat*, *Schizochytrium*, and *Tetraselmis chui*) commonly used for formulating animal feed include essential amino acids and polyunsaturated fatty acids (PUFAs) such as eicosapentaenoic acid (EPA), docosapentaenoic acid (DPA), and docosahexaenoic acid (DHA) that are not synthesized in animals' organism, as well as antioxidant compounds (such as carotenoids and flavonoids, trace elements, minerals, and vitamins). Extensive scientific evidence has been documented highlighting the impact of these molecules on animal health. However, microalgae also represent a source of unexploited bioactive compounds, which may have exceptional properties and significant applications, including, but not limited to, lipoproteins, sterols, and alkaloids [108].

Amongst the biomolecules present in microalgae, PUFAs have gained significant interest since these vital nutrients have been correlated with human health benefits [109]. The advantage of microalgae regarding PUFA formation and accumulation lies in an efficient elongase–desaturase network that is common in many algal species [110,111]. Indeed, long-chain polyunsaturated fatty acids (LCPUFAs) could regulate animals' pro-inflammatory response induced by farming conditions, especially in high-yielding individuals, through the production of eicosanoids with much less pro-inflammatory power or by inhibiting pro-inflammatory cytokines, resulting in the suppression of low-grade inflammation and stress. In this light, in a recent study by Mavrommatis et al. (2021) [112], the dietary supplementation with 20, 40, and 60 g *Schizochytrium* spp. downregulated the transcriptional profile of the TLR4 pathway in goats' blood monocytes. The former induced a cascade of downregulations in pro-inflammatory cytokines (*IFNG*, *IL1B*, *IL2*, *IL8*, and *TNF*) and chemokines (*CCL5* and *CXCL16*) in both blood monocytes and neutrophils. The mechanism that underlies this pro-inflammatory suppression may be attributed to the immunomodulatory effects of both DHA and ω6-DPA fatty acids contained in *Schizochytrium* biomass on immune cells' intramembrane receptors such as Toll-like receptors (TLRs) and G protein-coupled receptor 120 (GPR120) [112]. Another possible mode of action may be related to the metabolism of LCPUFAs in immune cells and their utilization to synthesize immunomodulatory mediators (eicosanoids). To further investigate these speculations, Kyriakaki et al. (2023) [113] assessed the expression of genes involved in eicosanoid production in monocytes and neutrophils of goats fed with *Schizochytrium* spp. and found a significant decrease in the expression of genes that regulate both cyclooxygenase (*COX2*) and lipoxygenase (*5-LOX*) pathways, indicating an overall suppression of pro-inflammatory response. These lines of evidence may have significant implications for dairy ruminants' resilience toward commercial farming since a low-grade pro-inflammatory response that activates the immune system can lead to severe competition for nutrient availability [114]. More specifically, animals' energy and nutrient requirements for maintenance and production can often be overlooked with regard to their role in immune function demands. It has been estimated that the activation of the immune system requires as much as 10–30% of metabolizable energy [114]. These energetic demands are supported by studies that have reported the utilization of 1 kg of glucose (within a 12 h period) for the immune system's activation in both cattle and swine [114]. Considering the topic of inflammation, Caroprese et al. (2012) [115] reported that a mixture of phytosterols from *Dunaliella tertiolecta* reduced the cytokine production in a sheep model of inflammation. Moreover, *Chlorella sorokiniana* suppressed the peripheral blood mononuclear cell proliferation and pro-inflammatory cytokine levels in sheep in vitro [116].

The bioactive compounds of microalgae possess several health benefits such as boosting the immune system, which will eventually reduce antibiotic dependence in livestock farming [108]. Indeed, Amaro et al. (2011) [117] summarized the potential antibacterial properties of microalgae, reporting that specific fatty acids, organic acids, and extra-metabolites could effectively counteract MRSA, *E. coli*, *Vibrio*, *Salmonella*, *Pseudomonas*, etc. Fries-Craft et al. (2021) [118] assessed the inclusion of 0.175% of a mixture consisting of Spirulina and *Chlorella* in the broilers' diet. Microalgae maintained intestinal integrity during the coccidiosis challenge and protected jejunal villus height. During the *Eimeria* challenge, splenic T cells in microalgae-fed broilers did not provide evidence of recruitment to peripheral tissues. These outcomes suggest that the ingredients in microalgae modified the immune response in a manner that reduced recruitment from secondary lymphoid organs in addition to protecting intestinal physiology. Dietary supplementation with 1% Spirulina or *Chlorella* microalgae as an alternative to antibiotic use was studied by Furbeyre et al. (2017) [104]. Diarrhea incidence was reduced in *Chlorella*-fed pigs compared with the control, Spirulina, and antibiotic (colistin) groups. Villus height at the jejunum was greater in microalgae-fed pigs compared with the control and antibiotic-fed pigs. This study reported a potential effect of both Spirulina and *Chlorella* supplementation on intestinal development and further advantages of *Chlorella* supplementation to manage mild digestive disorders. Similarly, the inclusion of 0.1% fermented *Chlorella* in growing pigs' diet improved growth performance, nutrient digestibility, and fecal microbial structure (higher *Lactobacillus* and lower *E. coli*), and decreased fecal noxious gas emission [119]. These sets of evidence demonstrate that microalgae and their derivatives could be useful ingredients in animal feed for minimizing the dependence on antibiotics, aiming to control antibiotic resistance and support animal health using natural bioactive compounds.

Besides its antibacterial, antiviral, and anthelmintic properties, microalgal biomass can be considered as a multi-component antioxidant system that in general is more effective through the interactions between the different antioxidant components. In this context, El-Bahr et al. (2020) [120] observed lower values of malondialdehydes (MDAs) and protein carbonyls (PCs) in the breast muscle of broilers, while the activity of superoxide dismutase (SOD) was increased due to dietary supplementation with microalgal species (*Arthrospira platensis*, *Chlorella vulgaris*, and *Halamphora coffeiformis* (formerly *Amphora coffeiformis*)). Remarkably, the dietary supplementation with microalgae-based antioxidants minimized the detrimental effect of mycotoxin-contaminated feed and partially improved the feed conversion ratio (FCR) in both heat-stressed and unstressed broiler chickens [121]. A recent study conducted by Christodoulou et al. (2022) [122] reported that the dietary supplementation with Spirulina significantly increased the antioxidant defense of sheep organisms through the higher activity of SOD, catalase (CAT), and glutathione peroxidase (GSH-Px), while the protein oxidative index (PC) was decreased. Interestingly, Spirulina used in the former study was derived through secondary sorting (no competition with the food and cosmetic industry), thus opening new valorization directions in the feed industry.

Even though microalgae and their bioactive compounds have been associated with many beneficial properties on animal health, the first principle of toxicology is that "all things are poisonous, and it is the dose that distinguishes between a drug and a poison" (Paracelsus C15th). More specifically, in an effort to fortify animal products with PUFAs through the supplementation of microalgae, the high propensity of PUFAs to oxidation could lead to a severe immune-oxidative burst. Dietary PUFA overload can activate cellular superoxide anion generators like xanthine and NADPH oxidases, resulting in superoxide anion formation and launching a cascade of pro-oxidant incidences [123]. Indeed, the high supplementation levels of *Schizochytrium* spp. (40 and 60 g/day) in goats' diet increased the activity of NADPH oxidase in blood plasma [124] and the mRNA levels of *NOX1* and *NOX2* in their monocytes and neutrophils [113], while a lower supplementation level (20 g) did not. Similarly, the dietary inclusion of *Schizochytrium* spp. significantly impaired sheep's oxidative system as reflected by the higher levels of MDAs and PCs. Another risk of incorporating microalgae into the human food chain is related to their ability to accumulate toxic metals. In semi-arid regions where the availability of fresh water is scarce, water wastes such as those obtained by mines are frequently used. Therefore, careful attention must be given before use as animal feed [125].

7. The Effect of Microalgae on Animal Product Quality

Even if the current cultivation technology does not allow for the substantial substitution of conventional feedstuffs by microalgae, and their inclusion in low levels in animal diets does not significantly improve animal performance, their application as feed additives for the promotion of animal product quality has been well documented with promising outcomes [110,126].

Laying hens fed with diets supplemented with *Microchloropsis gaditana* (formerly *Nannochloropsis gaditana*), containing ω-3 LCPUFAs such as EPA and DHA, resulted in the accumulation of these ω-3 FAs in the egg yolk [127]. Interestingly, a higher proportion of DHA than EPA accumulated in eggs when *Nannochloropsis oculata* was fed to laying hens, even though this microalga is richer in EPA [108]. In addition to the enrichment of eggs with beneficial LCPUFAs, the inclusion of 20% *Nannochloropsis oculata* in laying hens' diet increased the lutein and zeaxanthin content to 1.3 mg/egg [128]. In this context, the accumulation of carotenoids in egg yolk results in a darker orange color, which increases consumer acceptance [129]. Another important aspect of bioactive compounds in microalgae is related to their bioavailability. More specifically, it has been reported that lutein contained in *Chlorella* is incorporated more efficiently in eggs than synthetic carotenoids, also resulting in improved oxidative stability of yolk lipids [130]. Nevertheless, it should be mentioned that high levels of LCPUFAs in hens' diet can decrease tocopherol availability for proper egg yolk formation and induce pro-oxidant incidences with a further impact on birds' health and homeostasis [131]. Hence, the importance of supplementation levels should not be overlooked.

Considering the effect of dietary supplementation with microalgae on poultry and pigs' meat quality, the study of Martins et al. (2022) [126] has comprehensively summarized the latest insights. The dietary inclusion of Spirulina (4% or 8%) increased the yellow appearance of broilers' muscles, skin, fat, and liver, which increased the commercial value of the meat and consequently the consumers' acceptance [132]. Similarly, a high inclusion level (10%) of *Chlorella vulgaris* increased tenderness, yellowness, and total carotenoids in the breast and thigh meat of broilers [133]. Additionally, the dietary inclusion of *Schizochytrium* spp. rich in DHA increased the ω-3 content in the breast and thigh of broilers in numerous studies investigating a wide range of levels from 0.1% to 7.4% [100,101,119,134–137]. However, it should be highlighted that the higher dietary supplementation levels of *Schizochytrium* spp. also increased the concentration of lipid peroxidation metabolites [100,101]. In terms of pigs' meat quality, the effects of *Schizochytrium* spp. are quite similar to those observed in poultry [126]. Moreover, regarding the supplementation of protein-rich microalgae, the inclusion of *Chlorella vulgaris* at 5% increased the total carotenoid content in meat in weaned piglets and grower pigs [126,138].

In ruminants, the concept of enriching meat and milk with marine-origin fatty acids is more complicated. Due to the symbiotic microbiome colonizing the rumen, about 70–100% of PUFAs present in the feed are biohydrogenated, resulting in the formation of saturated fatty acids, mainly stearic acid, which is transferred to tissues and milk. However, there is another ruminal biochemical procedure that can be manipulated through dietary marine fatty acids, aiming to enrich products with PUFAs. The increased flow of LCPUFAs into the rumen changes biohydrogenation pathways, resulting in the accumulation of vaccenic acid due to the procedure's incompleteness. Vaccenic acid is desaturated through the activity of Δ9 desaturase to conjugated linoleic acid (CLA), a fatty acid with significant health benefits that humans receive through the consumption of milk and meat [139]. Thus, although the transfer efficiency of EPA, DPA, and DHA is quite low in ruminants due to their biohydrogenation [94,140], another important biomolecule can be formed. In this context, dietary supplementation with *Schizochytrium* spp., rich in ω6-DPA and DHA, enriched ovine [141] and caprine [94,142] milk with DHA, ω6-DPA, and CLA resulting in a two-fold increase in total milk PUFA content. More specifically, microalgae-fed goats and sheep produced milk fortified with up to four- and six-fold increased proportions of CLA, respectively. Additionally, in both goat and sheep milk, the ω6/ω3 ratio, the health-promoting index, and the atherogenic index were significantly improved, setting new horizons for the development of functional dairy products enriched with beneficial fatty acids for human health. The former constitutes an important aspect of the industry since ruminants' milk has been criticized for its high proportion of saturated fatty acids, which have been correlated with a high risk for human cardiovascular diseases [139].

On the contrary, the high accumulation of PUFAs in milk and dairy products increases their propensity to oxidation [124]. Indeed, the high inclusion level of *Schizochytrium* spp. in goats' diet impaired milk oxidative status through the accumulation of toxic aldehydes such as MDAs and protein oxidation products (PCs) [124,142]. Dairy sheep were found to be more prone to oxidation since MDAs in milk were increased even in the lowest supplementation level [141]. Nevertheless, Christodoulou et al. (2022) [122] reported that Spirulina supplementation in dairy sheep diet significantly increased the activity of antioxidant enzymes, namely SOD, CAT, and GSH-Px, in milk and its total antioxidant capacity. The former constitutes an important aspect since raw milk is frequently oxidized during its transportation to the industry; thus, receiving milk with a more stabilized oxidative status can improve its overall life span. The holistic consideration of the abovementioned trials allows us to highlight the potential benefits of combining microalgae rich in PUFAs and antioxidant compounds simultaneously, formulating feed additives aiming to fortify milk and dairy products with beneficial fatty acids for human health while concurrently controlling any side effects related to PUFAs' oxidation.

8. Microalgae in Monogastric Diets: The Use of Carbohydrate-Active Enzymes

The digestibility of microalgal biomass can be impaired by the complicated cellulosic cell walls of most microalgal species, distantly related to the architecture of plant cell walls. Little is known of the exact nature of the cell walls of microalgae in general. Most relevant studies are conducted in commonly used model organisms, such as *Chlorella* [143,144] and *Nanochloropsis* [145], but they are often contradictory, due to the significant variety observed depending on the species, the growth stage, and often the cultivation medium [146]. A relatively simple and straightforward method to determine the composition of algal cell walls is the hydrolysis of the material in harsh conditions and the compositional analysis of the resulting monomers (sugars or amino acids for example), as an indirect way to predict the polymers present and their proportions. Using this approach, Spain and Funk (2022) [143] characterized the composition of several Nordic species of microalgae, including *Chlorella vulgaris*, *Scenedesmus* sp., *Haematococcus lacustris*, and *Coelastrella* sp., and their changes according to growth phase. While for all strains, the same monosaccharides were present (arabinose, rhamnose, fucose, xylose, mannose, galactose, glucose, galacturonic, and glucuronic acids), the glucose content was found to considerably vary for *Coelastrella* sp. and *Scenedesmus* sp. in different growth phases. Accordingly, glycine, glutamic acid, aspartic acid, threonine, and alanine were found to be the most abundant amino acids in all strains, but the proportion between polar and non-polar amino acids shifted throughout the course of cultivation for *Scenedesmus* sp. Moreover, the protein-to-carbohydrate ratio also shifted during growth for all strains. Similarly, Weber et al. (2022) [144] studied the cell wall composition of *C. vulgaris* using alkaline or acidic extraction. Alkali extracts mainly contained glucosamine, indicating the presence of a chitin-like polymer, while acidic extracts mainly consisted of glucose, indicating the presence of cellulose or starch. Galactose, mannose, rhamnose, and uronic acids were also present, indicating the presence of pectin- and galactan-like polysaccharides, together with glycoproteins.

Nonetheless, most studies confirm that microalgal cells are enveloped in a thick and recalcitrant cell wall, containing various carbohydrate polymers, such as pectin, chitin, cellulose, β-glucan, β-galactan, mannan, and other hemicelluloses, as well as hydroxyproline-rich glycoproteins, but most importantly algaenan, a highly recalcitrant aliphatic lignin-like polymer consisting of long mono- or di-unsaturated fatty acids, connected with ester and ether bonds and substituted with amide and pyrrole groups [147]. Algaenan is considered an indigestible polymer and a major constituent of organic matter sediments in soil and marine environments [145].

Due to the complexity of the material, efficient treatment methodologies must be developed. Mechanical treatments have been previously applied, but the energy cost and the almost complete destruction of the algal cells are significant drawbacks of such methods, prohibiting their industrial use [148]. Enzyme treatment seems to be the optimal approach to improve the digestibility of algal biomass since it is an environmentally friendly alternative, it does not require the use of organic solvents, and most enzymes used in feed production are approved for animal consumption. To this end, there is a significant number of experimental studies targeting the formulation of optimal enzyme mixtures for increasing the digestibility of microalgae, but due to the complexity and heterogeneity of the material, the results are often contradictory. For example, Gerken et al. (2013) [149] studied the viability of microalgal cells after different enzymatic treatments, in *Nannochloropsis*, *Nannochloris*, and *Chlorella* strains. Their results revealed that no single enzyme, except lysozyme to a certain extent, impaired the viability of all the tested microalgae, but the application of enzyme mixtures achieved this effect. For *Chlorella*, the combination of lysozyme and sulfatase or trypsin resulted in almost complete cell permeability. However, chitinase, chitosanase, cellulase, pectinase, and phospholipase also induced an altered morphology of the cell wall. The necessity for combined enzyme action for the efficient extraction of fatty acids was also evident in the work of Liang et al. (2012) [150], where among the various proteases tested, as well as cellulose-acting enzymes, the best combination was found to be trypsin together with snailase, an enzyme mixture containing cellulase, hemicellulase, pectinase, and β-glucuronidase. The group of Zuorro et al. also studied the extraction of fatty acids from *C. sorokiniana*, testing commercial enzyme preparations including cellulase, pectinase, lysozyme, and hemicellulases [151]. The optimal enzyme mixture contained β-1,4-xylanase and β-1,4-mannanase, resulting in over 70% lipid recovery, highlighting the necessity of complementary enzyme specificities working in synergy to achieve optimal results. The disruption of algal cell walls of *C. zofingiensis* was studied using crude enzyme mixtures from several bacterial strains grown in wheat bran as an enzyme inducer [152]. The crude enzyme extracts were found to contain cellulase, xylanase, and laccase activities, and they significantly disrupted the algal cell wall, resulting in increased reducing sugars in the supernatant, as well as lipid extraction efficiency.

Regarding the enzymatic digestion of *Chlorella* cell walls, the group of Coelho et al. tested more than 200 carbohydrate-acting enzymes and sulfatases on *C. vulgaris* biomass, revealing 29 of them with a certain degree of activity [153]. The most effective candidates, including an exo-β-glucosaminidase, an alginate lyase, a peptidoglycan deacetylase, and a lysozyme, were tested as ternary mixtures for the optimization of reducing sugar release, leading up to 8-fold higher release of reducing sugars and 23-fold higher protein release, while the release of fatty acids was marginally improved. The same group used this approach to study the degradation of the cell walls of the microalga *Arthrospira platensis*, resulting in an efficient enzyme mixture containing only two enzymes, lysozyme and α-amylase [154]. The two-enzyme mixture resulted in 7-fold higher reducing sugars, 1.15-fold higher release in chlorophyl α, while the release of fatty acids was also facilitated.

Nannochloropsis is another microalgal species with significant potential as animal feed. The group of Lavecchia et al. have studied the enzymatic pretreatment of the biomass from this species in detail. The tested enzymes included cellulase, mannanase, glucanase, galactanase, xylanase, esterase, and lysozyme, and they all resulted in increased lipid recovery. The most effective enzymes were found to be cellulase, mannanase, glucanase, and galactanase, which were further studied in binary and ternary combinations in order to design an efficient enzyme cocktail for this strain, resulting in a maximum of 37.2 g of lipids per 100 g of biomass [155]. In a follow-up study, the same authors achieved over 70% of lipid extraction yield from the same strain, with the synergistic effect of cellulase and mannanase. Moreover, they showed that the crystallinity of cellulose was increased in the cell walls, indicating the degradation of amorphous cellulose [156].

Overall, it is evident that the cost-effective treatment of microalgal biomass for improving its digestibility can be ideally implemented with combinations of enzymes with different specificities, targeting the various constituents of the cell wall. However, significant research effort is required for the design of tailored enzymatic cocktails depending on the available material, since the heterogeneity of the microalgal biomass hinders the application of universally efficient enzyme formulations.

9. Environmental Aspects of Using Microalgae in Animal Nutrition

The Paris Agreement's aim of limiting the increase in global temperature to 1.5 °C above preindustrial levels demands rapid and ambitious mitigation strategies aiming to reduce global greenhouse gas (GHG) emissions while simultaneously attaining a significant reduction in the amount of methane (CH_4) produced by the agricultural sector [157]. Ruminants produce a significant amount of methane emitted through eructation as a normal biochemical function for the neutralization of CO_2 and H_2 formed in the rumen due to the microbial fermentation of the feed.

Green, brown, and red seaweeds are key marine habitats rich in bioactive compounds such as bromoform ($CHBr_3$), which inhibits methanogenesis. It is believed that $CHBr_3$, along with other halogenated volatile organic compounds (VOCs), competitively bind to the enzymes and reductases that facilitate the final steps of reducing CO_2 and H_2 by methanogens (Archaea) into CH_4. Bromoform is found within many seaweed species in low concentrations but has been found to accumulate in higher levels in the red seaweed *Asparagopsis taxiformis* [158]. Indeed, the inclusion of *A. taxiformis* (0.25% of organic matter (OM)) in beef diets reduced methane emission by 51 g/kg DMI, while a higher inclusion level (0.50% of OM) further reduced methane by 75 g/kg DMI [159]. Although *A. taxiformis* constitutes an important ally of ruminants against their high criticism as environmental polluters, dietary supplementation with this seaweed is negatively associated with one-health concept concerns mainly due to its bromoform content, a compound with potential carcinogenic properties.

Bromoform has been associated with ongoing health and environmental concerns even when included at low doses: The daily consumption of 67 g *A. taxiformis* (84.42 μg bromoform) resulted in rumenitis and residues in both urine and milk (10 and 9.1 μg bromoform, respectively) of cows [160]. The health and residue issues of bromoform should be considered with caution, as there are still very few studies published that define the long-term effects of feeding bromoform-rich seaweed on animal productivity, animal health, and residue deposition in milk and/or meat. Thus, Roskam et al. (2022) [161] investigated the antimethanogenic potential of bromoform-free brown and green seaweeds (*Pelvetia canaliculata*, *Ericaria selaginoides* (formerly *Cystoseira tamariscifolia*), *Bifurcaria bifurcata*, *Fucus vesiculosus*, *Himanthalia elongata*, *Ascophyllum nodosum*, and *Ulva intestinalis*) in vitro. The results showed that only *Fucus vesiculosus* reduced CH_4%; however, the absolute methane production was not significantly reduced. Notably, it is the bromoform content that effectively disrupts methanogen's function, while other bromoform-free seaweeds rich in tannins and phenolic compounds are incapable of mitigating methane formation before impairing the overall rumen habitat.

On the other hand, studies have also provided insights into methane mitigation properties in PUFA-rich microalgae involving a different mode of action. In the study of Mavrommatis et al. (2021) [162], the inclusion of *Schizochytrium* spp. in goats' diet decreased the abundance of total archaea and methanogens in the rumen particle-associated microbiota. Moreover, in the rumen liquid fraction the *Methanobrevibacter* spp., a dominant archaeon of the hydrogenotrophic pathway was significantly decreased [163]. The mode of action that resulted in the former changes is related to LCPUFAs contained in *Schizochytrium* spp. [164]. More specifically, it has been proposed that double bonds alter the shape of the molecule, such that kinked unsaturated fatty acids disrupt the bacterial lipid bilayer structure, resulting in chemiosmotic issues and imbalances in acyl CoA metabolism [165]. Nevertheless, recent evidence subverts the aforementioned assumptions, indicating that unsaturated fatty acids did not considerably affect bacterial growth of both Gram-negative and Gram-positive strains [166], while unsaturated fatty acids are involved in the prevention of biofilm formation in Gram-positive bacteria, even at very low levels [166]. Hence, rumen bacterial populations could return to a planktonic lifestyle if the biofilm is dispersed, making them prone to abiotic factors. In addition to methanogen suppression by marine fatty acids (EPA, DPA, and DHA), as reflected by their DNA footprint, methane mitigation was observed in an in vitro study [167].

10. Conclusions

Microalgae have the potential to revolutionize biotechnology in a number of areas, including feed, nutrition, pharmaceuticals, cosmeceuticals, and biofuels. The biological and chemical diversity of the microalgae has been the source of unique antioxidant molecules with the potential for industrial development as feed nutritional supplements. Microalgal biomass is an attractive alternative to traditional forms of biomass for the production of high value-added antioxidants due to high productivity, the ability to be cultivated on marginal lands, and the potential to utilize carbon dioxide. Since microalgal biomass is still largely unexplored, it represents a rich source for discovery in both academic and industrial sectors. Further research should be performed aiming to assess the potential of microalgae in substituting synthetic antioxidants (e.g., vitamin E) in animal feed since the synthetic ones have been linked with severe concerns for human health [168]. To bridge this scientific gap, targeted experimental trials should be designed in order to validate the equivalent of vitamin E antioxidant activity of specific microalgae in vivo. Additionally, although extensive evidence supports the antimicrobial potency of microalgae, scarce information exists about the in vivo antibacterial, antiprotozoal, and antihelminthic effects of biomolecules present in microalgae. This perspective should not only be investigated under pilot conditions but also should be validated at a commercial level, where the effect of other cofactors (e.g., pathogens, thermal stress, oxidative stress, social stress, welfare issues, etc.) are also concerned.

In addition to the direct effect of microalgae on improving both animals' health and performance, there are indirect aspects of their usage in livestock. Microalgae are sustainable natural bioresources that do not compete with terrestrial plants for arable land, freshwater, pesticides, fertilizers, and insecticides to grow, and yet they have high productivity [169]. Thus, substituting synthetic feed additives with microalgae could result in positive environmental outcomes in the livestock sector with higher-quality nutritional products [170–174].

Author Contributions: Conceptualization, N.E.L. and E.T. All authors have contributed to the original draft preparation. All authors have read and agreed to the published version of the manuscript.

Funding: This research was funded by the European Union (Grant Agreements: No 778263 and No 101086437).

Conflicts of Interest: The authors declare that the research was conducted in the absence of any commercial or financial relationships that could be construed as a potential conflict of interest.

References

1. Khan, M.I.; Shin, J.H.; Kim, J.D. The promising future of microalgae: Current status, challenges, and optimization of a sustainable and renewable industry for biofuels, feed, and other products. *Microb. Cell Factories* **2018**, *17*, 36. [CrossRef]
2. Coulombier, N.; Jauffrais, T.; Lebouvier, N. Antioxidant Compounds from Microalgae: A Review. *Mar. Drugs* **2021**, *19*, 549. [CrossRef] [PubMed]
3. Daneshvar, E.; Sik Ok, Y.; Tavakoli, S.; Sarkar, B.; Shaheen, S.M.; Hong, H.; Luo, Y.; Rinklebe, J.; Song, H.; Bhatnagar, A. Insights into upstream processing of microalgae: A review. *Bioresour. Technol.* **2021**, *329*, 124870. [CrossRef] [PubMed]
4. Jareonsin, S.; Pumas, C. Advantages of Heterotrophic Microalgae as a Host for Phytochemicals Production. *Front. Bioeng. Biotechnol.* **2021**, *9*, 628597. [CrossRef] [PubMed]
5. Available online: https://www.credenceresearch.com/report/algaeproducts-market (accessed on 3 March 2023).
6. Hempel, F.; Maier, U.G. Microalgae as Solar-Powered Protein Factories. *Adv. Exp. Med. Biol.* **2016**, *896*, 241–262. [CrossRef]
7. Madeira, M.S.; Cardoso, C.; Lopes, P.A.; Coelho, D.; Afonso, C.; Bandarra, N.M.; Prates, J.A.M. Microalgae as feed ingredients for livestock production and meat quality: A review. *Livest. Sci.* **2017**, *205*, 111–121. [CrossRef]
8. Correa, D.F.; Beyer, H.L.; Possingham, H.P.; García-Ulloa, J.; Ghazoul, J.; Schenk, P.M. Freeing land from biofuel production through microalgal cultivation in the Neotropical region. *Environ. Res. Lett.* **2020**, *15*, 094094. [CrossRef]
9. Carballo, C.; Chronopoulou, E.G.; Letsiou, S.; Maya, C.; Labrou, N.E.; Infante, C.; Power, D.M.; Manchado, M. Antioxidant capacity and immunomodulatory effects of a chrysolaminarin-enriched extract in Senegalese sole. *Fish Shellfish. Immunol.* **2018**, *82*, 1–8. [CrossRef]
10. Georgakis, N.; Ioannou, E.; Chatzikonstantinou, M.; Merino, M.; Chronopoulou, E.; Mullor, J.L.; Madesis, P.; Labrou, N.E. The cosmeceutical potential of the yellow-green alga *Trachydiscus minutus* aqueous extract: Development of a natural-based dermal formula as a proof of concept. *Cosmetics* **2023**, *10*, 75. [CrossRef]
11. Ioannou, E.; Labrou, N.E. Development of enzyme-based cosmeceuticals: Studies on the proteolytic activity of *Arthrospira platensis* and its efficient incorporation in a hydrogel formulation. *Cosmetics* **2022**, *9*, 106. [CrossRef]
12. Chatzikonstantinou, M.; Kalliampakou, A.; Gatzogia, M.; Flemetakis, E.; Katharios, P.; Labrou, N.E. Comparative analyses and evaluation of the cosmeceutical potential of selected *Chlorella* strains. *J. Appl. Phycol.* **2017**, *29*, 179. [CrossRef]
13. Chatzikonstantinou, M.; Vlachakis, D.; Papageorgiou, L.; Papageorgiou, A.C.; Labrou, N.E. The Glutathione Transferase Family of *Chlamydomonas reinhardtii*: Identification and characterization of novel sigma class-like enzymes. *Algal Res.* **2017**, *24*, 237–250. [CrossRef]

14. Yaakob, Z.; Ali, E.; Zainal, A.; Mohamad, M.; Takriff, M.S. An overview: Biomolecules from microalgae for animal feed and aquaculture. *J. Biol. Res.* **2014**, *21*, 6. [CrossRef] [PubMed]
15. Halliwell, B.; Gutteridge, J.M. *Free Radicals in Biology and Medicine*, 5th ed.; Oxford University Press: New York, NY, USA, 2015.
16. Givens, D.I.; Gibbs, R.A. Very long chain n-3 polyunsaturated fatty acids in the food chain in the UK and the potential of animal-derived foods to increase intake. *Nutr. Bull.* **2006**, *31*, 104–110. [CrossRef]
17. Dr, S.; Zaman, M.M.; Farooq, Z.; Hafeez, A.; Sajid, M.W.; Tariq, M.; Ali, S.W.; Ali, S.; Shafiq, M.; Iftikhar, M.; et al. Supplementation of PUFA extracted from microalgae for the development of chicken patties. *PeerJ* **2023**, *24*, e15355. [CrossRef]
18. Chen, W.; Li, T.; Du, S.; Chen, H.; Wang, Q. Microalgal polyunsaturated fatty acids: Hotspots and production techniques. *Front. Bioeng. Biotechnol.* **2023**, *11*, 1146881. [CrossRef]
19. Mourelle, M.L.; Gómez, C.P.; Legido, J.L. The potential use of marine microalgae and cyanobacteria in cosmetics and thalassotherapy. *Cosmetics* **2017**, *4*, 46. [CrossRef]
20. Bhalamurugan, G.L.; Valerie, O.; Mark, L. Valuable bioproducts obtained from microalgal biomass and their commercial applications: A review. *Environ. Eng. Res.* **2018**, *23*, 229–241. [CrossRef]
21. Pereira, A.G.; Otero, P.; Echave, J.; Carreira-Casais, A.; Chamorro, F.; Collazo, N.; Jaboui, A.; Lourenço-Lopes, C.; Simal-Gandara, J.; Prieto, M.A. Xanthophylls from the Sea: Algae as Source of Bioactive Carotenoids. *Mar. Drugs* **2021**, *19*, 188. [CrossRef]
22. Yao, R.; Fu, W.; Du, M.; Chen, Z.X.; Lei, A.P.; Wang, J.X. Carotenoids Biosynthesis, Accumulation, and Applications of a Model Microalga *Euglena gracilis*. *Mar. Drugs* **2022**, *20*, 496. [CrossRef]
23. Berthon, J.Y.; Nachat-Kappes, R.; Bey, M.; Cadoret, J.P.; Renimel, I.; Filaire, E. Marine algae as attractive source to skin care. *Free Radic. Res.* **2017**, *51*, 555–567. [CrossRef]
24. Odjadjare, E.C.; Mutanda, T.; Olaniran, A.O. Potential biotechnological application of microalgae: A critical review. *Crit. Rev. Biotechnol.* **2017**, *37*, 37–52. [CrossRef] [PubMed]
25. Tsai, C.F.; Lin, H.W.; Liao, J.M.; Chen, K.M.; Tsai, J.W.; Chang, C.S.; Chou, C.Y.; Su, H.H.; Liu, P.H.; Chu, Y.C.; et al. *Dunaliella salina* Alga Protects against Myocardial Ischemia/Reperfusion Injury by Attenuating TLR4 Signaling. *Int. J. Mol. Sci.* **2023**, *24*, 3871. [CrossRef] [PubMed]
26. Del Campo, J.A.; García-González, M.; Guerrero, M.G. Outdoor cultivation of microalgae for carotenoid production: Current state and perspectives. *Appl. Microbiol. Biotechnol.* **2007**, *74*, 1163–1174. [CrossRef]
27. Christaki, E.; Bonos, E.; Giannenas, I.; Florou-Paneri, P. Functional properties of carotenoids originating from algae. *J. Sci. Food Agric.* **2013**, *93*, 5–11. [CrossRef]
28. Yuan, J.P.; Peng, J.; Yin, K.; Wang, J.H. Potential health-promoting effects of astaxanthin: A high-value carotenoid mostly from microalgae. *Mol. Nutr. Food Res.* **2011**, *55*, 150–165. [CrossRef] [PubMed]
29. Lee, R.E. *Phycology*, 5th ed.; Cambridge University Press: Cambridge, UK, 2018.
30. Gouveia, L.; Veloso, V.; Reis, A.; Fernandes, H.; Novais, J.; Empis, J. Evolution of pigment composition in *Chlorella vulgaris*. *Bioresour. Technol.* **1996**, *57*, 157–163. [CrossRef]
31. Lemoine, Y.; Schoefs, B. Secondary ketocarotenoid astaxanthin biosynthesis in algae: A multifunctional response to stress. *Photosynth. Res.* **2010**, *106*, 155–177. [CrossRef]
32. Milledge, J.J. Commercial application of microalgae other than as biofuels: A brief review. *Rev. Environ. Sci. Biotechnol.* **2011**, *10*, 31–41. [CrossRef]
33. Rammuni, M.N.; Ariyadasa, T.U.; Nimarshana, P.H.V.; Attalage, R.A. Comparative assessment on the extraction of carotenoids from microalgal sources: Astaxanthin from *H. pluvialis* and β-carotene from *D. salina*. *Food Chem.* **2019**, *277*, 128–134. [CrossRef]
34. Borowitzka, M.A. Microalgae as sources of pharmaceuticals and other biologically active compounds. *J. Appl. Phycol.* **1995**, *7*, 3–15. [CrossRef]
35. Hamed, I. The evolution and versatility of microalgal biotechnology: A review. *Compr. Rev. Food Sci. Food Saf.* **2016**, *15*, 1104–1123. [CrossRef] [PubMed]
36. Khanra, S.; Mondal, M.; Halder, G.; Tiwari, O.N.; Gayen, K.; Bhowmick, T.K. Downstream processing of microalgae for pigments, protein and carbohydrate in industrial application: A review. *Food Bioprod. Process.* **2018**, *110*, 60–84. [CrossRef]
37. Molino, A.; Iovine, A.; Casella, P.; Mehariya, S.; Chianese, S.; Cerbone, A.; Rimauro, J.; Musmarra, D. Microalgae characterization for consolidated and new application in human food, animal feed and nutraceuticals. *Int. J. Environ. Res.* **2018**, *15*, 2436. [CrossRef] [PubMed]
38. Novoveská, L.; Ross, M.E.; Stanley, M.S.; Pradelles, R.; Wasiolek, V.; Sassi, J.F. Microalgal Carotenoids: A Review of Production, Current Markets, Regulations, and Future Direction. *Mar. Drugs* **2019**, *17*, 640. [CrossRef]
39. Lorenz, R.T.; Cysewski, G.R. Commercial potential for *Haematococcus* microalgae as a natural source of astaxanthin. *Trends Biotechnol.* **2000**, *18*, 160–167. [CrossRef]
40. Boussiba, S. Carotenogenesis in the green alga *Haematococcus pluvialis*: Cellular physiology and stress response. *Physiol. Plant.* **2000**, *108*, 111–117. [CrossRef]
41. Boussiba, S.; Vonshak, A. Astaxanthin accumulation in the green alga *Haematococcus pluvialis*. *Plant Cell Physiol.* **1991**, *32*, 1077–1082. [CrossRef]
42. Siqueira, S.F.; Queiroz, M.I.; Zepka, L.Q.; Jacob-Lopes, E. Introductory chapter: Microalgae biotechnology—A brief introduction. In *Microalgal Biotechnology*, 1st ed.; Jacob-Lopes, E., Zepka, L.Q., Queiroz, M.I., Eds.; IntechOpen: London, UK, 2018.

43. de Jesus Raposo, M.F.; De Morais, A.M.B.; De Morais, R.M.S.C. Marine polysaccharides from algae with potential biomedical applications. *Mar. Drugs* **2015**, *13*, 2967–3028. [CrossRef]
44. Pertiwi, H.; Nur-Mahendra, M.Y.; Kamaludeen, J. Astaxanthin as a Potential Antioxidant to Improve Health and Production Performance of Broiler Chicken. *Vet. Med. Int.* **2022**, *2022*, 4919442. [CrossRef]
45. Landrum, J.T.; Bone, R.A. Lutein, zeaxanthin, and the macular pigment. *Arch. Biochem. Biophys.* **2001**, *385*, 28–40. [CrossRef] [PubMed]
46. Shi, H.; Deng, X.; Ji, X.; Liu, N.; Cai, H. Sources, dynamics in vivo, and application of astaxanthin and lutein in laying hens: A review. *Anim. Nutr.* **2023**, *2*, 324–333. [CrossRef] [PubMed]
47. Shi, X.M.; Chen, F.; Yuan, J.P.; Chen, H. Heterotrophic production of lutein by selected *Chlorella* strains. *J. Appl. Phycol.* **1997**, *9*, 445–450. [CrossRef]
48. Zhang, X.W.; Shi, X.M.; Chen, F. A kinetic model for lutein production by the green microalga *Chlorella protothecoides* in heterotrophic culture. *J. Ind. Microbiol. Biotechnol.* **1999**, *23*, 503–507. [CrossRef]
49. Wu, Z.Y.; Shi, C.L.; Shi, X.M. Modeling of lutein production by heterotrophic *Chlorella* in batch and fed-batch cultures. *World J. Microbiol. Biotechnol.* **2007**, *23*, 1233–1238. [CrossRef]
50. Del Campo, J.A.; Rodriguez, H.; Moreno, J.; Vargas, M.A.; Rivas, J.; Guerrero, M.G. Accumulation of astaxanthin and lutein in *Chlorella zofingiensis* (Chlorophyta). *Appl. Microbiol. Biotechnol.* **2004**, *64*, 848–854. [CrossRef]
51. Wei, D.; Chen, F.; Chen, G.; Zhang, X.; Liu, L.; Zhang, H. Enhanced production of lutein in heterotrophic *Chlorella protothecoides* by oxidative stress. *Sci. China Life Sci.* **2008**, *51*, 1088–1093. [CrossRef]
52. Del Campo, J.A.; Moreno, J.; Rodríguez, H.; Vargas, M.A.; Rivas, J.; Guerrero, M.G. Carotenoid content of chlorophycean microalgae: Factors determining lutein accumulation in *Muriellopsis* sp. (Chlorophyta). *J. Biotechnol.* **2000**, *76*, 51–59. [CrossRef]
53. Sánchez, J.F.; Fernández-Sevilla, J.M.; Acién, F.G.; Cerón, M.C.; Pérez-Parra, J.; Molina-Grima, E. Biomass and lutein productivity of *Scenedesmus almeriensis*: Influence of irradiance, dilution rate and temperature. *Appl. Microbiol. Biotechnol.* **2008**, *79*, 719–729. [CrossRef]
54. Garbayo, I.; Cuaresma, M.; Vílchez, C.; Vega, J.M. Effect of abiotic stress on the production of lutein and β-carotene by *Chlamydomonas acidophila*. *Process Biochem.* **2008**, *43*, 1158–1161. [CrossRef]
55. Roberts, J.E.; Dennison, J. The photobiology of lutein and zeaxanthin in the eye. *J. Ophthalmol.* **2015**, *2015*, 687173. [CrossRef] [PubMed]
56. Molino, A.; Rimauro, J.; Casella, P.; Cerbone, A.; Larocca, V.; Karatza, D.; Hristoforou, E.; Chianese, S.; Musmarra, D. Microalgae valorisation via accelerated solvent extraction: Optimization of the operative conditions. *Chem. Eng. Trans.* **2018**, *65*, 835–840. [CrossRef]
57. Srivastava, A.; Kalwani, M.; Chakdar, H.; Pabbi, S.; Shukla, P. Biosynthesis and biotechnological interventions for commercial production of microalgal pigments: A review. *Bioresour. Technol.* **2022**, *352*, 127071. [CrossRef] [PubMed]
58. Khaw, Y.S.; Yusoff, F.M.; Tan, H.T.; Noor Mazli, N.A.I.; Nazarudin, M.F.; Shaharuddin, N.A.; Omar, A.R. The Critical Studies of Fucoxanthin Research Trends from 1928 to June 2021: A Bibliometric Review. *Mar. Drugs* **2021**, *19*, 606. [CrossRef]
59. Khaw, Y.S.; Yusoff, F.M.; Tan, H.T.; Noor Mazli, N.A.I.; Nazarudin, M.F.; Shaharuddin, N.A.; Omar, A.R.; Takahashi, K. Fucoxanthin Production of Microalgae under Different Culture Factors: A Systematic Review. *Mar. Drugs* **2022**, *20*, 592. [CrossRef]
60. Kratzer, R.; Murkovic, M. Food Ingredients and Nutraceuticals from Microalgae: Main Product Classes and Biotechnological Production. *Foods* **2021**, *10*, 1626. [CrossRef]
61. Sun, H.; Wang, Y.; He, Y.; Liu, B.; Mou, H.; Chen, F.; Yang, S. Microalgae-Derived Pigments for the Food Industry. *Mar. Drugs* **2023**, *21*, 82. [CrossRef]
62. Cuellar-Bermudez, S.P.; Aguilar-Hernandez, I.; Cardenas-Chavez, D.L.; Ornelas-Soto, N.; Romero-Ogawa, M.A.; Parra-Saldivar, R. Extraction and purification of high-value metabolites from microalgae: Essential lipids, astaxanthin and phycobiliproteins. *Microb. Biotechnol.* **2015**, *8*, 190–209. [CrossRef]
63. ElFar, O.A.; Billa, N.; Lim, H.R.; Chew, K.W.; Cheah, W.Y.; Munawaroh, H.S.H.; Balakrishnan, D.; Show, P.L. Advances in delivery methods of *Arthrospira platensis* (spirulina) for enhanced therapeutic outcomes. *Bioengineered* **2022**, *13*, 14681–14718. [CrossRef]
64. Kovaleski, G.; Kholany, M.; Dias, L.M.S.; Correia, S.F.H.; Ferreira, R.A.S.; Coutinho, J.A.P.; Ventura, S.P.M. Extraction and purification of phycobiliproteins from algae and their applications. *Front. Chem.* **2022**, *10*, 1065355. [CrossRef]
65. Sekar, S.; Chandramohan, M. Phycobiliproteins as a commodity: Trends in applied research, patents and commercialization. *J. Appl. Phycol.* **2008**, *20*, 113–136. [CrossRef]
66. Tan, H.T.; Yusoff, F.M.; Khaw, Y.S.; Ahmad, S.A.; Shaharuddin, N.A. Uncovering Research Trends of Phycobiliproteins Using Bibliometric Approach. *Plants* **2021**, *10*, 2358. [CrossRef] [PubMed]
67. Liu, R.; Qin, S.; Li, W. Phycocyanin: Anti-inflammatory effect and mechanism. *Biomed. Pharmacother.* **2022**, *153*, 113362. [CrossRef]
68. Bannu, S.M.; Lomada, D.; Gulla, S.; Chandrasekhar, T.; Reddanna, P.; Reddy, M.C. Potential Therapeutic Applications of C-Phycocyanin. *Curr. Drug Metab.* **2019**, *20*, 967–976. [CrossRef]
69. Ashaolu, T.J.; Samborska, K.; Lee, C.C.; Tomas, M.; Capanoglu, E.; Tarhan, Ö.; Taze, B.; Jafari, S.M. Phycocyanin, a super functional ingredient from algae; properties, purification characterization, and applications. *Int. J. Biol. Macromol.* **2021**, *193*, 2320–2331. [CrossRef]
70. Ji, H.; Liu, G.; Han, J.; Zhu, F.; Dong, X.; Li, B. C-phycocyanin inhibits epithelial-to-mesenchymal transition in Caski cells. *Cancer Cell Int.* **2020**, *20*, 292. [CrossRef] [PubMed]

71. Galasso, C.; Gentile, A.; Orefice, I.; Ianora, A.; Bruno, A.; Noonan, D.M.; Sansone, C.; Albini, A.; Brunet, C. Microalgal derivatives as potential nutraceutical and food supplements for human health: A focus on cancer prevention and interception. *Nutrients* **2019**, *11*, 1226. [CrossRef]
72. Santiago-Díaz, P.; Rivero, A.; Rico, M.; González González, A.; González-Dávila, M.; Santana-Casiano, M. Copper toxicity leads to accumulation of free amino acids and polyphenols in *Phaeodactylum tricornutum* diatoms. *Environ. Sci. Pollut. Res. Int.* **2023**, *30*, 51261–51270. [CrossRef]
73. León-Vaz, A.; León, R.; Vigara, J.; Funk, C. Exploring Nordic microalgae as a potential novel source of antioxidant and bioactive compounds. *New Biotechnol.* **2023**, *73*, 1–8. [CrossRef]
74. Parkes, R.; McGee, D.; McDonnell, A.; Gillespie, E.; Touzet, N. Rapid screening of phenolic compounds in extracts of photosynthetic organisms separated using a C18 monolithic column based HPLC-UV method. *J. Chromatogr. B Analyt. Technol. Biomed. Life Sci.* **2022**, *1213*, 123521. [CrossRef]
75. Scaglioni, P.T.; Pagnussatt, F.A.; Lemos, A.C.; Nicolli, C.P.; Del Ponte, E.M.; Badiale-Furlong, E. *Nannochloropsis* sp. and *Spirulina* sp. as a source of antifungal compounds to mitigate contamination by *Fusarium graminearum* species complex. *Curr. Microbiol.* **2019**, *76*, 930–938. [CrossRef] [PubMed]
76. Wang, P.; Li, D.; Ke, W.; Liang, D.; Hu, X.; Chen, F. Resveratrol-induced gut microbiota reduces obesity in high-fat diet-fed mice. *Int. J. Obes.* **2020**, *44*, 213–225. [CrossRef] [PubMed]
77. Espín, J.C.; González-Sarrías, A.; Tomás-Barberán, F.A. The gut microbiota: A key factor in the therapeutic effects of (poly)phenols. *Biochem. Pharmacol.* **2017**, *139*, 82–93. [CrossRef] [PubMed]
78. Tomás-Barberán, F.A.; Selma, M.V.; Espín, J.C. Interactions of gut microbiota with dietary polyphenols and consequences to human health. *Curr. Opin. Clin. Nutr. Metab. Care* **2016**, *19*, 471–476. [CrossRef]
79. Pang, J.; Liu, Y.; Kang, L.; Ye, H.; Zang, J.; Wang, J.; Han, D. *Bifidobacterium animalis* Promotes the Growth of Weaning Piglets by Improving Intestinal Development, Enhancing Antioxidant Capacity, and Modulating Gut Microbiota. *Appl. Environ. Microbiol.* **2022**, *88*, e0129622. [CrossRef]
80. Meng, Q.; Zhang, Y.; Li, J.; Shi, B.; Ma, Q.; Shan, A. Lycopene Affects Intestinal Barrier Function and the Gut Microbiota in Weaned Piglets via Antioxidant Signaling Regulation. *J. Nutr.* **2022**, *152*, 2396–2408. [CrossRef]
81. Martínez, Y.; Más, D.; Betancur, C.; Gebeyew, K.; Adebowale, T.; Hussain, T.; Lan, W.; Ding, X. Role of the Phytochemical Compounds like Modulators in Gut Microbiota and Oxidative Stress. *Curr. Pharm. Des.* **2020**, *26*, 2642–2656. [CrossRef] [PubMed]
82. Mirmohammadali, S.N.; Rosenkranz, S.K. Dietary phytochemicals, gut microbiota composition, and health outcomes in human and animal models. *Biosci. Microbiota Food Health* **2023**, *42*, 152–171. [CrossRef]
83. Novichkova, E.; Nayak, S.; Boussiba, S.; Gopas, J.; Zilberg, D.; Khozin-Goldberg, I. Dietary Application of the Microalga *Lobosphaera incisa* P127 Reduces Severity of Intestinal Inflammation, Modulates Gut-Associated Gene Expression, and Microbiome in the Zebrafish Model of IBD. *Mol. Nutr. Food Res.* **2023**, *67*, e2200253. [CrossRef]
84. Ran, L.; Yu, J.; Ma, R.; Yao, Q.; Wang, M.; Bi, Y.; Yu, Z.; Wu, Y. Microalgae oil from *Schizochytrium* sp. alleviates obesity and modulates gut microbiota in high-fat diet-fed mice. *Food Funct.* **2022**, *13*, 12799–12813. [CrossRef]
85. Zhou, J.; Wang, M.; Bäuerl, C.; Cortés-Macías, E.; Calvo-Lerma, J.; Carmen Collado, M.; Barba, F.J. The impact of liquid-pressurized extracts of Spirulina, *Chlorella* and *Phaedactylum tricornutum* on In Vitro antioxidant, antiinflammatory and bacterial growth effects, and gut microbiota modulation. *Food Chem.* **2023**, *401*, 134083. [CrossRef]
86. Sagaram, U.S.; Gaikwad, M.S.; Nandru, R.; Dasgupta, S. Microalgae as feed ingredients: Recent developments on their role in immunomodulation and gut microbiota of aquaculture species. *FEMS Microbiol. Lett.* **2021**, *368*, 71. [CrossRef]
87. de Jesus Raposo, M.F.; de Morais, A.M.; de Morais, R.M. Emergent Sources of Prebiotics: Seaweeds and Microalgae. *Mar. Drugs* **2016**, *14*, 27. [CrossRef]
88. Huang, Z.; Gao, J.; Peng, C.; Song, J.; Xie, Z.; Jia, J.; Li, H.; Zhao, S.; Liang, Y.; Gong, B. The Effect of the Microalgae *Chlorella vulgaris* on the Gut Microbiota of Juvenile Nile Tilapia (*Oreochromis niloticus*) Is Feeding-Time Dependent. *Microorganisms* **2023**, *11*, 1002. [CrossRef] [PubMed]
89. Lazado, C.C.; Nayak, S.; Khozin-Goldberg, I.; Zilberg, D. The gut mucosal barrier of zebrafish (*Danio rerio*) responds to the time-restricted delivery of *Lobosphaera incisa*-enriched diets. *Fish Shellfish Immunol.* **2019**, *89*, 368–377. [CrossRef] [PubMed]
90. Kulpys, J.; Paulauskas, E.; Pilipavicius, V.; Stankevicius, R. Influence of cyanobacteria *Arthrospira* (Spirulina) *platensis* biomaxx additives towards the body condition of lactation cows and biochemical milk indexes. *Agron. Res.* **2009**, *7*, 823–835.
91. El-Sabagh, M.R.; Abd Eldaim, M.A.; Mahboub, D.H.; Abdel-Daim, M. Effects of *Spirulina platensis* algae on growth performance, antioxidative status and blood metabolites in fattening lambs. *J. Agric. Sci.* **2014**, *6*, 92. [CrossRef]
92. Urrutia, O.; Mendizabal, J.A.; Insausti, K.; Soret, B.; Purroy, A.; Arana, A. Effects of addition of linseed and marine algae to the diet on adipose tissue development, fatty acid profile, lipogenic gene expression, and meat quality in lambs. *PLoS ONE* **2016**, *11*, 6. [CrossRef]
93. Díaz, M.T.; Pérez, C.; Sánchez, C.I.; Lauzurica, S.; Cañeque, V.; González, C.; De La Fuente, J. Feeding microalgae increases omega 3 fatty acids of fat deposits and muscles in light lambs. *J. Food Compos. Anal.* **2017**, *56*, 115–123. [CrossRef]
94. Mavrommatis, A.; Tsiplakou, E. The impact of the dietary supplementation level with *Schizochytrium* sp., on milk chemical composition and fatty acid profile of both blood plasma and milk of goats. *Small Rum. Res.* **2020**, *193*, 106252. [CrossRef]
95. Zahroojian, N.; Moravej, H.; Shivazad, M. Effects of dietary marine algae (*Spirulina platensis*) on egg quality and production performance of laying hens. *J. Agric. Sci. Technol.* **2013**, *15*, 1353–1360.

96. Evans, A.M.; Smith, D.L.; Moritz, J.S. Effects of algae incorporation into broiler starter diet formulations on nutrient digestibility and 3 to 21 d bird performance. *J. Appl. Poult. Res.* **2015**, *24*, 206–214. [CrossRef]
97. Bonos, E.; Kasapidou, E.; Kargopoulos, A.; Karampampas, A.; Nikolakakis, I.; Christaki, E.; Florou-Paneri, P. Spirulina as a functional ingredient in broiler chicken diets. *S. Afr. J. Anim. Sci.* **2016**, *46*, 94–102. [CrossRef]
98. Dlouha, G.; Sevcikova, S.; Dokoupilova, A.; Zita, L.; Heindl, J.; Skrivan, M. Effect of dietary selenium sources on growth performance, breast muscle selenium, glutathione peroxidase activity and oxidative stability in broilers. *Czech J. Anim. Sci.* **2008**, *53*, 265–269. [CrossRef]
99. Kang, H.K.; Salim, H.M.; Akter, N.; Kim, D.W.; Kim, J.H.; Bang, H.T.; Kim, M.J.; Na, J.C.; Hwangbo, J.; Choi, H.C.; et al. Effect of various forms of dietary *Chlorella* supplementation on growth performance, immune characteristics, and intestinal microflora population of broiler chickens. *J. Appl. Poult. Res.* **2013**, *22*, 100–108. [CrossRef]
100. Ribeiro, T.; Lordelo, M.M.; Alves, S.P.; Bessa, R.J.B.; Costa, P.; Lemos, J.P.C.; Ferreira, L.M.A.; Fontes, C.M.G.A.; Prates, J.A.M. Direct supplementation of diet is the most efficient way of enriching broiler meat with n-3 long-chain polyunsaturated fatty acids. *Br. Poult. Sci.* **2013**, *54*, 753–765. [CrossRef]
101. Ribeiro, T.; Lordelo, M.M.; Costa, P.; Alves, S.P.; Benevides, W.S.; Bessa, R.J.B.; Lemos, J.P.C.; Pinto, R.M.A.; Ferreira, L.M.A.; Fontes, C.M.G.A. Effect of reduced dietary protein and supplementation with a docosahexaenoic acid product on broiler performance and meat quality. *Br. Poult. Sci.* **2014**, *55*, 752–765. [CrossRef]
102. Grinstead, G.S.; Tokach, M.D.; Dritz, S.S.; Goodband, R.D.; Nelssen, J.L. Effects of *Spirulina platensis* on growth performance of weanling pigs. *Anim. Feed Sci. Tech.* **2000**, *83*, 237–247. [CrossRef]
103. Simkus, A.; Simkiene, J.; Cernauskiene, N.; Kvietkute, A.; Cernauskas, M.; Paleckaitis, S.; Kerziene, S. The effect of blue algae *Spirulina platensis* on pig growth performance and carcass and meat quality. *Vet. Med. Zoot.* **2013**, *6*, 83.
104. Furbeyre, H.; van Milgen, J.; Mener, T.; Gloaguen, M.; Labussière, E. Effects of dietary supplementation with freshwater microalgae on growth performance, nutrient digestibility and gut health in weaned piglets. *Animal* **2017**, *11*, 183–192. [CrossRef]
105. Banoch, T.; Svoboda, M.; Kuta, J.; Salakova, A.; Fajt, Z. The effect of iodine from iodine-enriched alga *Chlorella* spp. On the pork iodine content and meat quality in finisher pigs. *Acta Vet. Br.* **2012**, *81*, 339–346. [CrossRef]
106. Abril, R.; Garret, J.; Zeller, S.G.; Sander, W.J.; Mast, R.W. Safety assessment of DHA-rich microalgae from *Schizochytrium* sp. Part V: Target animal safety/toxicity study in growing swine. *Regul. Toxicol. Pharmacol.* **2003**, *37*, 73–82. [CrossRef] [PubMed]
107. Sardi, L.; Martelli, G.; Lambertini, L.; Parisini, P.; Mordenti, A. Effects of a dietary supplement of DHA-rich marine algae on Italian heavy pig production parameters. *Livest. Sci.* **2006**, *103*, 95–103. [CrossRef]
108. Saadaoui, I.; Rasheed, R.; Aguilar, A. Microalgal-based feed: Promising alternative feedstocks for livestock and poultry production. *J. Anim. Sci. Biotechnol.* **2021**, *12*, 76. [CrossRef]
109. Remize, M.; Brunel, Y.; Silva, J.L.; Berthon, J.Y.; Filaire, E. Microalgae n-3 PUFAs Production and Use in Food and Feed Industries. *Mar. Drugs* **2021**, *19*, 113. [CrossRef]
110. Georganas, A.; Giamouri, E.; Christodoulou, C.; Mavrommatis, A.; Zoidis, E.; Papadomichelakis, G.; Simitzis, P.; Tsiplakou, E.; Pappas, A.; Fegeros, F. Chapter: Impact of bioactive compounds on animals and the environment. In *Bioactive Compounds and Their Importance*; Nova Science Publisher: Hauppauge, NY, USA, 2021; pp. 61–103, ISBN 978-168507397-8/978-168507361-9.
111. Martins, D.A.; Custódio, L.; Barreira, L.; Pereira, H.; Ben-Hamadou, R.; Varela, J.; Abu-Salah, K.M. Alternative sources of n-3 long-chain polyunsaturated fatty acids in marine microalgae. *Mar. Drugs* **2013**, *11*, 2259–2281. [CrossRef]
112. Mavrommatis, A.; Theodorou, G.; Politis, I.; Tsiplakou, E. *Schizochytrium* sp. Dietary Supplementation Levels Modify Toll-like receptor-4 (TLR4) Signaling in Monocytes and Neutrophils of Dairy Goats. *Cytokine* **2021**, *148*, 155588. [CrossRef]
113. Kyriakaki, P.; Mavrommatis, A.; Tsiplakou, E. *Schizochytrium* spp. Dietary Supplementation Modulates Immune-Oxidative Transcriptional Signatures in Monocytes and Neutrophils of Dairy Goats. *Antioxidants* **2023**, *12*, 497. [CrossRef]
114. Burdick Sanchez, N.C.; Broadway, P.R.; Carroll, J.A. Influence of Yeast Products on Modulating Metabolism and Immunity in Cattle and Swine. *Animals* **2021**, *11*, 371. [CrossRef]
115. Caroprese, M.; Albenzio, M.; Ciliberti, M.G.; Francavilla, M.; Sevi, A. A mixture of phytosterols from *Dunaliella tertiolecta* affects proliferation of peripheral blood mononuclear cells and cytokine production in sheep. *Vet. Immunol. Immunopathol.* **2012**, *150*, 27–35. [CrossRef]
116. Ciliberti, M.G.; Albenzio, M.; Francavilla, M.; Neglia, G.; Esposito, L.; Caroprese, M. Extracts from Microalga *Chlorella sorokiniana* Exert an Anti-Proliferative Effect and Modulate Cytokines in Sheep Peripheral Blood Mononuclear Cells. *Animals* **2019**, *9*, 45. [CrossRef] [PubMed]
117. Amaro, H.M.; Guedes, A.C.; Malcata, F.X. Antimicrobial activities of microalgae: An invited review. In *Science against Microbial Pathogens: Communicating Current Research and Technological Advances*; Formatex Research Center: Badajoz, Spain, 2011; pp. 1272–1280.
118. Fries-Craft, K.; Meyer, M.M.; Bobeck, E.A. Algae-based feed ingredient protects intestinal health during *Eimeria* challenge and alters systemic immune responses with differential outcomes observed during acute feed restriction. *Poult. Sci.* **2021**, *100*, 101369. [CrossRef] [PubMed]
119. Yan, L.; Lim, S.U.; Kim, I.H. Effect of fermented *Chlorella* supplementation on growth performance, nutrient digestibility, blood characteristics, fecal microbial and fecal noxious gas content in growing pigs. *Asian-Aust. J. Anim. Sci.* **2012**, *25*, 1742–1747. [CrossRef] [PubMed]

120. El-Bahr, S.; Shousha, S.; Shehab, A.; Khattab, W.; Ahmed-Farid, O.; Sabike, I. Effect of Dietary Microalgae on Growth Performance, Profiles of Amino and Fatty Acids, Antioxidant Status, and Meat Quality of Broiler Chickens. *Animals* **2020**, *10*, 761. [CrossRef]
121. Bortoluzzi, C.; Schmidt, J.M.; Bordignon, H.L.F.; Fülber, L.M.; Layter, J.R.; Fernandes, J.I.M. Efficacy of yeast derived glucomannan or algae-based antioxidant or both as feed additives to ameliorate mycotoxicosis in heat stressed and unstressed broiler chickens. *Livest. Sci.* **2016**, *193*, 20–25. [CrossRef]
122. Christodoulou, C.; Kotsampasi, B.; Dotas, V.; Simoni, M.; Righi, F.; Tsiplakou, E. The effect of Spirulina supplementation in ewes' oxidative status and milk quality. *Anim. Feed Sci. Technol.* **2023**, *295*, 115544. [CrossRef]
123. Messina, C.M.; Manuguerra, S.; Arena, R.; Espinosa-Ruiz, C.; Curcuraci, E.; Esteban, M.A.; Santulli, A. Contaminant-induced oxidative stress underlies biochemical, molecular and fatty acid profile changes, in gilthead seabream (*Sparus aurata* L.). *Res. Vet. Sci.* **2023**, *159*, 244–251. [CrossRef]
124. Mavrommatis, A.; Chronopoulou, E.G.; Sotirakoglou, K.; Labrou, N.E.; Zervas, G.; Tsiplakou, E. The impact of the dietary supplementation level with *Schizochytrium* sp, on the oxidative capacity of both goats' organism and milk. *Livest. Sci.* **2018**, *218*, 37–43. [CrossRef]
125. Costa, D.F.A.; Castro-Montoya, J.M.; Harper, K.; Trevaskis, L.; Jackson, E.L.; Quigley, S. Algae as Feedstuff for Ruminants: A Focus on Single-Cell Species, Opportunistic Use of Algal By-Products and On-Site Production. *Microorganisms* **2022**, *10*, 2313. [CrossRef]
126. Martins, C.F.; Ribeiro, D.M.; Costa, M.; Coelho, D.; Alfaia, C.M.; Lordelo, M.; Almeida, A.M.; Freire, J.P.B.; Prates, J.A.M. Using Microalgae as a Sustainable Feed Resource to Enhance Quality and Nutritional Value of Pork and Poultry Meat. *Foods* **2021**, *10*, 2933. [CrossRef]
127. Bruneel, C.; Lemahieu, C.; Fraeye, I.; Ryckebosch, E.; Muylaert, K.; Buyse, J. Impact of microalgal feed supplementation on omega-3 fatty acid enrichment of hen eggs. *J. Funct. Foods* **2013**, *5*, 897–904. [CrossRef]
128. Fredriksson, S.; Elwinger, K.; Pickova, J. Fatty acid and carotenoid composition of egg yolk as an effect of microalgae addition to feed formula for laying hens. *Food Chem.* **2006**, *99*, 530–537. [CrossRef]
129. Kassis, N.; Drake, S.R.; Beamer, S.K.; Matak, K.E.; Jaczynski, J. Development of nutraceutical egg products with omega-3-rich oils. *LWT–Food Sci. Technol.* **2010**, *43*, 777–783. [CrossRef]
130. Englmaierová, M.; Skřivan, M.; Bubancová, I. A comparison of lutein, spraydried *Chlorella*, and synthetic carotenoids effects on yolk colour, oxidative stability, and reproductive performance of laying hens. *Czech. J. Anim. Sci.* **2013**, *58*, 412–419. [CrossRef]
131. Cherian, G.; Wolfe, F.W.; Sim, J.S. Dietary oils with added tocopherols: Effects on egg or tissue tocopherols, fatty acids, and oxidative stability. *Poult. Sci.* **1996**, *75*, 423–431. [CrossRef]
132. Toyomizu, M.; Sato, K.; Taroda, H.; Kato, T.; Akiba, Y. Effects of dietary Spirulina on meat colour in muscle of broiler chickens. *Br. Poult. Sci.* **2001**, *42*, 197–202. [CrossRef]
133. Alfaia, C.M.; Pestana, J.M.; Rodrigues, M.; Coelho, D.; Aires, M.J.; Ribeiro, D.M.; Major, V.T.; Martins, C.F.; Santos, H.; Lopes, P.A. Influence of dietary *Chlorella vulgaris* and carbohydrate-active enzymes on growth performance, meat quality and lipid composition of broiler chickens. *Poult. Sci. J.* **2021**, *100*, 926–937. [CrossRef]
134. Khan, I.A.; Parker, N.B.; Löhr, C.V.; Cherian, G. Docosahexaenoic acid (22:6 n-3)-rich microalgae along with methionine supplementation in broiler chickens: Effects on production performance, breast muscle quality attributes, lipid profile, and incidence of white striping and myopathy. *Poult. Sci.* **2021**, *100*, 865–874. [CrossRef]
135. Long, S.F.; Kang, S.; Wang, Q.Q.; Xu, Y.T.; Pan, L.; Hu, J.X.; Li, M.; Piao, X.S. Dietary supplementation with DHA-rich microalgae improves performance, serum composition, carcass trait, antioxidant status, and fatty acid profile of broilers. *Poult. Sci.* **2018**, *97*, 1881–1890. [CrossRef]
136. Baeza, E.; Chartrin, P.; Lessire, M.; Meteau, K.; Chesneau, G.; Guillevic, M.; Mourot, J. Is it possible to increase n-3 fatty acid content of meat without affecting its technological and/or sensory quality and the growing performance of chickens? *Br. Poult. Sci.* **2015**, *56*, 543–550. [CrossRef]
137. Moran, C.A.; Currie, D.; Keegan, J.D.; Knox, A. Tolerance of broilers to dietary supplementation with high levels of the DHA-rich microalga, *Aurantiochytrium limacinum*: Effects on health and productivity. *Animals* **2018**, *8*, 180. [CrossRef] [PubMed]
138. Coelho, D.; Pestana, J.; Almeida, J.M.; Alfaia, C.M.; Fontes, C.M.G.A.; Moreira, O.; Prates, J.A.M. A High Dietary incorporation level of *Chlorella vulgaris* improves the nutritional value of pork fat without impairing the performance of finishing pigs. *Animals* **2020**, *10*, 2384. [CrossRef] [PubMed]
139. McCrorie, T.A.; Keaveney, E.M.; Wallace, J.M.; Binns, N.; Livingstone, M.B.E. Human health effects of conjugated linoleic acid from milk and supplements. *Nutr. Res. Rev.* **2011**, *24*, 206–227. [CrossRef] [PubMed]
140. Farina, G.; Cattaneo, D.; Lecchi, C. A review on the role of EPA and DHA through goat nutrition to human health: Could they be effective both to animals and humans? *J. Dairy Vet. Anim. Res.* **2015**, *2*, 35–39. [CrossRef]
141. Zisis, F.; Kyriakaki, P.; Satolias, F.F.; Mavrommatis, A.; Simitzis, P.E.; Pappas, A.C.; Surai, P.F.; Tsiplakou, E. The Effect of Dietary Inclusion of Microalgae *Schizochytrium* spp. on Ewes' Milk Quality and Oxidative Status. *Foods* **2022**, *11*, 2950. [CrossRef]
142. Mavrommatis, A.; Sotirakoglou, K.; Kamilaris, C.; Tsiplakou, E. Effects of Inclusion of *Schizochytrium* spp. and Forage-to-Concentrate Ratios on Goats' Milk Quality and Oxidative Status. *Foods* **2021**, *10*, 1322. [CrossRef]
143. Spain, O.; Funk, C. Detailed characterization of the cell wall structure and composition of nordic green microalgae. *J. Agric. Food Chem.* **2022**, *70*, 9711–9721. [CrossRef]

144. Weber, S.; Grande, P.M.; Blank, L.M.; Klose, H. Insights into cell wall disintegration of *Chlorella vulgaris*. *PLoS ONE* **2022**, *17*, e0262500. [CrossRef]
145. Scholz, M.J.; Weiss, T.L.; Jinkerson, R.E.; Jing, J.; Roth, R.; Goodenough, U.; Posewitz, M.C.; Gerken, H.G. Ultrastructure and Composition of the *Nannochloropsis gaditana* Cell Wall. *Eukaryot. Cell* **2014**, *13*, 1450–1464. [CrossRef]
146. Domozych, D.S.; Ciancia, M.; Fangel, J.U.; Mikkelsen, M.D.; Ulvskov, P.; Willats, W.G.T. The Cell Walls of Green Algae: A Journey through Evolution and Diversity. *Front. Plant Sci.* **2012**, *3*, 82. [CrossRef]
147. Baudelet, P.H.; Ricochon, G.; Linder, M.; Muniglia, L. A new insight into cell walls of Chlorophyta. *Algal Res.* **2017**, *25*, 333–371. [CrossRef]
148. Dixon, C.; Wilken, L.R. Green microalgae biomolecule separations and recovery. *Bioresour. Bioprocess.* **2018**, *5*, 14. [CrossRef]
149. Gerken, H.G.; Donohoe, B.; Knoshaug, E.P. Enzymatic cell wall degradation of *Chlorella vulgaris* and other microalgae for biofuels production. *Planta* **2013**, *237*, 239–253. [CrossRef] [PubMed]
150. Liang, K.; Zhang, Q.; Cong, W. Enzyme-Assisted Aqueous Extraction of Lipid from Microalgae. *J. Agric. Food Chem.* **2012**, *60*, 11771–11776. [CrossRef]
151. Zuorro, A.; Malavasi, V.; Cao, G.; Lavecchia, R. Use of cell wall degrading enzymes to improve the recovery of lipids from *Chlorella sorokiniana*. *Chem. Eng. J.* **2019**, *377*, 120325. [CrossRef]
152. Guo, J.; Larabell, C.A. Soft X-ray tomography: Virtual sculptures from cell cultures. *Curr. Opin. Struct. Biol.* **2019**, *58*, 324–332. [CrossRef]
153. Coelho, D.; Lopes, P.A.; Cardoso, V.; Ponte, P.; Brás, J.; Madeira, M.S.; Alfaia, C.M.; Bandarra, N.M.; Gerken, H.G.; Fontes, C.M.G.A.; et al. Novel combination of feed enzymes to improve the degradation of *Chlorella vulgaris* recalcitrant cell wall. *Sci. Rep.* **2019**, *9*, 5382. [CrossRef]
154. Coelho, D.; Lopes, P.A.; Cardoso, V.; Ponte, P.; Brás, J.; Madeira, M.S.; Alfaia, C.M.; Bandarra, N.M.; Fontes, C.M.G.A.; Prates, J.A.M. A two-enzyme constituted mixture to improve the degradation of *Arthrospira platensis* microalga cell wall for monogastric diets. *J. Anim. Physiol. Anim. Nutr.* **2020**, *104*, 310–321. [CrossRef]
155. Zuorro, A.; Miglietta, S.; Familiari, G.; Lavecchia, R. Enhanced lipid recovery from *Nannochloropsis* microalgae by treatment with optimized cell wall degrading enzyme mixtures. *Bioresour. Technol.* **2016**, *212*, 35–41. [CrossRef]
156. Maffei, G.; Bracciale, M.P.; Broggi, A.; Zuorro, A.; Santarelli, M.L.; Lavecchia, R. Effect of an enzymatic treatment with cellulase and mannanase on the structural properties of *Nannochloropsis* microalgae. *Bioresour. Technol.* **2018**, *249*, 592–598. [CrossRef]
157. IPCC. Summary for policymakers. In *Climate Change 2021: The Physical Science Basis*; Contribution of Working Group I to the Sixth Assessment Report of the Intergovernmental Panel on Climate Change; Masson-Delmotte, V., Zhai, P., Pirani, A., Connors, S.L., Péan, C., Berger, S., Eds.; Cambridge University Press: Cambridge, UK, 2021.
158. Wasson, D.E.; Yarish, C.; Hristov, A.N. Enteric methane mitigation through *Asparagopsis taxiformis* supplementation and potential algal alternatives. *Front. Anim. Sci.* **2022**, *3*, 999338. [CrossRef]
159. Roque, B.M.; Venegas, M.; Kinley, R.D.; De Nys, R.; Duarte, T.L.; Yang, X. Red seaweed (*Asparagopsis taxiformis*) supplementation reduces enteric methane by over 80 percent in beef steers. *PLoS ONE* **2021**, *16*, e0247820. [CrossRef] [PubMed]
160. Muizelaar, W.; Groot, M.; van Duinkerken, G.; Peters, R.; Dijkstra, J. Safety and Transfer Study: Transfer of Bromoform Present in *Asparagopsis taxiformis* to Milk and Urine of Lactating Dairy Cows. *Foods* **2021**, *10*, 584. [CrossRef] [PubMed]
161. Roskam, E.; Kirwan, S.F.; Kenny, D.A.; O'Donnell, C.; O'Flaherty, V.; Hayes, M.; Waters, S.M. Effect of brown and green seaweeds on diet digestibility, ruminal fermentation patterns and enteric methane emissions using the rumen simulation technique. *Front. Anim. Sci.* **2022**, *3*, 1021631. [CrossRef]
162. Mavrommatis, A.; Skliros, D.; Simoni, M.; Righi, F.; Flemetakis, E.; Tsiplakou, E. Alterations in the rumen particle- associated microbiota of goats in response to dietary supplementation level of *Schizochytrium* sp. *Sustainability* **2021**, *13*, 607. [CrossRef]
163. Mavrommatis, A.; Sotirakoglou, K.; Skliros, D.; Flemetakis, E.; Tsiplakou, E. Dose and time response of dietary supplementation with *Schizochytrium* sp. in the rumen microbiota of dairy goats. *Livest. Sci.* **2021**, *247*, 104489. [CrossRef]
164. Mavrommatis, A.; Skliros, D.; Flemetakis, E.; Tsiplakou, E. Changes in the Rumen Bacteriome Structure and Enzymatic Activities of Goats in Response to Dietary Supplementation with *Schizochytrium* spp. *Microorganisms* **2021**, *9*, 1528. [CrossRef]
165. Maia, M.R.; Chaudhary, L.C.; Bestwick, C.S.; Richardson, A.J.; McKain, N.; Larson, T.R.; Wallace, R.J. Toxicity of unsaturated fatty acids to the biohydrogenating ruminal bacterium. *Butyrivibrio fibrisolvens*. *BMC Microbiol.* **2010**, *10*, 52. [CrossRef]
166. Yuyama, K.T.; Rohde, M.; Molinari, G.; Stadler, M.; Abraham, W.R. Unsaturated Fatty Acids Control Biofilm Formation of *Staphylococcus aureus* and Other Gram-Positive Bacteria. *Antibiotics* **2020**, *9*, 788. [CrossRef]
167. Vargas, J.E.; Andrés, S.; Snelling, T.J.; López-Ferreras, L.; Yáñez-Ruíz, D.R.; García-Estrada, C.; López, S. Effect of Sunflower and Marine Oils on Ruminal Microbiota, *In vitro* Fermentation and Digesta Fatty Acid Profile. *Front. Microbiol.* **2017**, *8*, 1124. [CrossRef]
168. Lourenço, S.C. Antioxidants of Natural Plant Origins: From Sources to Food Industry Applications. *Molecules* **2019**, *24*, 4132. [CrossRef] [PubMed]
169. Balina, K.; Romagnoli, F.; Blumberga, D. Seaweed biorefinery concept for sustainable use of marine resources. *Energy Procedia* **2017**, *128*, 504–511. [CrossRef]
170. Mlambo, V.; Mnisi, C.M.; Matshogo, T.B.; Mhlongo, G. Prospects of dietary seaweeds and their bioactive compounds in sustainable poultry production systems: A symphony of good things? *Front. Anim. Sci.* **2022**, *3*, 998042. [CrossRef]

171. Orzuna-Orzuna, J.F.; Chay-Canul, A.J.; Lara-Bueno, A. Performance, milk fatty acid profile and oxidative status of lactating small ruminants supplemented with microalgae: A meta-analysis. *Small Rumin. Res.* **2023**, *226*, 107031. [CrossRef]
172. Orzuna-Orzuna, J.F.; Hernández-García, P.A.; Chay-Canul, A.J.; Galván, C.D.; Ortiz, P.B.R. Microalgae as a dietary additive for lambs: A meta-analysis on growth performance, meat quality, and meat fatty acid profile. *Small Rumin. Res.* **2023**, *226*, 107072. [CrossRef]
173. Gebereyowhans, S.; Quan, Z.H.; Zhang, S.; Pang, X.; Workelul, K.; Lu, J.; Lv, J. Inclusion of microalgae in the caprine diet improves nutritional profile of milk and its Camembert cheese. *Int. J. Dairy Technol.* **2023**, *76*, 801–812. [CrossRef]
174. Vignaud, J.; Loiseau, C.; Hérault, J.; Mayer, C.; Côme, M.; Martin, I.; Ulmann, L. Microalgae Produce Antioxidant Molecules with Potential Preventive Effects on Mitochondrial Functions and Skeletal Muscular Oxidative Stress. *Antioxidants* **2023**, *12*, 1050. [CrossRef]

antioxidants

MDPI

Article

Partial Replacement of Synthetic Vitamin E by Polyphenols in Post-Weaning Piglets

Grazia Pastorelli [1,*], Rachida Benamri [2], Massimo Faustini [1], Roberta De Bellis [3], Valentina Serra [1], Lauretta Turin [1,*], Marc Haumont [4], Philippe Durand [4], Laura Bianchessi [1], Emmanuelle Prost-Camus [5], Thomas Pecqueur [2] and Michel Prost [4]

[1] Department of Veterinary Medicine and Animal Sciences, University of Milano, Via dell'Università 6, 26900 Lodi, Italy; massimo.faustini@unimi.it (M.F.); valentina.serra@unimi.it (V.S.); laura.bianchessi@unimi.it (L.B.)

[2] Cargill Animal Nutrition, Cargill Incorporated, Wayzata, MN 55391, USA; rachida_benamri@cargill.com (R.B.); thomas_pecqueur@cargill.com (T.P.)

[3] Department of Biomolecular Sciences, University of Urbino "Carlo Bo", 61029 Urbino, Italy; roberta.debellis@uniurb.it

[4] Laboratoire Lara-Spiral, 3 rue des Mardors, 21560 Couternon, France; laraspiral@laraspiral.com (M.H.); p.durand@laraspiral.com (P.D.); michelprost.spiral@wanadoo.fr (M.P.)

[5] Centre Europeen de Recherche et Analyses, 3 rue des Mardors, 21560 Couternon, France; centreeuropeen@orange.fr

* Correspondence: grazia.pastorelli@unimi.it (G.P.); lauretta.turin@unimi.it (L.T.)

Abstract: Vitamin E is an essential nutrient usually recommended in post-weaning piglets, when a decline in the serum vitamin E concentration is observed. Selected polyphenols have the potential to partially replace vitamin E in animal feed. The aim of this study was to investigate the effect of the dietary inclusion of some commercial polyphenol products (PPs) on the growth performance, antioxidant status and immunity of post-weaning piglets. A total of 300 piglets (BW 7.18 kg ± 1.18) were randomly assigned to six dietary groups: CON$^-$ (40 mg/kg vitamin E); CON$^+$(175.8 mg/kg vitamin E); and PP1, PP2, PP3 and PP4, in which 50% vitamin E of CON$^+$ was replaced with PP with equivalent vitamin E activity. The PP1 group exhibited lower performance ($p < 0.05$) than the other dietary groups, but a similar performance to that commonly registered in pig farms. Dietary polyphenols did not influence the IgG concentration or the IL-6, IL-10, IFN-γ and TNF-α cytokine concentrations. A lower IL-8 level was found in the PP4 group than in the other groups. The diets that affected the vitamin A content showed the highest value ($p < 0.05$) in the PP1 group, and a trend was noted for vitamin E with a higher content in PP4 and CON$^+$. The polyphenols-enriched diets, especially the PP3 diet, maintained an antioxidant capacity (whole blood KRL) similar to the CON$^+$ diet. In conclusion, the replacement of vitamin E with all PPs enables partial vitamin E substitution in post-weaning piglets.

Keywords: alpha-tocopherol; weaned piglets; dietary polyphenols; antioxidant status; immunity; cytokines

Citation: Pastorelli, G.; Benamri, R.; Faustini, M.; De Bellis, R.; Serra, V.; Turin, L.; Haumont, M.; Durand, P.; Bianchessi, L.; Prost-Camus, E.; et al. Partial Replacement of Synthetic Vitamin E by Polyphenols in Post-Weaning Piglets. *Antioxidants* **2023**, *12*, 1752. https://doi.org/10.3390/antiox12091752

Academic Editor: Volker Böhm

Received: 28 July 2023
Revised: 31 August 2023
Accepted: 8 September 2023
Published: 12 September 2023

1. Introduction

Vitamin E (tocopherol) is an essential lipid-soluble micronutrient known for its antioxidant, anti-inflammatory and immunomodulatory properties [1,2]. In feeding experiments with weaned pigs, several authors have observed a drop in vitamin E concentrations in the plasma in the first weeks after weaning [3,4]. Various factors influence the pig's vitamin E status both before and after weaning. Neonatal pigs are born with low tissue reserves of α-tocopherol [5]. Although colostrum contributes α-tocopherol to the neonate, its concentration is dependent on the dietary vitamin E level fed to the sow. In neonatal pigs administrated iron (Fe), serum α-tocopherol declines, as if vitamin E was used to curtail the pro-oxidant activity of injected Fe. Moreover, weaning causes a dramatic decrease in the activity of carboxyl ester hydrolase, the enzyme that cleaves the dl-α-tocopheryl acetate form of the vitamin used in standard commercial feeds. Adequate plasma vitamin E levels in weanling pigs may be attained with sufficient dl-α-tocopheryl acetate in the feed [6,7].

Furthermore, in order to prevent this deficiency in nursery pigs, the supplemental vitamin E has frequently been increased to 200 IU/kg, a level that is substantially higher than the 16 IU vitamin E/kg diet suggested by the National Research Council (NRC, ref. [8]).

Actually, the NRC (2012) recommends 16 and 11 UI/kg for piglets in the range of 7–11 and 11–25 kg, respectively. Supplementation with high levels of vitamin E is usually recommended for diets used during the post-weaning period, when piglets show reduced growth rates and are more susceptible to disease [9].

Improvements in the cell-mediated and humoral immune responses have been reported in various animal species by adjusting the dietary levels of vitamin E [2]. In particular, vitamin E supplementation resulted in increased lymphocyte proliferation, immunoglobulin levels, antibody responses, natural killer (NK) cell activity and interleukin (IL)-2 production. [2].

In recent years, there has been a growing level of awareness among food manufacturers regarding the origin of feed additives in livestock feed; consequently, the request for products of natural origin continues to increase. Polyphenols are substances of plant origin that have antioxidant properties [10], and may reduce the negative effects of oxidative stress [11,12]. Their antioxidant potential is comparable with that of the major biological antioxidants: tocopherol and ascorbic acid [13]. Polyphenols also have immunomodulatory, anti-inflammatory and bactericidal properties [14], making their use crucial in the post-weaning period. For example, polyphenolic compounds like curcumin and resveratrol have exhibited many beneficial effects in 21-day-old weaned piglets, by alleviating intestinal inflammation and improving intestinal immune function [15]. The dietary supplementation of grape pomace improved the intestinal microbiota and down-regulated the expression of pro-inflammatory cytokines in post-weaning piglets fed for 4 weeks [16]. Other studies reported beneficial effects of dietary polyphenols on the growth performance of weaned piglets [17,18].

This research postulated that polyphenols can partially replace vitamin E in terms of antioxidant activity based on the assumption of having a 50% equivalency to vitamin E (DL-α-tocopherol acetate).

The aim of the present study was to investigate the effects of partially replacing vitamin E with different commercial polyphenol products on the growth performance, antioxidant defenses and immune response of piglets during a 35-day post-weaning period.

2. Materials and Methods

2.1. Ethics Statement

Our in vivo trial complied with Italian regulations on animal experimentation and ethics (Legislative Decree 26/2014) [19], in accordance with European regulations (Directive 2010/63) [20], and was approved by the Animal Welfare Body of the University of Milan (number 140/2021). The study was performed between April and May 2022 at the Productive Pig Unit "Cascina Agrieffe" farm located at Gottolengo (BS) Lombardy (Italy).

2.2. Animals and Treatments

A total of 300 crossbred piglets (Large White x Landrace), (28 days old ± 2 days), with a mean body weight (BW) of 7.18 kg ± 1.18, were selected from 30 litters of contemporary sows. The piglets were individually ear-tagged and divided into six experimental dietary groups (5 pens per diet, 10 piglets per pen), balanced for sex and body weight. Each pen (2.0 × 1.5 m) was equipped with a self-feeder and nipple drinkers to allow ad libitum access to feed and water during the 35-day experimental period. The rooms had a forced-air ventilation system set at 60% relative humidity, and a temperature of 27 ± 2 °C. Each pen was provided with metal chains, with soft wooden bars hanging from the walls and straw-filled baskets as enrichment material.

The animals were assigned to six dietary treatments: a negative control diet (CON^-) corresponding to a diet with a low vitamin E content (40 mg/kg of vitamin E); a positive control diet (CON^+) corresponding to the standard diet used by farmers in which the vitamin E content (175.8 mg/kg) exceeds the nutritional requirements as recommended by NRC (2012 [8]); polyphenol product (PP) diets in which four different commercial polyphenol products (PP1, PP2, PP3, PP4) with equivalent vitamin E activity in terms of antioxidant capacity replaced 50% vitamin E of CON^+ (87.9 mg/kg vitamin E). The tested commercial polyphenol products contained the following:

PP1: Mix of citrus, grape and chestnut extracts and carrier;
PP2: Dried grape extract and carrier;
PP3: Freeze-dried melon juice and flesh palm oil and microcrystalline cellulose;
PP4: Grape and onion soluble, flavoring compounds and carrier.

The experimental diets were formulated to be isoenergetic and isoproteic on net energy, and were produced with the same batches of feeds by Tracciaverde S.R.L. (Bonemerse, Italy). None of the experimental diets contained any antimicrobial or growth promoter, and all were designed to meet or exceed the nutrient requirements of weaned piglets recommended by the NRC (2012 [8]; Table 1).

2.3. *Growth Performance*

The piglets were individually weighed on day 0 (d0) and on day 35 (d35), and the pen feed consumption (experimental unit for the feed intake evaluation) was recorded. The feed conversion ratio (FCR) was calculated by dividing the amount of feed consumed during the experimental period by the growth of the animals during that same time. The average daily gain was calculated from the measurements of weight and the number of experimental days. The feed intake of the pen was calculated by the difference between the offered feed and leftovers. The leftovers, if any, were weighed daily and considered for the final calculation of the feed consumed. The mortality was recorded daily throughout the trial.

2.4. *Collection of Blood Samples*

At the beginning (d0) and at the end of the dietary trial (d35), blood samples were collected from 2 randomly selected male piglets per pen via jugular vein puncture before the morning feeding (total number of specimens = 50).

Vacuum tubes (9 mL) containing K3EDTA (Vacuette® Tube, Cat. no. 455036; Greiner Bio-One, Kremsmünster, Austria) for plasma collection, and 9 mL vacuum tubes for serum (Vacuette® Tube, Cat. no. 455092; Greiner Bio-One, Kremsmünster, Austria) were used.

After collection, the blood samples were immediately transported to the laboratory. Plasma and serum were obtained from the blood samples by centrifugation ($3500\times g$ for 15 min at 4 °C), and were stored at –18 °C until subsequent analyses.

2.5. *Blood Analyses*

2.5.1. Vitamin A and Vitamin E Analysis

The retinol and tocopherol concentrations were measured via chromatography (HPLC) (Shimadzu, Japan) according to the protocol reported in Rettenmaier and Schüep, 1994 [21] and through an iCheck fluorometer–spectrophotometer (iCheck Vitamin E; BioAnalyt GmbH, Teltow, Germany) as described in Simoni et al., 2022 [22], respectively.

Table 1. Ingredients (%) and composition of basal diets (as-fed basis).

Ingredients as % of Feed Basis	CON⁻	CON⁺	PP
Barley meal	25.90	25.87	25.56
Wheat meal	22.45	22.45	22.45
Soy protein concentrate fermented	7.0	7.0	7.0
Flaked corn	7.0	7.0	7.0
Corn meal	5.0	5.0	5.0
Flaked wheat	8.89	8.89	8.89
Whey powder	4.0	4.0	4.0
Dextrose monohydrate	3.0	3.0	3.0
Hulled flaked barley	3.0	3.0	3.0
Soy protein concentrate	5.0	5.0	5.0
Soybean oil	1.5	1.5	1.5
Dried sugar beet pulp	1.5	1.5	1.5
Acidity regulators	1.0	1.0	1.0
Coconut oil	1.0	1.0	1.0
Plasma	1.0	1.0	1.0
L-Lysine	0.6	0.6	0.6
Fish meal	0.5	0.5	0.5
PP1, PP2, PP3, PP4	-	-	0.330
DL-methionine	0.23	0.23	0.23
Dicalcium phosphate	0.20	0.20	0.20
Calcium carbonate	0.2	0.2	0.2
L-treonine	0.2	0.2	0.2
Vitamin mineral premix [1]	0.20	0.20	0.20
Salt	0.15	0.15	0.15
L-valine	0.11	0.11	0.11
L-tryptophane	0.04	0.04	0.04
Aromas	0.08	0.08	0.08
Vit E supplement [2]	0.00	0.028	0.01
Chemical composition [3]			
Crude protein, %	17.20		
Ether extract, %	4.30		
Crude fiber, %	3.50		
Ash, %	5.00		
Sodium, %	0.20		
Calcium, %	0.50		
Phosphorus, %	0.60		
Lysine, %	1.30		
Methionine, %	0.50		
Net Energy (NE), kcal/kg	2340		

[1] Premix contained the following per kg nutrients of the diet: 6500.00 IU vitamin A, 1200.00 IU vitamin D3 (cholecalciferol), 47.50 mg betaine anhydrous, 40.0 mg betaine hydrochloride, 0.2 mg biotin, 0.50 mg folic acid, 20.0 mg niacinamide, 10.0 mg calcium pantothenate, 10.0 mg vitamin B1, 0.03 mg vitamin B12 (cobalamin), 5.0 mg vitamin B2 (riboflavin), 2.0 mg vitamin B6 (pyridoxine hydrochloride), 40 mg all-rac-atocopheryl-acetate; 1.5 mg vitamin K3, 15.0 mg Cu, 0.50 mg I, 40.0 mg Mn, 0.25 mg Se, 70.0 mg Zn oxide, 4680.00 L-lysine, 1970.00 mg L-threonine, 384.0 mg L-tryptophan. [2] Vitamin E provided per kilogram of premix: 485 mg. [3] Nutrient (expressed as fed basis) and net energy content were calculated using Plurimix software (Fabermatica, CR, Italy).

2.5.2. IgG and Cytokines Analysis

The serum IgG concentration was determined using a commercial enzyme-linked immunosorbent assay (ELISA; Porcine IgG (Immunoglobulin G) ELISA Kit; FineTest, Wuhan Fine Biotech Co., Ltd., Wuhan, China; cat. no. EP0084) based on sandwich binding. The colorimetric reaction was catalyzed with streptavidin-conjugated horseradish peroxidase (HRP), which produced a yellow product that was proportional to the target amount present in the sample. The serum samples were diluted at 1:200,000 before being analyzed in duplicate. The range of detection for this ELISA assay was 1.563–100 ng/mL.

The serum cytokines (IL-6, IL-8, IL-10, INF-γ) concentrations were determined with Luminex technology (Labospace S.r.l., Milan, Italy); the Luminex test allowed for quick and accurate measurements of the given targets using hundreds of internally colored plastic microbeads with a graduated mixture of red or infrared fluorescent dyes that emitted light at different wavelengths when struck by a laser [23].

The TNF-α levels were measured using a commercial sandwich ELISA assay (Cloud-Clone Corp., Katy, TX, USA; cat. no. SEA133Po) with a detection range of 15.6–1000 pg/mL. The wells of the microplate were pre-coated with antibodies specific to TNF-α and, after the addition of samples, avidin-conjugated to HRP, and the TMB chromogen substrate exhibited a color change that was measured spectrophotometrically at a wavelength of 450 nm.

2.5.3. Antioxidant Defenses

The activity of glutathione peroxidase (GPx) was determined using a commercial assay kit (Cayman Chemical Co., Ann Arbor, MI, USA, cat. no. 703102) according to the instructions provided by the manufacturer. Oxidized glutathione (GSSG), produced during the reduction of hydroperoxide by GPx, is recycled to its reduced state (GSH) by glutathione reductase (GR) and NADPH. The oxidation of NADPH to NADP+ is accompanied by a decrease in absorbance at 340 nm. Under conditions where GPx activity is limiting, the rate of decrease in absorbance is directly proportional to the GPx activity in the sample. One International Unit (U) is defined as the amount of GPx that will cause the oxidation of 1.0 nmol of NADPH to NADP+ per min at 25 °C.

Superoxide dismutases (SODs) catalyze the dismutation of the superoxide anion into hydrogen peroxide and molecular oxygen. There are three forms of human SOD: cytosolic Cu/Zn-SOD, mitochondrial Mn-SOD and extracellular SOD. The latter is present in plasma. The SOD activity was determined using a commercial assay kit (Cayman Chemical Co., cat. no. 706002) following the manufacturer's instructions. A water-soluble tetrazolium salt is used to detect the superoxide radical generated by the oxidation of xanthine into uric acid by xanthine oxidase. Superoxide ions that have not been removed by the superoxide dismutase present in the sample react with the tetrazolium salt to form a formazan dye. One International Unit (U) is defined as the amount of SOD needed to exhibit 50% dismutation of the superoxide radical. The 50% inhibitory activity by SOD can be determined using a colorimetric method, measuring the absorbance at 450 nm.

The overall antioxidant defense potential was determined with the biological KRL™ Test (M. Prost Patent) [24,25]. Whole blood and erythrocytes are submitted to oxidant stress, and the free radical-induced hemolysis is recorded via optical density decay with the KRL microplate reader. Inside the body, both the extracellular and intracellular antioxidant defense contribute to maintaining cellular integrity until hemolysis. The resistance of whole blood (KRLWB) and red blood cells (KRLRBC) to free radical attack is expressed as the time that is required to reach 50% of maximal hemolysis (half-hemolysis time, T1/2 in minutes).

2.6. *Antioxidant Activity of Diets*

2.6.1. Sample Treatment

Finely ground samples of all diets were extracted by adding 2 g of each sample to 10 mL of 70% (*v/v*) ethanol/double-distilled water. All samples were shaken in the dark for 1 h and then centrifuged at 13,000 rpm for 15 min at 4 °C. The recovered supernatants were stored at −20 °C until the analyses were carried out.

2.6.2. Antioxidant Activity (ORAC Assay)

The antioxidant capacity of the experimental diets was measured in terms of the Trolox equivalent antioxidant capacity, using the oxygen radical absorbance capacity (ORAC) assay. The assay was performed as in De Bellis et al., 2019 [26], detecting the fluorescence until total extinction (485 nm ex. and 520 nm em.) on a Fluostar Optima plate reader (BMG Labtech, Offenburg, Germany).

The results were compared to the Trolox antioxidant capacity, and, for this reason, the data are reported as the micromoles of Trolox equivalents (µmol TE/mg) of the samples.

All reaction mixtures were prepared in duplicate, and at least three independent assays were performed for each sample.

2.7. Statistical Analysis

Statistical analyses were performed using SPSS 26.0 (SPSS Inc., Chicago, IL, USA). Before hypothesis testing, all data were examined for normality and transformed where appropriate. The growth performance data were analyzed using the ANOVA procedure, with treatment and sex as the main effects, and the pen was considered as the experimental unit. Parameters for the blood samples (where multiple data were available for each animal) were subjected to repeated analyses. Vitamin E and TNF-α were natural-log-transformed to meet the normality assumption, and therefore subjected to analysis. The treatment effects were deemed significant at $p < 0.05$, and a trend was noted when the p-values were between 0.05 and 0.1.

3. Results

3.1. Antioxidant Activity of Diets and Growth Performance

The antioxidant activity levels of the six experimental diets measured using the ORAC assay are reported in Figure 1. The ORAC values of the experimental diets showed a significant difference between groups ($p < 0.001$), with concentrations equal to 67.8, 68.4, 70.5, 73.3, 78.6 and 80.0 µmol TE/mg in the CON$^-$, CON$^+$, PP1, PP2, PP3 and PP4 groups, respectively, with PP3 and PP4 showing the highest values.

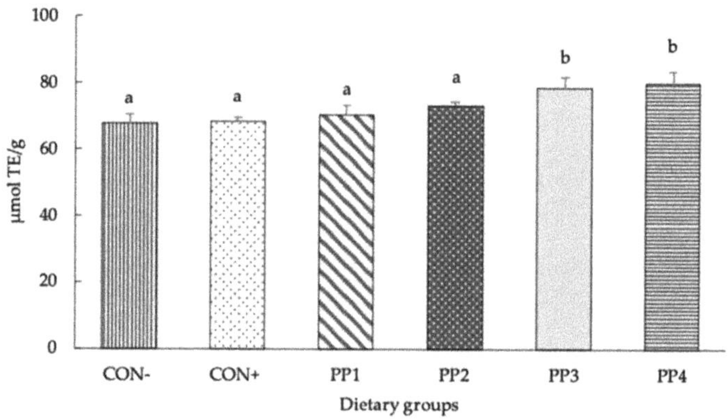

Figure 1. Antioxidant activity levels of experimental diets measured with ORAC assay. The data are reported as means ± SD. [a,b] Bars with different letters are significantly different ($p < 0.05$).

Concerning the growth performance, the average BW of the piglets at the beginning of the study was 7.18 kg, without differences due to dietary treatments and sex (7.12 kg female vs. 7.25 kg males) in accordance with the experimental design ($p > 0.05$) (Table 2).

However, at the end of the trial, after 35 days of differentiated feeding, the PP1 group exhibited a lower ($p < 0.05$) final body weight with a lower ($p < 0.05$) average daily gain (ADG) and worse feed conversion ratio (FCR) than the other six dietary treatments. No differences among the CON$^+$, CON$^-$, PP2, PP3 and PP4 groups for final body weight, ADG and FCR were found. At the end of the trials, there was no difference found between males and females (18.97 vs. 18.47). Moreover, all these results were not affected by health problems. In this regard, no mortality was registered throughout the entire period.

Table 2. Growth performance of piglets fed with different dietary polyphenol sources.

Groups	CON^-	CON^+	PP1	PP2	PP3	PP4	SEM	p-Value
BW (d0), kg	7.03	7.14	7.24	7.12	7.23	7.34	0.07	ns
BW (d35), kg	18.94 ^{a}	19.55 ^{a}	15.04 ^{b}	19.08 ^{a}	19.98 ^{a}	19.72 ^{a}	0.23	<0.05
ADG, g/d	340 ^{a}	354 ^{a}	223 ^{b}	342 ^{a}	364 ^{a}	354 ^{a}	0.005	<0.05
FCR, kg/kg	1.78 ^{a}	1.77 ^{a}	2.12 ^{b}	1.79 ^{a}	1.76 ^{a}	1.73 ^{a}	-	<0.05
Mortality, %	0	0	0	0	0	0	-	-

BW: body weight; ADG; average daily gain; FCR; feed conversion ratio; SEM: standard error of the mean ^{a,b}. Means within a row with different letters are significantly different at $p < 0.05$.

3.2. Blood Parameters

Table 3 reports the effects of dietary treatment at two different sampling times on the blood parameters of the piglets throughout the experiment.

Table 3. Effect of dietary treatment (D) and sampling time (T) on blood metabolites.

Groups	CON^-	CON^+	PP1	PP2	PP3	PP4	SEM	Diet	Treatment
Vit. A (ug/L)									
d 0	304.8	387.3	345.4	347.0	304.2	382.7	10.2		
d 35	310.3	353.6	395.4	329.6	308.0	337.3	9.35	0.045	ns
Vit. E * (mg/L)									
d 0	0.71	0.79	0.93	0.97	0.88	0.67	0.053		
d35	1.12	2.12	1.00	0.96	1.46	1.51	0.080	0.072	<0.001
IgG (mg/mL)									
d 0	1.57	2.15	2.45	1.98	2.42	1.88	0.09		
d 35	2.97	2.63	2.83	2.38	3.28	2.26	0.121	ns	0.001
IL-6 (ng/mL)									
d 0	0.13	0.01	0.45	0.06	0.05	0.05	0.017		
d35	0.04	0.04	0.04	0.04	0.11	0.05	0.013	ns	ns
IL-8 (ng/mL)									
d 0	0.302	0.201	0.181	0.282	0.354	0.312	0.022		
d35	0.708	0.724	0.787	0.390	0.404	0.358	0.038	0.022	0.001
IL-10 (ng/mL)									
d 0	0.495	0.108	0.158	0.281	0.204	0.279	0.06		
d 35	0.288	0.376	0.208	0.234	0.304	0.540	0.096	ns	ns
IFN-γ (ng mL)									
d 0	1.31	1.03	0.94	1.19	0.87	0.85	0.066		
d 35	0.55	0.54	1.34	0.54	1.20	0.65	0.120	ns	<0.001
TNF-α * (ng mL)									
d 0	0.019	0.015	0.012	0.005	0.010	0.031	0.000		
d35 **	0.003	0.000	0.001	0.003	0.004	0.009	0.000	ns	<0.01

SEM: standard error of the mean; * p-value calculated as ln-transformed; ** Although some pigs had values above the technique cut-off, the mean value was below the cut-off.

At d0, just before the pigs were assigned to receive experimental diets, there was no difference among the treatments for any studied variable ($p > 0.10$). At the end of the trial (d35), the plasma concentrations of vitamin A were greater in the PP1 and CON^+ groups compared to others ($p < 0.05$), with the highest positive variations observed in the PP1 group. Concerning the diet effect, the serum parameters evaluated were not significantly different among the treatments ($p > 0.05$), except for vitamin A and IL-8 ($p < 0.001$). For the vitamin E values, it was observed that piglets fed a CON^+ diet reached, on average (average of value at d0 and d35), higher values in comparison with other dietary treatments (1.45 vs. 0.92, 0.97, 0.97, 1.17 and 1.09 for CON^-, PP1, PP2, PP3 and PP4, respectively). The sampling time significantly affected the vitamin E content, with the lowest levels at day 0 for all groups. The vitamin E variation (%) result was significant ($p < 0.05$), with the highest positive value in the PP4 group vs. other the dietary treatments. The IL-8 results showed a time effect ($p < 0.001$) and a treatment effect ($p = 0.022$) with increased IL-8 at day 35. The comparison post hoc tests showed the following significances: CON^- vs. PP2, CON^- vs. PP4, CON^+ vs. PP2 and CON^+ vs. PP4.

3.3. Antioxidant Defenses

Table 4 shows the effects of the dietary treatment and sampling time on the antioxidant defenses of the piglets throughout the experiment.

Table 4. Effect of dietary treatment (D) and sampling time (T) on antioxidant defenses.

	CON^-	CON^+	PP1	PP2	PP3	PP4	D	T	DxT #
GPx	(U/mL)								
d 0	1.13 ± 0.19	0.99 ± 0.17	0.96 ± 0.20	1.13 ± 0.30	1.31 ± 0.30	1.24 ± 0.23	0.001	<0.001	<0.001
d 35	1.51 ± 0.26	1.34 ± 0.22	2.33 ± 0.39	1.42 ± 0.20	1.38 ± 0.32	1.44 ± 0.44			
SOD	(U/mL)								
d 0	24.2 ± 8.2	24.2 ± 10.5	20.8 ± 9.7	14.4 ± 2.0	28.1 ± 14.6	17.3 ± 7.7	0.112	<0.001	0.003
d35	10.5 ± 4.0	12.1 ± 6.8	16.1 ± 8.7	16.1 ± 6.0	19.1 ± 7.9	19.8 ± 6.5			
KRL_{WB} *	(min)								
d 0	125.4 ± 15.7	126.7 ± 8.5	125.9 ± 13.2	122.1 ± 16.9	134.4 ± 17.6	117.4 ± 15.0	0.033	<0.001	ns
d 35	105.8 ± 13.3	125.3 ± 17.7	106.2 ± 10.6	112.6 ± 17.2	120.9 ± 16.6	111.5 ± 9.9			
KRL_{RBC} **	(min)								
d 0	107.6 ± 17.1	107.5 ± 13.1	107.3 ± 17.4	96.0 ± 12.1	104.3 ± 17.7	93.6 ± 16.9	ns	ns	ns
d35	94.6 ± 18.1	102.1 ± 19.1	102.9 ± 10.6	101.3 ± 17.5	104.3 ± 17.3	97.3 ± 11.7			

* KRL_{WB}: KRL half-hemolysis time on whole blood; ** KRL_{RBC}: KRL half-hemolysis time on red blood cells. Results are expressed by means ± standard deviations of N = 10 piglets in each group. # p-values of repeated measures ANOVA: interactions between dietary treatment (D) and sampling time (T) main effects.

Concerning plasma antioxidant enzyme activities, the dietary sources significantly affected the plasma GPx activity, which was higher after 35 days ($p < 0.001$). A Tukey post hoc test revealed significant pairwise differences between group PP1 and other groups for GPx ($p < 0.043$). We found that the GPx activity was especially increased in this diet group, rising from 0.96 ± 0.20 to 2.33 ± 0.39 U/mL between d0 and d35.

The overall plasma SOD activity decreased significantly between d0 and d35. Although the lower initial SOD values in groups PP2 and PP4 make results interpretation difficult, we found that the SOD activity after 35 days was significantly higher with diets supplemented with polyphenols PP2, PP3 and PP4 than in the CON^- and CON^+ diets (Mann–Whitney $p < 0.05$).

Our results also show that the dietary source and time significantly affected the antioxidant defense potential measured in whole blood ($p < 0.05$). They show that whole blood KRL decreased significantly between d0 and d35 in the CON^- and PP1 groups, but was more stable (paired t test $p > 0.05$) in the other groups. At d35, whole blood KRL was higher in the CON^+ and PP3 diets than in the CON^- diet (Mann–Whitney $p < 0.05$). Moreover, a positive correlation was observed between the whole blood antioxidant defense potential and plasma SOD activity in piglets after 35 days (Pearson correlation, $r = 0.3248$ $p = 0.011$). No significant variations were found in the potential of antioxidant defenses measured at the level of red blood cells.

4. Discussion

Post-weaning is a critical stage for the health of piglets, as it affects the feed intake, immunity and redox status of the animals. The different diets tested were formulated to verify the effect of partially substituting the vitamin E content with different polyphenol products on the growth performance and some blood indicators mainly related to immunity and antioxidant status.

In the current study, the feed antioxidant capacity was measured using the ORAC assay, accepted as the current food industry standard for evaluating the antioxidant capacity of food additives, whole foods, juices and raw vitamins [27]. The higher ORAC values found both in the diets supplemented with PP3 and PP4 indicate a greater antioxidant activity compared to the other PPs, and even more than control groups; this is attributable to the presence of polyphenols such as flavanones (grapes), flavonols (onions), phenolic acids and flavonoids, which were mainly concentrated in grape seeds and skins [28] in the second group mentioned (PP4).

The zootechnical performance of piglets can be considered an indirect indicator of animal health. The mechanism of action of polyphenols as growth promoters lies in the fact that they can increase the secretion of endogenous enzymes, bile, mucus and salivary glands, retard the growth of pathogenic microorganisms in the gastrointestinal tract, and modulate gut morphology and architecture through their immunostimulatory, anti-inflammatory and antioxidant functions [29,30]. Furthermore, studies have shown that the utilization of polyphenolic compounds from aromatic plants in animals improved feed intake and growth performance, due to their enhancing effect on the flavor and palatability of feeds [31].

In our study, no differences were observed among the treatments (CON$^-$, CON$^+$, PP2, PP3 and PP4) regarding feed intake, ADG and FCR. The growth performance results of the PP1 group are probably attributable to the composition of the supplement that likely yielded unpleasant tastes and odors such as to modify the intake, presumably due to the presence of chestnuts that contain tannins, and citrus fruits. In fact, the bitter taste typical of citrus fruits seems to reduce palatability [32]; another study [33] also confirmed that bitterness is the main reason for the rejection of various food products.

Zhang et al. 2014 [34] reported no effect of diets in piglets fed for 21 days with a mixture of standardized plant extracts containing apple (16.5%), grape seed (27.5%), green tea (30%) and olive leaves (2.5%).

Furthermore, Gessner et al. [35] did not show significant differences in growth performance in a study on post-weaning piglets supplemented with 1% polyphenols derived from grape seeds and pomace meal [35]. Conversely, Rajković et al. [36], showed a final live weight (25.4 kg) that was higher ($p < 0.08$) than the control group (23.8 kg) in newly weaned piglets (6.9 kg) fed for 56 days with a diet based on corn supplemented with grape extracts (150 g/t) [36].

In showing similar results, the three vitamin E replacement strategies represented by PP2, PP3 and PP4 suggest that polyphenols can likely replace vitamin E; surprisingly, the CON$^-$ group reached a similar growth performance.

In general, the growth performance of piglets in our study reported values that are commonly registered in commercial pig farms, confirming good management, especially considering that the feed was not medicated. No mortality underlined good health conditions of the piglets; therefore, no metabolic shifts were needed to redistribute nutrients away from the growth processes toward immune system function, with a subsequent decrease in feed efficiency for growth.

Vitamins E and A are both antioxidant molecules; vitamin E is mainly related to hydroperoxyl radical scavenging [37], while vitamin A acts by donating hydrogen atoms [38]. The greater vitamin A concentration in pigs fed PP1 during post-weaning may be indicative of a lower inflammatory status, and in this study, it seems to be associated with greater GPX activity. The addition of PP1 to diets led to an increase (by 30% on average, $p = 0.022$) in serum retinol levels in comparison with the remaining groups.

The highest content of vitamin A in blood from piglets in the PP1 group at d35 could be related to the accumulation action from carotene content and antioxidant compounds from the dietary supplement; the citrus part contains polyphenols, primarily flavonoids. We speculated that the increased vitamin A level could be attributed to the ability of polyphenols of PP1 to strengthen and save the endogenous antioxidant system, as reported by Corbi et al. in rabbits [39].

Even though vitamin E and polyphenols are both added to swine diets for antioxidant purposes, they have a different mechanism of action. Vitamin E can be absorbed in the intestine and enter the systemic circulation, as literature reported that supplementing vitamin E in pig diets increased serum and tissue (loin muscle, liver, and fat) vitamin E concentrations [40–43]. On the other hand, there have been few studies conducted in vivo on the digestibility and bioavailability of polyphenols in pigs. Research studies suggest that only low percentages of dietary polyphenols may be absorbed in the small intestine, and have a low bioavailability due to their molecular structures [14,44,45]. The polyphenols are expected to have direct antioxidant effects in vivo in the intestinal lumen because of the higher concentration of polyphenols in the lumen compared to the systemic concentrations [46]. Moreover, the low amounts of absorbed polyphenols are then extensively bio-transformed in the liver and rapidly excreted in the urine and bile [47]. Subsequently, the colon microbiota's enzymes transform the bile-excreted polyphenol metabolites and the unabsorbed polyphenols into various metabolites [46,48]. The present study was not focused on the microbiota; therefore, we could not evaluate this aspect.

No clinical sign of vitamin E deficiency was noted in any of the piglets. In general, the raw data are in line with the concentrations detected in piglets in the same physiological stage [42], and also agree with the time effect found by Moreira and Mahan [49], who registered an increase from day 21 to day 42. Our trial ended 35 days after weaning. A vitamin E dose of 175.8 mg/kg increased the total tocopherol content in the blood serum, whereas the CON^{-} diet was the lowest on average, as expected.

There is no general agreement in the literature regarding the limits between adequate, marginal and deficient plasma vitamin E levels in pigs [50]. As reported by Sivertsen [50], the National Veterinary Institute has considered plasma levels of vitamin E below 1.0 μg/mL as distinctly deficient, while normal is considered equal to 2.0 mg/L [51]. It should be underlined that the proportion of pigs with a plasma vitamin E level below 1.5 μg vitamin E/mL is very common [52,53].

The IgG results showed a time effect ($p < 0.05$) that is in agreement with the literature. This immunoglobulin isotype, which is the most important and abundant defense molecule against infections, has been reported to decrease in concentration after weaning at the time of depletion of maternal immunity [54,55], and subsequently increases as an effect of the beginning of piglets' own synthesis of IgG, which starts at day 7, increases up to day 28 [56] and later up to day 45 [57], according to different studies. In the present study, which ended at day 35, an increase in IgG was observed already at such a time point. Dietary polyphenols did not influence the IgG concentration, confirming the data from Pistol et al. [58].

It is well-recognized that multiple factors that occur during weaning can lead to oxidative stress in post-weaned piglets.

Oxidative stress changes the structure of biomolecules, such as proteins, lipids and nucleic acids, leading to cell or organ injury; in addition, it may change gastrointestinal functions and induce cell apoptosis [59]. It may result in growth restriction, disease and even death [60].

Inflammation is the consequence of oxidative stress, and the pathways which produce the mediators of inflammation, such as cytokines, are all induced by oxidative stress [61]. Cytokines, including pro-inflammatory cytokines and anti-inflammatory cytokines, are necessary mediators of inflammatory responses. The balances of cytokines are central for protection against or susceptibility to infections [62]. Early life development is a stressful period for pigs, as the immature intestines of young animals are very vulnerable to invading pathogens, leading to inflammation [63]. Pro-inflammatory cytokines such as TNF-α, IL-6 and IL-8 trigger inflammatory responses and have negative effects on intestinal integrity and epithelial function [64]. In contrast, anti-inflammatory cytokines such as IL-10 hinder the inflammatory response [65]. Different studies found significant reductions in pro-inflammatory cytokines, such as TNF-α, NF-kB, IL-1β, IL-1ra, IL-2, IL-4, IL-6 or IL-8, when vitamin E [66] or polyphenols were fed to the nursery pigs [58,67,68].

In the current study, no statistically significant difference in anti-inflammatory cytokine (IL-10) was found among the dietary PP groups. Although no statistically significant differences were found, with the exception of IL-8 in the PP4 group, some pro-inflammatory cytokines showed a reduction at d35 in pigs fed diets formulated with PP4 (IFN-γ TNF-α), PP1 and PP2 (IL-6). A reduction in pro-inflammatory cytokines levels in healthy pigs indicates an improvement in immune status, which suggests that these pigs may be able to spend less energy on activating the immune defenses, which can potentially lead to improved energy utilization [41] and increased performance. However, in the present study, the growth performance results seem to be independent of the cytokines results.

All cytokine data are in the range found for piglets at the same age [69,70]. A significant time effect was detected for IL-8, TNF-α and IFN-γ ($p < 0.05$), showing higher values at Time 2 in comparison to Time 1, according to a previous study [71]. Conversely, the levels of IL-6 and IL-10 did not show any significant value in considering the time or the treatment effect. The pro-inflammatory cytokine IL-8 was significantly lower in the PP3 and PP4 groups than in the positive and negative controls. This may be explained by a diminished need for such chemoattractant cytokines to be functioning in the recruitment of neutrophils and other inflammatory cells.

Previous studies have shown a close relationship between post-weaning stress syndrome in piglets and oxidative stress that can last for weeks [72]. Whole blood KRL results emphasize that the overall potential of antioxidant defenses decreases after a 35-day post-weaning period, especially in piglets with a low vitamin E diet (CON^-). Although this is not significant, they also suggest that this decrease would be partly normalized by vitamin E or vitamin E/polyphenol supplementations, as in the CON^+ and PP3 groups. These restorations of antioxidant defense potential would be associated with an increase in superoxide dismutase activity that was observed in piglets supplemented with the PP2, PP3 and PP4 diets.

SOD and glutathione peroxidase (GPx) represent the main antioxidant enzymes in mammals, and protect the organism against prooxidants by reducing the accumulation of organic hydroperoxides and hydrogen peroxide. The activity of these enzymes is commonly used to monitor the antioxidant capability of the body [73].

Our analyses underline that plasma glutathione peroxidase activity increases in all piglets after 35 days of treatment, especially in animals with a PP1 diet. This increased plasma GPx activity in piglets suggests the induction of antioxidant defense mechanisms; in fact, it has been reported that antioxidant enzymes prevent the hosts from oxidative stress by increasing their activities [74]. This increase in plasma GPx activity is associated with a rapid increase in antioxidant demand, and has been described in other studies. [75].

Concerning the economic impact of this experimental trial, it should be underlined that apart from PP3, all tested commercial products have an inclusion price that is lower compared to synthetic vitamin E; therefore, the cost of the feed including the PP1, PP2 and PP4 products is lower than that of the CON^+ treatment. On the other hand, as the PP1 treatment decreased performance of piglets compared to the CON^+ treatment, the economic impact of this product is of negligible importance. Based on the obtained results, as PP4 treatment is less expensive than vitamin E and the results are positive, we can assume that the chances of obtaining a positive economic impact with this commercial product are very likely to occur.

5. Conclusions

The present study on the partial replacement of vitamin E in the diet of weaned piglets with different polyphenol products provides interesting results, since they come from a period characterized by high vitamin E requirements.

All experimental diets had no negative effects on the growth performance of piglets in the post-weaning period.

Replacing 50% of the vitamin E of the positive control diet with PP1 improved the vitamin A concentration and the antioxidant activity measured as glutathione peroxidase levels.

PP4 seems to be the most promising integration for the plasmatic increase in alpha tocopherol due to the spare action exerted by polyphenols, and contributed to a lower pro-inflammatory activity of IL-8.

Hence, under farming conditions among the polyphenol products, the PP4 treatment may be the most advisable integration as a result of its better growth performance observed compared to PP1.

The results of this study will have to be confirmed by further research, which should be extended to the subsequent growth phase for the evaluation of long-term effects.

Author Contributions: Conceptualization, G.P., R.B., T.P. and M.P.; validation, G.P., L.T. and V.S.; formal analysis, G.P., M.F., P.D. and M.H.; investigation, G.P., L.T., V.S., M.F., R.D.B., L.B., P.D. and E.P.-C.; resources, G.P., L.T., M.F. and M.P.; writing—original draft preparation, G.P., V.S., M.F. and L.T.; writing—review and editing, G.P., R.B.,T.P., L.T. V.S. P.D., R.D.B. and M.H.; supervision, G.P.; project administration, G.P.; funding acquisition, G.P. All authors have read and agreed to the published version of the manuscript.

Funding: This research was funded by Delacon company.

Institutional Review Board Statement: The study was approved by the Animal Welfare Committee of the University of Milan (OPBA_n 140_2021), according to the directive 2010/63/EU.

Informed Consent Statement: Not applicable.

Data Availability Statement: The data presented in the current study are available from the corresponding authors on reasonable request.

Acknowledgments: We thank Giancarlo Selmini for his precious assistance in formulating the diets.

Conflicts of Interest: R.B. was an employee of Delacon Biotechnik GmbH, which has sponsored the study. The sponsor had no role in the collection, analyses, or interpretation of data; in the writing of the manuscript or in the decision to publish the results. R.B. and T.P. are employees of Cargill Animal Nutrition. For ethical considerations, all polyphenolic additives were blinded during the study and until conclusions were drawn. The other authors declare no conflicts of interest.

References

1. Liao, S.; Omage, S.O.; Börmel, L.; Kluge, S.; Schubert, M.; Wallert, M.; Lorkowski, S. Vitamin E and metabolic health: Relevance of interactions with other micronutrients. *Antioxidants* **2022**, *11*, 1785. [CrossRef] [PubMed]
2. Lee, G.Y.; Han, S.N. The role of vitamin E in immunity. *Nutrients* **2018**, *10*, 1614. [CrossRef] [PubMed]
3. Håkansson, J.; Hakkarainen, J.; Lundeheim, N. Variation in vitamin E, glutathione peroxidase and retinol concentrations in blood plasma of primiparous sows and their piglets, and in vitamin E, selenium and retinol contents in sows' milk. *Acta Agric. Scand. Sect. A Anim. Sci.* **2001**, *51*, 224–234. [CrossRef]
4. Meyer, W.R.; Mahan, D.C.; Moxon, A.L. Value of dietary selenium and vitamin E for weanling swine as measured by performance and tissue selenium and glutathione peroxidase activities. *J. Anim. Sci.* **1981**, *52*, 302–311. [CrossRef]
5. Mahan, D.C.; Lepine, A.J. Effect of pig weaning weight and associated nursery feeding programs on subsequent performance to 105 kilograms body weight. *J. Anim. Sci.* **1991**, *69*, 1370–1378. [CrossRef]
6. Lauridsen, C.; Jensen, S.K. Influence of supplementation of all-rac-α-tocopheryl acetate preweaning and vitamin C postweaning on α-tocopherol and immune responses of piglets. *J. Anim. Sci.* **2005**, *83*, 1274–1286. [CrossRef]
7. Rey, A.I.; López-Bote, C.J.; Litta, G. Effects of dietary vitamin E (DL-α-tocopheryl acetate) and vitamin C combination on piglets oxidative status and immune response at weaning. *J. Anim. Feed Sci.* **2017**, *26*, 226–235. [CrossRef]
8. National Research Council. *Nutrient Requirements of Swine*, 11th ed.; The National Academies Press: Washington, DC, USA, 2012; p. 399.
9. Orengo, J.; Hernández, F.; Martínez-Miró, S.; Sánchez, C.J.; Peres Rubio, C.; Madrid, J. Effects of commercial antioxidants in feed on growth performance and oxidative stress status of weaned piglets. *Animals* **2021**, *11*, 266. [CrossRef]
10. Serra, V.; Salvatori, G.; Pastorelli, G. Dietary polyphenol supplementation in food producing animals: Effects on the quality of derived products. *Animals* **2021**, *11*, 401. [CrossRef]
11. Brenes, A.; Viveros, A.; Chamorro, S.; Arija, I. Use of polyphenol-rich grape by-products in monogastric nutrition. A review. *Anim. Feed. Sci. Tech.* **2016**, *211*, 1–17. [CrossRef]
12. Lipiński, K.; Mazur, M.; Antoszkiewicz, Z.; Purwin, C. Polyphenols in monogastric nutrition—A review. *Ann. Anim. Sci.* **2017**, *17*, 41–58. [CrossRef]
13. Surai, P.F. Polyphenol compounds in the chicken/animal diet: From the past to the future. *J. Anim. Physiol. Anim. Nutr.* **2014**, *98*, 19–31. [CrossRef]

14. Landete, J.M. Dietary intake of natural antioxidants: Vitamins and polyphenols. *Crit. Rev. Food. Sci. Nutr.* **2013**, *53*, 706–721. [CrossRef] [PubMed]
15. Gan, Z.; Wei, W.; Li, Y.; Wu, J.; Zhao, Y.; Zhang, L.; Wang, T.; Zhong, X. Curcumin and resveratrol regulate intestinal bacteria and alleviate intestinal inflammation in weaned piglets. *Molecules* **2019**, *24*, 1220. [CrossRef] [PubMed]
16. Wang, R.; Yu, H.; Fang, H.; Jin, Y.; Zhao, Y.; Shen, J.; Zhou, C.; Li, R.; Wang, J.; Fu, Y.; et al. Effects of dietary grape pomace on the intestinal microbiota and growth performance of weaned piglets. *Arch. Anim. Nutr.* **2020**, *74*, 296–308. [CrossRef] [PubMed]
17. Lu, Y.; Zhao, M.; Mo, J.; Lan, G.; Liang, J. Dietary supplementation ellagic acid on the growth, intestinal immune response, microbiota, and inflammation in weaned piglets. *Front. Vet. Sci.* **2022**, *9*, 980271. [CrossRef]
18. Galassi, G.; Battelli, M.; Verdile, N.; Rapetti, L.; Zanchi, R.; Arcuri, S.; Petrera, F.; Abeni, F.; Crovetto, G.M. Effect of a polyphenol-based additive in pig diets in the early stages of growth. *Animals* **2021**, *11*, 3241. [CrossRef]
19. Legislative Decree 26/2014—Implementation of Directive 2010/63/EU on the Protection of Animals Used for Scientific Purposes; GU n. 61 of 14 March 2014. Available online: https://www.gazzettaufficiale.it/eli/id/2014/03/14/14G00036/sg (accessed on 13 July 2023).
20. Directive 2010/63/EU of the European Parliament and of the Council of 22 September 2010 on the Protection of Animals Used for Scientific Purposes. OJ L276/33-79. Available online: https://eur-lex.europa.eu/legal-content/EN/TXT/PDF/?uri=CELEX:32010L0063&from=EN (accessed on 13 July 2023).
21. Schüep, W.; Rettenmaier, R. Analysis of vitamin E homologs in plasma and tissue: High-performance liquid chromatography. *Methods Enzymol.* **1994**, *234*, 294–302.
22. Simoni, M.; Goi, A.; Pellattiero, E.; Mavrommatis, A.; Tsiplakou, E.; Righi, F.; De Marchi, M.; Manuelian, C.L. Long-term administration of a commercial supplement enriched with bioactive compounds does not affect feed intake, health status, and growth performances in beef cattle. *Arch. Anim. Breed.* **2022**, *65*, 135–144. [CrossRef]
23. Khalifian, S.; Raimondi, G.; Brandacher, G. The use of luminex assays to measure cytokines. *J. Invest. Dermatol.* **2015**, *135*, e31. [CrossRef]
24. Prost, M. Method for Determining the Antiradical Defense Potential and Use Thereof, in Particular in Veterinary and Human Preventive Therapeutics. U.S. Patent 20,060,234,329, 22 October 2003.
25. Pastorelli, G.; Faustini, M.; Corino, C.; Rossi, R. Kit Radicaux Libres, a biological application for monitoring oxidative stress in pigs. *Ital. J. Anim. Sci.* **2013**, *12*, 432–435. [CrossRef]
26. De Bellis, R.; Piacentini, M.P.; Meli, M.A.; Mattioli, M.; Menotta, M.; Mari, M.; Valentini, L.; Palomba, L.; Desideri, D.; Chiarantini, L. In vitro effects on calcium oxalate crystallization kinetics and crystal morphology of an aqueous extract from Ceterach officinarum. Analysis of a potential antilithiatic mechanism. *PLoS ONE* **2019**, *14*, e0218734. [CrossRef]
27. Garrett, A.R.; Murray, B.K.; Robison, R.A.; O'Neill, K.L. Measuring antioxidant capacity using the ORAC and TOSC assays. *Methods Mol. Biol.* **2010**, *594*, 251–262.
28. Fundo, J.F.; Miller, F.A.; Garcia, E.; Santos, J.R.; Silva, C.L.; Brandão, T.R. Physicochemical characteristics, bioactive compounds and antioxidant activity in juice, pulp, peel and seeds of Cantaloupe melon. *J. Food Meas. Charact.* **2018**, *12*, 292–300. [CrossRef]
29. Costa, L.G.; Tait, L.; de Laat, R.; Dao, K.; Giordano, G.; Pellacani, C.; Cole, T.B.; Furlong, C.E. Modulation of paraoxonase 2 (PON2) in mouse brain by the polyphenol quercetin: A mechanism of neuroprotection? *Neurochem. Res.* **2013**, *38*, 1809–1818. [CrossRef] [PubMed]
30. Dhama, K.; Tiwari, R.; Chakraborty, S.; Saminathan, M.; Kumar, A.; Karthik, K.; Wani, M.Y.; Singh, S.V.; Rahal, A. Evidence based antibacterial potentials of medicinal plants and herbs countering bacterial pathogens especially in the era of emerging drug resistance: An integrated update. *Int. J. Pharmacol.* **2014**, *10*, 1–43. [CrossRef]
31. Christaki, E.; Giannenas, I.; Bonos, E.; Florou-Paneri, P. Chapter 2-Innovative uses of aromatic plants as natural supplements in nutrition. In *Feed Additives*; Florou-Paneri, P., Christaki, E., Giannenas, I., Eds.; Academic Press: London, UK, 2020; pp. 19–34.
32. Hasegawa, S.; Miyake, M. Biochemistry and biological functions of citrus limonoids. *Food Rev. Int.* **1996**, *12*, 413–435. [CrossRef]
33. Drewnowski, A.; Henderson, S.A.; Shore, A.B. Taste responses to naringin, a flavonoid, and the acceptance of grapefruit juice are related to genetic sensitivity to 6-n-propylthiouracil. *Am. J. Clin. Nutr.* **1997**, *66*, 391–397. [CrossRef]
34. Zhang, H.J.; Jiang, X.R.; Mantovani, G.; Lumbreras, A.E.V.; Comi, M.; Alborali, G.; Savoini, G.; Dell'Orto, V.; Bontempo, V. Modulation of plasma antioxidant activity in weaned piglets by plant polyphenols. *Ital. J. Anim. Sci.* **2014**, *13*, 3242. [CrossRef]
35. Gessner, D.K.; Fiesel, A.; Most, E.; Dinges, J.; Wen, G.; Ringseis, R.; Eder, K. Supplementation of a grape seed and grape marc meal extract decreases activities of the oxidative stress-responsive transcription factors NF-κB and Nrf2 in the duodenal mucosa of pigs. *Acta Vet. Scand.* **2013**, *55*, 18. [CrossRef]
36. Rajković, E.; Schwarz, C.; Tischler, D.; Schedle, K.; Reisinger, N.; Emsenhuber, C.; Ocelova, V.; Roth, N.; Frieten, D.; Dusel, G.; et al. Potential of grape extract in comparison with therapeutic dosage of antibiotics in weaning piglets: Effects on performance, digestibility and microbial metabolites of the ileum and colon. *Animals* **2021**, *11*, 2771. [CrossRef] [PubMed]
37. Navarrete, M.; Rangel, C.; Espinosa-García, J.; Corchado, J.C. Theoretical study of the antioxidant activity of vitamin E: Reactions of α-tocopherol with the hydroperoxy radical. *J. Chem. Theory Comput.* **2005**, *21*, 337–344. [CrossRef] [PubMed]
38. Dao, D.Q.; Ngo, T.C.; Thong, N.M.; Nam, P.C. Is Vitamin A an Antioxidant or a Pro-oxidant? *J. Phys. Chem. B* **2017**, *121*, 9348–9357. [CrossRef] [PubMed]

39. Corbi, G.; Conti, V.; Komici, K.; Manzo, V.; Filippelli, A.; Palazzo, M.; Vizzari, F.; Davinelli, S.; Di Costanzo, A.; Scapagnini, G.; et al. Phenolic plant extracts induce sirt1 activity and increase antioxidant levels in the rabbit's heart and liver. *Oxid. Med. Cell. Longev.* **2018**, *2018*, 2731289. [CrossRef]
40. Corino, C.; Oriani, G.; Pantaleo, L.; Pastorelli, G.; Salvatori, G. Influence of dietary vitamin E supplementation on "heavy" pig carcass characteristics, meat quality, and vitamin E status. *J. Anim. Sci.* **1999**, *77*, 1755–1761. [CrossRef]
41. Lauridsen, C. Evaluation of the effect of increasing dietary vitamin E in combination with different fat sources on performance, humoral immune responses and antioxidant status of weaned pigs. *Anim. Feed Sci. Technol.* **2010**, *158*, 85–94. [CrossRef]
42. Song, R.; Chen, C.; Johnston, L.J.; Kerr, B.J.; Weber, T.E.; Shurson, G.C. Effects of feeding diets containing highly peroxidized distillers dried grains with solubles and increasing vitamin E levels to wean–finish pigs on growth performance, carcass characteristics, and pork fat composition. *J. Anim. Sci.* **2014**, *92*, 198–210. [CrossRef]
43. Wang, D.; Jang, Y.D.; Kelley, M.; Rentfrow, G.K.; Azain, M.J.; Lindemann, M.D. Effects of multiple vitamin E levels and two fat sources in diets for swine fed to heavy slaughter weight of 150 kg: II. Tissue fatty acid profile, vitamin E concentrations, immune capacity, and antioxidant capacity of plasma and tissue. *Transl. Anim. Sci.* **2023**, *7*, txad087. [CrossRef]
44. Han, X.; Shen, T.; Lou, H. Dietary polyphenols and their biological significance. *Int. J. Mol. Sci.* **2007**, *8*, 950–988. [CrossRef]
45. Faria, A.; Fernandes, I.; Norberto, S.; Mateus, N.; Calhau, C. Interplay between anthocyanins and gut microbiota. *J. Agric. Food Chem.* **2014**, *62*, 6898–6902. [CrossRef]
46. Gessner, D.K.; Ringseis, R.; Eder, K. Potential of plant polyphenols to combat oxidative stress and inflammatory processes in farm animals. *J. Anim. Physiol. Anim. Nutr.* **2017**, *101*, 605–628. [CrossRef] [PubMed]
47. Hackman, R.M.; Polagruto, J.A.; Zhu, Q.Y.; Sun, B.; Fujii, H.; Keen, C.L. Flavanols: Digestion, absorption and bioactivity. *Phytochem. Rev.* **2008**, *7*, 195–208. [CrossRef]
48. Hein, E.M.; Rose, K.; van't Slot, G.; Friedrich, A.W.; Humpf, H.U. Deconjugation and degradation of flavonol glycosides by pig cecal microbiota characterized by fluorescence in situ hybridization (FISH). *J. Agric. Food Chem.* **2008**, *56*, 2281–2290. [CrossRef] [PubMed]
49. Moreira, I.; Mahan, D.C. Effect of dietary levels of vitamin E (all-rac-tocopheryl acetate) with or without added fat on weanling pig performance and tissue α-tocopherol concentration. *J. Anim. Sci.* **2002**, *80*, 663–669. [CrossRef]
50. Sivertsen, T.; Vie, E.; Bernhoft, A.; Baustad, B. Vitamin E and selenium plasma concentrations in weanling pigs under field conditions in Norwegian pig herds. *Acta Vet. Scand.* **2007**, *49*, 1. [CrossRef]
51. Hinson, R.B.; McCormick, K.A.; Moser, R.L.; Ackerman, M.A.; Main, R.G.; Mahoney, J.A. Reduced vitamin supplementation with fat-soluble vitamins A, D, and E added at National Research Council requirements may not be adequate for optimal sow and progeny performance. *J. Swine Health Prod.* **2022**, *30*, 79–94. [CrossRef]
52. Chung, Y.K.; Mahan, D.C.; Lepine, A.J. Efficacy of dietary D-α-tocopherol and DL-α-tocopheryl acetate for weanling pigs. *J. Anim. Sci.* **1992**, *70*, 2485–2492. [CrossRef]
53. Liu, P.; Kerr, B.J.; Weber, T.E.; Chen, C.; Johnston, L.J.; Shurson, G.C. Influence of thermally oxidized vegetable oils and animal fats on intestinal barrier function and immune variables in young pigs. *J. Anim. Sci.* **2014**, *92*, 2971–2979. [CrossRef]
54. García, G.R.; Dogi, C.A.; Ashworth, G.E.; Berardo, D.; Godoy, G.; Cavaglieri, L.R.; De Moreno de LeBlanc, A.; Greco, C.R. Effect of breast feeding time on physiological, immunological and microbial parameters of weaned piglets in an intensive breeding farm. *Vet. Immunol. Immunopathol.* **2016**, *176*, 44–49. [CrossRef]
55. Moeser, A.J.; Pohl, C.S.; Rajput, M. Weaning stress and gastrointestinal barrier development: Implications for lifelong gut health in pigs. *Anim. Nutr.* **2017**, *3*, 313–321. [CrossRef]
56. Rooke, J.A.; Carranca, C.; Bland, I.M.; Sinclair, A.G.; Ewen, M.; Bland, V.C.; Edwards, S.A. Relationships between passive absorption of immunoglobulin G by the piglet and plasma concentrations of immunoglobulin G at weaning. *Livest. Prod. Sci.* **2003**, *81*, 223–234. [CrossRef]
57. de Groot, N.; Fariñas, F.; Cabrera-Gómez, C.G.; Pallares, F.J.; Ramis, G. Weaning causes a prolonged but transient change in immune gene expression in the intestine of piglets. *J. Anim. Sci.* **2021**, *99*, skab065. [CrossRef] [PubMed]
58. Pistol, G.C.; Palade, L.M.; Marin, D.E.; Stancu, M.; Taranu, I. The effect of grape wastes, wine industry byproducts, on inflammatory and antioxidant biomarkers in post-weaning piglets. *Sci. Pap. Anim. Sci. Ser. Lucr. Ştiinţifice-Ser. Zooteh.* **2019**, *71*, 219–223.
59. Cao, S.T.; Wang, C.C.; Wu, H.; Zhang, Q.H.; Jiao, L.F.; Hu, C.H. Weaning disrupts intestinal antioxidant status, impairs intestinal barrier and mitochondrial function, and triggers mitophagy in piglets. *J. Anim. Sci.* **2018**, *96*, 1073–1083. [CrossRef]
60. Li, Y.; Zhao, X.; Jiang, X.; Chen, L.; Hong, L.; Zhuo, Y.; Lin, Y.; Fang, Z.; Che, L.; Feng, B.; et al. Effects of dietary supplementation with exogenous catalase on growth performance, oxidative stress, and hepatic apoptosis in weaned piglets challenged with lipopolysaccharide. *J. Anim. Sci.* **2020**, *98*, skaa067. [CrossRef]
61. Zheng, P.; Yu, B.; He, J.; Tian, G.; Luo, Y.; Mao, X.; Zhang, K.; Che, L.; Chen, D. Protective effects of dietary arginine supplementation against oxidative stress in weaned piglets. *Br. J. Nutr.* **2013**, *109*, 2253–2260. [CrossRef]
62. Xiong, S.; Guo, R.; Yang, Z.; Xu, L.; Du, L.; Li, R.; Xiao, F.; Wang, Q.; Zhu, M.; Pan, X. Treg depletion attenuates irradiation-induced pulmonary fibrosis by reducing fibrocyte accumulation, inducing Th17 response, and shifting IFN-γ, IL-12/IL-4, IL-5 balance. *Immunobiology* **2015**, *220*, 1284–1291. [CrossRef]
63. Lauridsen, C. From oxidative stress to inflammation: Redox balance and immune system. *Poult. Sci.* **2019**, *98*, 4240–4246. [CrossRef]

64. Al-Sadi, R.; Boivin, M.; Ma, T. Mechanism of cytokine modulation of epithelial tight junction barrier. *Front. Biosci.* **2009**, *14*, 2765. [CrossRef]
65. Opal, S.M.; DePalo, V.A. Anti-inflammatory cytokines. *Chest* **2000**, *117*, 1162–1172. [CrossRef]
66. Silva-Guillen, Y.V.; Arellano, C.; Boyd, R.D.; Martinez, G.; Van Heugten, E. Growth performance, oxidative stress and immune status of newly weaned pigs fed peroxidized lipids with or without supplemental vitamin E or polyphenols. *J. Anim. Sci Biotechnol.* **2020**, *11*, 2. [CrossRef]
67. Fiesel, A.; Gessner, D.K.; Most, E.; Eder, K. Effects of dietary polyphenol-rich plant products from grape or hop on pro-inflammatory gene expression in the intestine, nutrient digestibility and faecal microbiota of weaned pigs. *BMC Vet. Res.* **2014**, *10*, 196. [CrossRef] [PubMed]
68. Guo, M.; Wang, H.; Xu, S.; Zhuang, Y.; An, J.; Su, C.; Liu, X. Alteration in gut microbiota is associated with dysregulation of cytokines and glucocorticoid therapy in systemic lupus erythematosus. *Gut Microbes* **2020**, *11*, 1758–1773. [CrossRef] [PubMed]
69. Laskowska, E.; Jarosz, Ł.; Grądzki, Z. The effect of feed supplementation with effective microorganisms (EM) on pro-and anti-inflammatory cytokine concentrations in pigs. *Res. Vet. Sci.* **2017**, *115*, 244–249. [CrossRef] [PubMed]
70. Pomorska-Mól, M.; Wierzchosławski, K.; Włodarek, J.; Gogulski, M.; Pejsak, Z. Dynamics of pro-and anti-inflammatory cytokine changes in serum and assessment of their diagnostic utility during lactation impairment in pigs. *Res. Vet. Sci.* **2020**, *128*, 9–15. [CrossRef]
71. Turin, L.; Torinesi, R.; Pastorelli, G. Real-time PCR detection of the effect of postweaning on the expression of cytokines and NF-kB in piglets. *J. Biol. Regul. Homeost. Agents* **2019**, *33*, 1737–1745. [CrossRef]
72. Rossi, R.; Pastorelli, G.; Corino, C. Application of KRL test to assess total antioxidant activity in pigs: Sensitivity to dietary antioxidants. *Res. Vet. Sci.* **2013**, *94*, 372–377. [CrossRef]
73. Mao, X.; Lv, M.; Yu, B.; He, J.; Zheng, P.; Yu, J.; Wang, Q.; Chen, D. The effect of dietary tryptophan levels on oxidative stress of liver induced by diquat in weaned piglets. *J. Anim. Sci. Biotechnol.* **2014**, *5*, 49. [CrossRef]
74. Han, H.; Liu, Z.; Yin, J.; Gao, J.; He, L.; Wang, C.; Hou, R.; He, X.; Wang, G.; Li, T.; et al. D-galactose induces chronic oxidative stress and alters gut microbiota in weaned piglets. *Front. Physiol.* **2021**, *12*, 634283. [CrossRef]
75. Degroote, J.; Vergauwen, H.; Wang, W.; Van Ginneken, C.; De Smet, S.; Michiels, J. Changes of the glutathione redox system during the weaning transition in piglets, in relation to small intestinal morphology and barrier function. *J. Anim. Sci. Biotechnol.* **2020**, *23*, 45. [CrossRef]

antioxidants

Article

Antioxidant Activity of Sweet Whey Derived from Bovine, Ovine and Caprine Milk Obtained from Various Small-Scale Cheese Plants in Greece before and after In Vitro Simulated Gastrointestinal Digestion

Eleni Dalaka, Ioannis Politis and Georgios Theodorou *

Laboratory of Animal Breeding and Husbandry, Department of Animal Science, Agricultural University of Athens, 11855 Athens, Greece; elenidalaka@aua.gr (E.D.); i.politis@aua.gr (I.P.)
* Correspondence: gtheod@aua.gr; Tel.: +30-2105294450

Abstract: Whey-derived peptides have been associated with different biological properties, but most peptides are usually further hydrolyzed during the digestive process. In the present study, the antioxidant capacity of 48 samples of sweet whey (SW) derived from cheeses obtained from small-scale cheese plants made with bovine, ovine, caprine or a mixture of ovine/caprine milk was assessed using both cell-free and cell-based assays. SW digestates (SW-Ds) and a fraction (<3 kDa; SW-D-P3) thereof were obtained after in vitro digestion and subsequent ultrafiltration. Antioxidant properties using four different assays were evaluated before and after digestion. Our data showed higher values ($p < 0.05$) for ORAC, ABTS, FRAP and P-FRAP after in vitro digestion (SW-Ds and SW-D-P3) when compared with the corresponding values before digestion. In the non-digested SW, ORAC values were higher ($p < 0.05$) for the bovine SW compared with all the other samples. In contrast, the ABTS assay indicated a higher antioxidant activity for the ovine SW both before digestion and for SW-D-P3 compared with the bovine SW. The fraction SW-D-P3 of the ovine SW, using HT29 cells and H_2O_2 as an oxidizing agent, increased ($p < 0.05$) the cellular antioxidant activity. Furthermore, the same fraction of the ovine/caprine mixed SW increased, through the NF-κB pathway, the expression of SOD1 and CAT, genes implicated in the oxidative response in macrophage-like THP-1 cells. These findings indicate that SW, and particularly bovine and ovine SW, could be a candidate source for physical antioxidants in human and animal nutrition.

Keywords: cheese whey; in vitro digestion; bioactive peptides; antioxidant biochemical assays; cellular assays

check for updates

Citation: Dalaka, E.; Politis, I.; Theodorou, G. Antioxidant Activity of Sweet Whey Derived from Bovine, Ovine and Caprine Milk Obtained from Various Small-Scale Cheese Plants in Greece before and after In Vitro Simulated Gastrointestinal Digestion. *Antioxidants* **2023**, *12*, 1676. https://doi.org/10.3390/antiox12091676

Academic Editor: Volker Böhm

Received: 11 July 2023
Revised: 22 August 2023
Accepted: 25 August 2023
Published: 27 August 2023

1. Introduction

Sweet whey (SW), a by-product of the cheese-making process, is considered a pollutant, and its estimated annual production is about 200 million tons [1]. Considering the huge volumes of SW produced and the increasing consumer demand for more sustainable food production, constant efforts are being made to evaluate biologically relevant properties of whey [2].

The demand for functional foods has been growing continuously due to their nutritional value, health-promoting properties and their potential to reduce the risk of several diseases [3,4]. Whey proteins are a source of bioactive peptides with a range of biological properties including antihypertensive, antimicrobial, antidiabetic as well as antioxidant activity. Specifically, whey proteins and their derivatives are characterized as potent antioxidants by modulating a range of redox biomarkers and reactive oxygen species [5,6]. The antioxidant activity of whey protein peptides has been attributed to the synergistic action of sulfhydryl groups, free radical scavenging by specific amino acids and chelation of iron [7,8]. It should be noted that although there is a plethora of studies reporting milk whey antioxidant activity, most of them use as a source WPC and WPI of bovine origin and not fresh SW, while no comparative data with SW from small ruminants' milk are available.

During the last few years, much research has been dedicated to the processing and generation of bioactive peptides from food products [9]. Bioactive peptides are defined as components with biological activities over and above their nutritional value [10]. Furthermore, as it is known, whey proteins can release bioactive peptides by proteolysis in order to exert enhanced functions [11]. Polypeptides are degraded amongst other ways, by brush-border or cellular peptidases, whereas peptides with low molecular weight may remain intact and exert their activity at the tissue level [12].

Several recent studies have focused on cell-based assays as a more appropriate technique that serves as an intermediate between in vitro biochemical assays and in vivo clinical trials in animals or humans. Wolfe et al. developed a quantitative cellular antioxidant activity (CAA) assay based on human hepatocarcinoma HepG2 cells for quantifying the antioxidant activity of phytochemicals, food extracts and dietary supplements [13]. Liver cells, however, cannot be considered the ideal in vitro model for measuring the efficacy of dietary antioxidants. On the other hand, many studies mention a good correlation between the in vitro absorption in intestinal (Caco-2 and/or HT29) cellular models and in vivo intestinal absorption; hence, they are being widely used as effective tools for predicting the human intestinal absorption of food compounds and drugs [14–16]. For this reason, Wan et al. developed a Caco-2 CAA assay for the quantitative evaluation of antioxidants [17]. This was further validated by Kellett et al. who reported more robust results for CAA in a Caco-2 cell model compared to HepG2 cells and who verified that epithelial cell lines, known to be good models of the intestinal barrier, are more appropriate cell models to determine the antioxidant activity for phenolic antioxidants [18]. Many studies mention a good correlation between the in vitro absorption in intestinal (Caco-2 and/or HT29) cellular models and in vivo intestinal absorption; hence, they are being widely used as effective tools for predicting human intestinal absorption of food compounds and drugs [14–16].

The enzymatic hydrolysis of whey protein concentrate (WPC) and whey protein isolate (WPI) can produce antioxidant peptides, and hence, functional additives based on them could be produced. An advantageous approach for both dairy industries and livestock farmers should be the utilization of natural compounds, such as SW and its derivatives. Therefore, it is imperative to evaluate SW for its antioxidant capacity in vitro for the potential future increase of their use in animal nutrition [6,19].

The overproduction of ROS leads to oxidative stress and promotes inflammation by the NFE2L2 (NFE2 Like BZIP Transcription Factor 2) antioxidant pathway [20]. NFE2L2 is well recognized for its critical role in the response to oxidative stress and its binding to antioxidant responsive elements (AREs) and the subsequent regulation of their expression [21–23]. Enzymes such as catalase (CAT) and superoxide dismutase 1 (SOD1) are used as markers to reflect the antioxidant status of the cell [24]. More specifically, CAT and SOD1 represent the indirect antioxidant activity that reflects the removal of ROS. SOD1 decomposes superoxide anion to H_2O_2, which CAT then converts into water and oxygen [25], and the expression of both enzymes is regulated by AREs, which have been previously activated by NFE2L2 [26]. Another pathway that has been proposed as an alternative to NFE2L2 is that of the transcription nuclear factor kappa-light-chain-enhancer of activated B cells (NF-κB). The expression of antioxidant enzymes, among them CAT and SOD1, might also be regulated by the NF-κB pathway [27].

The aim of this study was to evaluate the antioxidant capacity of commercial SW from ovine, bovine, caprine and a mix of ovine/caprine milk origin using the harmonized static in vitro digestion protocol INFOGEST 2.0 [28]. The oxygen radical antioxidant capacity (ORAC), 2,2′-azinobis (3-ethylbenzothiazoline-6-sulfonic acid) radical scavenging assay (ABTS), ferric reducing antioxidant power (FRAP) and potassium ferricyanide reducing antioxidant power (P-FRAP) of SW before and after digestion were examined. Moreover, the effect of treatment with SW-D-P3 on CAA in HT29 cells and gene expression quantification of CAT, SOD and NFE2L2 in THP-1 cells, following induced oxidative stress by H_2O_2 and LPS, respectively, was assessed.

2. Materials and Methods

2.1. Chemicals and Reagents

The chemicals and enzymes used were all of high purity or an analytical reagent grade. Trolox (6-hydroxy-2,5,7,8-tetramethylchroman-2-carboxylic acid), 2,2′-azobis(2-methyl-propionamidine) dihydrochloride (AAPH), 2,4,6-tri (2-pyridyl)-*s*-triazine (TPTZ), fluorescein sodium salt (FL), ferric chloride ($FeCl_3$), potassium ferricyanide ($K_3Fe(CN)_6$), trichloroacetic acid (TCA) and 2′,7′-Dichlorofluorescein diacetate (DCFH-DA), hydrogen peroxide (H_2O_2—35% in water), 2,2′-azino-bis(3-ethylbenzothiazoline-6-sulfonic acid (ABTS), sodium persulfate, acetic acid (glacial), butylated hydroxytoluene (BHT), pepsin from porcine gastric mucosa (≥2.500 units/mg protein), porcine pancreatin (4 × USP, United States Pharmacopeia) bile extract porcine, phorbol 12-myristate 13-acetate (PMA) and lipopolysaccharides from *Escherichia coli* O111:B4 (LPS) were all purchased from Sigma-Aldrich (Saint Louis, MO, USA). The Amicon Ultra-4 Centrifugal Filter Devices (3kDa) and Millex-GP 33mm PES 0.22 um were from Merck Millipore (Burlington, MA, USA). The 96-well transparent flat-bottom plates were purchased from Kisker Biotech (Steinfurt, Germany), while the 96-well cell culture transparent flat-bottom plates were purchased from SPL Life Sciences (Pocheon, South Korea). 3-(4,5-dimethylthiazol2-yl)-2,5-diphenyltetrazolium bromide (MTT) was obtained from Cayman Chemical (Michigan, MI, USA). DMEM, RPMI 1640, L-glutamine, sodium pyruvate, non-essential amino acids and penicillin-streptomycin were purchased from Biosera (Cholet, France). Fetal bovine serum (FBS) was purchased from Gibco ThermoFisher Scientific (Waltham, MA, USA). Phosphate-buffered saline (PBS) and the PrimeScript RT reagent kit (Perfect Real Time) were purchased from Takara Bio (Shiga, Japan). NucleoZOL was obtained from Macherey-Nagel (Düren, Germany). DNase I (RNase-Free) was purchased from New England Biolabs (Ipswich, MA, USA). The FastGene 2× IC Green qPCR Universal Mix was purchased from Nippon Genetics (Tokyo, Japan).

2.2. Collection and Preparation of Samples

Sweet whey samples were obtained from several small-scale cheese plants in Greece. Cheeses were produced from bovine, ovine, caprine and a mixture of ovine/caprine milk. An equal number (12 each) of SW from each milk animal origin (bovine, ovine, caprine and a mix of ovine/caprine) was used. The ratio of ovine/caprine ranged from 80/20 to 70/30 in all samples used in this study. Freeze-drying of liquid sweet whey was carried out to remove water and other solvents, and then protein content of all 48 samples was determined by the Kjeldahl method [29]. Afterwards, all samples were resuspended in deionized water at the same protein concentration of 3% (v/v) before use in subsequent analyses.

2.3. In Vitro Digestion Protocol

The digestion procedure of the samples was modified using the protocol reported by the amended and improved digestion method INFOGEST 2.0 [28,29] with slight modifications. Electrolyte stock solutions of digestive fluids, including the simulated salivary fluid (SSF), simulated gastric fluid (SGF) and simulated intestinal fluid (SIF), were prepared, and the pH of the electrolyte stock solutions were adjusted with HCl (5 M) and NaOH (6 M). Enzyme and bile salt solutions were freshly prepared prior to use.

2.3.1. Simulated Oral Digestion

Briefly, 5 mL of each sample (30 mg protein/mL) was mixed with 4 mL of pre-warmed SSF followed by the addition of 25 μL of $CaCl_2$ (0.3 M) and 0.975 mL of distilled water and

NaOH (1 M) to reach a pH of 7. The oral solution was placed in a water bath at 37 °C and incubated with shaking for 2 min without the addition of α-salivary amylase.

2.3.2. Simulated Gastric Digestion

A total of 10 mL of the above oral bolus was mixed with 8 mL of pre-warmed SGF, 1 mL of porcine pepsin solution (2000 U/mL), 5 μL of $CaCl_2$ (0.3 M) and 0.995 mL of distilled water with HCl (5 M) to achieve a pH of 3. The final mixture was incubated for 2 h at 37 °C in a rotator at 300 rpm.

2.3.3. Simulated Intestinal Digestion

A total of 20 mL of the gastric chyme was mixed with 8 mL of pre-warmed SIF, 5 mL of pancreatin (100 U/mL), 2.87 mL of bile extract (5 mM), 40 μL of $CaCl_2$ (0.3 M) and 4.09 mL of distilled water with NaOH. The final pH of each mixture was adjusted to 7 to simulate the physiological intestinal digestion environment. Then, the final mixture was incubated for 2 h at 37 °C in a rotator at 300 rpm.

2.3.4. Digestates' Fractionation

SW digestates (SW-Ds) were immediately placed in a water bath for heat shock (85 °C, 10 min) to stop enzymatic activities and were then directly placed on ice. Undigested proteins were precipitated by centrifugation at 1200× *g* for 5 min. Then, aliquots of the digestates were filtered through a 0.22 μm polyvinylidene fluoride (PVDF) syringe filter. In order to remove high molecular weight peptides from SW-Ds and obtain the SW-D-P3 fraction, a membrane filter (Ultracel® low binding regenerated cellulose) with an MWCO of 3 kDa was used (Amicon Ultra-4 Centrifugal Filter Devices). The SW-D samples were centrifuged at 4000× *g* for 90 min in a swinging-bucket rotor, following the manufacturer's instructions. The SW-Ds and SW-D-P3 were then stored at −20 °C until needed for further analysis. Moreover, four replicates of blank digests were also prepared. Specifically, digests without SW (replaced with water) but with all required enzymes, electrolytes and bile salts were performed. The corresponding fractions after in vitro digestion were named BL-D for blank digest and BL-D-P3 for the fraction of digestate with peptides with a molecular weight below 3 kDa.

2.4. Biochemical Assays

2.4.1. Oxygen Radical Antioxidant Capacity (ORAC)

The ORAC method used with FL was first described by Ou et al. [30], and the improved method of Zulueta and colleagues [31,32] was used. Firstly, the SW, SW-Ds and SW-D-P3 were dissolved and diluted 1:30 in PBS (75 mM, pH 7.4) so as to be transparent and to avoid interferences. A total of 20 μL of each sample was added to a well in a 96-well microplate with 120 μL FL (117 nM in 75 mM PBS, pH 7.4). After incubation with shaking at 37 °C for 15 min, 60 μL AAPH (40 mM) was also added to each well. Fluorescence was read immediately every 2 min for a total of 80 min at 485 nm/535 nm (excitation/emission). The automated ORAC assay was carried out on a VICTOR 2030 multilabel counter (Perkin Elmer, Waltham, MA, USA), and a standard curve of Trolox was constructed with a final concentration range of 3.125–50 μM in 75 mM of PBS and a pH of 7.4. The antioxidant capacity was expressed as μmol Trolox Equivalents (TEs)/g protein based on the area under curve (AUC) for the fluorescence decay vs. time. SW before and after in vitro digestion were measured in triplicate, and the experiment was performed three independent times.

2.4.2. 2,2′-Azinobis (3-Ethylbenzothiazoline-6-Sulfonic Acid) Radical Scavenging Assay (ABTS)

The ABTS radical scavenging activity of the SW samples, before and after digestion, was measured according to the procedure of Ozgen et al. [33]. Firstly, 7 mM of an ABTS solution was mixed with 2.45 mM of a sodium persulfate solution in equal volumes, followed by incubation for 12–16 h at room temperature in the dark to generate ABTS radicals. After that, the above $ABTS^{\bullet+}$ solution was diluted with a sodium acetate buffer (20 mM, pH 4.5) until the absorbance provided nearly 0.7 at 734 nm. Then, 20 μL (3 mg protein/mL) of the sample and 230 μL of the $ABTS^{\bullet+}$ solution was mixed and incubated for 60 min at room temperature [4]. A standard curve of the antioxidant (Trolox) was constructed for a concentration range of 3.75–100 μM. The absorbance was measured at 734 nm using a 96-well Inifinite M200 Pro plate reader (Männedorf, Switzerland). The ABTS radical scavenging activity (%) was calculated as an ABTS radical scavenging activity (%) = [(Ac − As)/Ac] × 100 where Ac represents the absorbance of the control and As represents the absorbance of the samples after the reaction [34]. The SW samples before and after in vitro digestion were measured in triplicate, and the experiment was performed three independent times. The results are expressed as μmol TEs/g protein.

2.4.3. Ferric Reducing Antioxidant Power (FRAP)

The reducing activity of SW, SW-Ds and SW-D-P3 was determined according to the method of Benzie et al. [35] with appropriate modifications for 96-well microplates [36,37]. Firstly, the SW was suitably diluted at 1:3 and SW-Ds and SW-D-P3 at 1:2, respectively. A ferric-tripyridyltriazine (Fe^{III}-TPTZ) complex solution was prepared with 0.3 M sodium acetate with glacial acetic acid (pH 3,6), 20 mM $FeCl_3$ and 10 mM TPTZ (in 40 mM HCl) at a ratio of 10:1:1 (*v*/*v*/*v*) and was heated to 37 °C for 1 h. All stock solutions were prepared daily and kept in the dark. Consequently, 280 μL of the Fe^{III}-TPTZ solution was transferred to each well with 20 μL of the sample in a 96-well microplate, and the absorbance was read at 590 nm using the Epoch 2 microplate spectrophotometer (Biotek, Winooski, VT, USA). Trolox as standard (concentration range 0.18–5.88 μM) and methanol as blank were also added to each plate. Results are expressed as μmol TEs/g protein. Samples were measured in triplicate, and the experiment was performed three independent times.

2.4.4. Potassium Ferricyanide Reducing Power (P-FRAP)

The reducing power assay was performed according to the method described by Oyaizu [38] and Liang et al. [39] with some modifications. Briefly, the SW, SW-Ds and SW-D-P3 were appropriately diluted with water (1:6 for all samples) so that the final absorbance of 700 nm fell within the range of the calibration curve. Then, 50 μL of the sample solution were mixed with 50 μL of 0.2 M of PBS (pH 6.6) and 50 μL of a fresh (prepared daily) $K_3Fe(CN)_6$ (1% *w*/*v*) solution. The microplate was incubated for 20 min at 50 °C under agitation. Then, 50 μL of TCA (10% *w*/*v*) with 10 μL of $FeCl_3$ (0.1% *w*/*v*) were added, and the incubation was continued for an additional 10 min at 50 °C under agitation. The absorbance at 700 nm was measured, and a curve with a BHT was constructed (0–100 μM). Results are expressed as μmol BHT equivalents/g protein [40]. The reducing power assay measures the ability of a compound to reduce ferric ion (Fe^{3+}) to ferrous ion (Fe^{2+}) through electron or hydrogen donation [41]. Samples were measured in triplicate, and the experiment was performed three independent times.

2.5. Cellular Assays

2.5.1. Cell Culture and Cell Viability of HT29

Cells of the human colon adenocarcinoma cell line HT29 were cultured in DMEM supplemented with 100 U/mL of penicillin, 100 μg/mL of streptomycin, 10 U/mL of L-glutamine, 100 μM of non-essential amino acids, 1 mM of sodium pyruvate and 10% (*v*/*v*) FBS. Cells were incubated at 37 °C in humidified air containing 5% CO_2. Firstly, the effects of H_2O_2 on cell survival were measured using the MTT assay. The HT29 cells were seeded in 96-well plates at a concentration of 0.5×10^5 cells per well. After 24 h of culturing, the cells were treated with different concentrations of H_2O_2 (0–2 mM) at 37 °C. The following day, the cells were washed twice with PBS, incubated for an additional 2–3 h at 37 °C with MTT (0.5 mg/mL). Then, the supernatants were removed and 100 μL of DMSO was added. The absorbance was quantified at a 570 nm wavelength (Tecan Infinite® M200 PRO). Results are expressed as a percentage of the untreated control cells (without H_2O_2). Each point represents the mean of two experiments with each individual treatment being run in quadruplicate. Furthermore, different concentrations (0.75–6 mg protein/mL) of the SW-D-P3 fraction in the presence of 0,5 mM of H_2O_2 were evaluated for their cytotoxic effects in HT29 by the MTT assay.

2.5.2. Cellular Antioxidant Activity (CAA) Assay

CAA indicates the overall oxidative status by monitoring the decomposition of DCFH-DA in the cells and its oxidation by reactive oxygen species (ROS) into the fluorescent DCF [42,43]. Intracellular reactive oxygen species (ROS) levels were determined in the HT29 cells as described by Piccolomini et al. [43] and adapted by Feng et al. [44] with some modifications. Cells were seeded at 0.5×10^5 cells/well in 96-well plates for 24 h. Cells were washed with PBS and treated with 50 μL of BL-D-P3 or SW-D-P3 (6 mg protein/mL in DMEM) together with 50 μL of H_2O_2 (0.5 mM in DMEM) for 24 h. Afterwards, the cells were washed twice with PBS and treated with 100 μL of DCFH-DA (10 μM in PBS containing 0.2% methanol) for 30 min. Fluorescence at 485 nm/535 nm was recorded at 37 °C every 2 min for a total of 80 min using the VICTOR 2030 multilabel counter (Perkin Elmer, Waltham, MA, USA). CAA was measured in triplicate, and the experiment was performed three independent times. The results are expressed as a % of ROS generation to the untreated cells (without H_2O_2).

2.5.3. Cell Culture and Differentiation of THP-1

Cells of the human acute monocytic leukemia cell line THP-1 were maintained in RPMI 1640 supplemented with 10% (*v*/*v*) FBS, 10 U/mL of L-glutamine, 1 mM of sodium pyruvate, 100 U/mL of penicillin, 100 μg/mL of streptomycin and 100 μM of non-essential amino acids in a humidified incubator at 37 °C and 5% CO_2. To induce differentiation into macrophage-like ones, monocytes were placed into 12-well plates at a cell density of 0.8×10^6 cells/mL and incubated with 100 ng/mL of PMA for 48 h [45–48]. Then, the PMA-contained medium was removed, and cells were washed with PBS and subsequently incubated for 24 h in the supplemented PMA-free RPMI-1640. After the resting phase, macrophages were incubated for 24 h in the presence of 100 ng/mL of lipopolysaccharide (LPS) and SW-D-P3 (3 mg protein/mL) or BL-D-P3 for 24 h. Each sample was tested in triplicate.

2.5.4. Quantification of mRNA Transcripts Using Real Time-PCR (qPCR)

Total RNA extraction of the attached THP-1 cells was performed using the Nucleozol reagent according to the manufacturer's instructions. RNA samples were treated with DNase for the removal of the remaining DNA, and pure RNA was recovered by ethanol precipitation. The quantity and purity of RNA was calculated using a spectrophotometer (Q5000, Quawell Technology Inc., San Jose, CA, USA). Reverse transcription was performed from 500 ng of the total RNA with the PrimeScript RT reagent kit following the protocol of the manufacturer. A Real-Time thermal cycler (SaCycle–96, Sacace Biotechnologies, Como, Italy) was used for the qPCR using the FastGene 2× IC Green qPCR Universal Mix. Primers for target genes (*SOD1*, *CAT*, *NFE2L2*, *NFKB1* and *RELA*) and housekeeping genes (*RPS18*, *HPRT1*, *RPL37A* and *B2M*) were designed across intron/exon boundaries with an annealing temperature of 60 °C. Each cDNA sample was tested in duplicate. The relative gene expression was calculated using the $2^{-\Delta\Delta Ct}$ method. Primer details are listed in Table 1.

2.6. Statistical Analysis

The experimental results are reported as means ± the standard error of means (SEMs) of at least two biological replicates. All data were tested for normality using the Kolmogorov–Smirnov test and transformed in logarithmic or normalized form [49] where necessary until the data were normally distributed. Subsequently all data generated were compared using one-way ANOVA followed by Duncan's post hoc test. Differences between means were considered significant at $p < 0.05$. The statistical analysis was performed using the SPSS for Windows statistical package program, version 22.0.0. Graphs were generated using the GraphPad Prism 8 program.

Table 1. Oligonucleotide primer sequences, amplicon size and reaction efficiency in qPCR.

Gene (Accesion Number)	Primer Direction	Sequence (5′-3′)	Amplicon Size	Reaction Efficiency
SOD1 (NM_000454)	Forward Reverse	CGAGCAGAAGGAAAGTAATGG CCAAGTCTCCAACATGCC	194	95
CAT (NM_001752)	Forward Reverse	TGCCTATCCTGACACTCACC GAGCACCACCCTGATTGTC	137	92
NFE2L2 (NM_001145412)	Forward Reverse	GATCTGCCAACTACTCCCA GCCGAAGAAACCTCATTGTC	121	90
NFKB1 (NM_001165412)	Forward Reverse	GCACAAGGAGACATGAAACAG CCCAGAGACCTCATAGTTGTC	189	97
RELA (NM_001145138)	Forward Reverse	GGACTACGACCTGAATGCTG ACCTCAATGTCCTCTTTCTGC	228	105
RPS18 (NM_022551)	Forward Reverse	CTGAGGATGAGGTGGAACG CAGTGGTCTTGGTGTGCT	240	98
HPRT1 (NM_000194)	Forward Reverse	CTTTGCTTTCCTTGGTCAGG CAAATCCAACAAAGTCTGGCT	111	99
RPL37A (NM_000998)	Forward Reverse	AGTACACTTGCTCTTTCTGTGG GGAAGTGGTATTGTACGTCCAG	119	106
B2M (NM_004048)	Forward Reverse	GCTATCCAGCGTACTCCA CTTAACTATCTTGGGCTGTGAC	285	103

3. Results and Discussion

To evaluate whether gastrointestinal digestion affects the antioxidant activity of SW, the standardized INFOGEST static in vitro digestion model was applied. The antioxidant properties of SW, SW-Ds and SW-D-P3 were evaluated using a variety of methodological approaches. Due to the differences in the principle and mechanism of action of the different methods of antioxidant capacity evaluation, the use of only one antioxidant analysis can barely clarify the actual antioxidant status of samples [50]. Consequently, four different biochemical antioxidant methods are employed towards this aim, which can roughly be classified into two types, namely, assays based on hydrogen atom transfer (HAT) reactions and those based on single electron transfer (SET) [51]. HAT assays, such as ORAC, measure the ability of an antioxidant to inactivate a free radical (ROO·) by releasing a hydrogen atom in kinetic time. In contrast, assays that are dominated by SET-based reaction mechanisms, such as the end-point ABTS, measure the release of an electron to the (ROO·), converting it into an anion (ROO−) [52,53]. The latter causes a color change in the solution, indicating the concentration of the antioxidant. Thus, it was crucial to use at least one assay of each type in order to more completely evaluate the total antioxidant activity of a complex substrate such as SW. In addition, the cellular antioxidant activity of the SW-D-P3 fractions was evaluated using CAA in the HT29 intestinal cell line and expression of oxidative stress-related genes in the THP-1 monocytic cell line.

3.1. Assessment of Antioxidant Activity of SW before and after In Vitro Digestion Using ORAC, ABTS, FRAP and P-FRAP Biochemical Assays

The data shown in Table 2 indicate that the antioxidant activity of the samples after in vitro digestion was greatly increased, regardless of the biochemical method used. In all four assays, the antioxidant activity of SW was significantly increased SW-D>SW-D-P3>SW ($p < 0.05$). The digestive process had a positive effect on the antioxidant capacity of SW, represented by significant increases of around 2–4 fold on the ABTS, FRAP and P-FRAP assays after in vitro digestion, while an even greater augmentation was observed by the ORAC-FL assay (Table 2). In line with our results, there are other studies that report a significantly increased antioxidant activity of WPI after digestion, regardless of the assay used [37,54]. In addition, Garcia-Casas et al. [55] demonstrated that the bioaccessible fraction (corresponding to the SW-D samples of our study) of a SW-based beverage digestate exhibited an augmented antioxidant activity based on ABTS and FRAP. Since antioxidant compounds can use different mechanisms of action and each of the methods used evaluates their effect in a unique way, their evaluation of said compounds is complementary to each other. Namely, ABTS, FRAP and P-FRAP belong to the single-electron transfer (SET) assays while ORAC-FL belongs to the hydrogen atom transfer (HAT) assays. Interestingly, from our data in Table 2, the ORAC values showed a much-enhanced antioxidant activity for the SW samples after digestion, with an 8-fold increase, from 20.5 to 167.2 μmol TEs/g protein. Although Clausen et al. observed that the ORAC assay was more sensitive than ABTS in evaluating the scavenging of the peroxyl radicals of bovine whey proteins, it should be noted that this observation was made for intact whey proteins rather than fermented ones [56].

Table 2. Antioxidant activity by ABTS (2,2′-azinobis (3-ethylbenzothiazoline-6-sulfonic acid)), ORAC-FL (Oxygen Radical Absorbance Capacity-Fluorescence), FRAP (ferric reducing antioxidant power) and P-FRAP (potassium ferricyanide reducing power) of sweet whey before (SW) and after in vitro digestion (SW-Ds and SW-D-P3) regardless of milk origin.

Method (Units)	SW	SW-Ds	SW-D-P3
ABTS (μmol TEs/g protein)	20.3 ± 0.8 [a]	46.2 ± 0.8 [c]	38.7 ± 0.6 [b]
ORAC-FL (μmol TEs/g protein)	20.5 ± 1.0 [a]	167.2 ± 4.8 [c]	122.3 ± 3.9 [b]
FRAP (μmol TEs/g protein)	10.9 ± 0.5 [a]	31.4 ± 1.3 [c]	21.3 ± 0.6 [b]
P-FRAP (μmol BHT eqv/g protein)	7.5 ± 0.3 [a]	26.9 ± 0.8 [c]	18.7 ± 0.6 [b]

Values are means ± SEM (n = 48). Mean values in each row with different letters are significantly different (one-way ANOVA and Duncan test, $p < 0.05$).

In Table 2, the values of P-FRAP show a significant increase ($p < 0.05$) of antioxidant activity after simulated gastrointestinal digestion (7.51 for SW to 26.9 μmol BHT eqv/g protein for SW-Ds). The results of the increased P-FRAP values after gastrointestinal digestion are in accordance with a previous study by Shaukat et al. on buffalo milk-processed cheddar cheese [57].

The overall observed increase in antioxidant capacity after digestion could be attributed to the release of peptides and free amino acids by the simulated gastrointestinal digestion [11]. Moreover, since SW results from a fermentation procedure, this could further enhance the release of antioxidant bioactive peptides [58].

Peptides released during an enzymatic hydrolysis process have considerable variability in size and structural characteristics [59], with peptides of relatively low molecular weight tending to display a relatively high antioxidant capacity [60,61]. Besides gastrointestinal digestion, such fragments can be also produced technologically using a broad range of exogenously supplied enzymes such as alcalase, chymotrypsin and flavourzyme [11,62]. From the data in Table 2, it is evident that the SW-D-P3 fraction accounts for the majority (65 to 85%) of the antioxidant capacity of the SW-Ds, regardless of the assay used. This indicates that the peptides with antioxidant activity are predominantly of lower molecular weight. Consistent with our results, Athira et al. reported that the antioxidant activity of WPC alcalase hydrolysed permeate (3 kDa) was augmented compared to WPC [63]. Similarly, another study indicated that the peptides between 0.1 to 2.8 kDa of WPI alcalase hydrolysate displayed the strongest radical scavenging activity [64]. Furthermore, O'Keeffe and Fitzgerald showed a greater antioxidant activity of WPC hydrolysate fractions with low molecular mass peptides (<5 kDa and <1 kDa) than WPC as measured by ORAC [65].

Also, a recent study of Ballatore et al. reported the highest antioxidant activity for the <3 kDa fraction obtained from trypsin-hydrolyzed WPC [19].

Effect of Milk Animal Origin

The antioxidant activities with the ORAC and ABTS assays for bovine, ovine, caprine and a mix of ovine and caprine of SW, SW-Ds and SW-D-P3 are shown in Figure 1, panels a and b, respectively. The ORAC values of antioxidant activity were highest for bovine SW (26.0 μmol TEs/g protein) compared with the other three groups ($p < 0.05$), whereas after in vitro digestion (SW-Ds and SW-D-P3), no significant differences were observed ($p > 0.05$). In contrast, the ABTS values of ovine SW were higher, compared to their bovine counterparts, both before (23.2 μmol TEs/g protein) and after in vitro digestion (49.0 μmol TEs/g protein for the SW-Ds). There were no statistically significant differences ($p > 0.05$) in the FRAP and P-FRAP values between the SW samples derived from cheeses made from the four different milk sources (Figure 1c,d, respectively).

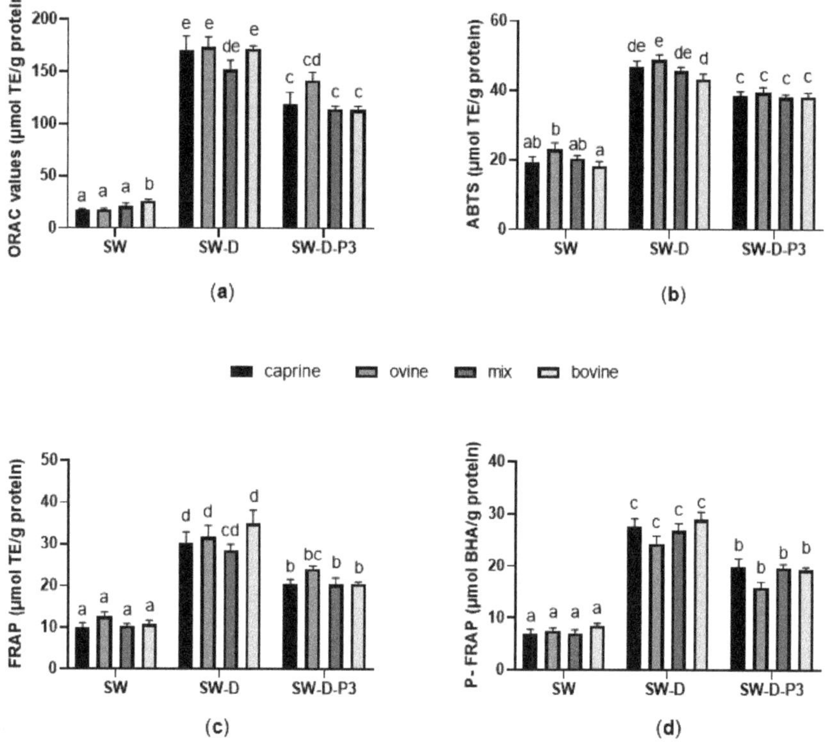

Figure 1. Antioxidant activity of sweet whey before (SW) and after in vitro digestion (digested SW (SW-Ds) and digested fraction below 3 kDa (SW-D-P3)) assessed with (**a**) oxygen radical absorbance capacity assay, expressed as μmol of TEs per gram of protein; (**b**) 2,2′-Azinobis (3- ethylbenzothiazoline-6-sulfonic acid), expressed as μmol of TE per gram of protein; (**c**) ferric reducing antioxidant power assay, expressed as μmol of TE per gram of protein; and (**d**) potassium ferricyanide reducing power, expressed as μmol of BHA per gram of protein. Results represent the mean of 3 experimental repetitions ± SEM ($n = 12$). Columns with different letters within the same panel are significantly different ($p < 0.05$).

3.2. Assessment of Cellular Antioxidant Activity of SW-D-P3

Food digestates are known to be cytotoxic to HT29 cells [66]. Thus, the HT29 cells were first exposed for 24 h to a range of concentrations of SW-D-P3 (0.75–6 mg protein/mL) to assess their cytotoxic effect. The MTT assay indicated that none of the concentration was cytotoxic. It was inferred that the concentration of the samples used in our study would not influence the activity of HT29 cells and thus would not affect the result of CAA. Therefore, the CAA assay was performed on the SW-D-P3 of a maximum concentration of 6 mg protein/mL. This concentration is in the range used in a previous study. In more detail, Kleekayai et al. evaluated the ROS generation in stressed HepG2 cells treated with hydrolysates of WPC at concentrations ranging from 0 to 10 mg/mL [67]. The highest concentration (10 mg/mL) exhibited the most potent cellular ROS generation reducing activity in the intracellular ROS generation in AAPH-stressed cells. Other studies reported a dose–response augmentation of CAA with the supplementation of whey proteins in different cell lines and are presented thoroughly in a review by Corrochano et al. [6].

The cell cytotoxicity of the oxidative stress inducer, H_2O_2, was also pre-evaluated at concentrations ranging from 0–2 mM in order to investigate their potential toxic effects on HT29 cells (Supplementary Data, Figure S1). A toxic effect yielding <80% cell viability was found at levels >1 mM. Due to the similar effects observed for H_2O_2 at concentrations below 1 mM, a concentration of 0.5 mM of H_2O_2 was selected to represent the oxidative stress inducer. Previous studies have reported the same range of H_2O_2 used as an oxidative stress inducer (0.25–0.7 mM) to evaluate the protective effect of food compounds [43,44,68,69]. Several studies have linked low-molecular size peptides with an improved antioxidant activity [64,70,71]. From our results in biochemical assays, it was concluded that the SW-D-P3 fraction was the one responsible for the majority of the antioxidant activity of digested SW. Therefore, in the present work, the HT29 epithelial cell line was used as a cellular model for the estimation of ROS% inhibition by SW-D-P3. The results presented in Figure 2 show that cell treatment with 6 mg protein/mL of SW-D-P3 significantly reduced radical formation in H_2O_2-treated cells, regardless of the SW milk origin, compared to the BL-D-P3 ($p < 0.05$). Furthermore, cell treatment with ovine SW-D-P3 significantly reduced the radical formation cells compared to the H_2O_2-treated group ($p < 0.05$), with a decrease of 14.97%, while the levels of radical formation for the cells treated with ovine and bovine SW-D-P3 did not differ significantly with those of the cells not treated with H_2O_2 ($p > 0.05$).

The results from previous studies regarding CAA determination in food and feed components after in vitro digestion using cell-based assays in various cell lines are scarce and are predominantly focused on non-dairy products. More specifically, CAA was found to be higher in digested whole grains compared to the free fraction in the HepG2 cell line [72]. Moreover, chickpea protein hydrolysate inhibited DCFH oxidation in a dose–response manner, and an increased CAA unit was noticed at higher hydrolysate protein concentrations [73]. A previous study by Zhang et al. [74] reported a decrease in ROS% generation in Caco-2 cells treated with soy protein hydrolysate compared to the H_2O_2 group. In another study [75], the digested and fractionated eggshell membrane (ESM) hydrolysate were evaluated in Caco-2 cells with AAPH used as a radical generator. In detail, the CAA of the digested ESM hydrolysate fraction with molecular weight of <5 kDa was higher compared with fractions of 5–10 kDa and >10 kDa. To the best of our knowledge, there are only two studies reporting CAA in eukaryotic cells in dairy-associated compounds after in vitro digestion. In the first one by Corrochano et al. [76], b-LG and a-LA, after in vitro digestion, effectively inhibited ROS and stimulated antioxidant enzymes in HT29 cells, while the same whey proteins were unable to reduce induced ROS formation in Caco-2 cells. Moreover, it should be noted that contrary to our results, they did not observe such an effect in WPI digestates. On the other hand, the second of these studies is in agreement with our results, as a better cellular antioxidant activity of digested WPI using the Caco-2 cell line was reported in response to H_2O_2 [77]. Interestingly, using a non-eukaryotic cell model, Ibrahim et al. reported a protective effect of camel milk protein hydrolysates (casein and whey hydrolysates separately) on yeast cells against H_2O_2-induced oxidative stress [78].

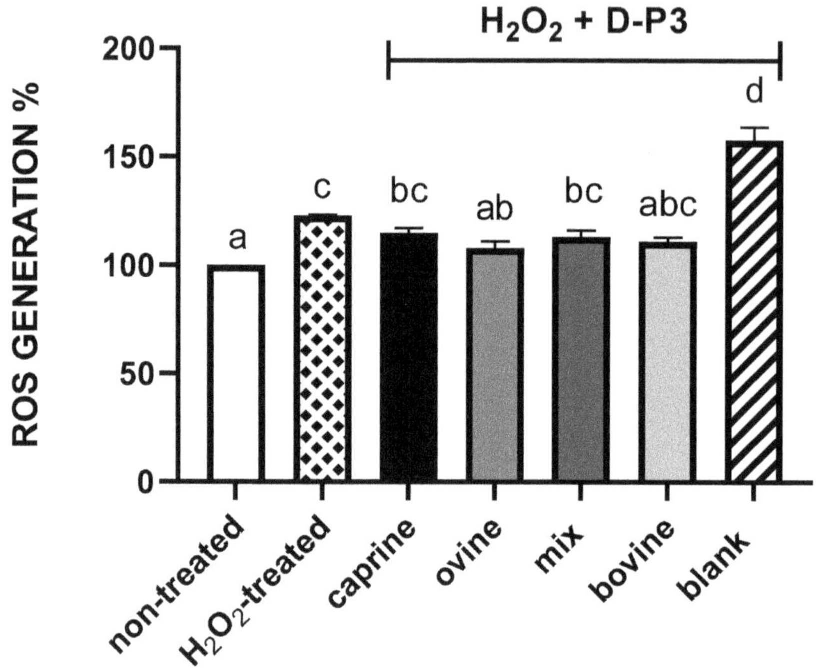

Figure 2. Reactive oxygen species (ROS)% generation in HT29 cells (seeded at 5×10^4 cells/well) after 24 h exposure to sweet whey after gastrointestinal digestion with peptides below 3 kDa (SW-D-P3) at a concentration of 6 mg protein/mL. Non-treated cells = cells without H_2O_2, H_2O_2-treated = cells treated with 0.5 mM of H_2O_2, caprine, ovine, bovine and a mix of caprine/ovine = SW-D-P3 and blank = BL-D-P3, all treated with 0.5 mM H_2O_2. Cell treatment was performed in triplicate on three different days, and the values reported are mean ± SEM. Columns with different letters are significantly different ($p < 0.05$).

3.3. Effect of SW-D-P3 on Expression of Antioxidant Genes

In the present study, the relative gene expression of NFE2L2, SOD1 and CAT, essential components of antioxidant signaling pathways, was measured in response to treatment with SW-D-P3 in LPS-challenged THP-1-derived macrophages. Additionally, the expression of *NFKB1* and *RELA* encoding the two main subunits of the NF-κB transcription factor was also quantified. The *CAT* expression was found to be higher in SW-D-P3-treated cells when compared to BL-D-P3, regardless of milk animal origin. This increase, however, was attenuated in the mixed and ovine samples when compared to the bovine and caprine ones ($p < 0.05$; Figure 3c). The *SOD1* expression in turn was higher ($p < 0.05$; Figure 3b) only in the mixed samples when compared with BL-D-P3 and with the ovine and bovine samples but not with the caprine ones. On the other hand, no statistically significant differences were observed between the samples or with the BL-D-P3 regarding the *NFE2L2* expression ($p > 0.05$; Figure 3a). Finally, the *NFKB1* and *RELA* expressions ($p < 0.05$; Figure 3d,e, respectively) were found to be higher in the bovine and mixed SW-D-P3 compared to BL-D-P3.

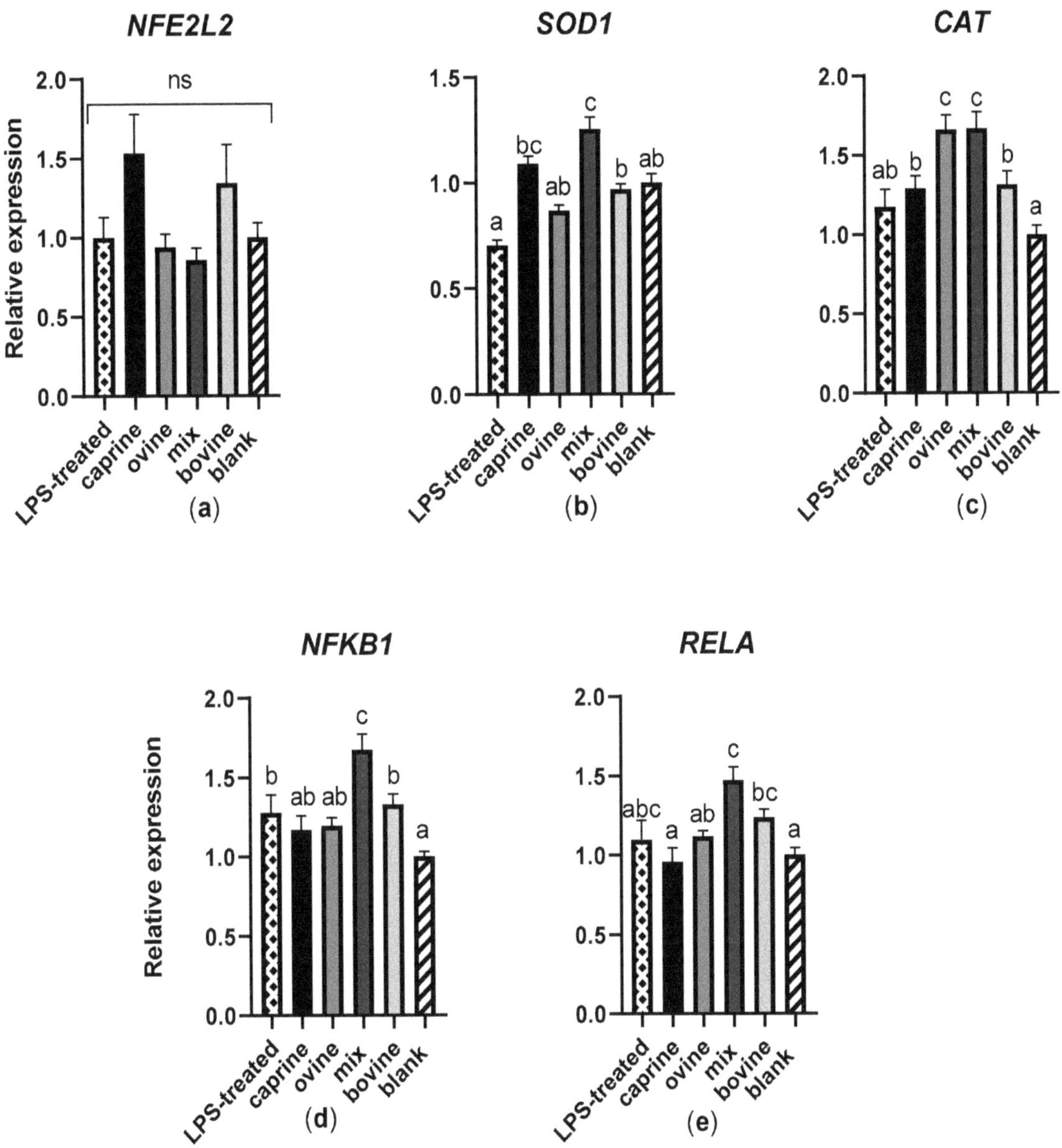

Figure 3. Effect of SW-D-P3 on LPS-induced mRNA expression in THP-1 cells. THP-1 cells were pretreated with PMA for 48 h (100 ng/mL), a 24 h rest and then were treated with LPS (100 ng/mL) with or without the presence of 3 mg protein/mL of SW-D-P3 (caprine, ovine, mix and bovine) or BL-D-P3 (blank) for 24 h. The expression levels of (**a**) *NFE2L2*, (**b**) *SOD1*, (**c**) *CAT*, (**d**) *NFKB1* and (**e**) *RELA* were measured using real-time PCR and were normalized to four housekeeping genes (*B2M*, *RPL37A*, *RPS18* and *HPRT1*). Data are represented as mean ± SEM of three independent determinations ($n = 3$). Columns with different letters within the same panel are significantly different ($p < 0.05$); ns = not significant ($p > 0.05$).

In line with our results, Xu et al. also reported that treatment with WPC increases the enzymatic activity of SOD1 and CAT, while, at the same time, no significant differences were found in the *NFE2L2* expression in the myoblast cell line C_2C_{12} following H_2O_2 oxidative stress [79]. Kerasioti et al. reported similar findings for the effect of WPC on SOD1 and CAT enzymatic activities and protein levels and NFE2L2 protein levels in the same cell line without the use of an oxidative agent, while using the same experimental parameter treatment with WPC resulted also in the increase of NFE2L2 protein levels in EA.hy926 endothelial cells [80]. In another study, HepG2 cells were treated with various concentrations of a glucose–WPC conjugate for 24 h followed by t-BHP oxidative stress to evaluate the role of NFE2L2 in the maintenance of the cellular redox status, reporting a dose-dependent increase of *NFE2L2* mRNA levels [81]. Furthermore, both the glutathione and catalase antioxidant systems are activated by WPC hydrolysate supplementation in human umbilical vein endothelial cells (HUVECs), resulting in an increase in cellular glutathione and CAT activity, albeit without the use of an oxidative stress factor [65]. Moreover, Corrochano et al. reported lower mRNA levels of *CAT* and *SOD1* in Caco-2 cells after a 4 h exposure to gastrointestinal digested samples of WPI when compared to the digestion control (equivalent to the BL-D-P3 of the present study), though, once more, no oxidative stress factor was used [76].

Based on the data produced by the present study, the induction of the expression of SOD1 and CAT by SW-D-P3 seems to be independent of the NFE2L2 pathway and is rather directed by NF-κB. The NF-κB pathway has been implicated in the regulation of many antioxidant and pro-oxidant targets [27]. More specifically, NF-κB is identified as a positive regulator of *SOD1* [82]. On the other hand, the role of NF-κB in regulation of the *CAT* expression is rather unclear as there are contradicting results supporting its role both as a negative [83] and as a positive [84] regulator of *CAT*. Our findings regarding the effect of SW-D-P3 in LPS-activated THP-1 support the latter and furthermore confirm the role of NF-κB in the induction of the *SOD1* expression, although further experimentation is needed to clarify the precise mechanism of their regulation. Corrochano et al. quantified mRNA transcripts of the antioxidant genes *SOD1* and *CAT* in intestinal cells that had been treated with 2.5 mg/mL of gastrointestinal bovine whey proteins (corresponding to the SW-Ds in our study) [76]. A recent study by Ishikawa evaluated some genes (e.g., *IL6* and *IL10*) implicated in inflammation and immunosuppression in THP-1 cells stressed with LPS and treated with 5 mg/mL of a whey protein hydrolysate [85].

Human or animal intervention trials with diets including whey products are the best way to assess their potential antioxidant benefit. However, only a small number of studies have evaluated the antioxidant effect of whey proteins/peptides using in vivo models. Ebaid et al. observed that dietary supplementation with whey proteins enhances the normal inflammatory responses during wound healing in diabetic mice by restoring the levels of oxidative stress [86]. Furthermore, Athira et al. reported the ameliorative potential of whey protein hydrolysates against paracetamol-induced oxidative stress in mice, compared with mice without whey protein administration. A significant increase in liver CAT and SOD levels and a reduction of the concentrations of oxidative biomarkers, such as alkaline phosphatase and creatinine, was observed [87].

In general, to consider the physiological benefits of sweet whey, it is important to know that whey proteins do not reach the intestine in their intact form. Sousa et al. compared the total protein digestibility of WPI between in vitro and in vivo situations which resulted in a good correlation, with a tendency toward an overestimation for the in vitro approach [88]. The antioxidant capacity in commercial whey products, which are commonly used as food ingredients especially in the sports nutrition sector, is well known [6,89]. To the best of our knowledge, this is the first report in which SW from small-scale cheese plants is used as a raw material for evaluating antioxidant capacity before and after in vitro digestion. The protection produced by SW against in vitro-induced oxidative stress (biochemical and cellular assays) with different methodologies reveals that SW could be used in animal/human nutrition. As with all in vitro and cell culture experiments, there remains a possibility that these results may not translate to in vivo situations. Also, to address the limitations concerning the suitable concentration related to the clinical efficacy of SW, its physiological efficacy is necessary to be investigated in vivo, either in animal nutrition and/or human clinical trials.

4. Conclusions

In this research, SW of different milk origins was digested using a standardized static in vitro digestion method for mimicking monogastric gastrointestinal digestion. The samples were evaluated by multiple assays to test antioxidant activity, since a single assay is not sufficient to test all relevant factors affecting antioxidant capacity. All antioxidant properties of the SW-Ds were significantly higher than those of the intact SW. Of special interest was the evaluation of the antioxidant activity of SW-D-P3 directly in mammalian cell lines. Intestinal cell lines are the models proposed to better assess the in vitro antioxidant potential of a dietary compound. However, there is only a relatively small number of studies that employ cell culture models to assess the antioxidant potential of food and feed components. In conclusion, the present study showed that SW-D-P3 exerts an antioxidant effect both in epithelial and derived activated macrophage cell lines. Furthermore, a slightly better antioxidant capacity seems to be associated with bovine and ovine when compared to caprine SW. A validation of the observed differences could be achieved by in vivo animal studies or human clinical trials. Further investigation is also deemed necessary in order to determine the peptides' sequences from SW with potential antioxidant activity in vivo.

Supplementary Materials: The following supporting information can be downloaded at: https://www.mdpi.com/article/10.3390/antiox12091676/s1, Figure S1: Effect of different concentrations of oxidative stress inducer, hydrogen peroxide (H_2O_2), on HT29 cell viability.

Author Contributions: Conceptualization, G.T. and E.D.; methodology and investigation, E.D.; writing—original draft preparation, E.D.; writing—review and editing, G.T.; supervision, G.T.; project administration, G.T.; funding acquisition, I.P., G.T. and E.D. All authors have read and agreed to the published version of the manuscript.

Funding: This research was co-financed by Greece and the European Union (European Social Fund-ESF) through the Operational Programme "Human Resources Development, Education and Lifelong Learning" in the context of the project "Strengthening Human Resources Research Potential via Doctorate Research" (MIS 5000432), implemented by the State Scholarships Foundation (IKY). The results of this study are part of a research project (MIS 5033108) funded by the Operational Program of the Region of Epirus co-financed by Greece and the European Union—European Regional Development Fund (ERDF).

Institutional Review Board Statement: Ethical approval was not required for this work, as the study did not involve human or animal participants or samples.

Informed Consent Statement: Not applicable.

Data Availability Statement: The data presented in this study are available on request from the corresponding author.

Acknowledgments: The THP-1 cell line was kindly provided by Kletsas Dimitrios from the National Centre of Scientific Research Demokritos (Athens, Greece). We would like to thank Sofia Mavrikou for kindly providing us with the HT29 cell line and giving us access to the Tecan Infinite M200 Pro plate reader (Laboratory of Cell Technology, Agricultural University of Athens). We would also like to thank Maria Kapsokefalou for giving us access to the instruments (Epoch 2 microplate spectrophotometer and VICTOR 2030 multilabel counter) of the Laboratory of Food Chemistry and Analysis (Agricultural University of Athens). Finally, we would like to thank all the small cheese plants for providing us with sweet whey samples.

Conflicts of Interest: The authors declare no conflict of interest. The funders had no role in the design of the study; in the collection, analyses, or interpretation of data; in the writing of the manuscript; or in the decision to publish the results.

References

1. Bintsis, T.; Papademas, P. Sustainable Approaches in Whey Cheese Production: A Review. *Dairy* **2023**, *4*, 249–270.
2. Barba, F.J. An integrated approach for the valorization of cheese whey. *Foods* **2021**, *10*, 564. [CrossRef] [PubMed]
3. Aspri, M.; Leni, G.; Galaverna, G.; Papademas, P. Bioactive properties of fermented donkey milk, before and after in vitro simulated gastrointestinal digestion. *Food Chem.* **2018**, *268*, 476–484. [CrossRef] [PubMed]
4. Akan, E. An evaluation of the in vitro antioxidant and antidiabetic potentials of camel and donkey milk peptides released from casein and whey proteins. *J. Food Sci. Technol.* **2021**, *58*, 3743–3751. [CrossRef]
5. Kleekayai, T.; O'neill, A.; Clarke, S.; Holmes, N.; O'sullivan, B.; Fitzgerald, R.J. Contribution of Hydrolysis and Drying Conditions to Whey Protein Hydrolysate Characteristics and In Vitro Antioxidative Properties. *Antioxidants* **2022**, *11*, 399. [CrossRef]
6. Corrochano, A.R.; Buckin, V.; Kelly, P.M.; Giblin, L. Invited review: Whey proteins as antioxidants and promoters of cellular antioxidant pathways. *J. Dairy Sci.* **2018**, *101*, 4747–4761. [CrossRef] [PubMed]
7. Tong, L.M.; Sasaki, S.; McClements, D.J.; Decker, E.A. Antioxidant activity of whey in a salmon oil emulsion. *J. Food Sci.* **2000**, *65*, 1325–1329. [CrossRef]
8. Tong, L.M.; Sasaki, S.; McClements, D.J.; Decker, E.A. Mechanisms of the antioxidant activity of a high molecular weight fraction of whey. *J. Agric. Food Chem.* **2000**, *48*, 1473–1478. [CrossRef]
9. Carrasco-Castilla, J.; Hernández-Álvarez, A.J.; Jiménez-Martínez, C.; Jacinto-Hernández, C.; Alaiz, M.; Girón-Calle, J.; Vioque, J.; Dávila-Ortiz, G. Antioxidant and metal chelating activities of peptide fractions from phaseolin and bean protein hydrolysates. *Food Chem.* **2012**, *135*, 1789–1795. [CrossRef]
10. Hayes, M.; Ross, R.P.; Fitzgerald, G.F.; Stanton, C. Putting microbes to work: Diary fermentation, cell factories and bioactive peptides. Part I: Overview. *Biotechnol. J.* **2007**, *2*, 426–434. [CrossRef]
11. Bustamante, S.Z.; González, J.G.; Sforza, S.; Tedeschi, T. Bioactivity and peptide profile of whey protein hydrolysates obtained from Colombian double-cream cheese production and their products after gastrointestinal digestion. *LWT* **2021**, *145*, 111334. [CrossRef]
12. Guo, Y.; Jiang, X.; Xiong, B.; Zhang, T.; Zeng, X.; Wu, Z.; Sun, Y.; Pan, D. Production and transepithelial transportation of angiotensin-I-converting enzyme (ACE)-inhibitory peptides from whey protein hydrolyzed by immobilized Lactobacillus helveticus proteinase. *J. Dairy Sci.* **2019**, *102*, 961–975. [CrossRef]
13. Wolfe, K.L.; Rui, H.L. Cellular antioxidant activity (CAA) assay for assessing antioxidants, foods, and dietary supplements. *J. Agric. Food Chem.* **2007**, *55*, 8896–8907. [CrossRef]
14. Faller, A.L.K.; Fialho, E.; Liu, R.H. Cellular antioxidant activity of Feijoada whole meal coupled with an in vitro digestion. *J. Agric. Food Chem.* **2012**, *60*, 4826–4832. [CrossRef]
15. Bothon, F.T.D.; Debiton, E.; Avlessi, F.; Forestier, C.; Teulade, J.C.; Sohounhloue, D.K.C. In vitro biological effects of two anti-diabetic medicinal plants used in Benin as folk medicine. *BMC Complement. Altern. Med.* **2013**, *13*, 51. [CrossRef]
16. Girard-Lalancette, K.; Pichette, A.; Legault, J. Sensitive cell-based assay using DCFH oxidation for the determination of pro- and antioxidant properties of compounds and mixtures: Analysis of fruit and vegetable juices. *Food Chem.* **2009**, *115*, 720–726. [CrossRef]
17. Wan, H.; Liu, D.; Yu, X.; Sun, H.; Li, Y. A Caco-2 cell-based quantitative antioxidant activity assay for antioxidants. *Food Chem.* **2015**, *175*, 601–608. [CrossRef]
18. Kellett, M.E.; Greenspan, P.; Pegg, R.B. Modification of the cellular antioxidant activity (CAA) assay to study phenolic antioxidants in a Caco-2 cell line. *Food Chem.* **2018**, *244*, 359–363. [CrossRef] [PubMed]
19. Ballatore, M.B.; Bettiol, M.D.R.; Vanden Braber, N.L.; Aminahuel, C.A.; Rossi, Y.E.; Petroselli, G.; Erra-Balsells, R.; Cavaglieri, L.R.; Montenegro, M.A. Antioxidant and cytoprotective effect of peptides produced by hydrolysis of whey protein concentrate with trypsin. *Food Chem.* **2020**, *319*, 126472. [CrossRef]
20. Nam, M.H.; Son, W.R.; Yang, S.Y.; Lee, Y.S.; Lee, K.W. Chebulic acid inhibits advanced glycation end products-mediated vascular dysfunction by suppressing ROS via the ERK/Nrf2 pathway. *J. Funct. Foods* **2017**, *36*, 150–161. [CrossRef]
21. Clementi, M.E.; Lazzarino, G.; Sampaolese, B.; Brancato, A.; Tringali, G. DHA protects PC12 cells against oxidative stress and apoptotic signals through the activation of the NFE2L2/HO-1 axis. *Int. J. Mol. Med.* **2019**, *43*, 2523–2531. [CrossRef]
22. Zhao, D.; Jiang, Y.; Sun, J.; Li, H.; Huang, M.; Sun, X.; Zhao, M. Elucidation of The Anti-Inflammatory Effect of Vanillin In Lps-Activated THP-1 Cells. *Nutrition* **2019**, *84*, 1920–1928. [CrossRef]
23. Saw, C.L.; Wu, Q.; Kong, A.N.T. Anti-cancer and potential chemopreventive actions of ginseng by activating Nrf2 (NFE2L2) anti-oxidative stress/anti-inflammatory pathways. *Chin. Med.* **2010**, *5*, 37. [CrossRef] [PubMed]
24. Dias, K.A.; da Conceição, A.R.; Pereira, S.M.S.; Oliveira, L.A.; da Silva Rodrigues, J.V.; Dias, R.S.; de Paula, S.O.; Natali, A.J.; da Matta, S.L.P.; Gonçalves, R.V.; et al. Curcumin-Added Whey Protein Positively Modulates Skeletal Muscle Inflammation and Oxidative Damage after Exhaustive Exercise. *Nutrients* **2022**, *14*, 4905. [CrossRef]
25. Jung, K.A.; Kwak, M.K. The Nrf2 system as a potential target for the development of indirect antioxidants. *Molecules* **2010**, *15*, 7266–7291. [CrossRef] [PubMed]
26. Treml, J.; Večeřová, P.; Herczogová, P.; Šmejkal, K. Direct and indirect antioxidant effects of selected plant phenolics in cell-based assays. *Molecules* **2021**, *26*, 2534. [CrossRef]
27. Morgan, M.J.; Liu, Z.-G. Crosstalk of reactive oxygen species and NF-κB signaling NF-κB. *Nat. Publ. Gr.* **2011**, *21*, 103–115. [CrossRef]

28. Brodkorb, A.; Egger, L.; Alminger, M.; Alvito, P.; Assunção, R.; Ballance, S.; Bohn, T.; Bourlieu-Lacanal, C.; Boutrou, R.; Carrière, F.; et al. INFOGEST static in vitro simulation of gastrointestinal food digestion. *Nat. Protoc.* **2019**, *14*, 991–1014. [CrossRef]
29. Minekus, M.; Alminger, M.; Alvito, P.; Ballance, S.; Bohn, T.; Bourlieu, C.; Carrière, F.; Boutrou, R.; Corredig, M.; Dupont, D.; et al. A standardised static in vitro digestion method suitable for food-an international consensus. *Food Funct.* **2014**, *5*, 1113–1124. [CrossRef]
30. Ou, B.; Hampsch-Woodill, M.; Prior, R.L. Development and validation of an improved oxygen radical absorbance capacity assay using fluorescein as the fluorescent probe. *J. Agric. Food Chem.* **2001**, *49*, 4619–4626. [CrossRef]
31. Zulueta, A.; Maurizi, A.; Frígola, A.; Esteve, M.J.; Coli, R.; Burini, G. Antioxidant capacity of cow milk, whey and deproteinized milk. *Int. Dairy J.* **2009**, *19*, 380–385. [CrossRef]
32. Zulueta, A.; Esteve, M.J.; Frígola, A. ORAC and TEAC assays comparison to measure the antioxidant capacity of food products. *Food Chem.* **2009**, *114*, 310–316. [CrossRef]
33. Ozgen, M.; Reese, R.N.; Tulio, A.Z.; Scheerens, J.C.; Miller, A.R. Modified 2,2-azino-bis-3-ethylbenzothiazoline-6-sulfonic acid (ABTS) method to measure antioxidant capacity of selected small fruits and comparison to ferric reducing antioxidant power (FRAP) and 2,2′-diphenyl-1- picrylhydrazyl (DPPH) methods. *J. Agric. Food Chem.* **2006**, *54*, 1151–1157. [CrossRef] [PubMed]
34. Feng, Y.; Yuan, D.; Cao, C.; Kong, B.; Sun, F.; Xia, X.; Liu, Q. Changes of in vitro digestion rate and antioxidant activity of digestion products of ethanol-modified whey protein isolates. *Food Hydrocoll.* **2022**, *131*, 107756. [CrossRef]
35. Benzie, I.F.F.; Strain, J.J. The Ferric Reducing Ability of Plasma (FRAP) as a Measure of "Antioxidant Power": The FRAP Assay. *Anal. Biochem.* **1996**, *239*, 70–76. [CrossRef]
36. González-Centeno, M.R.; Jourdes, M.; Femenia, A.; Simal, S.; Rosselló, C.; Teissedre, P.L. Proanthocyanidin composition and antioxidant potential of the stem winemaking byproducts from 10 different grape varieties (*Vitis vinifera* L.). *J. Agric. Food Chem.* **2012**, *60*, 11850–11858.
37. Corrochano, A.R.; Sariçay, Y.; Arranz, E.; Kelly, P.M.; Buckin, V.; Giblin, L. Comparison of antioxidant activities of bovine whey proteins before and after simulated gastrointestinal digestion. *J. Dairy Sci.* **2019**, *102*, 54–67. [CrossRef]
38. Oyaizu, M. Studies on Products of Browning Reaction Antioxidative Activities of Products of Browning Reaction Prepared from Glucosamine. *Jpn. J. Nutr. Diet.* **1986**, *44*, 307–315. [CrossRef]
39. Liang, T.; Yue, W.; Li, Q. Comparison of the phenolic content and antioxidant activities of apocynum venetum l. (Luo-Bu-Ma) and two of its alternative species. *Int. J. Mol. Sci.* **2010**, *11*, 4452–4464. [CrossRef]
40. Sánchez-Velázquez, O.A.; Mulero, M.; Cuevas-Rodríguez, E.O.; Mondor, M.; Arcand, Y.; Hernández-Álvarez, A.J. In vitro gastrointestinal digestion impact on stability, bioaccessibility and antioxidant activity of polyphenols from wild and commercial blackberries (*Rubus* spp.). *Food Funct.* **2021**, *12*, 7358–7378. [CrossRef] [PubMed]
41. Armstrong, D. *Advanced Protocols in Oxidative Stress III*; Elsevier: Amsterdam, The Netherlands, 2014; pp. 1–477. [CrossRef]
42. Lind, K.F.; Hansen, E.; Østerud, B.; Eilertsen, K.E.; Bayer, A.; Engqvist, M.; Leszczak, K.; Jørgensen, T.; Andersen, J.H. Antioxidant and Anti-Inflammatory Activities of Barettin. *Mar. Drugs* **2013**, *11*, 2655–2666. [CrossRef] [PubMed]
43. Piccolomini, A.F.; Iskandar, M.M.; Lands, L.C.; Kubow, S. High hydrostatic pressure pre-treatment of whey proteins enhances whey protein hydrolysate inhibition of oxidative stress and IL-8 secretion in intestinal epithelial cells. *Food Nutr. Res.* **2012**, *56*, 17549. [CrossRef]
44. Feng, L.; Peng, F.; Wang, X.; Li, M.; Lei, H.; Xu, H. Identification and characterization of antioxidative peptides derived from simulated in vitro gastrointestinal digestion of walnut meal proteins. *Food Res. Int.* **2019**, *116*, 518–526. [CrossRef] [PubMed]
45. Chanput, W.; Mes, J.; Vreeburg, R.A.M.M.; Savelkoul, F.J.; Wichers, H.J.; Savelkoul, H.F.J.; Wichers, H.J. Transcription profiles of LPS-stimulated THP-1 monocytes and macrophages: A tool to study inflammation modulating effects of food-derived compounds. *Food Funct.* **2010**, *1*, 254–261. [CrossRef] [PubMed]
46. Chanput, W.; Mes, J.J.; Wichers, H.J. International Immunopharmacology THP-1 cell line: An in vitro cell model for immune modulation approach. *Int. Immunopharmacol.* **2014**, *23*, 37–45. [CrossRef]
47. Baxter, E.W.; Graham, A.E.; Re, N.A.; Carr, I.M.; Robinson, J.I.; Mackie, S.L.; Morgan, A.W. Standardized protocols for differentiation of THP-1 cells to macrophages with distinct M(IFNγ+LPS), M(IL-4) and M(IL-10) phenotypes. *J. Immunol. Methods* **2019**, *478*, 112721. [CrossRef]
48. Daigneault, M.; Preston, J.A.; Marriott, H.M.; Whyte, M.K.B.; Dockrell, D.H. The identification of markers of macrophage differentiation in PMA-stimulated THP-1 cells and monocyte-derived macrophages. *PLoS ONE* **2010**, *5*, e8668. [CrossRef]
49. Templeton, G.F.; Templeton, G.F. A two-step approach for transforming continuous variables to normal: Implications and recommendations for IS research. *Commun. Assoc. Inf. Syst.* **2011**, *28*, 41–58. [CrossRef]
50. Cho, S.J. Changes in the antioxidant properties of rice bran protein isolate upon simulated gastrointestinal digestion. *LWT* **2020**, *126*, 109206. [CrossRef]
51. Huang, D.; Prior, R.L. The Chemistry behind Antioxidant Capacity Assays. *Agric. Food Chem.* **2005**, *53*, 841–1856. [CrossRef]
52. Di Meo, F.; Lemaur, V.; Cornil, J.; Lazzaroni, R.; Duroux, J.L.; Olivier, Y.; Trouillas, P. Free radical scavenging by natural polyphenols: Atom versus electron transfer. *J. Phys. Chem. A* **2013**, *117*, 2082–2092. [CrossRef]
53. Zou, T.B.; He, T.P.; Li, H.B.; Tang, H.W.; Xia, E.Q. The structure-activity relationship of the antioxidant peptides from natural proteins. *Molecules* **2016**, *21*, 72. [CrossRef]

54. Giromini, C.; Lovegrove, J.A.; Givens, D.I.; Rebucci, R.; Pinotti, L.; Maffioli, E.; Tedeschi, G.; Sundaram, T.S.; Baldi, A. In vitro-digested milk proteins: Evaluation of angiotensin-1-converting enzyme inhibitory and antioxidant activities, peptidomic profile, and mucin gene expression in HT29-MTX cells. *J. Dairy Sci.* **2019**, *102*, 10760–10771. [CrossRef] [PubMed]
55. García-Casas, V.E.; Seiquer, I.; Pardo, Z.; Haro, A.; Recio, I.; Olías, R. Antioxidant Potential of the Sweet Whey-Based Beverage Colada after the Digestive Process and Relationships with the Lipid and Protein Fractions. *Antioxidants* **2022**, *11*, 1827. [CrossRef]
56. Clausen, M.R.; Skibsted, L.H.; Stagsted, J. Characterization of major radical scavenger species in bovine milk through size exclusion chromatography and functional assays. *J. Agric. Food Chem.* **2009**, *57*, 2912–2919. [CrossRef] [PubMed]
57. Shaukat, A.; Nadeem, M.; Qureshi, T.M.; Kanwal, R.; Sultan, M.; Kashongwe, O.B.; Shamshiri, R.R.; Murtaza, M.A. Effect of in vitro digestion on the antioxidant and angiotensin-converting enzyme inhibitory potential of Buffalo milk processed cheddar cheese. *Foods* **2021**, *10*, 1661. [CrossRef]
58. Stobiecka, M.; Król, J.; Brodziak, A. Antioxidant Activity of Milk and Dairy Products. *Animals* **2022**, *12*, 245. [CrossRef]
59. Vogelsang-O'dwyer, M.; Sahin, A.W.; Bot, F.; O'mahony, J.A.; Bez, J.; Arendt, E.K.; Zannini, E. Enzymatic hydrolysis of lentil protein concentrate for modification of physicochemical and techno-functional properties. *Eur. Food Res. Technol.* **2023**, *249*, 573–586. [CrossRef]
60. Salami, M.; Moosavi-Movahedi, A.A.; Ehsani, M.R.; Yousefi, R.; Haertlé, T.; Chobert, J.M.; Razavi, S.H.; Henrich, R.; Balalaie, S.; Ebadi, S.A.; et al. Improvement of the antimicrobial and antioxidant activities of camel and bovine whey proteins by limited proteolysis. *J. Agric. Food Chem.* **2010**, *58*, 3297–3302. [CrossRef] [PubMed]
61. Peña-Ramos, E.A.; Xiong, Y.L.; Arteaga, G.E. Fractionation and characterisation for antioxidant activity of hydrolysed whey protein. *J. Sci. Food Agric.* **2004**, *84*, 1908–1918. [CrossRef]
62. Tavano, O.L. Protein hydrolysis using proteases: An important tool for food biotechnology. *J. Mol. Catal. B Enzym.* **2013**, *90*, 1–11. [CrossRef]
63. Athira, S.; Mann, B.; Saini, P.; Sharma, R.; Kumar, R.; Singh, A.K. Production and characterisation of whey protein hydrolysate having antioxidant activity from cheese whey. *J. Sci. Food Agric.* **2015**, *95*, 2908–2915. [CrossRef] [PubMed]
64. Peng, X.; Kong, B.; Xia, X.; Liu, Q. Reducing and radical-scavenging activities of whey protein hydrolysates prepared with Alcalase. *Int. Dairy J.* **2010**, *20*, 360–365. [CrossRef]
65. O'Keeffe, M.B.; FitzGerald, R.J. Antioxidant effects of enzymatic hydrolysates of whey protein concentrate on cultured human endothelial cells. *Int. Dairy J.* **2014**, *36*, 128–135. [CrossRef]
66. Le Maux, S.; Bouhallab, S.; Giblin, L.; Brodkorb, A.; Croguennec, T. Complexes between linoleate and native or aggregated β-lactoglobulin: Interaction parameters and in vitro cytotoxic effect. *Food Chem.* **2013**, *141*, 2305–2313. [CrossRef]
67. Kleekayai, T.; Le Gouic, A.V.; Deracinois, B.; Cudennec, B.; FitzGerald, R.J. In vitro characterisation of the antioxidative properties of whey protein hydrolysates generated under pH- And non pH-controlled conditions. *Foods* **2020**, *9*, 582. [CrossRef] [PubMed]
68. García Fillería, S.; Tironi, V. Intracellular antioxidant activity and intestinal absorption of amaranth peptides released using simulated gastrointestinal digestion with Caco-2 TC7 cells. *Food Biosci.* **2021**, *41*, 101086. [CrossRef]
69. García-Nebot, M.J.; Recio, I.; Hernández-Ledesma, B. Antioxidant activity and protective effects of peptide lunasin against oxidative stress in intestinal Caco-2 cells. *Food Chem. Toxicol.* **2014**, *65*, 155–161. [CrossRef]
70. Chen, M.; Li, B. The effect of molecular weights on the survivability of casein-derived antioxidant peptides after the simulated gastrointestinal digestion. Innov. *Food Sci. Emerg. Technol.* **2012**, *16*, 341–348. [CrossRef]
71. Kong, B.; Peng, X.; Xiong, Y.L.; Zhao, X. Protection of lung fibroblast MRC-5 cells against hydrogen peroxide-induced oxidative damage by 0.1–2.8 kDa antioxidative peptides isolated from whey protein hydrolysate. *Food Chem.* **2012**, *135*, 540–547. [CrossRef]
72. Gong, E.S.; Gao, N.; Li, T.; Chen, H.; Wang, Y.; Si, X.; Tian, J.; Shu, C.; Luo, S.; Zhang, J.; et al. Effect of in Vitro Digestion on Phytochemical Profiles and Cellular Antioxidant Activity of Whole Grains. *J. Agric. Food Chem.* **2019**, *67*, 7016–7024. [CrossRef]
73. Torres-Fuentes, C.; Contreras, M.D.M.; Recio, I.; Alaiz, M.; Vioque, J. Identification and characterization of antioxidant peptides from chickpea protein hydrolysates. *Food Chem.* **2015**, *180*, 194–202. [CrossRef] [PubMed]
74. Zhang, Q.; Tong, X.; Qi, B.; Wang, Z.; Li, Y.; Sui, X.; Jiang, L. Changes in antioxidant activity of Alcalase-hydrolyzed soybean hydrolysate under simulated gastrointestinal digestion and transepithelial transport. *J. Funct. Foods* **2018**, *42*, 298–305. [CrossRef]
75. Shi, Y.; Kovacs-Nolan, J.; Jiang, B.; Tsao, R.; Mine, Y. Antioxidant activity of enzymatic hydrolysates from eggshell membrane proteins and its protective capacity in human intestinal epithelial Caco-2 cells. *J. Funct. Foods* **2014**, *10*, 35–45. [CrossRef]
76. Corrochano, A.R.; Arranz, E.; De Noni, I.; Stuknytė, M.; Ferraretto, A.; Kelly, P.M.; Buckin, V.; Giblin, L. Intestinal health benefits of bovine whey proteins after simulated gastrointestinal digestion. *J. Funct. Foods* **2018**, *49*, 526–535. [CrossRef]
77. Ma, Y.; Xu, J.; Guo, R.; Teng, G.; Chen, Y.; Xu, X. In vitro gastrointestinal model for the elderly: Effect of high hydrostatic pressure on protein structures and antioxidant activities of whey protein isolate. *Food Biosci.* **2023**, *52*, 102452. [CrossRef]
78. Ibrahim, H.R.; Isono, H.; Miyata, T. Potential antioxidant bioactive peptides from camel milk proteins. *Anim. Nutr.* **2018**, *4*, 273–280. [CrossRef] [PubMed]
79. Xu, R.; Liu, N.; Xu, X.; Kong, B. Antioxidative effects of whey protein on peroxide-induced cytotoxicity. *J. Dairy Sci.* **2011**, *94*, 3739–3746. [CrossRef]
80. Kerasioti, E.; Stagos, D.; Tzimi, A.; Kouretas, D. Increase in antioxidant activity by sheep/goat whey protein through nuclear factor-like 2 (Nrf2) is cell type dependent. *Food Chem. Toxicol.* **2016**, *97*, 47–56. [CrossRef] [PubMed]
81. Pyo, M.C.; Yang, S.Y.; Chun, S.H.; Oh, N.S.; Lee, K.W. Protective Effects of Maillard Reaction Products of Whey Protein Concentrate against Oxidative Stress through an Nrf2-Dependent Pathway in HepG2 Cells. *Biol. Pharm. Bull.* **2016**, *39*, 1437–1447. [CrossRef]

82. Rojo, A.I.; Salinas, M.; Martín, D.; Perona, R.; Cuadrado, A. Regulation of Cu/Zn-Superoxide Dismutase Expression via the Phosphatidylinositol 3 Kinase/Akt Pathway and Nuclear Factor-κB. *J. Neurosci.* **2004**, *24*, 7324–7334. [CrossRef] [PubMed]
83. Schreiber, J.; Jenner, R.G.; Murray, H.L.; Gerber, G.K.; Gifford, D.K.; Young, R.A. Coordinated binding of NF-kappaB family members in the response of human cells to lipopolysaccharide. *Proc. Natl. Acad. Sci. USA* **2006**, *103*, 5899–5904. [CrossRef]
84. Zhou, L.Z.H.; Johnson, A.P.; Rando, T.A. NF kappa B and AP-1 mediate transcriptional responses to oxidative stress in skeletal muscle cells. *Free Radic. Biol. Med.* **2001**, *31*, 1405–1416. [CrossRef] [PubMed]
85. Ishikawa, F.; Matsubara, T.; Koyama, T.; Iwamoto, H.; Miyaji, K. Whey protein hydrolysate mitigates both inflammation and endotoxin tolerance in THP-1 human monocytic leukemia cells. *Immunity Inflamm. Dis.* **2022**, *10*, e737. [CrossRef] [PubMed]
86. Ebaid, H.; Salem, A.; Sayed, A.; Metwalli, A. Whey protein enhances normal inflammatory responses during cutaneous wound healing in diabetic rats. *Lipids Health Dis.* **2011**, *10*, 235. [CrossRef] [PubMed]
87. Athira, S.; Mann, B.; Sharma, R.; Kumar, R. Ameliorative potential of whey protein hydrolysate against paracetamol-induced oxidative stress. *J. Dairy Sci.* **2013**, *96*, 1431–1437. [CrossRef]
88. Sousa, R.; Recio, I.; Heimo, D.; Dubois, S.; Moughan, P.J.; Hodgkinson, S.M.; Portmann, R.; Egger, L. In vitro digestibility of dietary proteins and in vitro DIAAS analytical workflow based on the INFOGEST static protocol and its validation with in vivo data. *Food Chem.* **2023**, *404*, 134720. [CrossRef] [PubMed]
89. Manninen, A.H. Protein hydrolysates in sports nutrition. *Nutr. Metab.* **2009**, *6*, 38. [CrossRef]

antioxidants

MDPI

Article

Effect of Dietary Supplementation with a Mixture of Natural Antioxidants on Milk Yield, Composition, Oxidation Stability and Udder Health in Dairy Ewes

Agori Karageorgou, Maria Tsafou, Michael Goliomytis, Ariadni Hager-Theodorides, Katerina Politi and Panagiotis Simitzis *

Laboratory of Animal Breeding and Husbandry, Department of Animal Science, Agricultural University of Athens, Iera Odos 75, 11855 Athens, Greece; akarageorgou@aua.gr (A.K.); mgolio@aua.gr (M.G.); a.hager@aua.gr (A.H.-T.); katerinapoliti@aua.gr (K.P.)

* Correspondence: pansimitzis@aua.gr; Tel.: +30-2105-294-427

Abstract: Due to the limitations in the use of antibiotic agents, researchers are constantly seeking natural bioactive compounds that could benefit udder health status but also milk quality characteristics in dairy animals. The aim of the current study was therefore to examine the effects of a standardized mixture of plant bioactive components (MPBC) originated from thyme, anise and olive on milk yield, composition, oxidative stability and somatic cell count in dairy ewes. Thirty-six ewes approximately 75 days after parturition were randomly allocated into three experimental treatments, which were provided with three diets: control (C); without the addition of the mixture, B1; supplemented with MPBC at 0.05% and B2; supplemented with rumen protected MPBC at 0.025%. The duration of the experiment was 11 weeks, and milk production was weekly recorded, while individual milk samples for the determination of composition, oxidative stability, somatic cell count (SCC), pH and electric conductivity were collected. Every two weeks, macrophage, lymphocyte, and polymorphonuclear leukocyte counts were also determined in individual milk samples. It was observed that milk yield was the greatest in the B2 group, with significant differences within the seventh and ninth week ($p < 0.05$), whereas no significant differences were found for milk composition, with the exception of the seventh week, when protein, lactose and non-fat solid levels were lower in MPBC groups ($p < 0.05$). Oxidative stability was improved in the groups that received the MPBC, with significant differences at the third, seventh, tenth and eleventh week ($p < 0.05$). SCC was also significantly lower at the second, eighth and ninth week in B2 compared to the other groups ($p < 0.05$), while no significant effects on the macrophage, lymphocyte, and polymorphonuclear leukocyte counts were observed. In conclusion, the MPBC addition had a positive effect on sheep milk yield, oxidative stability and somatic cell count, without any negative effect on its composition.

Keywords: antioxidants; lymphocyte; macrophage; milk yield; plant bioactive compounds; polyphenols; oxidative stability; polymorphonuclear leukocytes; somatic cell count

check for updates

Citation: Karageorgou, A.; Tsafou, M.; Goliomytis, M.; Hager-Theodorides, A.; Politi, K.; Simitzis, P. Effect of Dietary Supplementation with a Mixture of Natural Antioxidants on Milk Yield, Composition, Oxidation Stability and Udder Health in Dairy Ewes. *Antioxidants* **2023**, *12*, 1571. https://doi.org/10.3390/antiox12081571

Academic Editors: Luciana Rossi and Matteo Dell'Anno

Received: 30 June 2023
Revised: 2 August 2023
Accepted: 4 August 2023
Published: 6 August 2023

1. Introduction

There is an increased public awareness of potential health hazards induced by the excessive use of in-feed antibiotics that is reflected in the legislation of many countries, such as the EU [1] and the efforts of animal scientists to find alternative safe natural feed additives [2,3]. Consumer concerns refer to toxicity, residues and metabolites in milk that can induce bacterial resistance in human infections. In general, farmers and industry comply with the legislation and several food safety controls are routinely carried out; however, these concerns remain possibly as a result of incorrect information [4,5]. Currently, research efforts are focused on the development of functional dairy products that fortify human health and are in harmony with the concept of sustainable production, green economy, environmental protection, and proper health and welfare status of dairy animals [2]. Bioactive compounds of plants, well known as phytobiotics, possess strong antioxidant and anti-inflammatory properties that depend upon their type and amount and are generally cheaper and safer compared to synthetic antibiotic agents [6]. Their dietary supplementation generally improves performance and ameliorates the health status of ruminants due to their multifaceted properties [6]. Milk Somatic Cell Count (SCC) is often used as an indirect index of mammary health, since high values are strongly related with changes in milk quality, poor udder health and inflammatory damage of mammary tissue leading to significant economic losses in modern dairy farms [7].

At the same time, the dietary inclusion of plants' bioactive compounds improves milk oxidative stability, resulting in dairy products of high-quality and safety [8,9]. An antioxidant is defined as any substance that when present in low concentrations compared to the oxidizable substrate (i.e., proteins, lipids, carbohydrates and DNA) significantly delays or prevents oxidation of this substrate through its radical scavenging, metal ion chelation, and singlet oxygen quenching properties [10]. A compound exerts its antioxidant activities by inhibiting the creation of reactive oxygen species (ROS), or directly cleaning free radicals [10]. ROS are small molecules that contain active oxygen and are produced as by-products in sub-cellular organelles such as mitochondria. A high concentration of ROS in any normal cell can turn it into a malignant cell [10]. Phytobiotic supplementation may be more efficient in animals that are under physiologic stress, such as the peak of milk production in dairy animals [11]. In addition, the combination of phytobiotics from different plants may exert an increased antioxidant activity due to their synergism [12]. However, several parameters, namely area of origin, period of harvesting within the year, used part of the plant (leaf, bark, seeds or root) and method of isolation (steam distillation, extraction with non-aqueous solvents, cold expression, etc.) modify the antioxidant capacity and the efficacy of each phytogenic compound. Discrepancies are also observed due to the type of the phytobiotic, its level of dietary supplementation, the composition and the digestibility of the basal diet, the level of feed intake and hygiene and environmental conditions [2].

The objective of the present study was to determine if the supplementation with a mixture of plant bioactive components (MPBC) to high-producing dairy ewes that are prone to subclinical udder health disorders may alleviate this stressful condition and, consequently, improve their lactation performance and health status.

2. Materials and Methods

2.1. Animals

Thirty-six 2-year-old Chios ewes with similar body condition scores (2.5–3.0) and a mean weight of 52.3 ± 1.9 kg that were at their second parity and thirty days after lamb weaning (75 ± 5 days after lamb birth) were randomly selected from the sheep herd of the experimental farm of the Agricultural University of Athens and allocated into three experimental groups of twelve ewes each based on their milk yield and body weight. The ewes of the flock were mated following estrus synchronization with Ovigest intravaginal progestogen sponges (Hipra S.A., Girona, Spain).

All animals initially consumed alfalfa hay and the same concentrate basal diet (Table 1) without the addition of MPBC for one week in order to become acclimatized to the experimental conditions. After this adaptive pre-experimental period, they received the three experimental diets for 11 weeks. One of the groups served as a control (C) and was fed with the previous concentrate diet, whereas the other two groups were offered the same concentrated diet further supplemented with MPBC (B1) at the level of 0.05% or with rumen protected MPBC at the level of 0.025% (B2). These levels were selected based on our preliminary studies. Furthermore, we attempted to evaluate whether a lower level of supplementation in a rumen protected form would return comparable results with the 0.05% level in the framework of precision livestock feeding. The MBPC used in this trial (NuPhoria, Nuevo S.A., Schimatari, Viotia, Greece) was a proprietary mixture of phytogenic substances originating from thyme (*Thymus vulgaris*), anise (*Pimpinella anisum*) and olive (*Olea europea*) at an approximate ratio of 20, 35 and 45% with a standardized active ingredient concentration of 100 g/kg. In detail, the levels of thymol, anethole and hydroxytyrosol in thyme, anise and olive were 24, 210 and 50 g/kg, respectively. Rumen-protected MPBC was obtained by freeze drying with maltodextrin at a rate of 50:50. Maltodextrins of different dextrose equivalents are commonly used as encapsulating agents due to their high-water solubility, low viscosity and colorless solutions [13,14]. Ewes in the present study consumed on average 2.0 kg of feed (concentrate and forage at a mean ratio of 50:50). Quantities of concentrates were constantly adjusted to the milk yield of each ewe and the additional demanded dry matter for high yielding animals was individually provided.

Table 1. Composition and analysis of dairy ewes' diet.

Components (g/kg)		
Corn	234	
Wheat	175	
Barley	175	
Soybean Meal (44%)	182.5	
Sunflower Meal (28%)	50	
Wheat Bran	150	
Sodium Chloride (NaCl)	10	
Limestone	18.5	
Monocalcium Phosphate	4	
Vitamins & Trace elements Premix *	1	
Calculated Analysis	Concentrates	Alfalfa hay
Dry Matter—DM (%)	86.0	93.5
Crude protein—CP (%)	17.0	10.2
Crude Fiber (%)	6.0	34.2
Ash (%)	6.5	7.4
Fat (%)	2.1	2.3
Calcium (%)	0.9	-
Phosphorus (%)	0.6	-
Sodium (%)	0.4	-

* Premix contained per kg: 25 g Mn, 30 g Fe, 45 g Zn, 0.10 g Se, 0.50 g Co, 1.75 g I, 10,000 kIU vitamin A, 2000 kIU vitamin D3, 20 kIU vitamin E (kIU: 1000 International Units).

Ewes were housed in 3 different pens (one pen per treatment) at the premises of the Agricultural University of Athens. Each pen consisted of an indoor and outdoor area and had the same direction and orientation, the same covered area (3 m^2/ewe) and was equipped with similar troughs for feeding (12 individual feeders indoors for concentrate and 1 feeder outdoors for alfalfa hay per pen). Water was available ad libitum and the diet that was formulated according to ewes' individual requirements, based on their body weight and milk yield, was provided twice daily at 8 a.m. and 15 p.m. Forage was offered to the animals after assuring that the concentrate was completely consumed. No refusals of forage and/or concentrates were observed.

2.2. *Determination of Milk Yield Composition and Oxidative Stability*

Ewes were milked twice per day (6:00 a.m. and 18:00 p.m.) in a 12 stall milking parlor (GEA Westfalia, Düsseldorf, Germany). A pulsation ratio of 50:50 was applied; pulsation rate was 150 cycles min^{-1} with 37.5 kPa vacuum level. Milk yield, determined as the sum of the morning and afternoon milking, was recorded on day 1 prior to and on week 1–11 after MPBC dietary supplementation. Fat corrected (FCM6%) milk yield was also calculated using the following formula:

Fat corrected milk (FCM) in 6% (FCM6%) = (0.28 + 0.12 × milk fat concentration (%)) × milk yield (kg/d).

Individual milk samples were also collected on acclimation week and on week 1–11 after MPBC dietary supplementation and analyzed for fat, protein, lactose, total solids-not-fat, pH, electric conductivity and somatic cell count by using the Lactoscan COMBO Milk Cell Analyser (Lactoscan, Nova Zagora, Bulgaria) in accordance with international standard protocol guidelines. Milk oxidative stability was evaluated by measuring the levels of malondialdehyde (MDA), a secondary lipid oxidation product formed by hydrolysis of lipid hydroperoxides. MDA concentration (ng/mL) was determined by applying a selective third-order derivative spectrophotometric method, previously developed by Botsoglou et al. [15].

2.3. Isolation of Milk Somatic Cells and Milk Somatic Cell Immunophenotyping

Every two weeks, 15 mL individual milk samples were also collected and kept on ice for the determination of macrophage, lymphocyte, and polymorphonuclear leukocytes count. Milk somatic cells (MSC) were isolated following a modified protocol of Koess and Hamann [16] optimized for sheep milk. Briefly, milk samples were centrifuged at 400× *g* for 15 min at 4 °C. Pellets were resuspended in 15 mL of dilution buffer, i.e., phosphate-buffered saline (PBS; pH 7.4) containing 0.01% sodium azide (NaN3) and 0.2% bovine serum albumin (BSA). Samples were centrifuged at 400× *g* for 10 min at 4 °C. Pellets were resuspended in 4 mL of dilution buffer and centrifuged at 400× *g* for 10 min at 4 °C. Cell pellets were resuspended in 1 mL of dilution buffer and filtered through 40 μm cell strainers.

Cell surface labelling was performed with anti-CD11b, anti-CD8 and anti-Cytokeratins 4 + 5 + 6 + 8 + 10 + 13 + 18 (anti-Pan Cytokeratins) for the identification of granulocytes and macrophages, T-cytotoxic and epithelial cells, respectively. In addition, propidium iodide (PI) staining was used to differentiate live from dead cells. Aliquots of MSC prepared as described above containing approximately 2×10^5 cells were centrifuged at 400× *g* for 5 min at 4 °C and cell pellets were resuspended in 50 μL ice cold antibody solutions containing combinations of 0.002 mg/mL anti-CD11b conjugated to Fluorescein isothiocyanate (FITC), 0.005 mg/mL anti-pan Cytokeratins conjugated to Allophycocyanin (APC) and 0.002 mg/mL CD8 R-PE antibodies. Cells in the staining solutions were incubated on ice and in the dark for 30 min, then a 2 mL dilution buffer was added, samples were centrifuged at 400× *g* for 5 min and cell pellets were resuspended in 100 μL dilution buffer. DNA staining was performed by addition of PI at a final concentration of 5 ng/μL. Following 10 min incubation at room temperature in the dark, 100 μL of PBS were added and samples were analyzed by flow cytometry (Cytomics FC 500, California, Beckman Coulter Inc., Fullerton, CA, USA).

Instrument voltage/gain for detectors FS, SS, FL1, FL2, FL3, FL4 and FL5 were set at 700/2.0, 680/20.0, 550/1.0, 650/1.0, 650/1.0, 614/1.0, and 601/1.0, respectively. The samples were run at medium speed and approximately 65,000 events were collected per sample. Data were stored as list mode files. Events that were identified as being of appropriate size and granularity based on their position on an FS/SS dotplot and were negative for PI were considered as live cells. PI stains nucleic acids only in cells with disrupted cell membranes, i.e., necrotic cells and thus cells that were negative for PI staining were considered as live cells. Dead cells were not included in further analysis as they often exhibit non-specific staining with cell surface antibodies and may be misclassified. Live cells that stained positive for CD11b (FL1) with higher SS values were classified as polymorphonuclear granulocytes (PMN) and live cells that stained positive for CD11b with lower SS values were classified as monocytes/macrophages (MPh). Epithelial cells were identified from the CD11b−/CD8− live cells that stained positive for pan-cytokeratins. Lymphocytes were identified as the live cells that were positioned in the FS/SS dotplot in the area identified by CD8+ cells. Proportions of each cell subset in the MSC were estimated as a percentage of the live cells.

2.4. Statistical Analysis

The experimental unit was the animal since it was the smallest unit upon which either the treatment was applied or the measurements were made. Data were subjected to repeated measures analysis of variance using the MIXED procedure of SAS software, with dietary treatment as the fixed factor and sampling week as the repeated factor. SCC was log transformed prior to statistical analysis in order to achieve normal distribution.

Significant differences were tested with Bonferroni adjustment at 0.05 significance level and the results are presented as least square means ± S.E.M.

3. Results

As shown, the milk yield (Figure 1A) and fat-corrected milk yield in 6% (Figure 1B) were generally higher in the MPBC supplemented groups than the controls with significant differences between the 7th and 9th week of the experiment for the B2 compared to the other groups ($p < 0.05$). In detail, the milk yield was 1335, 1021 and 1142 (±109) mL/day on week 7, 1253, 988 and 1042 (±99) mL/day on week 8 and 1262, 996 and 1179 (±121) mL/day on week 9 for the B2, C and B1 groups, respectively ($p < 0.05$; Figure 1A). The respective values for the fat-corrected milk yield were 1599, 1243 and 1367 (±142) mL/day on week 7, 1525, 1209 and 1230 (±104) mL/day on week 8 and 1543, 1187 and 1348 (±14,121) mL/day on week 9 for the B2, C and B1 groups, respectively ($p < 0.05$; Figure 1B).

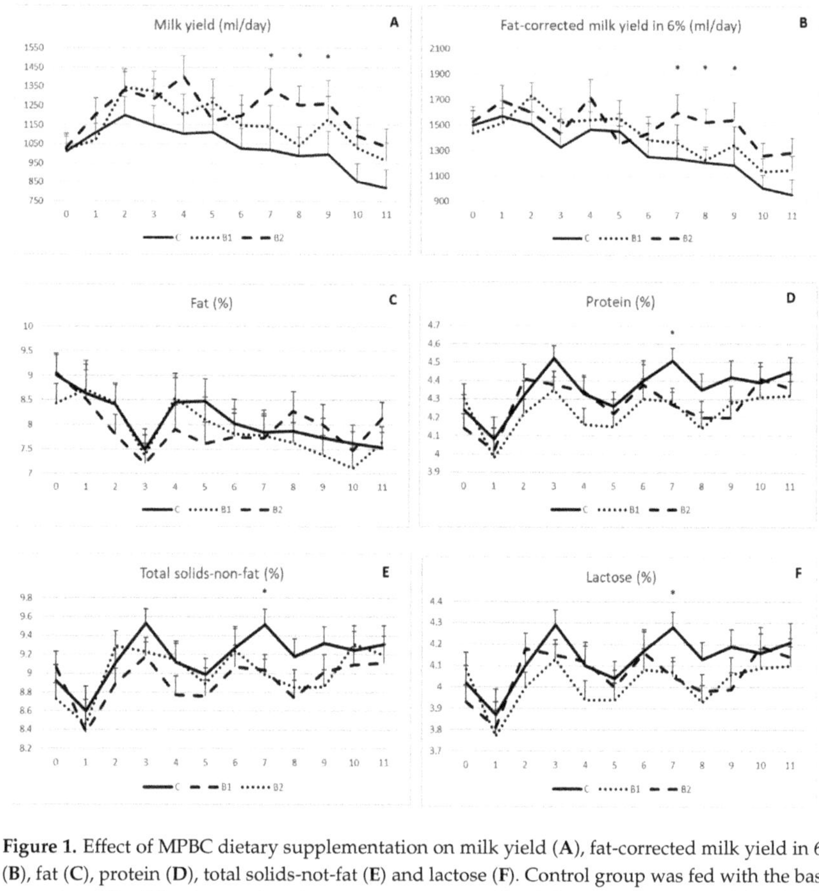

Figure 1. Effect of MPBC dietary supplementation on milk yield (**A**), fat-corrected milk yield in 6% (**B**), fat (**C**), protein (**D**), total solids-not-fat (**E**) and lactose (**F**). Control group was fed with the basal concentrate diet (C), whereas the other two groups were offered the same concentrated diet further with MPBC (B1) at the levels of 0.05% or with rumen protected MPBC (B2) at the levels of 0.025%. The use of (*) indicates significant difference at $p < 0.05$.

Milk composition was generally not affected by MPBC dietary supplementation. No significant differences were shown for milk fat throughout the experiment (Figure 1C). Similar findings were observed for milk protein (Figure 1D), total solids-non-fat (Figure 1E) and lactose (Figure 1F) with the only exception at the 7th week, when the values for these parameters were higher in the control compared with the MPBC supplemented groups ($p < 0.05$). In detail, milk protein (%) was 4.51, 4.29 and 4.27 (±0.07), milk total solids-non-fat (%) were 9.52, 9.04 and 8.99 (±0.16) and milk lactose was 4.28, 4.07 and 4.05 (±0.07) for the C, B1 and B2 groups, respectively.

As indicated in Figure 2A,B, milk pH and electrical conductivity were not influenced by MPBC dietary supplementation, since no significant differences were observed among the experimental groups during the 11-week experimental period. On the other hand, milk oxidative stability was in general improved in MPBC supplemented groups as shown by the reduced MDA values. Significant differences were shown on week 3, 7, 10 and 11 ($p < 0.05$; Figure 2C). The respective MDA values (ng/g) were 6.36 and 6.11 vs. 7.60 (±0.64) on week 3, 4.35 and 4.44 vs. 7.29 (±0.40) on week 7, 6.81 and 6.54 vs. 7.87 (±0.36) on week 10 and 5.89 and 5.52 vs. 6.72 (±0.29) on week 11 for the B1, B2 and C groups, respectively ($p < 0.05$; Figure 2C).

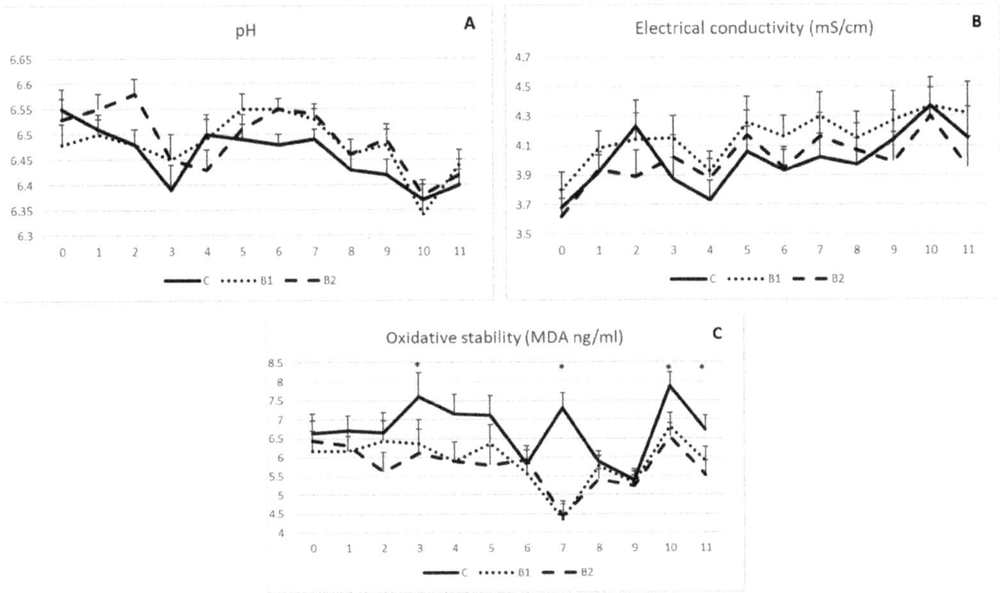

Figure 2. Effect of MPBC dietary supplementation on milk pH (**A**), electrical conductivity (**B**) and oxidative stability (MDA levels) (**C**). Control group was fed with the basal concentrate diet (C), whereas the other two groups were offered the same concentrated diet further supplemented with MPBC (B1) at the levels of 0.05% or with rumen protected MPBC (B2) at the levels of 0.025%. The use of (*) indicates significant difference at $p < 0.05$.

Rumen protected MPBC dietary supplementation decreased milk somatic cell count, as indicated in Figure 3A. However, significant differences were shown on week 2, 8 and 9. In detail, logSCC was 5.19 vs. 5.58 and 5.59 (±0.11) on week 2, 4.95 vs. 5.34 and 5.31 (±0.12) on week 8 and 4.87 vs. 5.30 and 5.23 (±0.12) on week 9 for B2, C and B1 group, respectively ($p < 0.05$; Figure 3A). On the other hand, proportions for lymphocyte (Figure 3B), macrophage (Figure 3C) and polymorphonuclear leucocytes (Figure 3D) were not significantly different among the experimental groups, although values for macrophage and polymorphonuclear leucocytes were numerically higher in controls compared to the MPBC dietary supplemented groups throughout the experiment.

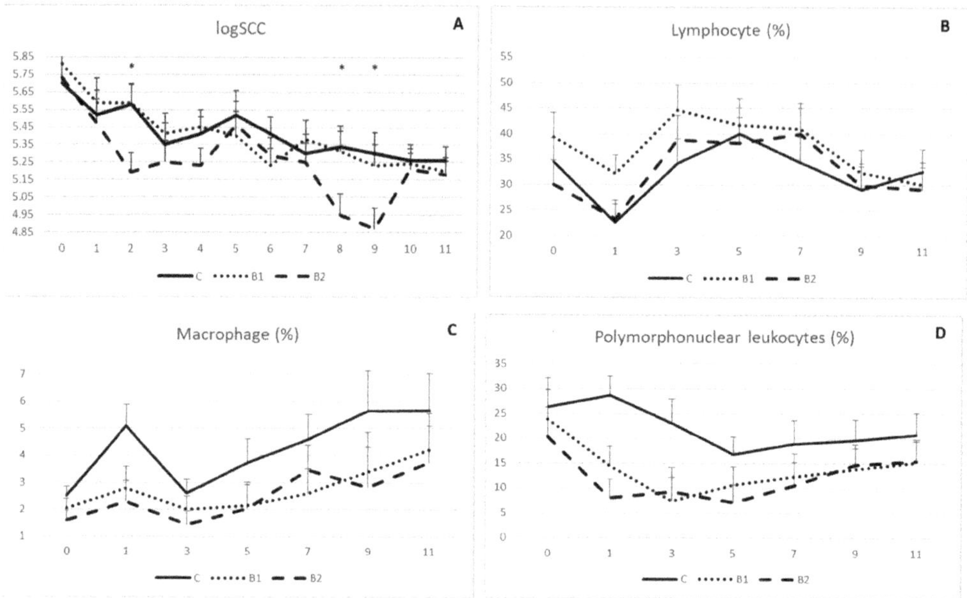

Figure 3. Effect of MPBC dietary supplementation on logSCC (**A**), lymphocyte (**B**), macrophage (**C**) and polymorphonuclear leukocytes (**D**) count of sheep milk. Control group was fed with the basal concentrate diet (C), whereas the other two groups were offered the same concentrated diet further with MPBC (B1) at the levels of 0.05% or with rumen protected MPBC (B2) at the levels of 0.025%. The use of (*) indicates significant difference at $p < 0.05$.

4. Discussion

As indicated by the findings of the present study, the milk yield and fat-corrected milk yield in 6% were generally higher in the MPBC supplemented groups than the controls with significant differences between the 7th and 9th week of the experiment for the B2 compared to the other groups. However, milk composition was not significantly affected by MPBC dietary supplementation, while oxidative stability was improved in B1 and B2 groups. According to the existing literature, thyme and/or celery seed mixture [17], thyme or celery essential oil [18], cornus extract enriched with EOs of oregano and thyme [19], orange peel essential oil (EO) [20] and EO components mixture (thymol, eugenol, vanillin, guaiacol and limonene) [21] induced an increase in the milk yield in dairy ewes. Moreover, Kholif et al. [22] reported that capsicum/thymus essential oils blend at 2 mL and/or enzymes cocktail at 4 g per day enhanced milk yield and milk fat levels in dairy ewes. On the other hand, no effect of citral oil [23] or anise, clove, and thyme EO [24] on milk yield was observed, while an increase in milk yield, protein and fat levels was observed in dairy goats as a result of Boswellia sacra resin [25] and rosemary or lemon grass [26] dietary supplementation. Feed efficiency, milk yield and levels of protein, fat, total solids were increased as a result of Lippia alba hay inclusion in the diet of dairy goats [27]. Choubey et al. [28] observed that the dietary supplementation with flowers, shoots and leaves of *Woodfordia fruticosa*, the whole plant of *Solanum nigrum* and the seeds of *Trigonella foenum-graecum* improves antioxidant status in adult goats. In contrast, Leparmarai et al. [29] showed that grape seed dietary supplementation did not influence milk yield, milk composition and blood antioxidant status in dairy sheep and goats.

Shabtay et al. [30] reported that the addition of pomegranate extract to dairy cow diets resulted in higher milk production. Enhanced daily outputs of milk, energy corrected milk and fat were also observed after coriander oil dietary supplementation without any negative effect on cow health [31]. Greater milk yield, total solids, protein, lactose and fat and decreased malondialdehyde values were observed in dairy cows supplemented with a phytogenic feed additives mixture that contained menthol, anethole and other terpinenes [32]. Moreover, according to a meta-analysis by Belanche et al. [33], long-term exposure to a commercial blend of EOs (Agolin) resulted in a slight increase in milk yield at the level of 4%, while no effects on feed intake and milk composition were evident. Braun et al. [34] suggested that the aforementioned effects on milk production in dairy cattle could be attributed to the improved rumen fermentation, feed efficiency, nutrients' absorption and utilization, and increased uptake of cations like calcium and ammonium as a result of phytobiotics' dietary supplementation. As indicated, herbs and their extracts can accelerate digestion by reducing residence time in the digestive tract [35], while the observation of similar values in feed intake among treatments may indicate that although the MPBC dietary supplementation improved nutrient digestibility, did not negatively affect feed palatability and acceptance. However, no effects of thyme oil and thymol [36], eucalyptus, thyme and anise oil [37], eugenol [38], cinnamaldehyde and eugenol [39], cinnamaldehyde and garlic oil [40], garlic or juniper berry EO [41], oregano leaves [42,43], EOs components mixture (thymol, eugenol, vanillin, guaiacol and limonene) [44–47], blend of oregano, cinnamon, thyme and orange peel EOs [48] and mixture of eugenol, geranyl acetate and coriander oil [49] on milk yield and composition of dairy cattle are observed. In a study carried out with a mixture of plant bioactive components, a decrease in milk fat content was observed, while the other milk components were not affected in dairy cows [50]. In water buffaloes, dietary supplementation with a phytogenic mix containing seeds of fennel, ajwain and fenugreek, tubers of ginger, leaves of *Swertia chirata*, roots of licorice, fruits of *Citrullus colocynthis*, *Terminalia chebula* and turmeric did not affect milk yield and rumen fermentation parameters, apart from pH [51].

Moreover, the improved milk oxidative stability as suggested by the reduced MDA levels indicating that the provision of polyphenols with potent antioxidant or co-antioxidant activity might be a beneficial strategy to protect mammary cells against the adverse effects of free radicals. Part of the molecular multifunctionality of the natural bioactive compounds is their antioxidant capacity, which improves the immune status and reduces the oxidative stress of animals [52–54]. Phenolic compounds have been found to in vitro modify immune status via the downregulation of the inflammatory response, since they reduce the production of cytokines and reactive oxygen species, as well as the functionality of cytotoxic T-lymphocytes and natural killer cells [55]. The anti-inflammatory and antioxidant activity of polyphenols has been confirmed by in vivo studies in dairy cows, goats and ewes [52,56,57]. The aforementioned properties are attributed to their ability to chelate with free radicals, inhibit the enzymes actions associated with the mechanisms of oxidative stress, reinforce the functionality of antioxidant mechanisms, and prevent the lipid oxidation [53,58–60].

As indicated, SCC was reduced in dairy ewes that were dietary supplemented with the rumen protected MPBC with significant differences on week 2, 8 and 9. The decreased levels of SCC in milk from ewes dietary supplemented with MPBC is associated with an ameliorated udder health status, since milk SCC is considered as an index of mammary health. This finding could be attributed to the provision of several hydrophylic and lipophylic phenols that are included in the MBPC and through their antioxidant properties fortify mammary cells against the adverse effects of free radicals produced as a result of oxidative stress. Similar findings were reported by Hashemzadeh-Cigari et al. [50] in dairy cows after their supplementation with a phytobiotics-rich herbal mixture (185 g/cow) that contained cinnamon bark, turmeric roots, rosemary leaves and clove buds. Moreover, supplementation of rosemary extract [59] or a mixture of essential oils [21] to lactating ewes and concentrated pomegranate extract to dairy cows [30] resulted in reduced milk SCC. Jaquezeski et al. [60] also found that curcumin dietary supplementation improved milk yield and antioxidant capacity, while a reduction in somatic cell count and protein oxidation was reported in dairy sheep. In dairy cows, thyme essential oil supplementation via esophageal tube decreased the standard plate count, while no differences were observed in the raw milk composition [61]. Moreover, although Rodrigues et al. [62] observed an increase in milk yield after dietary supplementation with a phytogenic mix, no effect on milk composition and incidence of clinical mastitis was observed in dairy cows.

Although SCC was decreased as an effect of rumen-protected MPBC dietary supplementation, no significant differences were observed among the treatments concerning the macrophage, lymphocyte, and polymorphonuclear leukocyte (PMN) counts. However, although there was not a clear tendency for the lymphocyte count, macrophages were increased and polymorphonuclear leukocytes were decreased from the beginning till the end of the experimental period. Although only numerical, MPBC supplemented groups had a lower PMN count compared to the controls, indicating a tendency for a healthier mammary gland, since PMN being the principal leucocytes that are increased during pathogen invasion, are closely correlated with high SCC [63] and oxidation stress in mammary gland [64]. On the other hand, macrophages represents 5–7% of leucocytes in ewe milk and are found to minimally contribute to the proteolytic activity in ewe milk [65]. Finally, lymphocytes represents approximately 40% of leucocyte population [66] and an absence of differences in the lymphocyte count as an effect of SCC suggests that in ewe milk this population is quite stable [63].

5. Conclusions

Plant extracts have been widely recognized as potential functional alternatives to antibiotics due to their green, safe, and efficient properties. Dietary supplementation with a mixture of phytogenic substances originated from thyme, anise and olive was effective in improving milk oxidative stability (third, seventh, tenth and eleventh week), enhancing performance (seventh–ninth week) and lowering SCC (second, eighth and ninth week) when provided in its rumen protected form in mid-lactation high-producing dairy ewes. Considering the positive outcome on milk yield, oxidative stability and SCC and the lack of any side effects on the other milk properties, the MPBC used in the present study and especially in its rumen protected form appeared as a promising candidate for a feed additive for dairy ewes. The use of the protected form of MPBC may be successfully adapted in a production system that incorporates precision livestock feeding. However, a further in-depth analysis is necessary regarding its production cost before establishing its regular use.

Author Contributions: Conceptualization, P.S. and M.G.; methodology, P.S. and A.H.-T.; validation, P.S., M.G. and A.H.-T.; formal analysis, M.G.; investigation, A.K., M.T., K.P. and P.S.; resources, P.S.; writing—original draft preparation, A.K. and P.S.; writing—review and editing, M.G., M.T., K.P. and A.H.-T.; visualization, P.S.; supervision, P.S.; project administration, P.S. All authors have read and agreed to the published version of the manuscript.

Funding: This research was funded by Nuevo SA, AUA grant no: 35.0054.

Institutional Review Board Statement: The animal study protocol was approved by Research Ethics Committee of the Agricultural University of Athens (protocol code 46/09.05.2022).

Informed Consent Statement: Not applicable.

Data Availability Statement: All data used to support the findings of this study are included within the article.

Acknowledgments: The authors would like to thank Athanasios Gelasakis from the Laboratory of Anatomy and Physiology of Farm Animals, Department of Animal Science, School of Animal Biosciences, Agricultural University of Athens for his technical assistance.

Conflicts of Interest: The authors declare no conflict of interest. The funders had no role in the design of the study, in the collection, analyses, or interpretation of data, in the writing of the manuscript and in the decision to publish the results.

References

1. EU Regulation 1831/2003/EC of 22 September 2003 on Additives for Use in Animal Nutrition, Replacing Directive 70/524/EEC on Additives in Feeding-Stuffs. Available online: https://eur-lex.europa.eu/legal-content/en/ALL/?uri=CELEX%3A32003R1831 (accessed on 15 June 2023).
2. Simitzis, P.E. Enrichment of animal diets with essential oils—A great perspective on improving animal performance and quality characteristics of the derived products. *Medicines* **2017**, *4*, 35. [CrossRef]
3. Serra, V.; Salvatori, G.; Pastorelli, G. Dietary polyphenol supplementation in food producing animals: Effects on the quality of derived products. *Animals* **2021**, *11*, 401. [CrossRef] [PubMed]
4. Bodas, R.; Prieto, N.; García-González, R.; Andrés, S.; Giráldez, F.J.; López, S. Manipulation of rumen fermentation and methane production with plant secondary metabolites. *Anim. Feed Sci. Technol.* **2012**, *176*, 78–93. [CrossRef]
5. Elghalid, O.A.; Kholif, A.E.; El-Ashry, G.M.; Matloup, O.H.; Olafadehan, O.A.; El-Raffa, A.M.; Abd El-Hady, A.M. Oral supplementation of the diet of growing rabbits with a newly developed mixture of herbal plants and spices enriched with special extracts and essential oils affects their productive performance and immune status. *Livest. Sci.* **2020**, *238*, 104082. [CrossRef]
6. Rochfort, S.; Parker, A.J.; Dunshea, F.R. Plant bioactives for ruminant health and productivity. *Phytochemistry* **2008**, *69*, 299–322. [CrossRef]
7. Schukken, Y.; Wilson, D.; Welcome, F.; Garrison-Tikofsky, L.; Gonzalez, R. Monitoring udder health and milk quality using somatic cell counts. *Vet. Res.* **2003**, *34*, 579–596. [CrossRef]
8. Salami, S.; Guinguina, A.; Agboola, J.; Omede, A.; Agbonlahor, E.; Tayyab, U. In vivo and postmortem effects of feed antioxidants in livestock: A review of the implications on authorization of antioxidant feed additives. *Animal* **2016**, *10*, 1375–1390. [CrossRef]
9. Nudda, A.; Cannas, A.; Correddu, F.; Atzori, A.S.; Lunesu, M.F.; Battacone, G.; Pulina, G. Sheep and goats respond differently to feeding strategies directed to improve the fatty acid profile of milk fat. *Animals* **2020**, *10*, 1290. [CrossRef]
10. Prasad, S.; Gupta, S.C.; Tyagi, A.K. Reactive oxygen species (ROS) and cancer: Role of antioxidative nutraceuticals. *Cancer Lett.* **2017**, *387*, 95–105. [CrossRef]
11. Gobert, M.; Martin, B.; Ferlay, A.; Chilliard, Y.; Graulet, B.; Pradel, P.; Bauchart, D.; Durand, D. Plant polyphenols associated with vitamin E can reduce plasma lipoperoxidation in dairy cows given n-3 polyunsaturated fatty acids. *J. Dairy Sci.* **2009**, *92*, 6095–6104. [CrossRef]
12. Burt, S. Essential oils: Their antibacterial properties and potential applications in foods—A review. *Int. J. Food Microbiol.* **2004**, *94*, 223–253. [CrossRef] [PubMed]
13. González, E.; Gómez-Caravaca, A.M.; Giménez, B.; Cebrián, R.; Maqueda, M.; Parada, J.; Martinez-Ferez, A.; Segura-Carretero, A.; Robert, P. Role of maltodextrin and inulin as encapsulating agents on the protection of oleuropein during in vitro gastrointestinal digestion. *Food Chem.* **2020**, *310*, 125976. [CrossRef] [PubMed]
14. Todorović, A.; Šturm, L.; Salević-Jelić, A.; Lević, S.; Osojnik Črnivec, I.G.; Prislan, I.; Skrt, M.; Bjekovic, A.; Poklar Ulrih, N.; Nedović, V. Encapsulation of bilberry extract with maltodextrin and gum arabic by freeze-drying: Formulation, characterisation, and storage stability. *Processes* **2022**, *10*, 1991. [CrossRef]
15. Botsoglou, N.A.; Fletouris, D.J.; Papageorgiou, G.E.; Vassilopoulos, V.N.; Mantis, A.J.; Trakatellis, A.G. A rapid, sensitive andspecific thiobarbituric acid method for measuring lipid peroxidation in animal tissues, food and feedstuff samples. *J. Agric. Food Chem.* **1994**, *42*, 1931–1937. [CrossRef]
16. Koess, C.; Hamann, J. Detection of mastitis in the bovine mammary gland by flow cytometry at early stages. *J. Dairy Res.* **2008**, *75*, 225–232. [CrossRef]
17. Khattab, M.S.A.; Kholif, A.E.; Abd El Tawab, A.M.; Shaaban, M.M.; Hadhoud, F.I.; El-Fouly, H.A.; Olafadehan, O.A. Effect of replacement of antibiotics with thyme and celery seed mixture on the feed intake and digestion, ruminal fermentation, blood chemistry, and milk lactation of lactating Barki ewes. *Food Funct.* **2020**, *11*, 6889–6898. [CrossRef]
18. Abd El Tawab, A.M.; Kholif, A.E.; Khattab, M.S.A.; Shaaban, M.M.; Hadhoud, F.I.; Mostafa, M.M.M.; Olafadehan, O.A. Feed utilization and lactational performance of Barki sheep fed diets containing thyme or celery. *Small Rumin. Res.* **2020**, *192*, 106249. [CrossRef]
19. Kalaitsidis, K.; Sidiropoulou, E.; Tsiftsoglou, O.; Mourtzinos, I.; Moschakis, T.; Basdagianni, Z.; Vasilopoulos, S.; Chatzigavriel, S.; Lazari, D.; Giannenas, I. Effects of cornus and its mixture with oregano and thyme essential oils on dairy sheep performance and milk, yoghurt and cheese quality under heat stress. *Animals* **2021**, *11*, 1063. [CrossRef]
20. Kotsampasi, B.; Tsiplakou, E.; Christodoulou, C.; Mavrommatis, A.; Mitsiopoulou, C.; Karaiskou, C.; Sossidou, E.; Fragioudakis, N.; Kapsomenos, I.; Bampidis, V.A.; et al. Effects of dietary orange peel essential oil supplementation on milk yield and composition, and blood and milk antioxidant status of dairy ewes. *Anim. Feed Sci. Technol.* **2018**, *245*, 20–31. [CrossRef]
21. Giannenas, I.; Skoufos, J.; Giannakopoulos, C.; Wiemann, M.; Gortzi, O.; Lalas, S.; Kyriazakis, I. Effects of essential oils on milk production, milk composition, and rumen microbiota in Chios dairy ewes. *J. Dairy Sci.* **2011**, *94*, 5569–5577. [CrossRef]
22. Kholif, A.E.; Kassab, A.Y.; Azzaz, H.H.; Matloup, O.H.; Hamdon, H.A.; Olafadehan, O.A.; Morsy, T.A. Essential oils blend with a newly developed enzyme cocktail works synergistically to enhance feed utilization and milk production of Farafra ewes in the subtropics. *Small Rumin. Res.* **2018**, *161*, 43–50. [CrossRef]
23. Canaes, T.S.; Zanferari, F.; Maganhe, B.L.; Takiya, C.S.; Silva, T.H.; Del Valle, T.A.; Rennó, F.P. Increasing dietary levels of citral oil on nutrient total tract digestibility, ruminal fermentation, and milk composition in Saanen goats. *Anim. Feed Sci. Technol.* **2017**, *229*, 47–56. [CrossRef]

24. El-Essawy, A.M.; Anele, U.Y.; Abdel-Wahed, A.M.; Abdou, A.R.; Khattab, I.M. Effects of anise, clove and thyme essential oils supplementation on rumen fermentation, blood metabolites, milk yield and milk composition in lactating goats. *Anim. Feed Sci. Technol.* **2021**, *271*, 114760. [CrossRef]
25. Soltan, Y.A.; Morsy, A.S.; Hashem, N.M.; Sallam, S.M. Boswellia sacra resin as a phytogenic feed supplement to enhance ruminal fermentation, milk yield, and metabolic energy status of early lactating goats. *Anim. Feed Sci. Technol.* **2021**, *277*, 114963. [CrossRef]
26. Kholif, A.E.; Matloup, O.H.; Morsy, T.A.; Abdo, M.M.; Abu Elella, A.A.; Anele, U.Y.; Swanson, K.C. Rosemary and lemongrass herbs as phytogenic feed additives to improve efficient feed utilization, manipulate rumen fermentation and elevate milk production of Damascus goats. *Livest. Sci.* **2017**, *204*, 39–46. [CrossRef]
27. da Silva, N.I.S.; de Moura, J.F.P.; de Lucena Nascimento, M.E.; Machado, F.C.G.; Costa, T.G.P.; de Araújo Filho, J.M.; Cordão, M.A.; Edvan, R.L.; Oliveira, R.L.; Bezerra, L.R. Effect of Lippia alba hay as phytogenic feed additive on the lactation performance, milk composition, and rumen and blood parameters of Alpine goats. *Small Rumin. Res.* **2022**, *215*, 106767. [CrossRef]
28. Choubey, M.; Pattanaik, A.K.; Baliyan, S.; Dutta, N.; Jadhav, S.E.; Sharma, K. Dietary supplementation of a novel phytogenic feed additive: Effects on nutrient metabolism, antioxidant status and immune response of goats. *Anim. Prod. Sci.* **2015**, *56*, 1612–1621. [CrossRef]
29. Leparmarai, P.T.; Sinz, S.; Kunz, C.; Liesegang, A.; Ortmann, S.; Kreuzer, M.; Marquardt, S. Transfer of total phenols from a grapeseed-supplemented diet to dairy sheep and goat milk, and effects on performance and milk quality. *J. Anim. Sci.* **2019**, *97*, 1840–1851. [CrossRef]
30. Shabtay, A.; Nikbachat, M.; Zenou, A.; Yosef, E.; Arkin, O.; Sneer, O.; Shwimmer, A.; Yaari, A.; Budman, E.; Agmon, G.; et al. Effects of adding a concentrated pomegranate extract to the ration of lactating cows on performance and udder health parameters. *Anim. Feed Sci. Technol.* **2012**, *175*, 24–32. [CrossRef]
31. Matloup, O.H.; Abd El Tawab, A.M.; Hassan, A.A.; Hadhoud, F.I.; Khattab, M.S.A.; Khalel, M.S.; Sallam, S.M.A.; Kholif, A.E. Performance of lactating Friesian cows fed a diet supplemented with coriander oil: Feed intake, nutrient digestibility, ruminal fermentation, blood chemistry, and milk production. *Anim. Feed Sci. Technol.* **2017**, *226*, 88–97. [CrossRef]
32. Kholif, A.E.; Hassan, A.A.; El Ashry, G.M.; Bakr, M.H.; El-Zaiat, H.M.; Olafadehan, O.A.; Matloup, O.H.; Sallam, S.M.A. Phytogenic feed additives mixture enhances the lactational performance, feed utilization and ruminal fermentation of Friesian cows. *Anim. Biotechnol.* **2021**, *32*, 708–718. [CrossRef] [PubMed]
33. Belanche, A.; Newbold, C.J.; Morgavi, D.P.; Bach, A.; Zweifel, B.; Yáñez-Ruiz, D.R. A meta-analysis describing the effects of the essential oils blend agolin ruminant on performance, rumen fermentation and methane emissions in dairy cows. *Animals* **2020**, *10*, 620. [CrossRef] [PubMed]
34. Braun, H.S.; Schrapers, K.T.; Mahlkow-Nerge, K.; Stumpff, F.; Rosendahl, J. Dietary supplementation of essential oils in dairy cows: Evidence for stimulatory effects on nutrient absorption. *Animal* **2019**, *13*, 518–523. [CrossRef] [PubMed]
35. Suresh, D.; Srinivasan, K. Studies on the in vitro absorption of spice principles—Curcumin, capsaicin and piperine in rat intestines. *Food Chem. Toxicol.* **2007**, *45*, 1437–1442. [CrossRef]
36. Benchaar, C. Diet supplementation with thyme oil and its main component thymol failed to favorably alter rumen fermentation, improve nutrient utilization, or enhance milk production in dairy cows. *J. Dairy Sci.* **2021**, *104*, 324–336. [CrossRef]
37. Giller, K.; Rilko, T.; Manzocchi, E.; Hug, S.; Bolt, R.; Kreuzer, M. Effects of mixed essential oils from eucalyptus, thyme and anise on composition, coagulation properties and antioxidant capacity of the milk of dairy cows. *J. Anim. Feed Sci.* **2020**, *29*, 3–10. [CrossRef]
38. Benchaar, C.; Lettat, A.; Hassanat, F.; Yang, W.Z.; Forster, R.J.; Petit, H.V.; Chouinard, P.Y. Eugenol for dairy cows fed low or high concentrate diets: Effects on digestion, ruminal fermentation characteristics, rumen microbial populations and milk fatty acid profile. *Anim. Feed Sci. Technol.* **2012**, *178*, 139–150. [CrossRef]
39. Tager, L.R.; Krause, K.M. Effects of essential oils on rumen fermentation, milk production, and feeding behavior in lactating dairy cows. *J. Dairy Sci.* **2011**, *94*, 2455–2464. [CrossRef]
40. Blanch, M.; Carro, M.D.; Ranilla, M.J.; Viso, A.; Vazquez-Anon, M.; Bach, A. 2016. Influence of a mixture of cinnamaldehyde and garlic oil on rumen fermentation, feeding behaviour and performance of lactating dairy cows. *Anim. Feed Sci. Technol.* **2016**, *219*, 313–323. [CrossRef]
41. Yang, W.Z.; Benchaar, C.; Ametaj, B.N.; Chaves, A.V.; He, M.L.; McAllister, T.A. Effects of garlic and juniper berry essential oils on ruminal fermentation and on the site and extent of digestion in lactating cows. *J. Dairy Sci.* **2007**, *90*, 5671–5681. [CrossRef]
42. Hristov, A.N.; Lee, C.; Cassidy, T.; Heyler, K.; Tekippe, J.A.; Varga, G.A.; Corl, B.; Brandt, R.C. Effect of Origanum vulgare L. leaves on rumen fermentation, production, and milk fatty acid composition in lactating dairy cows. *J. Dairy Sci.* **2013**, *96*, 1189–1202. [CrossRef]
43. Tekippe, J.A.; Hristov, A.N.; Heyler, K.S.; Cassidy, T.W.; Zheljazkov, V.D.; Ferreira, J.F.S.; Karnati, S.K.; Varga, G.A. Rumen fermentation and production effects of Origanum vulgare L. leaves in lactating dairy cows. *J. Dairy Sci.* **2011**, *94*, 5065–5079. [CrossRef] [PubMed]
44. Benchaar, C.; Petit, H.V.; Berthiaume, R.; Ouellet, D.R.; Chiquette, J.; Chouinard, P.Y. Effects of essential oils on digestion, ruminal fermentation, rumen microbial populations, milk production, and milk composition in dairy cows fed alfalfa silage or corn silage. *J. Dairy Sci.* **2007**, *90*, 886–897. [CrossRef] [PubMed]

45. Benchaar, C.; Petit, H.V.; Berthiaume, R.; Whyte, T.D.; Chouinard, P.Y. Effects of addition of essential oils and monensin premix on digestion, ruminal fermentation, milk production, and milk composition in dairy cows. *J. Dairy Sci.* **2006**, *89*, 4352–4364. [CrossRef] [PubMed]
46. Kung, L.; Williams, P.; Schmidt, R.J.; Hu, W. A blend of essential plant oils used as an additive to alter silage fermentation or used as a feed additive for lactating dairy cows. *J. Dairy Sci.* **2008**, *91*, 4793–4800. [CrossRef] [PubMed]
47. Joch, M.; Kudrna, V.; Hakl, J.; Božik, M.; Homolka, P.; Illek, J.; Tyrolova, Y.; Výborná, A. In vitro and in vivo potential of a blend of essential oil compounds to improve rumen fermentation and performance of dairy cows. *Anim. Feed Sci. Technol.* **2019**, *251*, 176–186. [CrossRef]
48. Spanghero, M.; Robinson, P.H.; Zanfi, C.; Fabbro, E. Effect of increasing doses of a microencapsulated blend of essential oils on performance of lactating primiparous dairy cows. *Anim. Feed Sci. Technol.* **2009**, *153*, 153–157. [CrossRef]
49. Santos, M.B.; Robinson, P.H.; Williams, P.; Losa, R. Effects of addition of an essential oil complex to the diet of lactating dairy cows on whole tract digestion of nutrients and productive performance. *Anim. Feed Sci. Technol.* **2010**, *157*, 64–71. [CrossRef]
50. Hashemzadeh-Cigari, F.; Khorvash, M.; Ghorbani, G.R.; Kadivar, M.; Riasi, A.; Zebeli, Q. Effects of supplementation with a phytobiotics-rich herbal mixture on performance, udder health, and metabolic status of Holstein cows with various levels of milk somatic cell counts. *J. Dairy Sci.* **2014**, *97*, 7487–7497. [CrossRef]
51. Hassan, F.U.; Ebeid, H.M.; Tang, Z.; Li, M.; Peng, L.; Peng, K.; Liang, X.; Yang, C. A Mixed Phytogenic Modulates the Rumen Bacteria Composition and Milk Fatty Acid Profile of Water Buffaloes. *Front. Vet. Sci.* **2020**, *7*, 569. [CrossRef]
52. Salem, A.Z.M.; López, S.; Robinson, P.H. Plant bioactive compounds in ruminant agriculture—Impacts and opportunities. *Anim. Feed Sci. Technol.* **2012**, *176*, 1–4. [CrossRef]
53. Tayengwa, T.; Mapiye, C. Citrus and winery wastes: Promising dietary supplements for sustainable ruminant animal nutrition, health, production, and meat quality. *Sustainability* **2018**, *10*, 3718. [CrossRef]
54. Piao, M.; Tu, Y.; Zhang, N.; Diao, Q.; Bi, Y. Advances in the Application of Phytogenic Extracts as Antioxidants and Their Potential Mechanisms in Ruminants. *Antioxidants* **2023**, *12*, 879. [CrossRef] [PubMed]
55. Provenza, F.D.; Villalba, J.J. The role of natural plant products in modulating the immune system: An adaptable approach for combating disease in grazing animals. *Small Rumin. Res.* **2010**, *89*, 131–139. [CrossRef]
56. Correddu, F.; Lunesu, M.F.; Buffa, G.; Atzori, A.S.; Nudda, A.; Battacone, G.; Pulina, G. Can agro-industrial by-products rich in polyphenols be advantageously used in the feeding and nutrition of dairy small ruminants? *Animals* **2020**, *10*, 131. [CrossRef]
57. Niderkorn, V.; Jayanegara, A. Opportunities offered by plant bioactive compounds to improve silage quality, animal health and product quality for sustainable ruminant production: A review. *Agronomy* **2021**, *11*, 86. [CrossRef]
58. Soldado, D.; Bessa, R.J.B.; Jerónimo, E. Condensed tannins as antioxidants in ruminants—Effectiveness and action mechanisms to improve animal antioxidant status and oxidative stability of products. *Animals* **2021**, *11*, 3243. [CrossRef]
59. Chiofalo, V.; Liotta, L.; Fiumanò, R.; Riolo, E.B.; Chiofalo, B. Influence of dietary supplementation of Rosmarinus officinalis L. on performances of dairy ewes organically managed. *Small Rumin. Res.* **2012**, *104*, 122–128. [CrossRef]
60. Jaguezeski, A.M.; Perin, G.; Bottari, N.B.; Wagner, R.; Fagundes, M.B.; Schetinger, M.R.C.; Morsch, V.M.; Stein, C.S.; Moresco, R.N.; Barreta, D.A.; et al. Addition of curcumin to the diet of dairy sheep improves health, performance and milk quality. *Anim. Feed Sci. Technol.* **2018**, *246*, 144–157. [CrossRef]
61. Castelani, L.; Pfrimer, K.; Giglioti, R.; van Cleef, E.H.C.B.; Salles, M.S.V.; Júnior, L.C.R. Effects of thyme (*Thymus vulgaris* L.) essential oil supplementation on the microbiological quality of raw milk of lactating dairy cows. *Res. Vet. Sci.* **2023**, *161*, 118–121. [CrossRef]
62. Rodrigues, R.O.; Cooke, R.F.; Firmino, F.C.; Moura, M.K.; Angeli, B.F.; Ferreira, H.A.; Brandao, A.P.; Gex-Farby, M.; Ostrensky, A.; Vasconcelos, J.L. Productive and physiological responses of lactating dairy cows supplemented with phytogenic feed ingredients. *Transl. Anim. Sci.* **2019**, *3*, 1133–1142. [CrossRef] [PubMed]
63. Albenzio, M.; Caroprese, M. Differential leucocyte count for ewe milk with low and high somatic cell count. *J. Dairy Res.* **2011**, *78*, 43–48. [CrossRef] [PubMed]
64. Hamed, H.; El Feki, A.; Gargouri, A. Total and differential bulk cow milk somatic cell counts and their relation with antioxidant factors. *Comptes Rendus Biol.* **2008**, *331*, 144–151. [CrossRef]
65. Caroprese, M.; Marzano, A.; Schena, L.; Marino, R.; Santillo, A.; Albenzio, M. Contribution of macrophages to proteolysis and plasmin activity in ewe bulk milk. *J. Dairy Sci.* **2007**, *90*, 2767–2772. [CrossRef] [PubMed]
66. Albenzio, M.; Santillo, A.; Caroprese, M.; d'Angelo, F.; Marino, R.; Sevi, A. Role of endogenous enzymes in proteolysis of sheep milk. *J. Dairy Sci.* **2009**, *92*, 79–86. [CrossRef]

antioxidants

MDPI

Review

Antioxidant Activity, Metabolism, and Bioavailability of Polyphenols in the Diet of Animals

Drago Bešlo [1,*], Nataša Golubić [1], Vesna Rastija [1], Dejan Agić [1], Maja Karnaš [1], Domagoj Šubarić [1] and Bono Lučić [2]

[1] Faculty of Agrobiotechnical Sciences Osijek, J. J. Strossmayer University Osijek, Vladimira Preloga 1, HR-31000 Osijek, Croatia; golubicnatasa@gmail.com (N.G.); vrastija@fazos.hr (V.R.); dagic@fazos.hr (D.A.); mkarnas@fazos.hr (M.K.); dsubaric@fazos.hr (D.Š.)

[2] NMR Center, Ruđer Bošković Institute, Bijenička cesta 54, HR-10000 Zagreb, Croatia; lucic@irb.hr

* Correspondence: dbeslo@fazos.hr; Tel.: +385-3155-4915

Abstract: As the world's population grows, so does the need for more and more animal feed. In 2006, the EU banned the use of antibiotics and other chemicals in order to reduce chemical residues in food consumed by humans. It is well known that oxidative stress and inflammatory processes must be combated to achieve higher productivity. The adverse effects of the use of pharmaceuticals and other synthetic compounds on animal health and product quality and safety have increased interest in phytocompounds. With the use of plant polyphenols in animal nutrition, they are gaining more attention as a supplement to animal feed. Livestock feeding based on a sustainable, environmentally friendly approach (clean, safe, and green agriculture) would also be a win–win for farmers and society. There is an increasing interest in producing healthier products of animal origin with a higher ratio of polyunsaturated fatty acids (PUFAs) to saturated fatty acids by modulating animal nutrition. Secondary plant metabolites (polyphenols) are essential chemical compounds for plant physiology as they are involved in various functions such as growth, pigmentation, and resistance to pathogenic organisms. Polyphenols are exogenous antioxidants that act as one of the first lines of cell defense. Therefore, the discoveries on the intracellular antioxidant activity of polyphenols as a plant supplement have contributed significantly to the improvement of antioxidant activity, as polyphenols prevent oxidative stress damage and eliminate excessively produced free radicals. To achieve animal welfare, reduce stress and the need for medicines, and increase the quality of food of animal origin, the addition of polyphenols to research and breeding can be practised in part with a free-choice approach to animal nutrition.

Keywords: animal nutrition; supplementation of polyphenols; bioavailability of polyphenols; biotransformation; metabolism; antioxidant/pro-oxidant activity; diversity of microbiota; gut microbiota composition; immunomodulation; animal health; animal product quality

Citation: Bešlo, D.; Golubić, N.; Rastija, V.; Agić, D.; Karnaš, M.; Šubarić, D.; Lučić, B. Antioxidant Activity, Metabolism, and Bioavailability of Polyphenols in the Diet of Animals. *Antioxidants* **2023**, *12*, 1141. https://doi.org/10.3390/antiox12061141

Academic Editors: Luciana Rossi and Matteo Dell'Anno

Received: 7 April 2023
Revised: 4 May 2023
Accepted: 15 May 2023
Published: 23 May 2023

1. Introduction

Plant-based dietary supplements for domestic animals have increased significantly in the last decade. Polyphenols are secondary metabolites of plants that contain bioactive compounds and have beneficial effects on animal organisms. Polyphenols in plants play an important role in growth and reproduction and provide protection against pathogens and herbivores [1]. There is great interest in these phytochemicals for their health benefits and effects on animals. [2,3]. Phenolic compounds, also known as polyphenols, are a class of compounds found in various plant species. They are built from one or more aromatic rings on which one or two hydroxyl groups are present. There are three main groups of polyphenols: flavonoids, non-flavonoids, and tannins. Their biological function depends mainly on the chemical structure [4]. More than 10,000 compounds have been identified that have anti-inflammatory, immunomodulatory, and antimutagenic properties. The positive effects of phenolic compounds are attributed to their antioxidant activity [5]. Their role as antioxidants is comparable to important biological and well-known antioxidants: vitamins E and C. Despite these benefits, polyphenols are characterized as substances with low bioavailability, and additional research is needed to investigate their efficacy in the diet of domestic animals [2]. The higher level of unsaturated fatty acids compared to saturated fatty acids arouses interest in the production of healthier animal products in order to change the diet of animals. This dietary strategy has been linked to increased lipid peroxidation, a process in which free radicals "steal" electrons from lipids in cell membranes, leading to cell damage. It is important to maintain the quality of meat and dairy products by mitigating oxidative decay. The addition of antioxidant molecules to food or to the final product controls and reduces the occurrence of oxidative stress.

It is extremely important to understand and relate the inflammatory processes in animals to prevent oxidative stress. Oxidative stress is the result of an imbalance between reactive oxygen species (ROS) and antioxidants. These include enzymatic antioxidants (e.g., carotenoids, tocopherols, polyphenols, and glutathione) and antioxidant enzymes (e.g., superoxide dismutase (SOD), catalase (CAT), and glutathione peroxidase (GPX)) [6]. The accumulation of ROS in cells can seriously damage macromolecules and continuously (in a chain) stimulate the production of ROS.

Oxidative stress contributes to the development of various diseases and chronic pathological conditions. It is known to be involved in the development of pneumonia [7] and sepsis [8] in pigs. It also occurs, for example, in piglets that are switched from mother's milk to solid dry feed [9], because the digestive tract and the body's own system are not yet mature, resulting in reduced appetite and stunted growth [10]. In animals exposed to stressful conditions such as transport, starvation, and low and high temperatures, glucocorticoids (CTC) are released, leading to an increase in free radicals and an increased rate of oxidation of microfibrillar proteins, resulting in the inhibition of muscle growth and muscle wasting [11,12].

Antioxidant compounds are needed to prevent the formation of new free radicals, and polyphenols are prominent compounds that react with free radicals. The formation of free radicals ultimately leads to oxidative stress, and the addition of antioxidant compounds can stop the further spread of ROS. In addition, some experimental studies have confirmed positive results in vivo with the addition of polyphenols, indicating their potential as natural antioxidants [13–15].

Interest in the use of natural antioxidants in food production instead of synthetic antioxidants has increased in recent years, as they are less harmful to the environment and are also used for economic reasons. In addition, natural antioxidants are better for the end consumer as they are considered safer [4].

Recently, there has been growing interest in the use of by-products in animal nutrition because they contain high levels of unsaturated fatty acids (PUFA) as well as a high concentration of polyphenols [16]. This diet may contribute to a greater stability of poultry meat to fatty acid oxidation and a greater stability of meat products for human consumption [17,18]. In pig diets, the use of grape pomace (GP) with a high PUFA content (60.9–64.4%) and a high PUFA/SFA ratio (2.80–3.0) resulted in improved effective growth, altered the composition of fatty acids in adipose tissue, and resulted in better meat quality [19,20].

A review of the literature on the supplementation of polyphenols in animal diets was done to follow the changes (metabolism and biotransformations) of polyphenols in the digestive tract of monogastric animals. Some monogastric animals may consume foods of animal origin in addition to a plant-based diet, such as pigs, dogs or birds, while horses or rabbits consume plant-based diets. Results from the literature on the bioavailability, biotransformation and metabolism of polyphenols in relation to the gut microbiota and its diversity are considered and reviewed. Biological activities such as antioxidant and pro-oxidant activities are analysed as well as the effects of polyphenols on animal growth and the immune system by modulating the diversity of the gut microbiota in animals.

Finally, the results of polyphenol supplementation on animal health and the quality of animal products used in human nutrition will be analysed.

2. Classification of Polyphenols

The classification of polyphenols is based on the number of phenolic rings and the structural elements that connect the rings. They are divided into four classes: phenolic acids, flavonoids, stilbenes, and lignans (Figure 1). The phenolic acids are further subdivided into hydroxybenzoic acid and hydroxycinnamic acid. About one-third of the polyphenolic compounds are phenolic acids and are found in almost all plant materials. Caffeic acid, gallic acid, and ferulic acid are some of the most common phenolic acids.

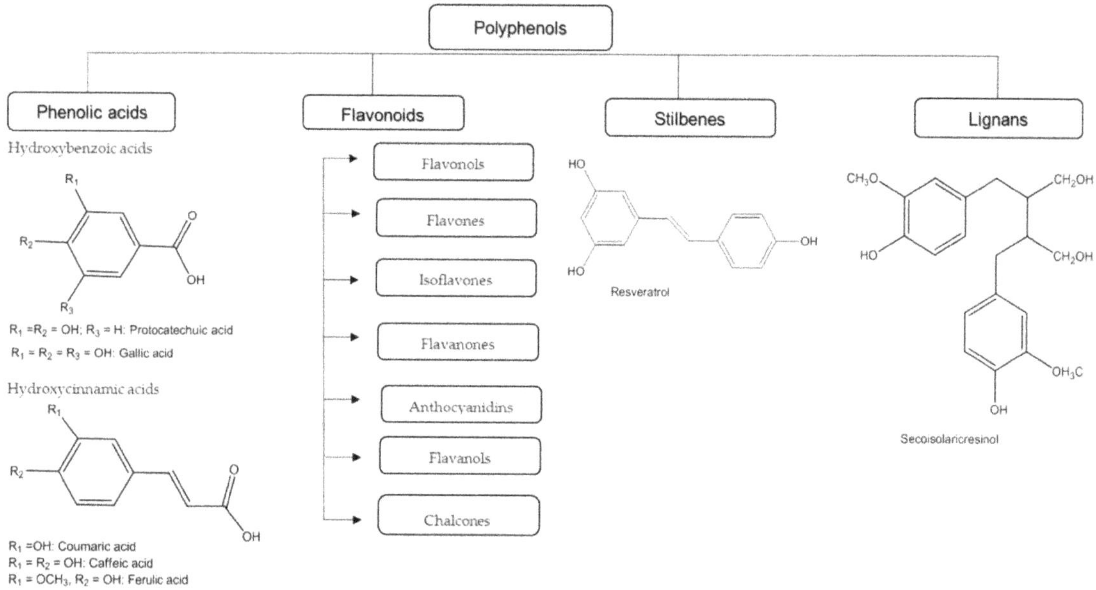

Figure 1. Classification of polyphenols based on their chemical structure (arranged according to Bešlo et al., 2022, [21]).

The most common polyphenols are flavonoids, which have a common basic structure. They consist of two aromatic rings joined by three carbon atoms, forming an oxygenated heterocycle.

Flavonoids have been classified into different subclasses based on the substitution content of the C ring, the oxidation state of the heterocyclic ring, and the position of the B ring (Figure 2). For these reasons, flavonoids are divided into seven main subclasses: flavanols, flavonones, isoflavones, anthocyanins, chalcones, flavones, and flavonols (Figure 2). In flavanones, flavones, flavonols, flavanols, and anthocyanins ring B is located at position 2 of the heterocyclic ring, while in isoflavonoids at position 3. Flavanones and flavonols have a saturated central heterocyclic ring and, in this case, one or more chiral centers are present. On the other hand, anthocyanins, isoflavones, flavones, and flavonols have an unsaturated central heterocyclic ring and the molecule is achiral [22,23].

Stilbenes contain two phenyl residues linked by a two-carbon methylene bridge. Most stilbenes in plants act as antifungal compounds that are synthesized only in response to infection or injury. The most extensively studied stilbene is resveratrol. Lignans are diphenolic compounds with the structure of 2,3-dibenzylbutane formed by the dimerization of two cinnamic acid residues [21,24].

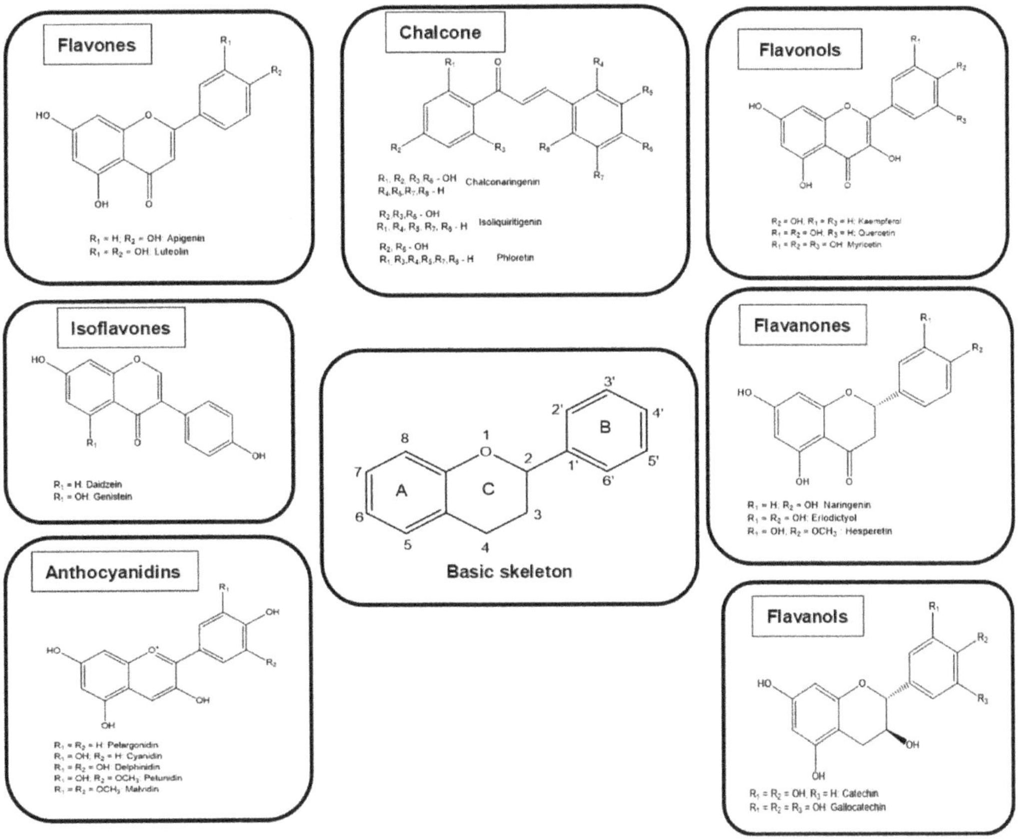

Figure 2. Chemical structures of flavonoids.

3. Reactive Oxygen Species and Reactive Nitrogen Species

All aerobic organisms have a need for molecular oxygen, and at the same time it is ironic that they have to defend themselves against the dangerous oxidase [25]. The basis of aerobic life is the oxidation of organic compounds, namely carbohydrates, proteins, and fats, which provide the metabolic energy necessary for life functions. Free radicals and antioxidants have become common topics in today's discussion of disease mechanisms [26]. It is normal for various metabolic processes to produce free radicals. The formation of free radicals is well regulated by the physiological process in the aerobic cell. In homeostasis, the formation of free radicals and their elimination are in equilibrium because during the formation of free radicals in the cell, there is an antioxidant defense system that eliminates ROS and reactive nitrogen radicals (RNS, reactive nitrogen species). When an imbalance occurs, oxidative stress results [27].

3.1. Sources of ROS

The production of ROS is a natural component of aerobic life, responsible for various cellular functions, from oxygen transport pathways to defense against microbial invasion and gene expression to growth and death promotion [28–31]. ROS can be produced in the mitochondria, where O_2 is reduced to $O_2{\cdot}^-$ in the ETC (electron transport chain) process, where the transfer of electrons from NADH and $FADH_2$ to O_2 can lead to a "leakage" of electrons at complexes I and III. However, the enzymes monoamine oxidase,

2-ketoglutarate dehydrogenase, and glycerol phosphate dehydrogenase, which are located in the mitochondrion, can additionally contribute to production [32–34].

A reaction takes place at the inner membrane of the mitochondria in the respiratory chain in which partial intermediates O_2 can be formed [35].

$$O_2 \xrightarrow{+e^-} \cdot O_2^- \xrightarrow{+e^-(+2H^+)} H_2O_2 \xrightarrow{+e^-} \cdot HO + OH^- \xrightarrow{+e^-(+2H^+)} 2H_2O$$

The three primary species, i.e., the superoxide anion ($O_2{}^-$), the hydrogen peroxide (H_2O_2), and the hydroxyl radical ($\cdot$OH), are called oxygen with reactive properties. $O_2{}^-$ and $\cdot$OH are usually referred to as "free radicals" and H_2O_2 and $^1O_2$ as non-radicals. An excessive amount of ROS is very toxic to the cell. A usually high concentration of ROS in a cell is called "oxidative stress" and can damage various molecules in the cell, e.g., lipids, proteins and DNA. In particular, in the cell membrane, it can act as a double bond and cause lipid peroxidation, increasing the permeability and fluidity of the membrane. In damaging proteins ROS can cause site-specific modifications, fragmentation of protein chain, as well as susceptibility to proteolysis [36]. Finally, ROS can damage DNA through deoxyribose oxidation, strand scission, nucleotide removal, base modification and DNA–protein cross-linking [13–15,37]. The protective effect against ROS is exerted by several enzymes, namely superoxide dismutase (SOD), catalase (CAT) and glutathione peroxidase (GPx), as well as non-enzymatic compounds such as vitamin E, β-carotene, ascorbic acid, glutathione (GSH) and exogenously supplied polyphenols [38–41].

In addition to the production of ROS in organelles, several enzymes, including cytochrome P450 (CYP) 2E1, NADPH oxidase (NOX), cyclooxygenases, xanthine oxidases, and lipoxygenases, produce ROS in the plasma membrane and cytosol [42].

Similarly, sources of the formation of oxidants are unwanted substances in food, such as pesticides, organic solvents or mycotoxins. These compounds induce the xenobiotic system of the liver, which produces oxidants as by-products [43]. Oxidative stress is directly related to inflammation because oxidants activate NF-kB, an important regulator of inflammation [44]. NF-kB is a protein complex found in almost all types of animal cells and i bound in an inactive state to inhibitory proteins in the cytosol. Upon stimulation by oxidants and various other stimuli such as cytokines, free radicals, heavy metals, bacterial stimuli and viruses, inhibitory proteins are released from nuclear factor kappa B (NF-kB), facilitating the translocation of active NF-kB into the nucleus and activation of transcription of a large number of genes involved in all aspects of inflammation (e.g., vasodilation, chemotaxis, cell adhesion and phagocytosis) [45]. Typical proteins encoded by NF-kB target genes include pro-inflammatory cytokines, chemokines, inflammatory enzymes, adhesion molecules and various receptors. Several NF-kB-regulated proteins such as cytokines and chemokines stimulate the production of oxidants by activated neutrophils (respiratory burst) and their mitochondria, thereby promoting oxidative stress [46]. If this cycle in which oxidative stress develops cannot be broken due to long-term and excessive production of oxidants, the inflammatory process becomes chronic and cells and tissues are damaged [47].

3.2. Nitrosative Stress

RNS are nitric oxide (NO$\cdot$), peroxynitrite (ONOO$\cdot$) and nitric oxide NO_2 [42,48]. NO is formed from L-arginine by three major isoforms of nitric oxide synthase (NOS). It is a very versatile molecule with numerous functions and mechanisms of action. It is described as a diffusing radical that leads to vasodilation and plays a key role in the vascular system. Post-translational modifications mediated by NO lead to the formation of another reactive nitrogen with superoxide, resulting in peroxynitrate ONOO$\cdot$ [49]. In an aqueous environment, the peroxynitrous acid HONO is formed, which can dissociate with NO_2 and the highly reactive hydroxyl radical ($\cdot$OH) [50,51]. The hydroxyl radical is reactive enough to remove electrons from almost any biological molecule. For example, it forms a tyrosyl radical with tyrosine, which can react with NO_2 to form nitrotyrosine.

The phagocyte immune system uses the phagocyte's ability to modify proteins and other molecules. By swallowing and killing bacteria due to the activation of macrophages and neutrophils, ROS are released in a "burst of breath". After phagocytosis of antigenic particles, NADPH oxidase produces large amounts of superoxide [52–57]. Furthermore, the bactericidal function of activated macrophages depends on the presence of peroxynitrite and the fact that inhibition of superoxide or nitric oxide generation inhibits this function [54]. While peroxynitrite serves to protect the body from pathogens, its overproduction or dysregulation of these metabolic pathways is also harmful.

3.3. Free Radicals and Internal Defense

Oxidative stress means a disturbance in the balance between the formation of free radicals and the ability of the body's defense systems to eliminate them [58]. Oxidants are free radicals, which are atoms, ions or molecules that contain an unpaired electron in the outer electron shell. The most common are reactive oxygen radicals (ROS) and reactive nitrogen radicals (RNS). ROS are superoxide radicals (O_2^-), hydroxyl radicals (OH), peroxyl radicals (RO_2), alkyl radicals (RO), hypochlorous acid (HClO), ozone (O_3), singlet oxygen ($^1\Delta gO_2$) and hydrogen peroxide (H_2O_2). The hydroxyl radical is the most reactive, i.e., it is characterized by low substrate specificity and a short half-life. Figuratively speaking, it can most easily split an electron from the surrounding molecules. It is an important trigger of lipid peroxidation.

RNS are reactive species that contain both nitrogen and oxygen, e.g., nitric oxide (NO), nitrogen dioxide (NO_2), peroxynitrite ($ONOO^-$), etc. NO produces various isoforms of the enzyme NOS (nitric oxide synthase) [42]. At low concentrations, it is essential as a neurotransmitter and hormone that causes vasodilation. In high concentrations, it combines with an oxygen molecule or a superoxide radical to produce RNS, whose effect on cells is similar to that of ROS [59,60].

There are physiological mechanisms in cells to suppress ROS when they become out of control under conditions of oxidative stress. These negative effects can be significantly attenuated by antioxidants, a heterogeneous class of chemicals whose common feature is the ability to interrupt radical chain reactions and thus prevent or limit cell damage. In cells, there are defense mechanisms that attempt to maintain homeostasis. These include the enzymes superoxide dismutase (SOD), catalase, and glutathione peroxidase.

Superoxide dismutase catalyzes the following reaction:

$$O_2\cdot + O_2\cdot + 2H^+ \rightarrow H_2O_2 + O_2$$

Catalase:

$$2H_2O_2 \rightarrow 2\,H_2O + O_2$$

$$ROOH + AH_2 \rightarrow H_2O + ROH + A$$

Glutathione peroxidase:

$$ROOH + 2GSH \rightarrow ROH + GSSG + H_2O$$

3.4. Exogenous Molecules in Defense against Oxidative Stress

Plants are a tremendous source of antioxidants. Many of them are already known and widely used, and many others are still waiting to be discovered. Plant antioxidants generally have the chemical character of polyphenols and can break radical chain reactions by forming phenoxy radicals, which are generally less reactive and/or converted into relatively stable dimers, quinones, etc. For a number of natural polyphenols, some studies on the structure-activity relationship are available [61–64], in which it was shown that the number and arrangement of OH groups in the structure of polyphenols are the most important factors for their antioxidant activity. In general, these arrangements of the OH groups reflect the dissociation energy of the OH bonds through which the polyphenols exert their activity as antioxidants.

Due to their proven therapeutic importance in prevention and treatment, bioactive molecules of plant origin have attracted much attention, whether whole plants, plant extracts, or even isolated compounds are used. Larger randomized studies are needed to provide clear evidence of the benefits/risks of antioxidant supplementation. Antioxidants are also sensitive to oxidation, so their use as foods (or food supplements) should be carefully considered as oxidation and reduction reactions do not take place in them in isolation. Taking high doses of antioxidants is coming under increasing criticism, as evidence of some adverse effects is accumulating. The investigation of their chemical constituents as future prophylactic and therapeutic agents is of particular interest as they are more effective and safer than those commonly available [65].

The cause of cell damage can be lipid peroxidation induced by free radicals, DNA chain cleavage, and protein oxidation [66–68]. ROS and RNS have important physiological effects in certain concentrations. It is known that living organisms can regulate the concentration of ROS and RNS through dietary intake of antioxidants and endogenous production of antioxidants and, for example, systems for inactivating excess radicals in order to maintain a balance between antioxidants and ROS/RNS that allows the body to function normally. An imbalance that favours the accumulation of ROS/RNS is defined as oxidative/nitrosative stress. Current indications suggest that oxidative/nitrosative stress is involved in several diseases, including liver disease [69,70].

Free radicals include not only ROS and RNS but also other radicals [71,72]—details are summarized in (Table 1).

Table 1. Reactive species (ROS and RNS) contributing to the occurrence of oxidative stress [a].

Name of the Substance		Symbol	Half-Life [b]	Characteristic
Radicals	Superoxide	$O_2\cdot$	10^{-6} s	unstable molecule, signaling molecule
	Hydroxyl	$OH\cdot$	10^{-10} s	very reactive and unstable species, it is created in the Fenton and Haber–Weiss reaction with an iron catalyst
	Alkoxyl radical	$RO\cdot$	10^{-6} s	organic (lipid) radical
	Peroxyl Radical	$ROO\cdot$	17 s	it is formed from organic hydroperoxide (ROOH), by removing hydrogen
	Nitric oxide	$NO\cdot$	s	environmental toxin, endogenous signaling molecule
	Nitrogen dioxide	$NO_2\cdot$	s	highly reactive species, environmental toxin
Non radical	Hydrogen peroxide	H_2O_2	Stable	a cellular toxin, signaling role, generation of other free radicals
	Singlet oxygen	$^1O_2$	10^{-6} s	the first excited form of oxygen
	Ozone	O_3	s	environmental toxin
	Organic peroxide	ROOH	Stabile	it easily decomposes into radicals, so it serves as a catalyst for radical reactions
	Peroxynitrous	$ONOO^-$	Stabile	highly reactive species, environmental toxin
	nitrogen oxides	NO_x	s	environmental toxin, including NO and NO_2 derivatives formed in the combustion process

[a] Arranged from references [71,72]. [b] The half-life of some radicals depends on the surrounding medium, for example, the half-life of NO· in an air-saturated solution can be a few minutes; s seconds.

The enzymatic antioxidant defense system includes antioxidant enzymes such as superoxide dismutase, glutathione reductase, catalase, and other enzymes that play a key role in detoxifying radicals into non-reactive molecules [69,70] (Figure 3). Under physiological conditions, these enzymes keep the radical concentration in the cell low and their activity is regulated by precise mechanisms at the molecular level. All these enzymes are essential for maintaining homeostasis between oxidation and antioxidant capacity and for the survival of all aerobic organisms.

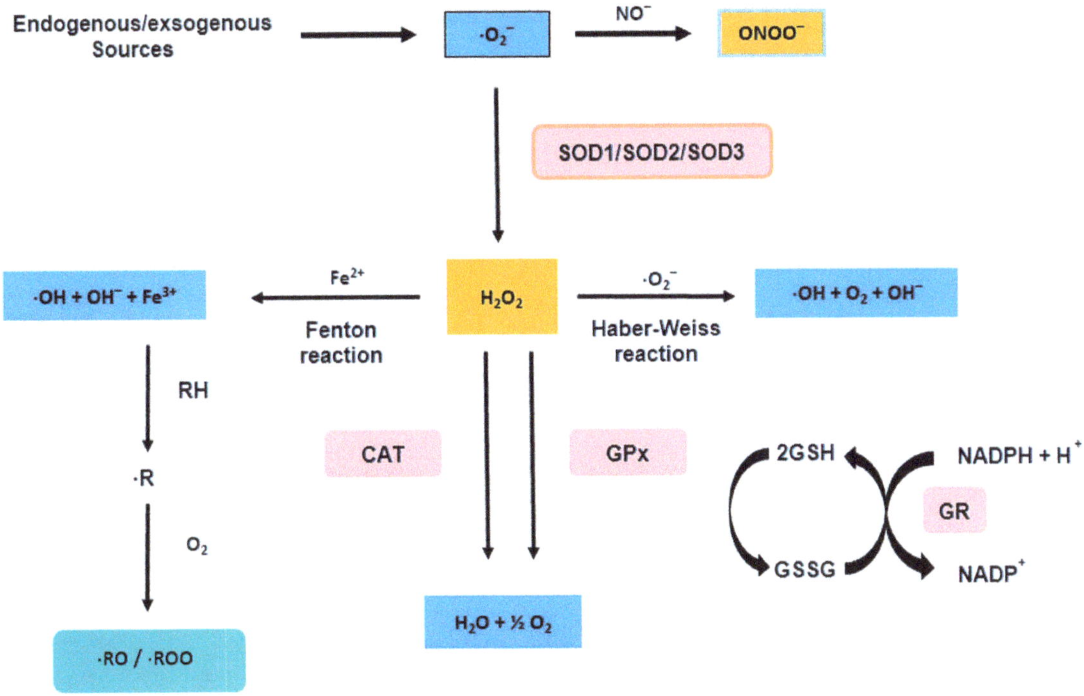

Figure 3. Sources and generation of reactive oxygen species (ROS). Abbreviations: CAT (catalase), GR (glutathione reductase), SOD (superoxide dismutase), RH (lipid membrane), R (alkyl radical).

4. Distribution of Polyphenols in Nature

Polyphenols are found in almost all plants, They have been confirmed to be present in their various parts such as roots, leaves, flowers, fruits and seeds, which, among other things, protect plants from various pests and UV radiation [73]. Their distribution is variable at the tissue level. Thus, the outer layers of plants contain higher amounts of polyphenols than the inner layers, or the insoluble polyphenolic compounds are associated with the cell wall, while the soluble ones are found in the vacuoles [74]. In particular, prolonged scalding of the fruits of the plant, decreases the concentration of polyphenols decreases, especially at higher temperatures, which is a consequence of the sensitivity of polyphenols to oxidation [21]. Fruits, vegetables, legumes, nuts, and herbs are typical sources rich in polyphenols, and these foods are also consumed by animals [75,76]. Antioxidant properties are attributed to polyphenols, because they can act as chain breakers or radical scavengers, depending on their chemical structure [77,78].

Plants synthesize an incredible variety of metabolites with different properties. Secondary plant metabolites are generally associated with plant defense, but also with a variety of biological properties. The concentration of secondary metabolites and their activities in the biological system vary depending on the maturity of the plant and its parts, as well as the soil conditions, the availability of water and light, and other environmental conditions under which the plant grows [79].

In order to ensure the desired high productivity of livestock production, it is necessary to control oxidative stress and inflammatory processes, which are closely linked. Oxidative stress has become a major challenge in animal husbandry because it affects the growth of animals. Oxidative stress reduces the globulin concentration in plasma and thus reduces the immune status of poultry [80]. Many studies have been conducted to solve this problem and one of the safest solutions is the use of polyphenols. Polyphenols are well known as exogenous antioxidants that act as one of the first lines of cellular defense. The discovery of the intracellular antioxidant activity of polyphenols as a plant food supplement contributes significantly to the improvement of antioxidant activity because polyphenols prevent oxidative stress damage and eliminate excessively produced free radicals [80].

5. Digestion, Absorption, and Metabolism of Polyphenols

The biological effects of certain compounds contained in plant foods for animals or other plant preparations depend on their uptake, metabolism, distribution, and excretion from the organism, i.e., on the bioavailability after their uptake into the organism as well as on the reducing properties of the resulting metabolites. Understanding the processes involved in the uptake and distribution of polyphenols is crucial for determining their potential bioactive effects in vivo as well as their overall importance in the prevention of a number of diseases associated with oxidative stress [81–84].

5.1. Digestion/Metabolism/Biotransformation and Bioavailability of Polyphenols in Animals

Recent studies on the bioabsorption of polyphenols indicate their low bioavailability after after intake of relatively high doses. The main obstacle to their pharmacological use is their poor bioavailability, which is related to the interactions of polyphenols in the different phases of digestion, absorption, and distribution that alter their molecular structure, especially the interactions with food, digestive enzymes, and transporters in the intestine and blood proteins [85].

In explaining the biological effects of polyphenols, it was assumed that they are bioavailable and can effectively reach the target tissue. It is very important to understand the processes by which they are absorbed, metabolized, and excreted from the body. The study on absorption is complicated because of the molecular complexity of foods rich in polyphenols and other factors such as the degree of polymerization and conjugation with other compounds and phenols. Most polyphenols are present in food in the form of esters, glycosides or polymers, and cannot be absorbed in these forms. Once absorbed, polyphenols are recognized by the body as xenobiotics, so their bioavailability is relatively low compared to micro- and macronutrients.

The metabolism of polyphenols is similar to metabolic detoxification to reduce their potential cytotoxic effects by increasing their hydrophilicity and facilitating excretion in urine or bile [23]. The structure of polyphenols, not their concentration, determines the rate and also the extent of absorption as well as the type of circulating metabolites in plasma. Depending on their degree of structural complexity and polymerization, these compounds can be readily absorbed in the small intestine (monomeric and dimeric polyphenols) or almost unchanged reach the large intestine (oligomeric and polymeric polyphenols) [86].

According to some estimates, only 5–10% of ingested polyphenols can be absorbed in the small intestine. After absorption, less complex polyphenolic compounds can be hydrolyzed and biotransformed in enterocytes and then in hepatocytes. The result is a series of hydrophilic conjugated metabolites (methyl, glucuronide, and sulphate derivatives) that rapidly enter the bloodstream and are further distributed to organs or excreted in the urine [24]. It is difficult to track the fate of each compound in the body because ten new compounds may be formed when each phenolic compound is metabolised [87]. Very often it is not possible to detect the parent phenol, because in most cases metabolic processes convert phenolic antioxidants into completely different molecules [87]. As xenobiotics, they may first undergo oxidation, reduction or hydrolysis reactions (phase I metabolism) in enterocytes, introducing or exposing a functional group, such as a hydroxyl group, especially for conjugation (phase II metabolism) [87,88]. The aromatic structures in enterocytes usually remain intact, but the hydroxyl groups on the phenolic aromatic ring are successfully conjugated to glucuronide, sulphate, and/or methylated metabolites by the action of uridine-5-diphosphate glucuronosyltransferase (UGT), sulphotransferase (SULT) and catechol-*O*-methyltransferase. (COMT) (Figure 4) [89–91].

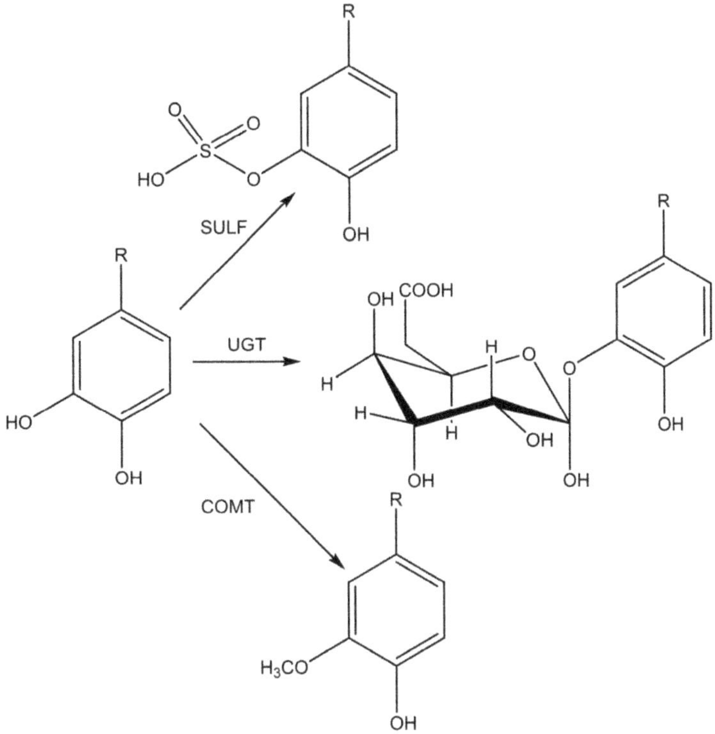

Figure 4. Conjugation reaction of phenolic hydroxy groups: glucuronidation, sulphonation, and/or methylation by uridine-5-diphosphate glucuronosyltransferase (UGT), sulphotransferase (SULT), and catechol-O-methyltransferase (COMT).

The conjugation of catechol antioxidants usually occurs mainly at the meta position [84,92,93]. However, conjugation with a glucuronic acid or sulphate group can also occur at the para position, usually following conjugation at the meta position with a methyl group.

5.2. Biotransformation of Polyphenols by the Gut/Colon Microbiota

5.2.1. Fate of Polyphenols in the Stomach and Small Intestine

When food passes from the oral cavity to the esophagus, the ingested polyphenols enter the stomach. Due to low pH and contact with bacteria, some flavonoids can be degraded into phenolic acids, but in general no significant changes occur under the acidic conditions of the stomach [68]. Flavonoid glycosides can hydrolyze under acidic conditions in the stomach and pass unchanged into the small intestine.

Some of the metabolites from the enterocytes enter the duodenum via bile and are then hydrolyzed in the large intestine by bacterial enzymes (mainly β-glucuronidase). 95% of the body's microbiota is found in the intestines, primarily in the colon, where microbial cells outnumber the total number of cells in the body [4]. With the sequencing methods available today, it is possible to determine the presence of a healthy or disturbed microbiome [86]. The intestinal microbiota of the digestive system plays an important role in health and disease, but our understanding of the composition, dynamics, and functionality of the intestinal ecosystem is still rudimentary [87].

5.2.2. Polyphenols and the Gut Microbiome

This enterohepatic recycling can result in a longer retention of polyphenols in the body. The remaining polyphenols (90–95% of the total ingested polyphenols) may accumulate in the lumen of the large intestine, from where they are excreted under the action of enzymes together with conjugates into the lumen of the intestine. The intestinal microbial community produces metabolites such as aromatic acids (hydroxyphenyloctene, phenylpropionic acid, phenylbutyric acid, phenylvalerolactones, etc.) [86,94]. All these phenolic metabolites produced by microbes can be absorbed or excreted in the feces. Once absorbed, they enter the liver via the portal vein, where they can undergo extensive metabolism (including glucuronidation, methylation, sulphation, or a combination thereof) until they finally enter the systemic circulation and are distributed to organs or excreted in the urine. The intestinal microbiota is responsible for the extensive degradation of the original structures of polyphenols into a series of molecules of low molecular weight, which, because they can be absorbed, may be responsible for the biological activity resulting from the consumption of foods rich in polyphenols, rather than the original compounds in the food. After consumption, the concentration of polyphenols that is achieved varies considerably depending on the type of polyphenols and the food source. With normal consumption, the concentration of intact flavonoids in plasma rarely exceeds 1 μM, and peak concentrations of 1 μM are usually reached 1–2 h after ingestion [86]. Polyphenols and their derivatives are mainly excreted through urine and bile.

There are very few studies on the interaction of polyphenolics with the gut microbiota in animal diets. *In vivo,* studies have shown that resveratrol has great potential as an antibiotic alternative to reverse the negative effects of weaning stress in piglets infected with *E. coli* and *Salmonella* [95–98] on their growth, immunity, and microbiota [99]. Similarly, grape pomace supplementation in the diet of pigs reduces diarrhea caused by *E. coli* [100]. Fiesel et al. [100] showed that feeding pomace flour extract and grape pomace alters the microbial composition and leads to a reduction in *Streptococcus* spp. and *Clostridium* in the fecal microbiota.

5.2.3. Reaction Phases of the Biotransformation of Polyphenols

The biotransformation reactions that take in place during phase I metabolism are oxidation, reduction, and hydrolysis. The main purpose of these reactions is to increase the polarity of heterogeneous phenolic compounds to facilitate their excretion [101,102]. Day et al. [91] found that these reactions significantly affect the antioxidant activity of flavonoids and their cross-reaction with proteins.

The second stage of biotransformation involves the incorporation of various chemical radicals into exogenous compounds. The free radicals transported in the body originate from endogenous, polar, and easily accessible molecules. The main goal of this phase is to increase the polarity of exogenous compounds [103].

The liver and small intestine (especially the jejunum and ileum) are the key organs responsible for various biotransformations leading to the formation of different conjugated forms of flavonoids; however, the kidney and other organs and tissues are also involved in flavonoid metabolism [24]. The most important metabolic reactions of flavonoid biotransformation are glucuronidation, sulphation, O-methylation, oxidation, reduction, and hydrolysis [103].

5.2.4. The Most Important Metabolic Reactions in the Biotransformation of Polyphenols

The most important metabolic reaction to which flavonoids are subjected is the glucuronidation reaction. It consists of the transfer of glucuronic acid, which is bound to a specific substrate as UDP-glucuronic acid by the microsomal enzymes uridine-5′-diphosphate-glucuronisyltransferases (UGTs). The enzyme family UGT exhibits exceptional diversity in substrate recognition and catalyzes the glucuronidation of a large number of functional groups (e.g., -OH, -COOH, -NH_2, and -SH). The reaction takes place on the luminal side of the endoplasmic reticulum, and UGT1A9 and UGT1A3, which are found in the intestine and liver, are thought to play the most important role [104]. The glucuronidation reaction allows the organism to make the substrates of the individual metabolic pathways more water-soluble. It is assumed that 80% of the metabolic pathways of flavonoids are due to the action of the enzyme UGT [105,106]. The chemical reaction of sulphation is mediated by sulphotransferases (SULT), enzymes responsible for catalyzing the sulphation of flavonoids and many drugs. In studies dealing with the sulphation of flavonoids, it was found that the sulphation process takes place mainly at the C7- OH position and that removal of the hydroxyl group reduces or inhibits the sulphation process [107–109].

Manach et al. [110] found that sulphate esters and glucuronides retain some of their antioxidant activity and still retard the in vitro oxidation of low-molecular-weight lipoproteins. However, Zhang et al. [111] have shown that glucuronidation of flavonoids reduces their biological activity. The glucuronides of daidzein and genistein have 10- and 40-fold lower affinity for estrogen receptors than the corresponding aglycones, respectively [112]. O-methylation of flavonoids is a common xenobiotic transformation in micro-organisms, plants, and mammals catalyzed by O-methyltransferases (OMTs) [113]. Compared to other flavonoids, flavan-3-ols are more susceptible to methylation in the jejunum, as evidenced by the specificity of catechol O-methyltransferase (COMT) for these compounds. O-methylation may reduce the biological activity of polyphenols: their antioxidant activity and effect on endothelial function [114].

Methylation may also affect the reduction of toxicity of flavonoids and polyphenols in general. Indeed, most polyphenols contain catechin groups that can be oxidized to toxic quinones in vivo. Similar quinones formed from endogenous estrogens and catecholamines lead to the formation of superoxide radicals through a reaction with nucleophilic molecules in cells. Even the smallest changes in the structure of flavonoids can significantly affect their activities in the body. Most often, biological activity is reduced and excretion accelerated, but there is also evidence of the formation of more biologically active compounds [115]. However, methylation of free hydroxyl groups in quercetin leads to more metabolically stable derivatives that readily penetrate membranes, resulting in better bioavailability. All these effects enhance the biological effects of quercetin, and the same applies to other flavones [115].

5.2.5. Uptake of Polyphenol Metabolites into the Body

After the metabolism of polyphenols and their absorption in the small intestine, the metabolites of polyphenols enter the systemic circulation, where they are transported to the liver via the portal vein. The main metabolites found in the portal vein are mainly glucuronides and methylated glucuronides. These polar conjugates have also been found to cross the hepatocyte membrane and be further modified in other cell types, with most metabolites being excreted via the kidneys.

After the glucuronidation, sulphation, and methylation reactions, there are two types of metabolic pathways for polyphenol metabolites. One pathway leads to polyphenol metabolites in the plasma, which are then excreted into the urine via the kidneys, and the other pathway leads to transport into the colon.

A large proportion of ingested polyphenols reach the colon, including polyphenols that have not been absorbed in the small intestine (80–90%) and polyphenols that have been absorbed and metabolized (in the liver or small intestine) and then transported via membrane transporters or via bile directly back into the lumen of the colon. The colon contains a rich microbiota (10^{12} micrroorganisms/cm^3) that has a high catalytic and hydrolytic capacity, leading to the degradation of polyphenols and the formation of a large number of new metabolites (second phase of metabolism) [116].

Enzymatic degradation of flavonoids in the colon leads to the formation of a large number of new metabolites because bacterial enzymes, unlike host enzymes, can catalyze many reactions, including hydrolysis, dehydroxylation, demethylation, ring cleavage and decarboxylation, and rapid deconjugation [115]. Enzymes of the intestinal microflora catalyze the degradation of the flavonoid scaffold into simpler molecules such as phenolic acids. The extent of absorption of flavonoid metabolites in the intestine is not yet fully understood. Therefore, it is necessary to determine the role of the gut microflora in the overall distribution and potential bioactivity of flavonoids in the diet [114,117–119].

5.3. The Relationship between the Intake of Food Rich in Polyphenols and the Change in Intestinal Microbiota

From today's perspective, consumers of certain products demand that safe foods be used to feed animals. This is a challenge for today's science and points to the use of natural food supplements in animal husbandry, with increasing attention being paid to the use of polyphenols. Polyphenols are not only known for their antioxidant effects, but they also benefit animals by improving their immunity. Phenols have anti-inflammatory and antimicrobial properties in the gut [120].

Multicellular organisms exist as metaorganisms consisting of both macroscopic hosts and symbiotic commensals [121]. The gut microbiota is a natural inhabitant of the gastrointestinal tract that resides in the host [122]. Its composition depends on the host's homeostasis [123]. The host microbiota is a very living composition and is characterized by diversity and constant change over time due to the microbiological composition of the gastrointestinal tract. Changes in the microbiota also occur during breast milk consumption and during the transition to solid food. The absence of breast milk leads to changes in the gut microbiota, which may be the main cause of diarrhea [124]. The role of the gut microbiota is to digest impervious nutrients [125]. It is argued that polyphenols from foods, in general, influence the growth of probiotics and the gut microbiota in pigs and other animals, which could have an impact on gut health. It was originally claimed that food polyphenols alter the supply of biochemicals to the gut microbiome, which enables fermentation in the gut [126]. Changes in the distribution of certain bacterial species in the intestinal microbiome of the organism contribute to the antioxidant and anti-inflammatory effects, and further reduction of pathogenic species by polyphenols may stimulate the growth of probiotics [127,128]. It is assumed that polyphenols are most active in the intestines, which means that they can exert antioxidant, anti-inflammatory and antibiotic effects. At one point in the structure, hydrolyzed glycosidic compounds can react with enzymes of the gut microbiome in the form of aglycones before undergoing conjugation reactions [45,128]. Whole plants or their parts are a great asset when they are an integral part of the diet. Thus, plant leaf extracts inhibited human hepatocellular carcinoma (HepG2) cell growth [129] and induced cytotoxic activity against HepG2 cancer cells via oxidative stress in vitro [130], although the manifestation of activity will be markedly different under in vivo conditions due to the metabolism of polyphenols. Reference is also made to possible negative effects if polyphenols are consumed in excess and if they are taken alone and not as a limited supplement to the basic diet [131,132]. After ingestion, they are usually extensively metabolized in the gut microbiome of humans and animals before being absorbed into the bloodstream, where they are treated as xenobiotics and eventually excreted. Some polyphenols are known to interfere with the absorption of essential nutrients (e.g., iron and other metals formed from polyphenols), but their health benefits outweigh the potential negative effects [128,133,134]. Differences in the metabolism of polyphenols (i.e., their biotransformation and absorption) in the intestines may be due to mutual individual differences in the intestinal microflora and differences in the structures of polyphenols [135]. For example, many phytoestrogen flavonoids (e.g., genistein, apigenin, kaempferol, and naringenin) are more easily metabolized and absorbed in the form of aglycones, but the efficiency of their hydrolysis and absorption of glycosides (the most abundant form in plants) depends on the type of microflora [136,137]. The effects of polyphenol-rich food additives on the microbiome measured in various studies are listed in Table 2.

Table 2. Presentation of the effect of polyphenols on intestinal microbiota.

Polyphenols and Source	Sample	Impact on Microbiome	References
Red wine extract	Fecal (human) in vitro	Increase: *Bifidobacterium* spp. *Lactobacillus/Enterococcus* spp. *Bacteroides* spp. Inhibit: *Clostridin hystolyticum*	[138]
Grape pomace	Fecal (human) in vitro	*Increase: Bifidobacterium* spp. *Clostrida* *Campylobacter* *Inhibit: Escherichia coli* *Salmonella*	[139]
Blueberry extract	Fecal (human) in vitro	*Increase: Lactobacillus* *Bifidobacterium* spp.	[140]
Grape seed extract fraction	Fecal (human) in vitro	Increase: *Lactobacillus/Enterococcus* spp. Inhibit: *Clostridin hystolyticum*	[141]
Tea polyphenols	Fecal (pigs) in vitro	Increase: *Lactobacillus* Inhibit: *Bacteroidaceae* *Clostridium perfringens*	[142]
Red wine polyphenols powder	Fecal (rats) in vitro	Increase: *Lactobacilli* *Bifidobacteria* Inhibit: *Propionibacteria* *Bacteriodes, Clostridia*	[143]
Grape pomace concentrate, Grape seed extract	Fecal (broiler chicks) in vitro	Increase: *E. Coli, Enterococcus* spp., *Lactobacillus* spp.	[144]
Lowbush wild blueberries	Fecal (rats) in vitro	Incerase: *Thermonospora* spp., *Corynebacteria spp.* *Slackia* spp. Inhibit: *Lactobacillus* spp. *Enterococcus* spp.	[145]
Apple pomace	In vivo (rats)	*Increase: Lactobacillus* *Bifidobacterium* *Bacteriodoceae* *Eubacterium* Inhibit: *Bacteroides* spp.	[146]
Tannin supplementation	In vivo (mouse)	Increase: *Bacteroides, Lactobacillus* Inhibit: *Clostridium*	[147]

The individual sources of polyphenols used as supplements to animal feed in several studies are listed in Table 2. The microorganisms were determined from feces or in vivo, and the good bacteria whose concentration increased and the pathogens whose concentration decreased were listed. From the data presented, it can be concluded that polyphenols have a stimulating effect on the modulation (in a positive direction) of the intestinal microbiome of the animals tested.

6. Antioxidant and Pro-Oxidant Activity of Polyphenols

6.1. Antioxidant Activity of Polyphenols

Oxidative stress has become a major challenge for animal husbandry due to the impairment of animal growth. Discoveries about the intracellular antioxidant action of polyphenols as a plant additive in animal feed contributed significantly to the improvement of antioxidant activity. Due to the positive effect of polyphenols in preventing oxidative stress damage and eliminating excessively produced free radicals, polyphenols are receiving more and more attention [136].

The antioxidant effect of polyphenols in neutralizing free radicals involves the transfer of a hydrogen atom from the active hydroxyl group of the polyphenolic compound to a free radical (Ar-OH +R → Ar-O + RH) [113]. Di Meo et al. [135] have proposed four mechanisms for the reaction shown, which are illustrated in Figure 5.

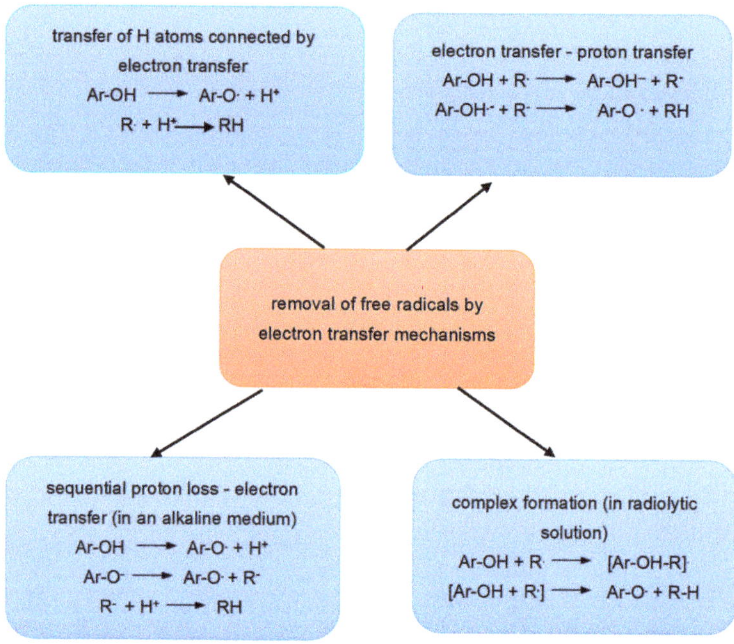

Figure 5. Polyphenol free radical scavenging activity by the hydrogen transfer mechanism. Abbreviations: R·—free radical, Ar-OH—phenolic compound.

In the first example, hydrogen atom transfer (HAT) occurs, a mechanism in which a hydrogen atom is transferred from an antioxidant (ArOH) to a radical (R·), reducing the radical (RH) and forming an antioxidant radical (AO). The mechanism identified with HAT in biological systems is called PCET (proton-coupled electron transfer) and is characterized by the pre-reaction complex in which the proton transfer occurs along the H-bond to one of the free O-atom pairs of the free radical.

The second mechanism involves the transfer of an electron (one-electron transfer-proton transfer, SET-PT), which occurs in two steps. In the first step the electron is transferred and the first radical cation reaction ($ArOH^{\cdot+}$) formed from antioxidants and anions (R^-), takes place. In the second step, an exchange of protons takes place between the obtained products of the first reaction.

In the third mechanism, proton loss by electron transfer takes place (Sequential Proton Loss-Electron Transfer, SPLET), in which two processes take place: the elimination of protons from antioxidants and the formation of anions (AO·) and electron transfer.

In the fourth mechanism (adduct formation), the radical forms a hydrogen bond with the HO group of the antioxidant. In a radiolytic solution, a radical that takes up an H ion can lead to the formation of a double bond in an aromatic ring.

6.2. Pro-oxidative Activity of Polyphenols

Polyphenols act as antioxidants at low concentrations, while at higher concentrations they have a pro-oxidant effect. This depends on many factors. One of the most important factors is the "seasonal type", where weather conditions during the growing season of the plants, such as high air temperature and the amount of UV radiation, lead to the formation of higher concentrations of polyphenols in the plants. Therefore, the concentration of polyphenols is higher in food produced for animals during the growing season of plants, which may have a pro-oxidant effect. Polyphenols may have a pro-oxidative effect on other cells in environments where the partial pressure and concentration of oxygen are increased [136].

The same flavonoids can have both negative and positive effects, depending on the source and concentration of free radicals and the concentration of polyphenols in the administered dose [148]. The pro-oxidant properties of food polyphenols may deteriorate due to the lack of glutathione (GSH) in cells, the lack of chemical stability, and the activation of cellular copper ions (Cu^+) [123] formed during autoxidation together with the semiquinone radical (oxidized flavonoid). The cytotoxic semiquinone radical reduces NADH, which is involved in the formation of ROS in redox reactions [149].

Copper ions contribute to the increase in superoxide radical concentration and the formation of hydrogen peroxide and hydroxyl ions. However, the pro-oxidative action of polyphenols may also have beneficial effects by inducing mild oxidative stress, generating antioxidants and xenobiotic-degrading enzymes, and contributing to a general cytoprotective effect [149,150]. Polyphenols are very susceptible to autoxidation [151].

The most studied property is the ability to protect the organism from free radicals and oxygenated reactive species formed during oxygen metabolism [152], which are also the main cause of radical damage in the cell [153]. Radical damage to cells leads to changes in net charge, which alters their osmotic pressure and leads to an increase in volume and death. Lipid peroxidation products derived from dead cells can also promote carcinogenesis [154]. Flavonoids prevent the damage caused by free radicals, in part by directly scavenging the radicals. Simply put, flavonoids are oxidized by radicals (R^-) and converted into a more stable and less reactive form according to the following mechanism [155]:

$$\text{flavonoid (OH)} + R^\bullet \rightarrow \text{flavonoid } (O^\bullet) + RH$$

The newly formed flavonoxy radical (O^-) is stabilized by resonance, as the unpaired electron can and will be displaced throughout the aromatic ring, but it can also participate in reactions of dimerization, dismutation, recombination with other radicals, or oxidation to quinone, whether it reacts with radicals, other antioxidants, or perhaps biomolecules. The reaction of flavonoxy radicals with oxygen produces quinone and superoxide anions. This reaction is responsible for the undesirable pro-oxidant effect of flavonoids in healthy tissue, but may be helpful in combating tumor tissues. This confirms the assumption that the antioxidant capacity of flavonoids depends not only on the redox potential of the cell itself, i.e., the O^-/OH potential, but also on the reactivity of the flavone oxyradical formed [152].

7. The Influence of Polyphenols on Animals

7.1. Effect of the Addition of Polyphenols on Animals

The addition of polyphenols to animal feed had no clear effect on animal growth. Inhibited secretion of digestive enzymes, increased protein excretion, and reduced digestibility of proteins and amino acids may have a detrimental effect on metabolism, as evidenced by a decrease in body weight and nutritional efficiency [152–157].

Fiesel et al. [100] reported a significant decrease in digestibility of total proteins and cellulose in weaners whose diets were supplemented with spent hops, a source of natural polyphenols. The supplement had no effect on performance and the gain-to-feed ratio was better in the experimental group than in the control group pigs (638 g/kg versus 579 g/kg).

The addition of Moringa oleifera, a source of quercetin and kaempferol, to the diet of chickens contributed to a significant increase in body weight at 21 days of age (low: 928 g; median: 932.5 g; high: 954.6 g) compared to chickens in positive and negative control groups (887.6 g; 918.7 g) and improved feed conversion (1.47; 1.44; 1.45) compared to 1.53 for positive control group during the experiment [158]. Supplementation of the chicken diet with 0.2% plant extracts of *Chelidonium majus*, *Lonicera japonica* and *Saposhnikovia divaricata* (sources of flavonoids, tannins, phenolic compounds, saponins, terpenoids and essential oils) resulted in a significant increase in final body weight (1949 g; 1930 g; 1930 g vs. 1845 g) and increased daily gain of about 3 g compared to the control chickens [143,159].

In a study by El-Iraqi et al. [160], chickens under heat stress (32–40 °C) were fed a diet enriched with dried peppermint and *Ginkgo biloba* (a source of flavonoids). This resulted in a significant increase in final body weight and a decrease in feed conversion compared to chickens whose diets were enriched with single herbs or vitamin C.

Viveros et al. [143] observed a significant decrease in body weight of 21-day-old chickens whose diets were enriched with grape seed extract (7.2 g/kg diet) compared to chickens in the other groups (486 g vs. 553; 557; 542 g). A significant reduction in feed conversion rate was also observed in chickens fed grape pomace concentrate (60 g/kg feed) or the antibiotic avoparcin (50 mg/kg feed) compared to chickens in the other groups (1.43; 1.43 vs. 1.51). The conversion rate of the feed mixture of pigs whose diet was enriched with polyphenols (from grape seeds and grape marc) increased compared to the control group (652 vs. 624 g/kg; $p < 0.05$, which is a significant correlation) [161,162]. In the study by Lipiński et al. (2015) [163], dietary supplementation with polyphenols (grape seeds and onions) had no effect on growth performance, percentage carcass weight loss, breast muscle yield, or meat composition in broilers.

In the study by Flis et al. [164], phenolic compounds in oat grains fed to pigs at the end of fattening (45% of the diet) had no effect on animal growth or carcass leanness. Dietary supplementation with cranberry extract, a rich source of phenolic compounds, had no effect on body weight or feed efficiency in poultry [165]. Simitzis et al. [166] showed that dietary supplementation with hesperidin and tocopherol acetate had no significant effects on growth or slaughter weight of broilers. The addition of grape by-products did not improve efficiency parameters, while herbs rich in flavonoids gave positive results, especially in animals under heat stress [4].

7.2. Immunomodulatory Effect of Polyphenols on Intestinal Health of Animals

The body's immune system consists of innate and adaptive immunity. The first line of defense is innate immunity, which forms barriers between the organism and the environment [167]. The innate system can be divided into cellular and non-cellular systems. The cellular system consists of several subgroups of cells, namely, dendritic cells (DC), monocytes, macrophages, granulocytes, and natural killer (NK) cells. The non-specific systems are very diverse and range from mucosal barriers to complex protein molecules. The basic function is to prevent the invasion of pathogens and destroy them by phagocytosis [168]. The adaptive immune system consists of T and B cells. B cells secrete antibodies, while T cells synthesize cytokinins that destroy infected or malignant tissue. Many studies have confirmed the positive relationship between microflora and host health. The bioefficacy of polyphenols in interaction with the gut microbiome can promote the development of beneficial micro-organisms in the host gut and positively modulate the gut microbiome [169]. Previous studies in mice have confirmed that polyphenols positively modulate the gut microbiota by altering the composition of certain bacteria in the gut, such as *Bifidobacterium, Proteobacteria, Actinobacteria, Deferribacteres, Lactobacillus, Helicobacter, Desulfovibrio, Adlercreutzia, Prevotella,* and *Flexispira* [170]. For example, it was found that the addition of resveratrol to the diet of mice at a dose of 200 mg/kg per day improved the dysbiosis of the intestinal microbiota caused by a high-fat diet, i.e., it resulted in an inhibition of the growth of *Enterococcus faecalis* and an increase in the growth of *Lactobacillus and Bifidobacterium* [171]. Similar results and positive effects were obtained with the addition of blueberry polyphenol extract to the diet of mice, i.e., weight gain was prevented and fat metabolism returned to normal [172]. It was found that the addition of blueberry polyphenol extract modulated the composition of the microbiome—the populations of Proteobacteria and Deferribacteres increased and the population of Actinobacteria in the gut of mice decreased. Specifically, populations of *Bifidobacterium, Desulfovibrio, Adlercreutzia, Helicobacter*, and *Flexispira* were found to increase, and *Adlercreutzia* and *Prevotella* decreased, compared to mice fed a high-fat diet without polyphenols [167]. Experimental evidence suggests that polyphenols play an important role in the proliferation of beneficial gut bacteria in mice, such as *Akkermansia muciniphila*, and the reduction in pathogenic gut microbiota [168].

The interactions between the gut microbiota and polyphenols in humans are crucial as they modulate the gut microbiota and not only influence the composition of gut bacteria, but also improve the bioavailability of polyphenols through the formation of more bioactive metabolites, enhance their health-promoting effects, protect epithelial cells, and prevent inflammatory processes in the gut. Polyphenols suppress the development of pathogenic bacteria and promote the growth of beneficial bacteria that act as prebiotics [169]. The gut microbiota enhances the bioavailability of polyphenols by producing more bioactive metabolites that increase health effects and influence gut morphology [169]. The influence of polyphenols in stress during weaning of piglets and calves improves digestion and absorption of nutrients, improves the function of the intestinal barrier, improves the function of the intestinal microbiota, and thus provides positive effects [170]. The study of the concentration of polyphenols in body tissues is not directly related to their concentration in the diet. In pigs whose diets were supplemented with 50 mg/kg quercetin for 4 weeks, quercetin levels were higher in the kidneys (6.31 nmol/g) and colon (13.92 nmol/g) than in the liver (2.83 nmol/g) or plasma (0.67 mol/L). In contrast, quercetin concentrations in tissues remained unchanged (3.78 nmol/g in liver, 1.84 nmol/g in kidney), while they increased in plasma (1.1 mol/L) when pigs' diets were supplemented with higher doses of quercetin (500 mg/kg) for 3 days [4]. This indicates that increased concentrations of polyphenols were found only in organs involved in the metabolism of polyphenolic compounds. Concentrations of unabsorbed dietary phenolic compounds have significant effects on the gut environment by suppressing or stimulating the growth of some gut microbiota. Dietary polyphenols have prebiotic properties and act antimicrobially by promoting the growth of certain strains of beneficial bacteria (*Bacillus* spp. and *Lactobacillus* spp.) in the digestive tract, while competitively excluding certain pathogenic bacteria and stabilizing the gut microbiota and strengthening the immune system of animals [171].

7.3. The Effectiveness of Polyphenols on the Quality of Products of Animal Origin

Feeding animals diets rich in n-3 PUFA has been shown to improve the nutritional quality of fats in animal products. However, it has been shown that a dietary strategy with increased PUFA content makes the products of these animals susceptible to lipoperoxidation. For this reason, the use of antioxidants in the diet has been recommended to limit lipid peroxidation, maintain animal health, and achieve better product quality [172]. Thus, vitamin E was resorted to as a synthetic antioxidant. However, high doses of vitamin E have been reported to have a pro-oxidant effect. Therefore, the use of a natural antioxidant has been recommended to optimize antioxidant protection in animals fed a PUFA-rich diet. Polyphenols are of interest among natural antioxidants. Various herbs, spices and plant residues are rich in polyphenols and could prevent lipid oxidation. Studies on poultry, for example, have confirmed that plant extracts or dietary hesperidin supplementation prevent lipoperoxidation [101,166,173].

From the detailed review by Serra et al. (2021) (Table 3) [4], it appears that the enrichment of animal feed with polyphenols improves the products of animal origin (based on meat or milk) without significant negative effects on animal health. Food-derived polyphenols have been shown to undergo no significant metabolic changes, thus enriching meat [174,175] and dairy products [176–180]. It has also been shown that the addition of 5% dried GP, a rich PUFA source, to the diet of pigs improves the concentration of total n-3 PUFA (especially α-linolenic acid) in muscle after 12 days of feeding [181]. Polyphenols have been found to contribute to the regeneration of tocopherol in broiler plasma due to the potential reduction of one electron, as well as protecting vitamin E from oxidation and thus slowing lipid oxidation [182].

The addition of polyphenols makes the animals' diet more diverse, leading to greater diversity of gut flora and a better immune status for the animals. Combined with the ability for the animals to freely choose additional feed of plant origin containing polyphenols, this leads to a reduction in stress in the animals [183,184]. It has been shown that animals that are free to choose their feed optimise their individual macronutrient requirements, which has a positive effect on their health [183], whereas a monotonous and uniform diet causes stress [184]. Modern genetically bred pigs have not lost the ability to optimally select their feed in terms of optimising individual nutrient requirements [185]. The animals' free choice of feed improves their lives, which contributes to a better health status of the animals.

In the future, more precise studies need to be defined that combine research on the addition of polyphenols to feed with the partial freedom of animals in intensive farming to choose the type and amount of feed [186]. In this way, a better quality of food of animal origin used for human consumption will be achieved. So there is a great opportunity to use natural antioxidants in the future to protect animal health and improve food of animal origin for human consumption.

8. Conclusions

Medicinal herbs and preparations made from medicinal herbs are one of the oldest ways of treating various diseases. Plants are known to contain many bioactive chemical compounds attributed to health effects. One of the most common secondary metabolites of plants are polyphenols. The process of absorption, metabolism, distribution and excretion from the organism is crucial for the determination of potentially bioactive effects in vivo and for the prevention of a number of diseases associated with oxidative stress. The bioavailability of polyphenols is related to the inter-reactions of polyphenols in the different phases of digestion. Most of the ingested polyphenols (90–95%) reach the large intestine. It is known that the intestinal microbiota is rich (it contains about 10^{12} microorganisms/cm^3). Polyphenols are degraded by the microbiota and a large number of metabolites are formed. The intestinal microbiota improves the bioavailability of polyphenols, whereby metabolites are formed that affect the modulation of the intestinal microbiota in a positive direction, increasing the concentration of beneficial bacteria, while decreasing pathogenic ones. Previous studies on pigs and poultry have confirmed that plant polyphenols alleviate inflammatory conditions and improve intestinal microbiota. Polyphenols are known to have pronounced antioxidant activity in vitro, such as ROS removal, metal ion chelation, and an effect on increased transcription of antioxidant enzymes. However, due to low absorption, polyphenols are degraded in a xenobiotic manner, so it is necessary to compare them with other antioxidants such as vitamin C, vitamin E, albumins, uric acid and glutathione. It has also been confirmed that plant polyphenols can lower the level of ROS in animals with intestinal inflammation. Feeding pigs with grape pomace improved growth performance and changed the composition of fatty acids in finished products, which is important for human nutrition and such finished products. In addition to the positive effects, polyphenols also have a pro-oxidative effect and form complexes with metals, making them less available.

Further research and experimentation with polyphenolic compounds as feed additives for animals is a potential that is certainly worthy of much more research, as a relatively small number of experiments have provided positive evidence of an effect on the health status of animals. In addition to finding the right ratio of polyphenol preparations and plants in which they can best be used, and depending on the characteristics of the animal we want to improve, it is also important to focus research on maintaining the quality of polyphenol preparations so that they can be used as efficiently as possible. In order to achieve better overall results in terms of animal health, the quality of their lives and the quality of food of animal origin, the addition of polyphenols to the diet can be combined in research and in practice with the free choice approach to feed.

Author Contributions: Conceptualisation, D.B. and B.L.; resources, N.G., V.R., D.A., M.K. and D.Š. Writing—original draft preparation, D.B. and B.L.; Reviewing and editing, D.B., B.L., V.R., D.A., M.K. and D.Š; visualization, D.B. and B.L. All authors have read and agreed to the published version of the manuscript.

Funding: This publication was supported by the Croatian Ministry of Science and Education (D. Bešlo, D.; V. Rastija, D. Agić, M. Karnaš, D. Šubarić and B. Lučić) through basic grants given to their institutions. B. Lučić was also supported by the Croatian Government and the European Union through grant KK.01.1.1.01.0002 (Bioprospecting of the Adriatic Sea, Scientific Centre of Excellence BioProCro).

Conflicts of Interest: The authors declare no conflict of interest.

Abbreviations

PUFA	polyunsaturated fatty acid
ROS	reactive oxygen species
RNS	reactive nitrogen species
SOD	superoxide dismutase
CAT	catalase
GTx	glutathion peroxidase
SFA	saturation fatty acid
CTC	glucocorticoide
GP	grape pomace
ETC	electron transport chain
CYP	cytochrome P450
NOX	NADPH oxidase
NF-kB	nuclear factor kappa B
UGT	uridine-5-diphosphate glucuronosyltranspherase
SULT	sulphotransferase
COMT	catechol-O-methyltransferase
OMTs	O-methyltransferases
HAT	hydrogen atom transfer
PCET	one-electron transfer-proton transfer
SPLET	sequential proton loss electron transfer

References

1. Bravo, L. Polyphenols: Chemistry, Dietary Sources, Metabolism, and Nutritional Significance. *Nutr. Rev.* **1998**, *56*, 317–333. [CrossRef] [PubMed]
2. Lipiński, K.; Mazur, M.; Antoszkiewicz, Z.; Purwin, C. Polyphenols in monogastric nutrition. *Ann. Anim. Sci.* **2017**, *17*, 41–58. [CrossRef]
3. Abdel-Moneim, A.M.E.; Shehata, A.M.; Alzahrani, S.O.; Shafi, M.E.; Mesalam, N.M.; Taha, A.E.; Swelum, A.A.; Arif, M.; Fayyaz, M.; Abd El-Hack, M. The role of polyphenols in poultry nutrition. *J. Anim. Physiol. Anim. Nutr.* **2020**, *104*, 1851–1866. [CrossRef] [PubMed]
4. Serra, V.; Salvatori, G.; Pastorelli, G. Dietary polyphenol supplementation in food producing animals: Effects on the quality of derived products. *Animals* **2021**, *11*, 401. [CrossRef]
5. Heim, K.E.; Tagliaferro, A.R.; Bobilya, J.D. Flavonoid antioxidants: Chemistry, metabolism and structure-activity Relationships. *J. Nutr. Biochem.* **2002**, *13*, 572–584. [CrossRef]
6. Sies, H. Biochemistry of Oxidative Stress. *Angew. Chem. Int. Ed. Engl.* **1986**, *25*, 1058–1071. [CrossRef]
7. Lauridsen, C.; Hojsgaard, S.; Sørensen, M.T. Influence of dietary rapeseed oil, vitamin E and copper on the performance and the antioxidative and oxidative status in pigs. *J. Anim. Sci.* **1999**, *77*, 906–916. [CrossRef]
8. Basu, S.; Eriksson, M. Oxidative injury and survival during endotoxemia. *FEBS Lett.* **1998**, *438*, 159–160. [CrossRef]
9. Campbell, J.M.; Crenshaw, J.D.; Polo, J. The biological stress of early weaned piglets. *J. Anim. Sci. Biotechn.* **2013**, *4*, 19. [CrossRef]

10. Wijtten, P.J.; van der Meulen, J.; Verstegen, M.W. Intestinal barrier function and absorption in pigs after weaning: A review. *Brit. J. Nutr.* **2011**, *105*, 967–981. [CrossRef]
11. Ohtsuka, A.; Ohtani, T.; Horiguchi, H.; Kojima, H.; Hayashi, K. Vitamin E reduces glucocorticoidinduced growth inhibition and lipid peroxidation inrats. *J. Nutr. Sci. Vitaminol.* **1998**, *44*, 237–247. [CrossRef] [PubMed]
12. Tanigucha, N.; Ohtsuka, A.; Hayashi, K. Effect of dietary corticosterone and vitamin E on growth and oxidative stress in broiler chickens. *Anim. Sci. J.* **1999**, *70*, 195–200. [CrossRef]
13. Sharma, P.; Jha, A.B.; Dubey, R.S.; Pessarakli, M. Reactive oxygen species, oxidative damage, and antioxidative defense mechanism in plants under stressful conditions. *J. Bot.* **2012**, *2012*, 26. [CrossRef]
14. Cadet, J.; Wagner, J.R. DNA base damage by reactive oxygen species, oxidizing agents, and UV radiation. *Cold Spring Harb. Perspect. Biol.* **2013**, *5*, a012559. [CrossRef]
15. Cadet, J.; Davies, K.J.A.; Medeiros, M.H.G.; Di Mascio, P.; Wagner, J.R. Formation and repair of oxidatively generated damage in cellular DNA. *Free Radic. Biol. Med.* **2017**, *107*, 13–34. [CrossRef]
16. Chamorro, S.A.; Viveros, A.A.; Rebolé, A.; Rica, B.D.I.; Arija, I.; Brenes, A. Influence of dietary enzyme addition on polyphenol utilization andmeat lipid oxidation of chicks fed grape pomace. *Food Res. Int.* **2015**, *73*, 197–203. [CrossRef]
17. Rymer, C.; Given, D.I. Effects of vitamin E and fish oil inclusion in broiler diets on meat fatty acid composition and on the flavour of a composite sample of breast meat. *J. Sci. Food Agric.* **2010**, *90*, 1628–1633. [CrossRef]
18. Aziza, A.E.; Quezada, N.; Cherian, G. Antioxidative effect of dietary camelina meal in fresh, stored, or cooked broiler chicken meat. *Poult. Sci.* **2010**, *89*, 2711–2718. [CrossRef]
19. Brenes, A.A.; Viveros, I.; Goñi, C.; Centeno, S.G.; Sáyago-Ayerdi, I.; Saura-Calixto, A.F. Effect of grape pomace concentrate and vitamin E on digestibility of polyphenols and antioxidant activity in chickens. *Poult. Sci.* **2008**, *87*, 307–316. [CrossRef]
20. Pascual, J.V.M.; Rafecas, M.A.; Canela, J.; Boatella, R.; Bou, A.C.; Barroeta, C.; Codony, R. Effect of increasing amounts of a linoleic-rich dietary fat on the fat composition of four pig breeds. Part II: Fatty acid composition in muscle and fat tissues. *Food Chem.* **2007**, *100*, 1639–1648. [CrossRef]
21. Bešlo, D.; Došlić, G.; Agić, D.; Rastija, V.; Šperanda, M.; Gantner, V.; Lučić, B. Polyphenols in Ruminant Nutrition and Their Effects on Reproduction. *Antioxidants* **2022**, *11*, 970. [CrossRef] [PubMed]
22. Dretcanu, G.; Stirbu, I.; Leoplold, N.; Cruceriu, D.; Danciu, C.; Stănilă, A.; Fărcas, A.; Monica Borda, I.; Iuhas, C.; Diaconeasa, Z. Chemical Structure, Sources and Role of Bioactive Flavonoids in Cancer Prevention: A Review. *Plants* **2022**, *11*, 1117. [CrossRef] [PubMed]
23. Pandey, K.B.; Rizv, S.I. Plant polyphenols as dietary antioxidants in human health and disease. *Oxid. Med. Cell. Longev.* **2009**, *2*, 270–278. [CrossRef] [PubMed]
24. Manach, C.; Scalbert, A.; Morand, C.; Rémésy, C.; Jiménez, L. Polyphenols: Food sources and bioavailability. *Am. J. Clin. Nutr.* **2004**, *79*, 727–747. [CrossRef]
25. Gomes, M.N.; Muratov, E.N.; Pereira, M.; Peixoto, J.C.; Rosseto, L.P.; Cravo, P.V.L.; Andrade, C.; Neves, B.J. Chalcone Derivatives: Promising Starting Points for Drug Design. *Molecules* **2017**, *22*, 1210. [CrossRef]
26. Halliev, B. Reactive species and antioxidants. Redox biology is a fundamental theme of aerobic life. *Plant. Physiol.* **2006**, *141*, 312–322. [CrossRef]
27. Aruoma, O.I. Nutrition and health aspects of free radicals and antioxidants. *Food Chem. Toxicol.* **1994**, *32*, 671–683. [CrossRef]
28. Li, S.; Tan, H.-Y.; Wang, N.; Zhang, Z.-J.; Lao, L.; Wong, C.-W.; Feng, Y. The Role of Oxidative Stress and Antioxidants in Liver Diseases. *Int. J. Mol. Sci.* **2015**, *16*, 26087–26124. [CrossRef] [PubMed]
29. Ramos-Tovar, E.; Muriel, P. Free radicals, antioxidants, nuclear factor-E2-related factor-2 and liver damage. *J. Appl. Toxicol.* **2020**, *40*, 151–168. [CrossRef]
30. Valko, M.; Leibfritz, D.; Moncol, J.; Mark, T.D.; Cronin, M.T.D. Free radical and antioxidants in normal physiological functions and human disease. *Int. J. Biochem. Cell. Biol.* **2007**, *39*, 44–84. [CrossRef]
31. Lee, J.; Giordano, S.; Zhang, J. Autophagy, mitochondria and oxidative stress: Cross-talk and redox signalling. *Biochem. J.* **2012**, *441*, 523–540. [CrossRef] [PubMed]
32. Ramos-Tovar, E.; Muriel, P. Molecular Mechanisms That Link Oxidative Stress, Inflammation, and Fibrosis in the Liver. *Antioxidants* **2020**, *9*, 1279; [CrossRef]
33. Cichoz-Lach, H. Oxidative stress as a crucial factor in liver diseases. *World J. Gastroenterol.* **2014**, *20*, 8082–8091. [CrossRef] [PubMed]
34. Giacco, F.; Brownlee, M. Oxidative Stress and Diabetic Complications. *Circ. Res.* **2010**, *107*, 1058–1070. [CrossRef] [PubMed]
35. Pham-Huy, L.A.; He, H.; Pham-Huy, C. Free Radicals, Antioxidants in Disease and Health. *Int. J. Biomed. Sci.* **2008**, *4*, 89–96.
36. Halliwell, B.; Gutteridge, J.M.C. *Free Radicals in Biology and Medicine*, 3rd ed.; Oxford Science Publications: Oxford, UK, 1999.
37. Juan, C.A.; Pérez de la Lastra, J.M.; Plou, F.J.; Pérez-Lebeña, E. The Chemistry of Reactive Oxygen Species (ROS) Revisited: Outlining Their Role in Biological Macromolecules (DNA, Lipids and Proteins) and Induced Pathologies. *Int. J. Mol. Sci.* **2021**, *22*, 4642. [CrossRef]
38. Liang, X.; Wang, S.; Wang, L.; Ceylan, A.F.; Ren, J.; Zhang, Y. Mitophagy inhibitor liensinine suppresses doxorubicin-induced cardiotoxicity through inhibition of drp1-mediated maladaptive mitochondrial fission. *Pharmacol. Res.* **2020**, *157*, 104846. [CrossRef]

39. Edeas, M.; Attaf, D.; Mailfert, A.S.; Nasu, M.; Joubet, R. Maillard reaction, mitochondria and oxidative stress: Potential role of antioxidants. *Pathol. Biol.* **2010**, *58*, 220–225. [CrossRef]
40. Majima, H.J.; Indo, H.P.; Suenaga, S.; Matsui, H.; Yen, H.C.; Ozawa, T. Mitochondria as possible pharmaceutical targets for the effects of vitamin E and its homologues in oxidative stressrelated diseases. *Curr. Pharm. Des.* **2011**, *17*, 2190–2195. [CrossRef] [PubMed]
41. Mao, G.; Kraus, G.A.; Kim, I.; Spurlock, M.E.; Bailey, T.B.; Beitz, D.C. Effect of a mitochondria-targeted vitamin E derivative on mitochondrial alteration and systemic oxidative stress in mice. *Br. J. Nutr.* **2011**, *106*, 87–95. [CrossRef] [PubMed]
42. Snezhkina, A.V.; Kurdymon, O.L.; Savvateeva, M.V.; Melnikova, N.V.; Krasnov, G.S.; Dmitriev, A.A. ROS Generation and Antioxidant Defense Systems in Normal and Malignant Cells. *Oxid. Med. Cell Longev.* **2019**, *2019*, 6175804. [CrossRef] [PubMed]
43. Banerjee, B.D.; Seth, V.; Ahmed, R.S. Pesticide-Induced Oxidative Stress: Perspectives and Trends. *Rev. Environ. Health* **2001**, *16*, 1–40. [CrossRef] [PubMed]
44. Pantano, C.; Reynaert, N.L.; van der Vliet, A.; Janssen-Heininger, Y.M. Redox-sensitive kinases of the nuclear factor-kappaB signaling pathway. *Antioxid. Redox Signal.* **2006**, *8*, 1791–1806. [CrossRef] [PubMed]
45. Gessner, D.K.; Ringseis, R.; Eder, K. Potential of Plant Polyphenols to Combat Oxidative Stress and Inflammatory Processes in Farm Animals. *J. Anim. Physiol. Anim. Nutr.* **2017**, *101*, 605–628. [CrossRef] [PubMed]
46. Pillarisetti, S.; Saxena, U. Role of oxidative stress and inflammation in the origin of Type 2 diabetes—A paradigm shift. *Expert. Opin. Ther. Targets.* **2004**, *8*, 401–408. [CrossRef]
47. Lykkesfeldt, J.; Svendsen, O. Oxidants and antioxidants in disease: Oxidative stress in farm animals. *Vet. J.* **2007**, *173*, 502–511. [CrossRef]
48. Bedard, K.; Krause, K.H. The NOX family of ROS- generating NADPH oxidases: Physiology and pathophysiology. *Physiol. Rev.* **2007**, *87*, 245–313. [CrossRef]
49. Klebanoff, S.J. Oxygen metabolism and the toxic properties of phagocytes. *Ann. Intern. Med.* **1980**, *93*, 480–489. [CrossRef]
50. Nauser, T.; Koppenol, W.H. The rate constant of the reaction of superoxide with nitrogen monoxide: Approaching the diffusion limit. *J. Phys. Chem. A* **2002**, *106*, 4084. [CrossRef]
51. Francis, S.H.; Busch, J.L.; Corbin, J.D.; Sibley, D. cGMP-dependent protein kinases and cGMP phosphodiesterases in nitric oxide and cGMP action. *Pharmacol. Rev.* **2010**, *62*, 525–563. [CrossRef]
52. McDonald, L.J.; Murad, F. Nitric oxide and cGMP signaling. *Adv. Pharmacol.* **1995**, *34*, 263–275. [PubMed]
53. Gilbert, R.S.; Herschman, H.R. Transforming growth factor beta differentially modulates the inducible nitric oxide synthase gene in distinct cell types. *Biochem. Biophys. Res. Commun.* **1993**, *195*, 380–384. [CrossRef] [PubMed]
54. Carreras, M.C.; Pargament, G.A.; Catz, S.D.; Poderoso, J.J.; Boveris, A. Kinetics of nitric oxide and hydrogen peroxide production and formation of peroxynitrite during the respiratory burst of human neutrophils. *FEBS Lett.* **1994**, *341*, 65–68. [CrossRef]
55. Allen, R.G.; Lafuse, W.P.; Powell, N.D.; Webster Marketon, J.I.; StinerJones, L.M.; Sheridan, J.F.; Bailey, M.T. Stressor-induced increase in microbicidal activity of splenic macrophages is dependent upon peroxynitrite production. *Infect. Immun.* **2012**, *80*, 3429–3437. [CrossRef] [PubMed]
56. Dahlgren, C.; Karlsson, A. Respiratory burst in human neutrophils. *J. Immunol. Methods* **1999**, *232*, 3–14. [CrossRef] [PubMed]
57. Adams, L.; Franco, M.C.; Estevez, A.G. Reactive nitrogen species in cellular signaling. *Exp. Biol. Med.* **2015**, *240*, 711–717. [CrossRef] [PubMed]
58. Lobo, V.; Patil, A.; Phatak, A.; Chandra, N. Free radicals, antioxidants and functional foods: Impact on human health. *Pharm. Rev.* **2010**, *4*, 118–126. [CrossRef]
59. Beeridge, M.V.; Tan, A.S.; McCoy, K.D.; Wang, R. The biochemical and cellular basis of cell proliferation assay that use tetrazolium salts. *Biochemica* **1996**, *4*, 15–20.
60. Forrester, S.J.; Kikuchi, D.S.; Hernandes, M.S.; Xu, Q.; Griendling, K.K. Reactive Oxygen Species in Metabolic and Inflammatory Signaling. *Circ. Res.* **2018**, *122*, 877–902. [CrossRef]
61. Valko, M.; Rhodes, C.J.; Moncola, J.; Izakovic, M.; Mazura, M. Free radicals, metals and antioxidants in oxidative stress-induced cancer. *Chem. Biol. Interact.* **2006**, *160*, 1–40. [CrossRef]
62. Lučić, B.; Amić, D.; Trinajstić, N. Antioxidant QSAR Modeling as Exemplified on Polyphenols. In *Advanced Protocols in Oxidative Stress I*; Armstrong, D., Ed.; Humana Press: Totowa, NJ, USA, 2008; Volume 477, pp. 207–218. ISBN 978-1-60327-218-6.
63. Filipović, M.; Marković, Z.; Đorović, J.; Marković, J.D.; Lučić, B.; Amić, D. QSAR of the Free Radical Scavenging Potency of Selected Hydroxybenzoic Acids and Simple Phenolics. *Comptes Rendus Chim.* **2015**, *18*, 492–498. [CrossRef]
64. Amić, D.; Davidović-Amić, D.; Bešlo, D.; Rastija, V.; Lučić, B.; Trinajstić, N. SAR and QSAR of the antioxidant activity of flavonoids. *Curr. Med. Chem.* **2007**, *14*, 827–845. [CrossRef]
65. Altemimi, A.; Lakhssassi, N.; Baharlouei, A.; Watson, D.G.; Lightfoot, D.A. Phytochemicals: Extraction, Isolation, and Identification of Bioactive Compounds from Plant Extracts. *Plants* **2017**, *6*, 42. [CrossRef] [PubMed]
66. Genestra, M. Oxyl radicals, redox-sensitive signalling cascades and antioxidants. *Cell. Signal. 2007 19*, 1807–1819. [CrossRef]
67. Muriel, P. Role of free radicals in liver diseases. *Hepatol. Int.* **2009**, *3*, 526–536. [CrossRef] [PubMed]
68. Casas-Grajales, S.; Muriel, P. The Liver, Oxidative Stress, and Antioxidants. In *Liver Pathophysiology*; Elsevier BV: Berlin, Germany, 2017; pp. 583–604.
69. Pacher, P.; Beckman, J.S.; Liaudet, L. Nitric Oxide and Peroxynitrite in Health and Disease. *Physiol. Rev.* **2007**, *87*, 315–424. [CrossRef]

70. Abrams, G.A.; Trauner, M.; Nathanson, M.H. Nitric oxide and liver disease. *Gastroenterologist* **1995**, *3*, 220–233.
71. Phaniendra, A.; Jestadi, D.B.; Periyasamy, L. Free Radicals: Properties, Sources, Targets, and Their Implication in Various Diseases. *Ind. J. Clin. Biochem.* **2015**, *30*, 11–26. [CrossRef]
72. Venza, I.; Venza, M.; Visalli, M.; Lentini, G.; Teti, D.; d'Alcontres, F.S. ROS as Regulators of Cellular Processes in Melanoma. *Oxid. Med. Cell. Longev.* **2021**, *2021*, 1–19. [CrossRef]
73. Di Ferdinandoa, M.; Brunetti, C.; Agati, G.; Tattini, M. Multiple functions of polyphenols in plants inhabiting unfavorable Mediterranean areas. *Environ. Exp. Bot.* **2014**, *103*, 107–116. [CrossRef]
74. Amrit, B.K.; Ponnampalam, E.N.; Macwan, S.; Wu, H.; Aziz, A.; Muir, S.; Dunshea, F.R.; Suleria, H.A.R. Comprehensive screening and characterization of polyphenol compounds from pasture grasses used for livestock production under temperate region. *Anim. Feed. Sci. Technol.* **2023**, *300*, 115657. [CrossRef]
75. Scalbert, A.; Morand, C.; Manach, C.; Rémésy, C. Absorption and metabolism of polyphenols in the gut and impact on health. *Biomed. Pharmacother.* **2002**, *56*, 276–282. [CrossRef] [PubMed]
76. Duda-Chodak, A.; Tar, T. Possible Side Effects of Polyphenols and Their Interactions with Medicines. *Molecules* **2023**, *28*, 2536. [CrossRef] [PubMed]
77. Wink, M. Compartmentation of secondary metabolites and xenobiotics in plant vacuoles. *Adv. Bot. Res.* **1997**, *25*, 141–169. [CrossRef]
78. Tsao, R. Chemistry and Biochemistry of Dietary Polyphenols. *Nutrients* **2010**, *2*, 1231–1246. [CrossRef]
79. Asfaw, D.E. Antioxidant Properties of Phenolic Compounds to Manage Oxidative Stress—A Review. *J. Adv. Agron. Crop. Sci.* **2022**, *1*, 1–16.
80. Rice-Evans, C. Flavonoid antioxidants. *Curr. Med. Chem.* **2001**, *8*, 797–807. [CrossRef]
81. Chen, D.; Mubeen, B.; Hasnain, A.; Rizwan, M.; Adrees, M.; Naqvi, S.A.H.; Iqbal, S.; Kamran, M.; El-Sabrout, A.M.; Elansary, H.O.; et al. Role of Promising Secondary Metabolites to Confer Resistance Against Environmental Stresses in Crop Plants: Current Scenario and Future Perspectives. *Front. Plant. Sci.* **2022**, *13*, 881032. [CrossRef]
82. Rein, M.J.; Renouf, M.; Cruz-Hernandez, C.; Actis-Goretta, L.; Thakkar, S.K.; da Silva Pinto, M. Bioavailability of bioactive food compounds: A challenging journey to bioefficacy. *Br. J. Clin. Pharmacol.* **2013**, *75*, 588–602. [CrossRef]
83. Eseberri, I.; Trepiana, J.; Léniz, A.; Gómez-García, I.; Carr-Ugarte, H.; González, M.; María, P.; Portillo, M.P. Variability in the Beneficial Effects of Phenolic Compounds: A Review. *Nutrients.* **2022**, *14*, 1925. [CrossRef]
84. Kumar, S.; Pandey, A.K. Chemistry and Biological Activitiesof Flavonoids: An Overview. *Sci. World J.* **2013**, *16*, 162750. [CrossRef]
85. Rice-Evans, C.; Spencer, J.P.E.; Schroeter, H.; Rechner, A.R. Bioavailability of flavonoids and potential bioactive forms in vivo. *Drug. Metab. Drug. Interact.* **2000**, *17*, 291–310. [CrossRef] [PubMed]
86. Scalbert, A.; Williamson, G. Dietary Intake and Bioavailability of Polyphenols 1. *J. Nutr.* **2000**, *130*, 2073S–2085S. [CrossRef] [PubMed]
87. Qiao, J.P.; Kong, X.Y.; Kong, A.Y.; Han, M. Pharmacokinetics and biotransformation of tea polyphenols. *Curr. Drug Metab.* **2014**, *15*, 30–36. [CrossRef]
88. Gonzales, G.B.M.; Smagghe, G.; Grootaert, C.; Zotti, M.; Raes, K.; Van Camp, J. Flavonoid interactions during digestion, absorption, distribution and metabolism: A sequential structure -activity/property relationship-based approach in the study of bioavailability and bioactivity. *Drug Metab. Rev.* **2015**, *47*, 175–190. [CrossRef]
89. Monagas, M.; Urpi-Sarda, M.; Sánchez-Patán, F.; Lorach, R.; Garrido, I.; Gómez-Cordovés, C.; Andrés-Lacueva, C.; Bartolomé, B. Insights into the metabolism and microbial biotransformation of dietary flavan-3-ols and the bioactivity of their metabolites. *Food Funct.* **2010**, *1*, 233–253. [CrossRef]
90. Kay, C.D.; Pereira-Caro, G.; Ludwig, I.A.; Clifford, M.N.; Crozier, A. Anthocyanins and Flavanones Are More Bioavailable than Previously Perceived: A Review of Recent Evidence. *Annu. Rev. Food Sci. Technol.* **2017**, *8*, 155–180. [CrossRef]
91. Day, A.J.; Cañada, F.J.; Díaz, J.C.; Kroon, P.A.; McLauchlan, R.; Faulds, C.B.; Plumb, G.W.; Morgan, M.R.; Williamson, G. Dietary flavonoid and isoflavone glycosides are hydrolysed by the lactase site of lactase phlorizin hydrolase. *FEBS Lett.* **2000**, *468*, 166–170. [CrossRef]
92. Costa, M.; Sezgin-Bayindir, Z.; Losada-Barreiro, S.; Paiva-Martins, F.; Saso, L.; Bravo-Díaz, C. Polyphenols as Antioxidants for Extending Food Shelf-Life and in the Prevention of Health Diseases: Encapsulation and Interfacial Phenomena. *Biomedicines* **2021**, *9*, 1909. [CrossRef]
93. Jiamboonsri, P.; Pithayanukul, P.; Bavovada, R.; Leanpolchareanchai, J.; Gao, S.; Hu, M. In Vitro Glucuronidation of Methyl Gallate and Pentagalloyl Glucopyranose by Liver Microsomes. *Drug Metab. Pharmacokinet.* **2016**, *31*, 292–303. [CrossRef]
94. Zong, L.; Inoue, M.; Nose, M.; Kojima, K.; Sakaguchi, N.; Isuzugawa, K.; Takeda, T.; Ogihara, Y. Metabolic fate of gallic acid orally administered to rats. *Biol. Pharm. Bull.* **1999**, *22*, 326–329. [CrossRef] [PubMed]
95. Serreli, G.; Deiana, M. Biological Relevance of Extra Virgin Olive Oil Polyphenols Metabolites. *Antioxidants.* **2018**, *7*, 170. [CrossRef] [PubMed]
96. Selma, M.V.; Espin, J.C.; Tomás-Barberán, F.A. Interaction between phenolics and gut microbiota: Role in human health. *J. Agric. Food Chem.* **2009**, *57*, 6485–6501. [CrossRef] [PubMed]
97. Brenes, A.; Viveros, A.; Chamorro, S.; Arija, I. Use of Polyphenol-Rich Grape by-Products in Monogastric Nutrition: A Review. *Anim. Feed. Sci. Technol.* **2016**, *211*, 1–17. [CrossRef]

98. Sánchez-Patán, F.; Cueva, C.; Monagas, M.; Walton, G.E.; Gibson, G.R.; Martín-Álvarez, P.J.; Moreno-Arribas, M.V.; Bartolomé, B. Gut microbial catabolism of grape seed flavan-3-ols by human faecal microbiota. Targetted analysis of precursor compounds, intermediate metabolites and end-products. *Food Chem.* **2012**, *131*, 337–347. [CrossRef]
99. Ahmed, S.T.; Hossain, M.E.; Kim, G.M.; Hwang, J.A.; Ji, H.; Yang, C.J. Effects of resveratrol and essential oils on growth performance, immunity and microbial shedding in challenged piglets. *Asian-Aust. J. Anim. Sci.* **2013**, *26*, 683–690. [CrossRef]
100. Fiesel, A.; Gessner, D.K.; Erika Most, E.; Eder, K. Effects of dietary polyphenol-rich plant products from grape or hop on pro-inflammatory gene expression in the intestine, nutrient digestibility and faecal microbiota of weaned pigs. *BMC Vet. Res.* **2014**, *10*, 196–206. [CrossRef]
101. Qui, N.H.; Thu, N.T.A. Recent Advances of Using Polyphenols to Extenuate Oxidative Stress in Animal Production: Evidence from Poultry. *Kafkas Univ. Vet. Fak. Derg.* **2022**, *28*, 535–541. [CrossRef]
102. Parkinson, A.; Ogilvie, B.W. Biotransformation of xenobiotics. In *Casarett & Doull's Essentials of Toxicology*; Klaassen, C.D., Watkins, J.B., Eds.; McGraw-Hill Medical: Chicago, IL, USA, 2010.
103. Kroon, P.A.; Clifford, M.N.; Crozier, A.; Day, A.J.; Donovan, J.L.; Manach, C.; Williamson, G. How should we assess the effects of exposure to dietary polyphenols in vitro? Am. *J. Clin. Nut.* **2004**, *80*, 15–21. [CrossRef]
104. Chen, Y.; Xie, S.; Chen, S.; Zeng, S. Glucuronidation of flavonoids by recombinant UGT1A3 and UGT1A9. *Biochem. Pharmacol.* **2008**, *76*, 416–425. [CrossRef]
105. Wu, B.; Kulkarni, K.; Basu, S.; Zhang, S.; Hu, M. First-pass metabolism via UDP glucuronosyltransferase: A barrier to oral bioavailability of phenolics. *J. Pharm. Sci.* **2011**, *100*, 3655–3681. [CrossRef] [PubMed]
106. Wu, B.; Morrow, J.K.; Singh, R.; Zhang, S.; Hu, M. Three-dimensional quantitative structure-activity relationship studies on UGT1A9- mediated 3-O-glucuronidation of natural flavonols using a pharmacophore- based Comparative Molecular Field Analysis model. *J. Pharm. Exp. Ther.* **2011**, *336*, 403–413. [CrossRef] [PubMed]
107. Wu, B.; Basu, S.; Meng, S.; Wang, X.; Hu, M. Regioselective sulfation and glucuronidation of phenolics: Insights into the structural basis. *Curr. Drug Metab.* **2011**, *12*, 900–916. [CrossRef] [PubMed]
108. Meng, S.; Wu, B.; Singh, R.; Yin, T.; Morrow, J.K.; Zhang, S.; Hu, M. SULT1A3-mediated regiospecific 7-O-sulfation of flavonoids in Caco-2 cells can be explained by the relevant molecular docking studies. *Mol. Pharm.* **2012**, *9*, 862–873. [CrossRef] [PubMed]
109. Yang, H.L.; Chen, S.C.; Senthil Kumar, K.J.; Yu, K.N.; Lee Chao, P.D.; Tsai, S.Y.; Hou, Y.C.; Hseu, Y.C. Antioxidant and anti-inflammatory potential of hesperetin metabolites obtained from hesperetin-administered rat serum: An ex vivo approach. *J. Agric. Food Chem.* **2012**, *60*, 522–532. [CrossRef]
110. Manach, C.; Williamson, G.; Morand, C.; Scalbert, A.; Rémésy, C. Bioavailability and bioefficacy of polyphenols in humans. I. Review of 97 bioavailability studies. *Am. J. Clin. Nutr.* **2005**, *81*, 230S–242S. [CrossRef]
111. Zhang, L.; Lin, G.; Zuo, Z. Position preference on glucuronidation of mono-hydroxylflavones in human intestine. *Life Sci.* **2006**, *78*, 2772–2780. [CrossRef]
112. Papuc, C.; Gheorghe, V.; Goran, G.V.; Predescu, C.N.; Nicorescu, V.; Stefan, G. Plant Polyphenols as Antioxidant and Antibacterial Agents for Shelf-Life Extension of Meat and Meat Products: Classification, Structures, Sources, and Action Mechanisms. *Compr. Rev. Food Sci. Food Saf.* **2017**, *16*, 1243–1267. [CrossRef]
113. Hosny, M. Microbial transformation of 2,5 dihydroxycinnamic acid by Aspergillus niger and Rhizopus oryzae. *Az. J. Pharm. Sci.* **2014**, *49*, 104–116. [CrossRef]
114. Spencer, J.P. Metabolism of tea flavonoids in the gastrointestinal tract. *J. Nutr.* **2003**, *133*, 3255–3261. [CrossRef]
115. Walle, T. Methylation of Dietary Flavones Increases Their Metabolic Stability and Chemopreventive Effects. *Int. J. Mol. Sci.* **2009**, *10*, 5002–5019. [CrossRef] [PubMed]
116. Zhang, W.; Qi, S.; Xue, X.; Al Naggar, Y.; Wu, L.; Wang, K. Understanding the Gastrointestinal Protective Effects of Polyphenols Using Foodomics-Based Approaches. *Front. Immunol.* **2021**, *12*, 671150. [CrossRef] [PubMed]
117. Aura, A.M. Microbial metabolism of dietary phenolic compounds in the colon. *Phytochem. Rev.* **2008**, *7*, 407–429. [CrossRef]
118. Cardona, F.; Andrés-Lacueva, C.; Tulipani, S.; Tinahones, F.J.; Queipo-Ortuño, M.I. Benefits of polyphenols on gut microbiota and implications in human health. *J. Nutr. Biochem.* **2013**, *24*, 1415–1422. [CrossRef]
119. Ozdal, T.; Sela, D.A.; Xiao, J.; Boyacioglu, D.; Chen, F.; Capanoglu, E. The Reciprocal Interactions between Polyphenols and Gut Microbiota and Effects on Bioaccessibility. *Nutrients* **2016**, *8*, 78. [CrossRef]
120. Mahfuz, S.; Shang, Q.; Piao, X. Phenolic compounds as natural feed additives in poultry and swine diets: A review. *J. Anim. Sci. Biotechnol.* **2021**, *12*, 48. [CrossRef]
121. Belkaid, Y.; Hand, W.T. Role of the Microbiota in Immunity and Inflammation. *Cell* **2014**, *157*, 121–141. [CrossRef]
122. Kim, H.B.; Isaacson, R.E. The pig gut microbial diversity: Understanding the pig gut microbial ecology through the next generation high throughput sequencing. *Vet. Microbiol.* **2015**, *177*, 242–251. [CrossRef]
123. Makarewicz, M.; Drożdż, I.; Tarko, T.; Duda-Chodak, A. The Interactions between Polyphenols and Microorganisms, Especially Gut Microbiota. *Antioxidants* **2021**, *10*, 188. [CrossRef]
124. Duda-Chodak, A.; Tarko, T.; Satora, P.; Sroka, P. Interaction of dietary compounds, especially polyphenols, with the intestinal microbiota: A review. *Eur. J. Nutr.* **2015**, *54*, 325–341. [CrossRef]
125. Kaulmann, A.; Bohn, T. Bioactivity of Polyphenols: Preventive and Adjuvant Strategies toward Reducing Inflammatory Bowel Diseases—Promises, Perspectives, and Pitfalls. *Oxid. Med. Cell. Longev.* **2016**, *2016*, 9346470. [CrossRef]

126. Ed Nignpense, B.; Francis, N.; Blanchard, C.; Bommannan Santhakumar, A. Bioaccessibility and Bioactivity of Cereal Poly-phenols: A Review. *Foods* **2021**, *10*, 1595. [CrossRef] [PubMed]
127. Axling, U.; Olsson, C.; Xu, J.; Fernandez, C.; Larsson, S.; Ström, K.; Ahrné, S.; Holm, C.; Molin, G.; Berger, K. Green Tea Powder and Lactobacillus Plantarum Affect Gut Microbiota, Lipid Metabolism and Inflammation in High-Fat Fed C57BL/6J Mice. *Nutr. Metab.* **2012**, *9*, 105. [CrossRef] [PubMed]
128. Scott, M.B.; Styring, A.K.; McCullagh, J.S.O. Polyphenols: Bioavailability, Microbiome Interactions and Cellular Effects on Health in Humans and Animals. *Pathogens* **2022**, *11*, 770. [CrossRef]
129. Rummun, N.; Rondeau, P.; Bourdon, E.; Pires, E.; McCullagh, J.; Claridge, T.D.W.; Bahorun, T.; Li, W.-W.; Neergheen, V.S. Terminalia Bentzoë, a Mascarene Endemic Plant, Inhibits Human Hepatocellular Carcinoma Cells Growth In Vitro via G0/G1 Phase Cell Cycle Arrest. *Pharmaceuticals* **2020**, *13*, 303. [CrossRef] [PubMed]
130. Rummun, N.; Pires, E.; McCullagh, J.; Claridge, T.W.D.; Bahorun, T.; Li, W.-W.; Neergheen, V.S. Methyl Gallate—Rich Fraction of Syzygium Coriaceum Leaf Extract Induced Cancer Cell Cytotoxicity via Oxidative Stress. *S. Afr. J. Bot.* **2021**, *137*, 149–158. [CrossRef]
131. Cory, H.; Passarelli, S.; Szeto, J.; Tamez, M.; Mattei, J. The Role of Polyphenols in Human Health and Food Systems: A Mini-Review. *Front. Nutr.* **2018**, *5*, 87. [CrossRef]
132. D'Archivio, M.; Filesi, C.; Varì, R.; Scazzocchio, B.; Masella, R. Bioavailability of the Polyphenols: Status and Controversies. *Int. J. Mol. Sci.* **2010**, *11*, 1321–1342. [CrossRef]
133. Simons, A.L.; Renouf, M.; Hendrich, S.; Murphy, P.A. Human Gut Microbial Degradation of Flavonoids: Structure-function Relationships. *J. Agric. Food Chem.* **2005**, *53*, 4258–4263. [CrossRef]
134. Rastmanesh, R. High Polyphenol, Low Probiotic Diet forWeight Loss Because of Intestinal Microbiota Interaction. *Chem. Biol. Interact.* **2011**, *189*, 1–8. [CrossRef]
135. Di Meo, F.; Anouar, E.; Podloucka, P.; Fabre, G.; Trouillas, P. Understanding antioxidant properties of natural compounds at the atomic scale. *J. Serb. Soc. Comput.* **2013**, *7*, 58–70.
136. Majewska, M.; Czeczot, H. Flavonoids in prevention and therapy of diseases. *Pol. Merkur. Lekarski.* **2012**, *32*, 50–54. (In Polish)
137. Sánchez-Patán, F.; Cueva, C.; Monagas, M.; Walton, E.G.; Gibson, M.R.G.; Quintanilla-López, J.E.; Lebrón-Aguilar, R.P.; Martín-Álvarez, P.; Moreno-Arribas, J.M.V.; Bartolomé, B. In Vitro Fermentation of a Red Wine Extract by Human Gut Microbiota: Changes in Microbial Groups and Formation of Phenolic Metabolites. *J. Agric. Food Chem.* **2012**, *60*, 2136–2147. [CrossRef]
138. Kafantaris, I.; Kotsampasi, B.; Christodoulou, V.; Kokka, E.; Kouka, P.; Terzopoulou, Z.; Gerasopoulos1, K.; Stagos, D.; Mitsagga, C.; Giavasis, I.; et al. Grape pomace improves antioxidant capacity and faecal microflora of lambs. *J. Anim. Physiol. Anim. Nutr.* **2016**, *101*, 1–14. [CrossRef] [PubMed]
139. Molan, A.L.; Lila, M.A.J.; Mawson, J.; De, S. In vitro and in vivo evaluation of the prebiotic activity of watersoluble blueberry extracts. *World J. Microbiol. Biotechnol.* **2009**, *25*, 1243–1249. [CrossRef]
140. Cueva, C.; Sánchez-Patán, F.; Monagas, M.; Walton, G.E.; Gibson, G.R.; Martín-Álvarez, P.J.; Bartolomé, B.; Moreno-Arribas, M.V. In vitro fermentation of grape seed flavan-3-ol fractions by human faecal microbiota: Changes in microbial groups and phenolic metabolites. *FEMS Microbiol. Ecol.* **2013**, *83*, 792–805. [CrossRef]
141. Xu, Q.; Cheng, M.; Jiang, R.; Zhao, X.; Zhu, J.; Liu, M.; Chao, X.; Zhang, C.; Zhou, B. Effects of dietary supplement with a Chinese herbal mixture on growth performance, antioxidant capacity, and gut microbiota in weaned pigs. *Front. Vet. Sci.* **2022**, *9*, 971647. [CrossRef]
142. Dolara, P.; Luceri, C.; De Filippo, C.; Femia, A.P.; Giovannelli, L.; Caderni, G.; Cecchini, C.; Silvi, S.; Orpianesi, C.; Cresci, A. Red wine polyphenols influence carcinogenesis, intestinal microflora, oxidative damage and gene expression profiles of colonicmucosa in F344 rats. *Mutat. Res.* **2005**, *591*, 237–246. [CrossRef]
143. Viveros, A.; Chamorro, S.; Pizarro, M.; Arija, I.; Centeno, C.; Brenes, A. Effects of dietary polyphenol-rich grape products on intestinal microflora and gut morphology in broiler chicks. *Poult. Sci.* **2011**, *90*, 566–578. [CrossRef]
144. Lacombe, A.; Li, R.W.; Klimis-Zacas, D.; Kristo, A.S.; Shravani Tadepalli, S.; Krauss, E.; Young, R.; Vivian, C.; Wu, H. Lowbush wild blueberries have the potential to modify gut microbiota and xenobiotic metabolism in the rat colon. *PLoS ONE* **2013**, *8*, e67497. [CrossRef]
145. Koutsos, A.; Lima, M.; Conterno, L.; Gasperotti, M.; Bianchi, M.; Fava, F.; Vrhovsek, U.; Lovegrove, J.U.; Tuohy, K.M. Effects of Commercial Apple Varieties on Human Gut Microbiota Composition and Metabolic Output Using an In Vitro Colonic Model. *Nutrients* **2017**, *9*, 533; [CrossRef]
146. Gasaly, N.; Gottelan, M. Interference of dietary polyphenols with potentially toxic amino acid metabolites derived from the colonic microbiota. *Amino Acids* **2022**, *54*, 311–324. [CrossRef] [PubMed]
147. Martemucci, G.; Costagliola, C.; Mariano, M.; D'andrea, L.; Napolitano, P.; D'Alessandro, A.G. Free Radical Properties, Source and Targets, Antioxidant Consumption and Health. *Oxygen* **2022**, *2*, 48–78. [CrossRef]
148. Hu, L.; Huang, T.; Liu, X.J.; Cai, Y.D. Predicting protein phenotypes based on protein-protein interaction network. *PLoS ONE* **2011**, *6*, e17668. [CrossRef]
149. Procházková, D.; Boušová, I.; Wilhelmová, N. Antioxidant and prooxidant properties of flavonoids. *Fitoterapia* **2011**, *82*, 513–523. [CrossRef] [PubMed]
150. Halliwell, B. Are polyphenols antioxidants or prooxidants? What do we learn from cell culture and in vivo studies? *Arch. Biochem. Biophys.* **2008**, *476*, 107–112. [CrossRef]

151. Halliwell, B. Dietary polyphenols: Good, bad, or indifferent for your health? *Cardiovasc. Res.* **2007**, *73*, 341–347. [CrossRef]
152. Irina, I.; Mohamed, G. Biological Activities and Effects of Food Processing on Flavonoids as Phenolic Antioxidants. In *Advances in Applied Biotechnology*; Petre, M., Ed.; University of Pitesti: Pitesti, Romania, 2012; ISBN 978-953-307-820-5.
153. Ren, W.; Qiao, Z.; Wang, H.; Zhu, L.; Zhang, L. Flavonoids: Promising anticancer agents. *Med. Res. Rev.* **2003**, *23*, 519–534. [CrossRef]
154. Rohn, S.; Petzke, K.J.; Rawel, H.M.; Kroll, J. Reactions of chlorogenic acid and quercetin with a soy protein isolate–Influence on the in vivo food protein quality in rats. *Mol. Nutr. Food. Res.* **2006**, *50*, 696–704. [CrossRef]
155. Brenes, A.; Viveros, A.; Goñi, I.; Centeno, C.; Saura-Calixto, F.; Arija, I. Effect of grape seed extract on growth performance, protein and polyphenol digestibilities, and antioxidant activity in chickens. *Span. J. Agric. Res.* **2010**, *8*, 326–333. [CrossRef]
156. Chamorro, S.; Viveros, A.; Centeno, C.; Romero, C.; Arija, I.; Brenes, A. Effects of dietary grape seed extract on growth performance, amino acid digestibility and plasma lipids and mineral content in broiler chicks. *Animal* **2013**, *7*, 555–561. [CrossRef] [PubMed]
157. Nkukwana, T.T.; Muchenje, V.; Pieterse, E.; Masika, P.J.; Mabusela, T.P.; Hoffman, L.C.; Dzama, K. Effect of Moringa oleifera leaf meal on growth performance, apparent digestibility, digestive organ size and carcass yield in broiler chickens. *Livest. Sci.* **2014**, *161*, 139–146. [CrossRef]
158. Park, J.H.; Kang, S.N.; Chu, G.M.; Jin, S.K. Growth performance, blood cell profiles, and meat quality properties of broilers fed with Saposhnikovia divaricata, Lonicera japonica, and Cheli-donium majus extracts. *Livest. Sci.* **2014**, *165*, 87–94. [CrossRef]
159. Surai, P.F. Polyphenol compounds in the chicken/animal diet: From the pastto the future. *J. Anim. Physiol. Anim. Nutr.* **2014**, *98*, 19–31. [CrossRef] [PubMed]
160. El-Iraqi, K.G.; Abdelgawad, E.M.; Ibrahim, H.M.; El Sawe, A.E. Effect of Ginkgo biloba, dry peppermint and vitamin C as anti-stress on broiler welfare during summer heat stress. *Glob. Vet.* **2013**, *10*, 770–778.
161. Gessner, D.K.; Fiesel, A.; Most, E.; Dinges, J.; Wen, G.; Ringseis, R.; Eder, K. Supplementation of a grape seed and grape marc meal extract decreases activities of the oxidative stress-responsive transcription factors NF-κB and Nrf2 in the duodenal mucosa of pigs. *Acta Vet. Scand.* **2013**, *55*, 18–28. [CrossRef] [PubMed]
162. Chedea, V.S.; Palade, L.M.; Pelmus, R.S.; Dragomir, C.; Taranu, I. Red Grape Pomace Rich in Polyphenols Diet Increases the Antioxidant Status in Key Organs—Kidneys, Liver, and Spleen of Piglets. *Animals* **2019**, *9*, 149. [CrossRef]
163. Lipiński, K.; Korniewicz, D.; Antoszkiewicz, Z.; Mazur, M. Effect of onion and grape seed extracts on performance and the vitamin E and antioxidant status in sows. In Proceedings of the XLIV Scientific Session: Nutrition of Livestock, Companion and Wild Animals, Warsaw, Poland, 16–17 June 2015; pp. 16–17.
164. Flis, M.; Sobotka, W.; Antoszkiewicz, Z.; Lipiński, K.; Zduńczyk, Z. Effect of husked and naked oat used in the diets supplemented with linseed oil on the growth performance of pigs, carcass and meat quality. *Arch. Spec. Issue.* **2007**, *50*, 161–171. [CrossRef]
165. Leusink, G.; Rempel, H.; Skura, B.; Berkyto, M.; White, W.; Yang, Y.; Fitzpatrick, S. Growth performance, meat quality, and gut microflora of broiler chickens fed with cranberry extract. *Poult. Sci.* **2010**, *89*, 1514–1523. [CrossRef]
166. Simitzis, P.E.; Symeon, G.K.; Charismiadou, M.A.; Ayoutanti, A.G.; Deligeorgis, S.G. The effects of dietary hesperidin supplementation on broiler performance and chicken meat characteristics. *Can. J. Anim. Sci.* **2011**, *91*, 275–282. [CrossRef]
167. Clark, R.; Kupper, T. Old meets new: The interaction between innate and adaptive immunity. *J. Invest. Dermatol.* **2005**, *125*, 629–637. [CrossRef] [PubMed]
168. Hachimura, S.; Totsuka, M.; Hosono, A. Immunomodulation by food: Impact on gut immunity and immune cell function. *Biosci. Biotechnol. Biochem.* **2018**, *82*, 584–599. [CrossRef] [PubMed]
169. Serra, D.; Almeida, L.M.; Dinis, T.C. Dietary polyphenols: A novel strategy to modulate microbiota-gut-brain axis. *Trends Food Sci. Technol.* **2018**, *78*, 224–233. [CrossRef]
170. Jiao, X.; Wang, Y.; Lin, Y.; Lang, Y.; Li, E.; Zhang, X.; Zhang, Q.; Feng, Y.; Meng, X.; Li, B. Blueberry Polyphenols Extract as a Potential Prebiotic with Anti-Obesity Effects on C57BL/6 J Mice by Modulating the Gut Microbiota. *J. Nutr. Biochem.* **2019**, *64*, 88–100. [CrossRef]
171. Qiao, Y.; Sun, J.; Xia, S.; Tang, X.; Shi, Y.; Le, G. Effects of Resveratrol on Gut Microbiota and Fat Storage in a Mouse Model with High-Fat-Induced Obesity. *Food Funct.* **2014**, *5*, 1241–1249. [CrossRef]
172. Rodríguez, M.; Rebollar, P.G.; Mattioli, S.; Castellini, C. N-3 PUFA Sources (Precursor/Products): A Review of Current Knowledge on Rabbit. *Animals* **2019**, *9*, 806. [CrossRef]
173. Shimizu, M. Multifunctions of dietary polyphenols in the regulation of intestinal inflammation. *J. Food Drug Anal.* **2017**, *25*, 93–99. [CrossRef]
174. Corrêa, T.A.F.; Rogero, M.M.; Hassimotto, N.M.A.; Lajolo, F.M. The Two-Way Polyphenols-Microbiota Interactions and Their Effects on Obesity and Related Metabolic Diseases. *Front. Nutr.* **2019**, *6*, 188. [CrossRef]
175. Iqbal, Y.; Cottrell, J.J.; Suleria, H.A.R.; Dunshea, F.R. Gut Microbiota-Polyphenol Interactions in Chicken: A Review. *Animals* **2020**, *10*, 1391. [CrossRef]
176. Scislowski, V.; Bauchart, D.; Gruffat, D.; Laplaud, P.M.; Durand, D. Effects of dietary n−6 or n−3 polyunsaturated fatty acids protected or not against ruminal hydrogenation on plasma lipids and their susceptibility to peroxidation in fattening steers. *J. Anim. Sci.* **2005**, *83*, 2162–2174. [CrossRef]
177. Wood, J.D.; Enser, M. Factors influencing fatty acids in meat and the role of antioxidants in improving meat quality. *Br. J. Nutr.* **1997**, *78*, S49–S60. [CrossRef] [PubMed]

178. Gladine, C.; Morand, C.; Rock, E.; Bauchart, D.; Durand, D. Plant extracts rich in polyphenols (PERP) are efficient antioxidants to prevent lipoperoxidation in plasma lipids from animals fed n-3 PUFA supplemented diets. *Anim. Feed Sci. Technol.* **2007**, *136*, 281–296. [CrossRef]
179. Branciari, R.; Ranucci, D.; Trabalza-Marinucci, M.; Codini, M.; Orru, M.; Ortenzi, R.; Forte, C.; Ceccarini, M.R.; Valiani, A. Evaluation of the antioxidant properties and oxidative stability of Pecorino cheese made from the raw milk of ewes fed *Rosmarinus officinalis* L. leaves. *Int. J. Food Sci. Technol.* **2015**, *50*, 558–565. [CrossRef]
180. Todaro, M.; Alabiso, M.; Scatassa, M.L.; Di Grigoli, A.; Mazza, F.; Maniaci, G.; Bonanno, A. Effect of the inclusion of fresh lemon pulp in the diet of lactating ewes on the properties of milk and cheese. *Anim. Feed Sci. Tech.* **2017**, *225*, 213–223. [CrossRef]
181. Habeanu, M.; Lefter, N.A.; Ropota, M.; Chedea, V.S.; Gheorghe, A.; Toma, S.M.; Ciurescu, G.; Dragomir, C. Dried grape pomace influenced fatty acids composition of Longissimus dorsi muscle and plasma polyphenols spectrum in finishing pigs. *Indian J. Anim. Sci.* **2015**, *85*, 786–789.
182. Muñoz-González, I.; Chamorro, S.; Pérez-Jiménez, J.; López-Andrés, P.; Álvarez-Acero, I.; Herrero, A.M.; Nardoia, M.; Brenes, A.; Viveros, A.; Arija, I.; et al. Phenolic metabolites in plasma and thigh meat of chickens supplemented with grape byproducts. *J. Agric. Food Chem.* **2019**, *67*, 4463–4471. [CrossRef]
183. Atwood, S.B.; Provenza, F.D.; Wiedmeier, R.D.; Banner, R.E. Influence of Free-Choice vs Mixed-Ration Diets on Food Intake and Performance of Fattening Calves. *J. Anim. Sci.* **2001**, *79*, 3034. [CrossRef]
184. Catanese, F.; Obelar, M.; Villalba, J.J.; Distel, R.A. The Importance of Diet Choice on Stress-Related Responses by Lambs. *Appl. Anim. Behav. Sci.* **2013**, *148*, 37–45. [CrossRef]
185. Pichler, J.; Schwarz, C.; Gierus, M.; Schedle, K. Choice Feeding in Fattening Pigs: Effect of Diets Differing in Nutrient Density on Feeding Behaviour and Fattening Performance. *Czech J. Anim. Sci.* **2020**, *65*, 247–257. [CrossRef]
186. Meier, J.S.; Kreuzer, M.; Marquardt, S. Design and Methodology of Choice Feeding Experiments with Ruminant Livestock. *Appl. Anim. Behav. Sci.* **2012**, *140*, 105–120. [CrossRef]

antioxidants

Article

Stevioside Improves Antioxidant Capacity and Intestinal Barrier Function while Attenuating Inflammation and Apoptosis by Regulating the NF-κB/MAPK Pathways in Diquat-Induced Oxidative Stress of IPEC-J2 Cells

Qinglei Xu, Mingzheng Liu, Xiaohuan Chao, Chunlei Zhang, Huan Yang, Jiahao Chen and Bo Zhou *

College of Animal Science and Technology, Nanjing Agricultural University, Nanjing 210095, China; 2019205004@njau.edu.cn (Q.X.); 2020205018@stu.njau.edu.cn (M.L.); 2021205020@stu.njau.edu.cn (X.C.); 2020105039@stu.njau.edu.cn (C.Z.); 2021105038@stu.njau.edu.cn (H.Y.); 2021105039@stu.njau.edu.cn (J.C.)
* Correspondence: zhoubo@njau.edu.cn

Abstract: As a natural sweetener, stevioside is extracted from *Stevia rebaudiana* Bertoni and possesses potent antioxidant activity. However, little information is known about its protective role in maintaining the intestinal epithelial cells health under oxidative stress. The aim of this study was to investigate the protective effects and underlying mechanisms of stevioside on alleviating inflammation, apoptosis, and improving antioxidant capacity in intestinal porcine epithelial cells (IPEC-J2) under oxidative stress by diquat. The results demonstrated that the pretreatment with stevioside (250 μM) for 6 h increased cell viability and proliferation and prevented apoptosis induced by diquat at 1000 μM for 6 h in IPEC-J2 cells, compared with the diquat alone-treated cells. Importantly, stevioside pretreatment significantly reduced ROS and MDA production as well as upregulated T-SOD, CAT, and GSH-Px activity. Moreover, it also decreased cell permeability and improved intestinal barrier functions by significantly upregulating the tight junction protein abundances of claudin-1, occludin, and ZO-1. At the same time, stevioside significantly down-regulated the secretion and gene expression of IL-6, IL-8, and TNF-α and decreased the phosphorylation levels of NF-κB, IκB, and ERK1/2 compared with the diquat alone group. Taken together, this study demonstrated that stevioside alleviated diquat-stimulated cytotoxicity, inflammation, and apoptosis in IPEC-J2 cells, protecting cellular barrier integrity and mitigating oxidative stress by interfering with the NF-κB and MAPK signaling pathways.

Keywords: stevioside; antioxidant; oxidative stress; inflammation; IPEC-J2 cells; NF-κB/MAPK signaling pathway

Citation: Xu, Q.; Liu, M.; Chao, X.; Zhang, C.; Yang, H.; Chen, J.; Zhou, B. Stevioside Improves Antioxidant Capacity and Intestinal Barrier Function while Attenuating Inflammation and Apoptosis by Regulating the NF-κB/MAPK Pathways in Diquat-Induced Oxidative Stress of IPEC-J2 Cells. *Antioxidants* **2023**, *12*, 1070. https://doi.org/10.3390/antiox12051070

Academic Editors: Luciana Rossi and Matteo Dell'Anno

Received: 22 April 2023
Revised: 8 May 2023
Accepted: 8 May 2023
Published: 10 May 2023

1. Introduction

Oxidative stress (OS) arises as a consequence of an imbalance between oxidation and antioxidant levels in the body [1]. Oxidative stress refers to the excessive production of reactive oxygen species (ROS) and reactive nitrogen species (RNS) relative to antioxidant defense, resulting in the disorder of free radical metabolism in the body [2]. The physiological function of piglets is not fully developed at weaning, and their ability to provide antioxidant defense is weak [3]. The intestinal tract of pigs is composed of monolayer intestinal epithelial cells, which play an important role in the digestion and absorption of nutrients, immune barrier function, and amino acid metabolism [4]. When piglets are subjected to stress, the degree of the oxidation exceeds the clearance rate of oxides, and the imbalance of oxidation system and antioxidant system causes oxidative stress and then leads to intestinal epithelial damage [5]. Oxidative stress causes the intestinal barrier function of piglets to be damaged, which affects the digestion and absorption of nutrients, and the immune function is decreased, which leads to the decline of disease resistance and even diarrhea [6]. In addition, oxidative stress is common in the intensive production of pigs, which poses greater risks to animal health and seriously affects the economic benefits of the pig industry. Therefore, it is very important to develop effective methods to protect pigs against oxidative injury caused by oxidative stress.

Stevioside is a natural sweetener extracted from *Stevia rebaudiana* Bertoni that is often used as a sweetener additive [7]. Natural noncaloric sweeteners that can substitute for sucrose in the food industry have attracted more and more attention with the rising incidence of obesity and diabetes [8]. In addition to the application of stevioside as a sweetener in food, stevioside also has the pharmacological effects of hypoglycemic [9], antioxidant [10], anti-inflammatory activities [11] and immune regulation [12]. In addition, the abuse of antibiotics causes antibiotic residues in foodstuffs, which led to the ban of antibiotics as feed additives in many countries [13]. Stevioside, as a natural feed additive, has attracted widespread attention from researchers. Previous studies have shown that stevioside as a feed additive has positive potential to improve goat feed intake [14]. Furthermore, it should be noted that the administration of stevioside has been shown to yield favorable outcomes in terms of animal health and growth. This is exemplified by the fact that the provision of stevioside to maternal subjects resulted in an increase in innate immunity and a notable improvement in the physiological state of subsequent generations of chicken offspring [15]. Stevioside alleviated the intestinal mucosal damage of broilers caused by lipopolysaccharide (LPS) through its anti-inflammatory and antioxidant activities [16]. At the same time, supplementary stevioside in cattle diets improved growth performance, carcass traits, meat quality, and decreased the lipid oxidation of beef during storage [17]. The incorporation of stevioside into the diet of broiler chickens resulted in improved growth performance. As such, stevioside may be considered a viable option for use as a growth-promoting feed additive in broiler chickens [18]. Our previous study showed that 0.5% stevioside in acidified water improved the GSH activity of weaned piglets and showed an antioxidant effect [19]. However, the underlying molecular mechanism of how stevioside exerts an antioxidant effect in pigs remains unknown.

Diquat, as a classical inducer of oxidative stress, is widely used to establish an oxidative stress model in animals [20]. Intestinal epithelial cells participate in digestion, absorption, secretion, barrier integrity, and the stress response of the intestine [21]. The IPEC-J2 cell line was derived from the jejunum epithelium of newborn piglets and is often be used as an in vitro model of small intestinal epithelial cells in pigs [22]. Despite the potential benefits attributed to stevioside as an antioxidant, its molecular mechanism of action and the extent to which it may be applied in the context of oxidative stress remain ambiguous, particularly within the confines of IPEC-J2 cell lines. Thus, the purposes of this study were to investigate the antioxidant effects of stevioside against oxidative damage induced by diquat in IPEC-J2 cells and to explore the underlying molecular mechanism.

2. Materials and Methods

2.1. Chemicals

Diquat (purity $\geq$ 99.9%) was purchased from TMRM (Tan-Mo Technology Co., Ltd., Changzhou, China). Stevioside standards (purity $\geq$ 98% by high performance liquid chromatography, CAS: 57817-89-7) were purchased from Shanghai Yuanye Bio-Technology Co., Ltd. (Shanghai, China). Dulbecco's modified Eagle's medium/nutrient mixture F12 (DMEM/F12) medium, fetal bovine serum (FBS), trypsin, and penicillin-streptomycin were purchased from Gibco (Grand Island, NY, USA). The antibodies used here were obtained from ABclonal Technology (ABclonal, Wuhan, China).

2.2. Cell Culture and Treatments

The IPEC-J2 cells were kindly provided by Prof. Chunmei Li (Nanjing Agricultural University, Nanjing, China), cultured in DMEM/F12 medium containing 10% fetal bovine serum (FBS) and 1% penicillin-streptomycin, and maintained at 37 °C with a 5% CO_2 atmosphere. Cells were processed as follows: control group (CON), stevioside group (ST), diquat group (DQ), and stevioside + diquat group (ST + DQ). Specifically, the control cells were cultivated in normal DMEM/F12 containing 10% FBS (CON). Cells were exposed to optimum concentration diquat alone for optimum treatment durations in the diquat treatment group (DQ), while those in the stevioside treatment group (ST) were exposed to optimum concentration stevioside for optimum treatment durations. Cells were pretreated with optimum concentration stevioside for optimum treatment durations, followed by co-incubation with optimum concentration diquat for optimum treatment durations in the stevioside + diquat treatment group (ST + DQ).

2.3. Cell Viability Assay

IPEC-J2 cells were cultured in 96-well plates (Costar, Corning Inc., Corning, NY, USA) for diquat or stevioside treatment. Firstly, IPEC-J2 cells were treated with diquat to establish an oxidative stress model in vitro. The cytotoxicity of diquat was evaluated using the methyl thiazolyl tetrazolium (MTT) assay (Nanjing Jiancheng Bioengineering Institute, Nanjing, China). IPEC-J2 cells were seeded in 96-well culture plates at 1×10^4 cells/well, and 100 μL of complete culture medium was added to culture overnight. After 4, 6, or 8 h of treatment with diquat at final concentrations of 0, 100, 250, 500, 750, 1000, 1250, and 1500 μM, incubated cells with 1× MTT working solution for 4 h at 37 °C. The absorbance at 570 nm was measured using a microplate reader (Tecan, Austria GmbH, Grödig, Austria). Using the 50% inhibitory concentration (IC 50) as a standard, we selected the optimal concentration of diquat to simulate oxidative stress in IPEC-J2 cells.

After growing to 80% confluence in 96-well plates, the cells were pretreated with 0, 50, 100, 250, and 500 μM stevioside for 4, 6, or 8 h. After incubation with optimum stevioside for optimum treatment durations, cells were treated with stevioside + diquat for optimum treatment durations. To analyze cell viability, cells were incubated with CCK-8 solution (Vazyme, Nanjing, China), and the absorbance at 450 nm was measured using a microplate reader (Tecan, Austria GmbH, Grödig, Austria). Cell viability was shown as a percentage of viable cells compared to the wells containing control cells not exposed to stevioside or diquat.

2.4. Measurement of Cell Proliferation

To assess cell proliferation, the Cell-light™ EdU Apollo® 567 In Vitro Imaging Kit (manufactured by Ribobio, Guangzhou, China) was employed and implemented in accordance with the recommended guidelines. Cellular staining was observed with fluorescence microscopy (Zeiss, LSM 700; Oberkochen, Germany). Proliferation was quantified and shown as the percentage of EdU-positive cells.

2.5. Flow Cytometric Determination of Cell Apoptosis

Cells were treated as described above (CON, ST, DQ, and ST + DQ). Apoptosis analysis was performed using an Annexin V-FITC/PI Apoptosis Detection Kit (Vazyme, Nanjing, China). In brief, the cells were collected from a 12-well culture plate, and then they were incubated with 5 μL annexin V-FITC and 5 μL propidium iodide at room temperature for 20 min in the dark. Then, the stained cells were analyzed by a flow cytometer (BD Biosciences, San Jose, CA, USA). Apoptotic cells were expressed as a percentage of total cells.

2.6. Measurement of Intracellular ROS Production

IPEC-J2 cells (10^4 cells per well) were seeded in 96-well plates with 8 parallel holes in each group. Intracellular reactive oxygen species (ROS) production was monitored using a ROS assay kit (Nanjing Jiancheng Bioengineering Institute, Nanjing, China). In brief, IPEC-J2 cells were stained with 10 μM 2′,7′-dichlorohydro-fluorescein diacetate (DCFH-DA) for 30 min at 37 °C, then washed with FBS-free media and re-suspended in PBS. Intracellular ROS production was determined with excitation at 480 nm and emission at 530 nm using a microplate reader (Tecan, Austria GmbH, Grödig, Austria). Its fluorescence signal intensity is proportional to the ROS levels in IPEC-J2 cells. The intracellular ROS levels were expressed as relative fluorescence signals and then normalized to the control group.

2.7. Determination of Oxidative Stress Parameters

After the IPEC-J2 cells were treated as described above, the cells were carefully washed twice with PBS and lysed with RIPA Lysis Buffer (containing PMSF) (Solarbio, Beijing, China) for 10 min. Centrifuge at 1000× *g* for 10 min at 4 °C and collect the supernatants. Then, the activities of total superoxide dismutase (T-SOD), malondialdehyde (MDA), catalase (CAT), glutathione peroxidase (GSH-Px) and total antioxidant capacity (T-AOC) were measured using commercial kits (Nanjing Jiancheng Bioengineering Institute, Nanjing, China) according to the manufacturer. It is expressed as U/mg protein, and experiments were performed in triplicate.

2.8. Detection of Cytokines

The IPEC-J2 cells were treated with one of the four treatments previously described (CON, DQ, ST + DQ, and ST groups). After stimulation, we collected cell supernatants for the detection of cytokines. The concentrations of tumor necrosis factor-α (TNF-α), interleukin-6 (IL-6), and interleukin-8 (IL-8) were detected with enzyme-linked immunosorbent assay (ELISA) kits (Nanjing Jiancheng Bioengineering Institute, Nanjing, China) in accordance with the manufacturer's instructions. In short, the competitive inhibition ELISA method was used to detect the content of IL-6, IL-8, and TNF-α in these four groups. Data were collected by reading optical absorption at 450 nm wavelength using a microplate reader (Tecan, Austria GmbH, Grödig, Austria).

2.9. Quantitative Reverse-Transcription Polymerase Chain Reaction (qRT-PCR)

IPEC-J2 cells were prepared in 12-well culture plates, cultured at a density of 5×10^5 cells per well and adhered for 24 h. After four treatments (CON, DQ, ST + DQ, ST), total RNA was extracted using TRIzol total RNA isolation reagent (Invitrogen, Carlsbad, CA, USA) according to the manufacturer's instructions. RNA quality and concentration were assessed using a NanoDrop ND-2000 Spectrophotometer (NanoDrop Technologies, Wilmington, DE, USA). Total RNA integrity was verified by electrophoresis on a 1.5% agarose gel. Approximately 1 μg total RNA per sample was used to generate cDNA by reverse transcription using the HiScript ® III RT SuperMix for qPCR (+gDNA wiper) kit (Vazyme, Nanjing, China). Real-time quantitative PCR was conducted using ChamQ Universal SYBR qPCR Master Mix (Vazyme, Nanjing, China) on a CFX Connect™ Real-Time PCR Detection System (Bio-Rad, Hercules, CA, USA) with a total volume of 20 μL per reaction. All the primers used in this study are listed in Supplementary Table S1. GAPDH was applied as a housekeeping gene for normalizing gene levels. The RT-qPCR data were analyzed using the $2^{-\Delta\Delta Ct}$ method [23] to calculate the relative fold changes of target genes.

2.10. Western Blotting

The IPEC-J2 cells treated above were harvested, the total proteins were extracted, and the concentrations were detected. Loaded an equal amount of protein (30 μg) into each lane and separated them at 80 V for an initial 30 min, 120 V for 1 h. Then, the gel was transferred onto a polyvinylidene fluoride membrane (PVDF, Millipore, Merck KgaA, Darmstadt, Germany) at 330 mA for 1 h at 4 °C. After blocking with QuickBlock™ Western Blot Blocking Buffer (Beyotime, Shanghai, China) for 30 min, the blots were incubated with primary antibodies overnight at 4 °C. After washing three times with Tris-buffered saline + 0.1% Tween20 (TBST), they were incubated with secondary antibodies for 1 h at room temperature. Chemiluminescence detection was performed using a high-sensitivity ECL chemiluminescence detection kit (Vazyme, Nanjing, China) according to the manufacturer's instructions. Immunoreactive bands were imaged using the ChemiDoc™ Imaging System (Bio-Rad, Hercules, CA, USA) and quantified using ImageJ software (National Institutes of Health, Bethesda, MD, USA).

2.11. Statistical Analysis

All results are expressed as the mean ± standard error of the mean (SEM). Data were visualized and statistical analyses were conducted using GraphPad Prism version 9 (GraphPad Software, Inc., San Diego, CA, USA). Means were compared using one-way analysis of variance (ANOVA), followed by Duncan's post hoc tests in SPSS (version 26.0, SPSS Inc., Chicago, IL, USA). The means of the two groups were compared using a two-tailed Student's t-test. The 50% inhibitory concentration (IC 50) was calculated using SPSS 26.0. $p < 0.05$ was considered a significant difference.

3. Results

3.1. Stevioside Attenuated Diquat-Induced Cytotoxicity in IPEC-J2 Cells

To establish a model of oxidative stress in IPEC-J2 cells, we measured the cell viability of IPEC-J2 cells after treatment with diquat by the MTT method. Compared with the control group, diquat treatment decreased cell viability in a dose- and time-dependent manner (Figure 1A). The cell viability of IPEC-J2 was significantly decreased by 50% after 6 h of treatment with 1000 μM diquat ($p < 0.05$) as compared with the control group (Figure 1A). Therefore, diquat was used to induce oxidative stress at a concentration of 1000 μM for 6 h treatment time in subsequent experiments.

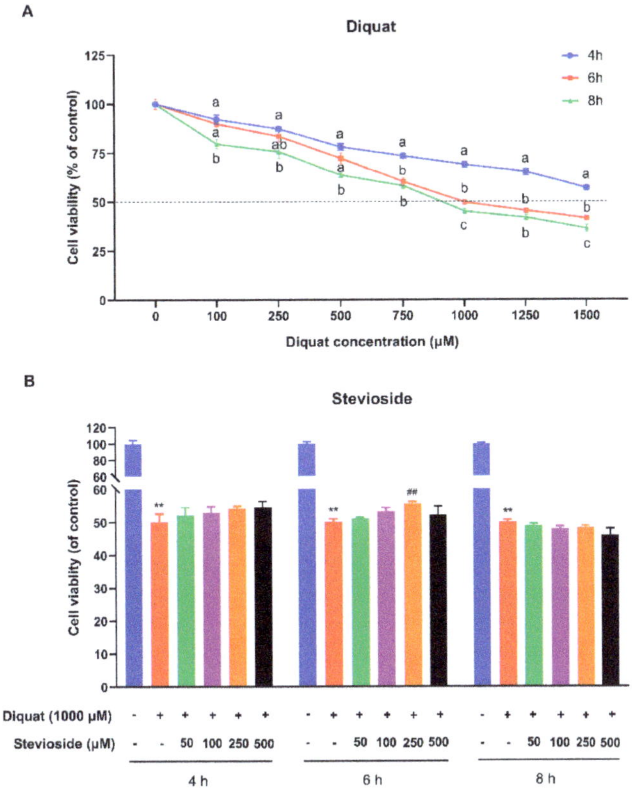

Figure 1. Effects of stevioside and/or diquat on the viability of IPEC-J2 cells. (**A**) IPEC-J2 cells incubated with diquat (0, 100, 250, 500, 750, 1000, 1250, or 1500 μM) for 4, 6, or 8 h. (**B**) Stevioside protects against diquat-induced cell damage of IPEC-J2 cells. IPEC-J2 cells were incubated with or without stevioside at the different concentrations (50, 100, 250, or 500 μM) for 4 h, 6 h, or 8 h, after which the medium was replaced with fresh medium containing 1000 μM diquat. After incubation

for 6 h, the cell viability of IPEC-J2 cells was measured by the CCK-8 (Vazyme, Nanjing, China) assay. All data are presented as the mean ± standard error of the mean (SEM) of three independent experiments. Differences among multiple groups were compared using one-way ANOVA followed by Tukey-Kramer's post hoc tests. Means without a common letter differ, $p < 0.05$. ** $p < 0.01$ vs. nontreated control cells; ## $p < 0.01$ vs. diquat-treated cells.

As shown in Figure 1B, a 4 h or 6 h pretreatment with stevioside increased IPEC-J2 cell viability after diquat treatment, indicating that short-term stevioside pretreatment attenuates diquat-induced cell damage. Pretreatment with stevioside for 6 h more effectively protected cell viability than 4 h or 8 h pretreatment. Lower and higher concentrations of stevioside were less effective after pretreatment for 6 h, while 500 μM stevioside directly decreased IPEC-J2 cell viability and had no protective effect on diquat-induced injury. It is worth noting that diquat-attenuated cell viability was significantly reversed by the pretreatment of 250 μM stevioside for 6 h ($p < 0.05$) (Figure 1B). So, we selected a 250 μM stevioside concentration and a 6 h pretreatment time to conduct further research.

3.2. *Stevioside Promoted Cell Proliferation in Diquat-Treated IPEC-J2 Cells*

IPEC-J2 cell proliferation was first determined using a commercially available Cell-Light EdU DNA cell proliferation kit. The result revealed that the percentages of EdU-positive cells in DQ-treated cells were significantly decreased (5.85% ± 1.39% vs. 41.51% ± 0.77%) ($p < 0.05$) compared with the control group (Figure 2A,B). In contrast, pre-treatment with stevioside significantly increased the EdU-positive cells in the ST + DQ treatment group compared with those exposed only to DQ (40.17 ± 2.20% vs. 5.85 ± 1.39%) ($p < 0.05$) (Figure 2A,B). The results of EdU incorporation experiments indicated a profound decline in fluorescence intensity upon IPEC-J2 cell exposure to diquat (1000 μM, 6h), thus signifying a noticeable reduction in DNA replication. Meanwhile, significant restoration of the DNA replication activity following diquat-induced damage was observed with the administration of stevioside at a concentration of 250 μM. To further verify the effect of stevioside on cell proliferation, the mRNA expression levels of proliferating cell nuclear antigen (*PCNA*) and Cyclin D1 (*CCND1*) were detected (Figure 2C,D). The mRNA expression levels of *PCNA* and *CCND1* decreased significantly after the 1000 μM diquat treatment (Figure 2C,D). However, the expression of *PCNA* and *CCND1* was up-regulated by stevioside pre-treatment in the IPEC-J2 cells (Figure 2C,D). These results indicated that stevioside could extremely increase the ratio of cells in the proliferation phase, dramatically promote the expression of cell proliferation-related genes, and thus promote cell proliferation.

3.3. *Stevioside Alleviated Diquat-Induced Apoptosis in IPEC-J2 Cells*

The IPEC-J2 cell apoptosis was measured by flow cytometry, as shown in Figure 3. Compared with the CON group, diquat incubation led to increased apoptosis, which was rescued by stevioside pretreatment (Figure 3A,B). The expression of the apoptosis-associated genes, *BCL-2*, *BAX*, and *BCL-2/BAX* ratio, was evaluated by qRT-PCR to further understand the antiapoptotic effect of the stevioside. qRT-PCR analysis demonstrated that diquat treatment significantly increased the mRNA abundance of *BAX* ($p < 0.05$) (Figure 3C) and reduced the abundance of *BCL-2* ($p < 0.05$) (Figure 3D) as compared with the control. Expression of the antiapoptotic factor Bax was decreased by stevioside pretreatment, whereas no obvious decrease was observed after diquat treatment alone (Figure 3C). The stevioside pretreatment dramatically increased the mRNA abundance of *BCL-2* and the ratio of *BCL-2/BAX* compared with the diquat treatment ($p < 0.05$) (Figure 3D,E). These results indicated that stevioside pretreatment significantly ameliorated diquat-induced apoptosis.

3.4. *Stevioside Regulated Barrier Function in Diquat-Induced IPEC-J2 Cells*

To assess the role of stevioside or diquat on tight junction permeability, we determined the abundance of tight junction proteins in IPEC-J2 cells. The results showed that diquat treatment significantly decreased the abundance of tight-junction proteins (Figure 4A) and the gene expression levels of claudin-1, occludin, and ZO-1 (Figure 4B–D) as compared with the control group ($p < 0.05$). The pretreatment of 250 μM stevioside to the culture medium enhanced the gene expression abundance of claudin-1, occludin, and ZO-1 as compared with the diquat treatment ($p < 0.05$) (Figure 4B–D).

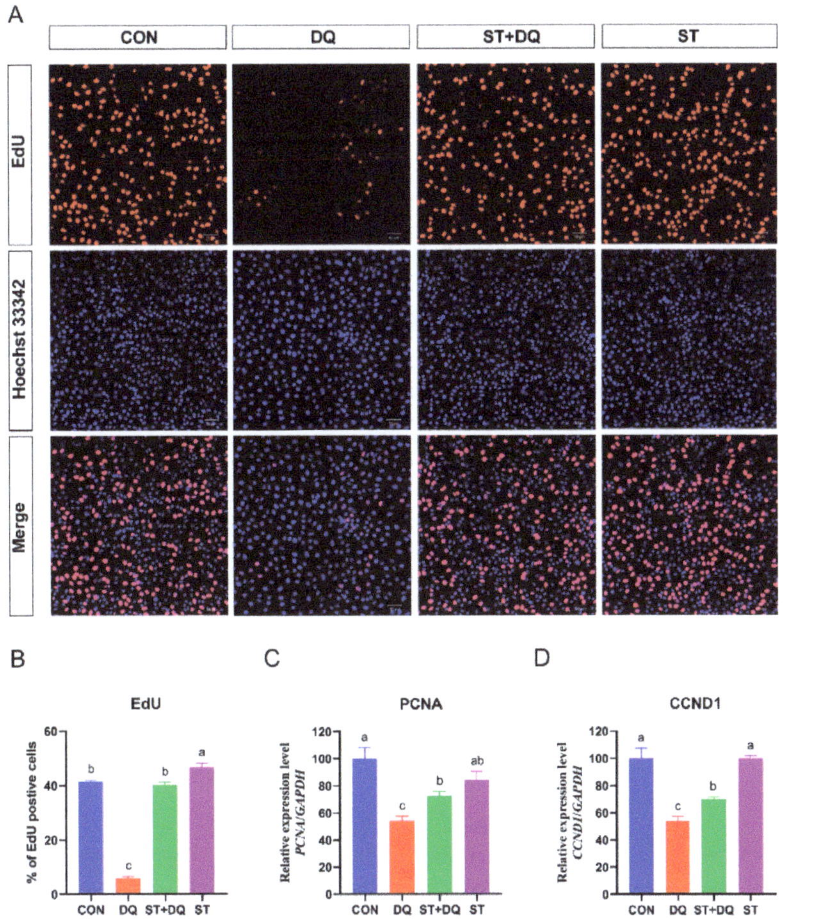

Figure 2. Effects of stevioside and/or diquat on the cell proliferation of IPEC-J2 cells. IPEC-J2 cells were incubated for 6 h in the presence or absence of 250 μM stevioside. Replace the medium with a fresh medium containing 1000 μM diquat and incubate for 6 h. (**A**) Cell proliferation detection was determined by the EdU assay. Proliferating cells were stained with EdU positive (red fluorescence), and cells were stained with Hoechst 33,342 (nuclear blue fluorescence). The red and blue images were merged to produce images with plink or purple fluorescence. The images were taken under a fluorescence microscope (magnification 10×, scale bar = 50 μm). (**B**) Statistical results of the proportion of EdU-positive cells. The mRNA expression of the proliferating cell nuclear antigen (*PCNA*) gene (**C**) and the Cyclin D1 (*CCND1*) gene (**D**) were measured by qRT-PCR. CON, control group, cells without being treated; DQ, diquat group, cells were only treated by diquat; ST + DQ, stevioside+diquat group, cells were pretreated with stevioside and were then treated by diquat; ST, stevioside group, cells were only treated by stevioside; EdU, 5-ethynyl-2′-deoxyuridine. Values are expressed as the mean ± standard error of the mean (SEM) of three independent experiments. Differences among multiple groups were compared using one-way ANOVA followed by Tukey-Kramer's post hoc tests. Means without a common letter differ, $p < 0.05$.

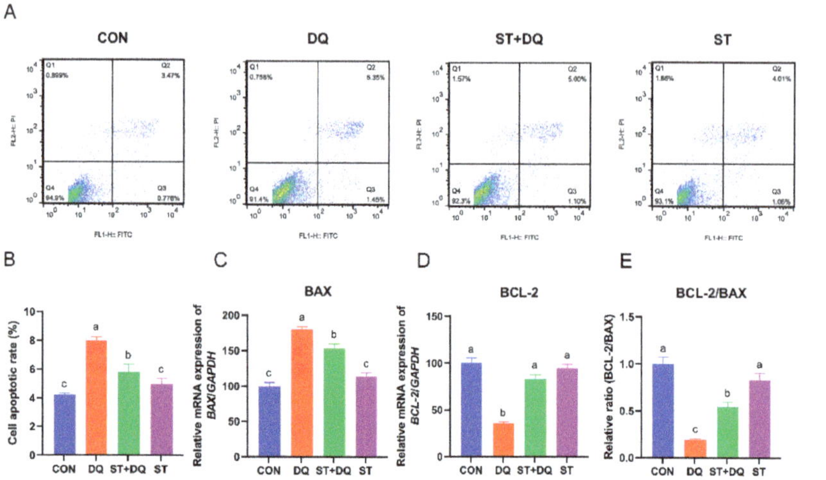

Figure 3. Effects of stevioside and/or diquat on the cell apoptosis of IPEC-J2 cells. IPEC-J2 cells were subjected to incubation conditions consisting of the presence or absence of 250 μM stevioside for a duration of 6 h. Subsequent to this step, the pre-existing medium was replaced with fresh medium containing 1000 μM of diquat, and the incubation was extended for an additional 6 h. (**A**) Apoptosis was determined by flow-cytometric analysis. Q1: dead cells; Q2: late apoptotic cells; Q3: early apoptotic cells; Q4: live cells. (**B**) Quantification of apoptotic cells based on flow cytometry data. Relative mRNA expression levels of BAX (**C**) and BCL-2 (**D**) were determined by qRT-PCR analysis. (**E**) Relative BCL-2/BAX mRNA expression ratio. CON, control, cells without being treated; DQ, diquat, cells were only treated by diquat; ST + DQ, stevioside+diquat, cells were pretreated with stevioside and were then treated by diquat; ST, stevioside, cells were only treated by stevioside; BCL-2, B-cell leukemia/lymphoma-2; BAX, BCL-2-associated X, apoptosis regulator; and GAPDH, glyceraldehyde-3-phosphate dehydrogenase. GAPDH was used as a normalizer. Values are expressed as the mean ± standard error of the mean (SEM) of three independent experiments. Differences among multiple groups were compared using one-way ANOVA followed by Tukey-Kramer's post hoc tests. Means without a common letter differ, $p < 0.05$.

3.5. Stevioside Regulated the Cellular Redox State in Diquat-Induced IPEC-J2 Cells

To evaluate the regulatory effect of stevioside on the redox state of diquat-induced IPEC-J2 cells, we measured intracellular ROS production capacity using a fluorescent ROS probe (DCFH-DA) (Figure 5A). The results showed that the ROS accumulation level was significantly increased after exposure to diquat ($p < 0.05$) (Figure 5A), indicating that diquat enhances the intracellular ROS burst in IPEC-J2 cells. However, pretreatment with stevioside significantly suppressed the diquat-induced ROS burst in IPEC-J2 cells.

To further investigate the redox state, the MDA was determined, which can be affected by diquat-induced ROS release. After diquat treatment alone, the level of MDA was significantly increased in IPEC-J2 cells ($p < 0.05$) (Figure 5B), indicating the occurrence of lipid peroxidation. After stevioside pretreatment, MDA content was significantly decreased ($p < 0.05$) (Figure 5B), which indicated significantly inhibited lipid peroxidation in IPEC-J2 cells. Next, we detected the levels of T-SOD, CAT, GSH-Px, and T-AOC, which serve as important indicators of the cellular redox state. After diquat treatment, intracellular T-SOD and GSH-Px levels were significantly decreased ($p < 0.05$) (Figure 5C,E), while the levels of CAT and T-AOC were decreased by 8.2%, and 4.7%, respectively, compared to the CON group (Figure 5D,F). However, pretreatment with stevioside markedly increased the T-SOD, CAT, and GSH-Px levels compared to those in the DQ group ($p < 0.05$) (Figure 5C–E), indicating the positive effect of stevioside on the redox state in IPEC-J2 cells.

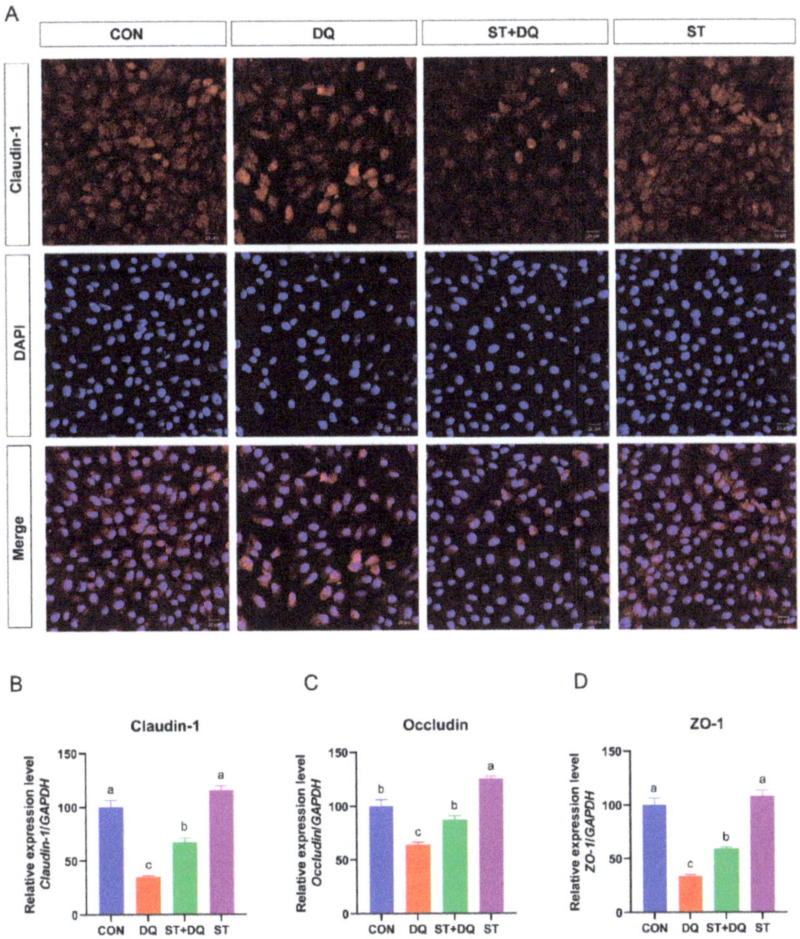

Figure 4. Effects of stevioside on the abundances of tight junction proteins in diquat-challenged IPEC-J2 cells. (**A**) Immunofluorescence staining localizes the tight-junction protein claudin-1. IPEC-J2 cells were seeded into 12-well plates at a density of 2.5×10^5 cells/well (n = 3). Immunofluorescence staining of Claudin-1 (red) and DAPI (blue) in IPEC-J2 cells. The red and blue images were merged to produce images with plink or purple fluorescence. Magnification 20×, and scale bars representing 20 µm. (**B**) The mRNA expression level of claudin-1. (**C**) The mRNA expression level of occludin. (**D**) The mRNA expression level of ZO-1. CON, control, cells without being treated; DQ, diquat, cells were only treated by diquat; ST + DQ, stevioside+diquat, cells were pretreated with stevioside and were then treated by diquat; ST, stevioside, cells were only treated by stevioside. All data are presented as the mean ± standard error of the mean (SEM) of three independent experiments. Differences among multiple groups were compared using one-way ANOVA followed by Tukey-Kramer's post hoc tests. Means not sharing a common letter are significantly different ($p < 0.05$).

3.6. Effects of Stevioside on Inflammatory Cytokines

To further verify the protective effects of stevioside against intestinal inflammation, the secretion and gene expression of several inflammatory cytokines were determined in the IPEC-J2 cells. Compared with the CON group, diquat treatment caused a significant increase in the secretion of pro-inflammatory cytokines, including tumor necrosis factor-α (TNF-α), interleukin-6 (IL-6), and interleukin-8 (IL-8) ($p < 0.05$) (Figure 6A–C). In contrast to the DQ group, stevioside pretreatment dramatically decreased the secretion of pro-inflammatory cytokines, including TNF-α, IL-6, and IL-8 in the IPEC-J2 cell culture supernatant ($p < 0.05$) (Figure 6A–C). In addition, the relative mRNA abundances of *TNF-α*, *IL-6*, and *IL-8* were significantly upregulated by diquat alone treatments for 6 h in IPEC-J2 cells ($p < 0.05$) (Figure 6D,F), compared with the control group. However, they were dramatically down-regulated by stevioside pretreatment ($p < 0.05$) (Figure 6D,F). These results preliminarily inferred that stevioside could alleviate diquat cytotoxicity and decrease inflammation.

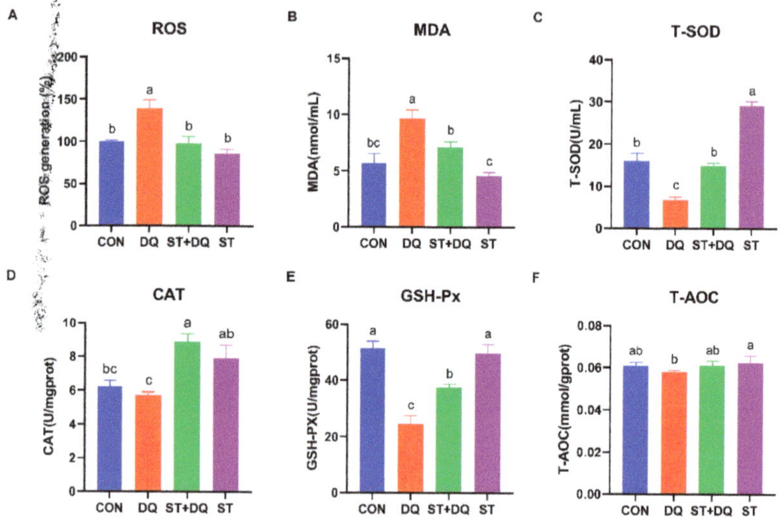

Figure 5. Stevioside scavenged intracellular ROS and improved the antioxidant capacity of diquat-treated IPEC-J2 cells. (**A**) The intracellular ROS content was confirmed by ELISA analysis. IPEC-J2 cells that were not subjected to DCFH2-DA incubation were utilized as blank controls. The fluorescence intensity of the cells was calculated relative to that of the control. (**B**) Cellular MDA levels; (**C**) Cellular T-SOD levels; (**D**) Cellular CAT levels; (**E**) Cellular GSH-PX levels; and (**F**) Cellular T-AOC levels. CON, control, cells without being treated; DQ, diquat, cells were only treated by diquat; ST + DQ, stevioside + diquat, cells were pretreated with stevioside and were then treated by diquat; ST, stevioside, cells were only treated by stevioside. All data are presented as the mean ± standard error of the mean (SEM) of three independent experiments. Differences among multiple groups were compared using one-way ANOVA followed by Tukey-Kramer's post hoc tests. Means not sharing a common letter are significantly different ($p < 0.05$).

3.7. Stevioside Regulated NF-κB/MAPK Signaling Pathways in IPEC-J2 Cells

To investigate whether the antioxidative effects of stevioside were mediated via inhibition of the nuclear factor kappa B (NF-κB)/mitogen-activated protein kinase (MAPK) signaling pathways, we evaluated the mRNA and protein abundance of NF-κB, p-NF-κB, IκB, p-IκB, ERK1/2, and p-ERK in IPEC-J2 cells. As shown in Figure 7, the diquat treatment significantly upregulated the phosphorylation levels of IκB and the mRNA abundance compared with the control and stevioside treatments in the IPEC-J2 cells ($p < 0.05$) (Figure 7A,D). In contrast, the phosphorylation levels of IκB were significantly reduced in the stevioside pretreatment group than in the diquat group ($p < 0.05$) (Figure 7A,D). In addition, the abundance level of NF-κB p65 subunit, p-NF-κB and related genes in the DQ group were significantly higher expressed compared with the CON group ($p < 0.05$) (Figure 7B,E), indicating the presence of NF-κB activation. However, stevioside pretreatment significantly reduced the mRNA and protein abundances of phosphorylated IκB and NF-κB in the diquat-challenged IPEC-J2 cells ($p < 0.05$) (Figure 7A,B,D,E). Besides, stevioside significantly decreased the mRNA and phosphorylation levels of ERK1/2 as compared with the DQ group ($p < 0.05$) (Figure 7C,F).

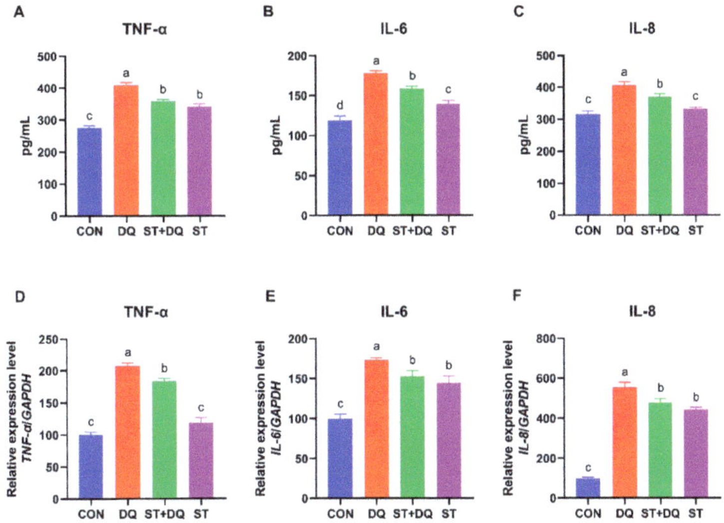

Figure 6. Effects of stevioside on the secretion and gene expression of inflammatory cytokines in diquat-challenged IPEC-J2 cells. The levels of TNF-α (**A**), IL-6 (**B**), and IL-8 (**C**) secretion. The gene expression levels of TNF-α (**D**), IL-6 (**E**), and IL-8 (**F**). CON, control, cells without being treated; DQ, diquat, cells were only treated by diquat; ST + DQ, stevioside+diquat, cells were pretreated with stevioside and were then treated by diquat; ST, stevioside, cells were only treated by stevioside. Values are expressed as the mean ± standard error of the mean (SEM) of three independent experiments. Differences among multiple groups were compared using one-way ANOVA followed by Tukey-Kramer's post hoc tests. Means without a common letter differ, $p < 0.05$.

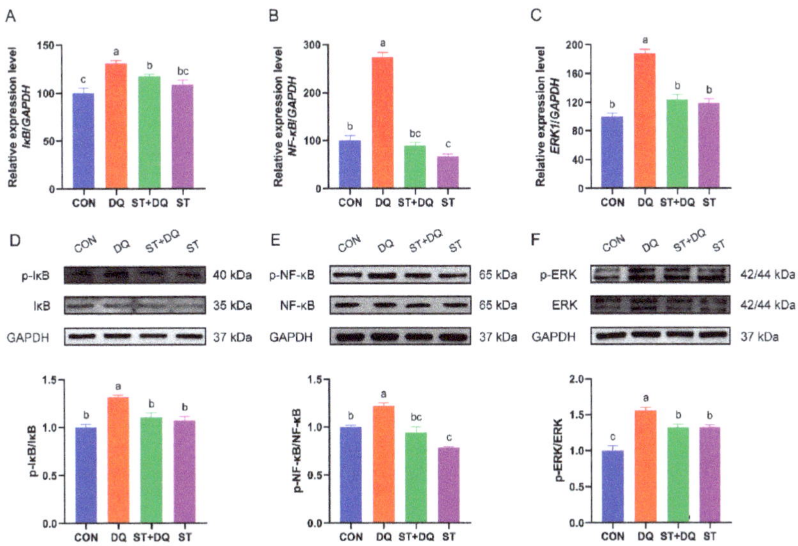

Figure 7. Effects of stevioside on the activation of the NF-κB/MAPK signaling pathways in diquat-challenged IPEC-J2 cells. Relative mRNA levels of (**A**) IκB, (**B**) NF-κB and (**C**) ERK1 were detected

by qRT-PCR. GAPDH was used as a control. Western-blot analysis of the phosphorylation levels of (**D**) IκB and (**E**) NF-κB. (**F**) The phosphorylation level of extracellular signal-regulated kinase (ERK) 1/2. CON, control, cells without being treated; DQ, diquat, cells were only treated by diquat; ST + DQ, stevioside+diquat, cells were pretreated with stevioside and were then treated by diquat; ST, stevioside, cells were only treated by stevioside. The values presented are the means ± SEM of three independent experiments. Means without a common letter differ, $p < 0.05$.

4. Discussion

In pig production, factors such as birth stress [24], weaning stress [25], feed mycotoxin pollution [26], environment [27] and social factors [28] induce the body to produce a large number of free radicals, resulting in oxidative stress. Oxidative stress has a negative impact on the production performance, health, and reproductive performance of pigs, which seriously affects the economic benefits of the pig industry [29]. Therefore, it is of great significance for animal and human health to study the prevention and treatment measures and intervention programs for porcine oxidative stress. Stevioside, a natural sweet compound from *Stevia rebaudiana* Bertoni, has been widely used as a non-nutritive sweetener in food to combat obesity and hyperglycemia [30]. Recently, stevioside has been widely reported to have antioxidant effects, which can alleviate the damage of oxidative stress on the intestinal epithelium [31]. However, whether stevioside plays an antioxidant role in IPEC-J2 cells under oxidative stress and the underlying molecular mechanism remain unclear. Therefore, this research aimed to detect the toxic effect of diquat on IPEC-J2 cell line in this study and to investigate the potential protective effect and molecular mechanism of stevioside in alleviating the oxidative stress injury of the IPEC-J2 cells induced by diquat.

Previous studies have shown that stevioside extract improves neural cell viability and proliferation [32]. Interestingly, stevioside also has the potential to be an anticancer drug because it induces a dose-dependent decline in the cell viability of breast cancer cells but promotes the proliferation of successfully surviving cells [33]. EdU is a thymidine analog that is incorporated into cells only during the S-phase of the cell cycle and thus can be used to assess cellular proliferation. Consistent with these results, DQ exposure inhibited the IPEC-J2 cells viability and proliferation according to the CCK-8 and EdU staining assays. Similarly, our results showed that stevioside pretreatment attenuated cell damage by improving cell viability and proliferation in DQ-treated IPEC-J2 cells. We also found that higher concentrations and a longer treatment time of stevioside directly decreased the viability of IPEC-J2. The present study suggests that optimal concentration and stimulation time of stevioside have the ability to enhance the viability of IPEC-J2 cells by promoting cell proliferation.

Oxidative stress is widely involved in small intestinal epithelium apoptosis [34]. Excessive oxidative stress even induces cell death [35]. A previous study determined that diquat treatment induced apoptosis and increased the level of intracellular ROS in IPEC-1 cells [36]. Similarly, the results of apoptosis detection indicated that pre-treatment with stevioside significantly decreased diquat-induced apoptosis rate and increased cell viability in IPEC-J2 cells. The present results revealed that stevioside alleviated the oxidative stress in the small intestine by inhibiting intestinal epithelial apoptosis. This indicated that stevioside is considered an effective antioxidant, which protects IPEC-J2 cells against oxidative stress-induced apoptosis.

Oxidative stress is caused by an imbalance between pro-oxidants and antioxidants [37], which often results in the disruption of the intestinal barrier integrity in the small intestinal epithelium [38]. Oxidative stress leads to impairment of the intestinal epithelial barrier, affects the digestion and absorption of nutrients, and may lead to various diseases [39]. The intestinal barrier's integrity is crucial for maintaining intestinal homeostasis and protecting the intestinal epithelium from toxins and pathogens [40]. Tight junction proteins, including claudin-1, occludin, and ZO-1, are markers of intestinal integrity and play a crucial role in maintaining the intestinal epithelial barrier's function [41]. Therefore, we evaluated the effects of diquat and/or stevioside on the integrity and barrier function of small intestinal epithelial cells by detecting the expression of claudin-1, occludin, and ZO-1. As expected, diquat treatment significantly decreased the expression of claudin-1, occludin, and ZO-1 compared with the CON group, while stevioside pretreatment upregulated the expression of tight junction related genes in the diquat-stimulated IPEC-J2 cells. A previous study demonstrated that stevioside improved intestinal barrier integrity, protected intestinal barrier function, and reduced inflammation in mice [42]. Consistent with these findings, our study demonstrated that stevioside prevented the increase in diquat-induced cell permeability, maintaining intestinal barrier integrity in pigs.

Reactive oxygen species (ROS) are important markers of oxidative stress, and oxidative stress can be directly assessed by measuring ROS levels [43]. A previous study demonstrated that DQ exposure increased the level of intracellular ROS and induced oxidative stress in IPEC-J2 cells [44]. Stevioside extract is recognized as an antioxidant that scavenges free radicals and exerts antioxidant activity [45]. The present results indicated that pre-treatment with stevioside significantly inhibited DQ-induced intracellular ROS accumulation, which proved that it had antioxidant properties in IPEC-J2 cells. As an indicator of lipid peroxidation [46], MDA showed a similar trend to ROS levels in this study. Diquat treatment promoted the accumulation of MDA, while stevioside pretreatment inhibited the production of MDA induced by diquat. These results indicated that the protective effect of stevioside involves antioxidant enzyme activity.

To further investigate the antioxidant mechanisms of stevioside, we evaluated the activities of some antioxidant-associated enzymes in IPEC-2 cell lysates. T-AOC normally reflects the capacity of the nonen-zymatic antioxidant defense system and is often used as a biomarker to investigate oxidative status [47]. The elevation in T-AOC after pretreatment with stevioside demonstrated that stevioside suppresses oxidative stress at least in part via the nonenzymatic antioxidant defense system. Catalase (CAT) is one of the main antioxidant enzymes, which mainly catalyzes the decomposition of hydrogen peroxide in cells to detoxify, thus playing a functional role in protecting the antioxidant system [48]. Antioxidant enzymes, such as T-SOD and CAT, can detoxify ROS into safe molecules, thus protecting cells against ROS damage [49]. In the present study, we observed that stevioside pretreatment increased the activity of T-SOD and CAT compared with diquat treatment, which also indirectly explains that stevioside could protect IPEC-2 cells against oxidative damage. GSH-Px is an important antioxidant enzyme in the cellular antioxidant defense system. The increase in its activity can eliminate the increased reactive oxidative species [50]. In the present study, we observed inhibition of GSH-Px activity by diquat, which was counteracted by stevioside pretreatment. Stevioside treatment alone increased the activity of GSH-Px to increase the antioxidant capacity. Similar results have been obtained in weaned piglets fed stevia residue extract [51]. These above results confirmed the antioxidant effect of stevioside in IPEC-2 cells.

Oxidative stress can trigger an inflammatory response, which in turn can directly induce oxidative stress [52]. The production of pro-inflammatory cytokines, such as IL-6, IL-8, and TNF-α, which are often representative hallmarks of an inflammatory response [53]. In the present study, the diquat treatment promoted the secretion of IL-6, IL-8, and TNF-α compared with the CON group and also caused a significant increase in gene expression (IL-6, IL-8, and TNF-α genes) in the IPEC-J2 cells. However, the stevioside treatment downregulated pro-inflammatory gene and protein expressions in the diquat-stimulated IPEC-J2 cells and cell supernatant. These results suggested that stevioside treatment attenuated the diquat-induced cellular inflammatory response by decreasing the production of pro-inflammatory cytokines. Consistent with our results, stevioside played an anti-inflammatory and immunomodulatory role in the human colon carcinoma cell line (Caco-2) by potentially suppressing lipopolysaccharide-induced pro-inflammatory cytokine TNF-α, IL-1β, and IL-6 productions [54]. Similarly, the latest research shows that dietary stevioside supplementation improves immunity in broilers [55]. The above results indicated that stevioside had certain anti-inflammatory and immunomodulatory effects in IPEC-J2 cells.

Pro-inflammatory cytokine production is dependent on activation of the transcription factor nuclear factor kappa B (NF-κB) signaling pathway, which regulates the expression of genes and proteins involved in the inflammatory response and immune system [56]. Therefore, we next investigated the activation of the NF-κB pathway and found that the diquat treatment significantly increased the phosphorylation of NF-κB and IκB in the IPEC-J2 cells, whereas their expression was markedly suppressed in the stevioside pretreatment group. This attenuated activation of the NF-κB pathway in turn reduced the production of pro-inflammatory cytokines, which was consistent with the changes in the cytokine levels in our results. A study has reported that the activation of the NF-κB pathway is highly correlated with the activation of the mitogen-activated protein kinase (MAPK) pathway [57]. ERK1/2 is upstream of NF-κB and the ERK1/2 MAPK pathway enhances NF-κB transcriptional activity [58]. Research suggests that stevioside attenuates LPS-stimulated inflammation by downregulating the phosphorylation levels of proteins related to the MAPK signaling pathway in mouse macrophage cell lines [59]. Similarly, our study also revealed that stevioside pretreatment significantly inhibited diquat-induced ERK1/2 phosphorylation in IPEC-J2 cells. A report demonstrated that inhibition of the MAPK signaling pathway reduced the production of pro-inflammatory cytokines [60]. Meanwhile, the NF-κB/MAPK pathway is critical for regulating inflammatory genes [61]. This means that the decrease in ERK1/2 phosphorylation in the current study may be one of the reasons for decreased IL-6, IL-8, and TNF-α expression. On the other hand, the activation of the NF-κB and MAPK signaling pathways decreased the expression of tight junction proteins in dextran sulphate sodium-induced acute colitis, which increased intestinal permeability and enhanced the inflammatory response [62]. Therefore, we accordingly speculated that stevioside improved intestinal barrier integrity and alleviated the inflammatory response by inhibiting the NF-κB and MAPK pathways in diquat-induced IPEC-J2 cells.

In this study, we confirmed the anti-oxidation and anti-inflammatory effects of stevioside in a piglet-derived cell line. This provides basic experimental data for adding stevioside to feed to improve animal growth and health. Additionally, the latest research shows that adding 400 mg/kg stevia residue extract to the diet had no significant effect on the average daily feed intake but improved the antioxidant capacity of weaned piglets and increased the relative abundance of beneficial bacteria to improve piglet health [51]. Our present study found that the pretreatment of IPEC-J2 cells with 250 μM stevioside for 6 h effectively enhanced antioxidant capacity. However, it is important to note that differences may exist between in vivo and in vitro studies. Meanwhile, adding high concentrations of stevia glycosides to feed could potentially alter the feed's palatability and impact feed intake, highlighting the need for further research to determine the optimal dosage of stevia glycosides for use in feed. Furthermore, our findings suggest that steviosides play antioxidant and anti-inflammatory roles in cells. However, since steviosides are metabolized into steviol alcohols, stevioside metabolites are also active and have anti-inflammatory and immunomodulatory activities in colonic epithelial cells [54]. Steviosides can also directly interact with the gut microbiota [63], indicating that their mechanisms of action are multifaceted, possibly exerting antioxidant effects through multiple synergistic mechanisms. However, this study has certain limitations, including the need for validation in multiple cell lines and further in vivo studies to corroborate these findings.

5. Conclusions

Taken together, this is the first study to confirm that stevioside alleviated the inflammatory response, apoptosis, and exerted an antioxidant effect to alleviate diquat-induced oxidative stress through the attenuation of the activation of NF-κB/MAPK pathways. However, further research is necessary to validate these findings in vivo as well as determine the optimal percentage of stevioside to be added to pig feed.

Supplementary Materials: The following supporting information can be downloaded at: https://www.mdpi.com/article/10.3390/antiox12051070/s1, Table S1: Primer sequences used for real-time quantitative PCR.

Author Contributions: Conceptualization, B.Z. and Q.X.; data curation, investigation, and methodology, Q.X.; visualization, M.L.; validation, X.C. and H.Y.; software, C.Z. and J.C.; writing—original draft preparation, Q.X.; writing—review and editing, B.Z.; formal analysis and supervision, B.Z.; project administration, B.Z.; funding acquisition, B.Z. All authors have read and agreed to the published version of the manuscript.

Funding: This research was funded by Fundamental Research Funds for the Central Universities (funding number KYLH202008).

Institutional Review Board Statement: Not applicable.

Informed Consent Statement: Not applicable.

Data Availability Statement: All data are presented in the article and Supplementary Materials.

Acknowledgments: We thank Chunmei Li (Nanjing Agricultural University, China) for kindly providing the IPEC-J2 cell line.

Conflicts of Interest: The authors declare no competing financial interest or conflict of interest.

References

1. Yu, W.; Chen, G.; Yan, J.; Wang, X.; Zhu, Y.; Zhu, L. Single-cell sequencing analysis reveals gastric cancer microenvironment cells respond vastly different to oxidative stress. *J. Transl. Med.* **2022**, *20*, 250. [CrossRef] [PubMed]
2. Zhang, X.F.; Zhao, W.Y.; Li, B.; Li, W.Q.; Zhang, C.X.; Hou, X.C.; Jiang, J.; Dong, Y.Z. Ratiometric fluorescent probes for capturing endogenous hypochlorous acid in the lungs of mice. *Chem. Sci.* **2018**, *9*, 8207–8212. [CrossRef] [PubMed]
3. Lauridsen, C. Effects of dietary fatty acids on gut health and function of pigs pre- and post-weaning. *J. Anim. Sci.* **2020**, *98*, skaa086. [CrossRef] [PubMed]
4. Ji, F.J.; Wang, L.X.; Yang, H.S.; Hu, A.; Yin, Y.L. Review: The roles and functions of glutamine on intestinal health and performance of weaning pigs. *Animal* **2019**, *13*, 2727–2735. [CrossRef] [PubMed]
5. Qin, W.; Xu, B.; Chen, Y.; Yang, W.; Xu, Y.; Huang, J.; Duo, T.; Mao, Y.; Zhou, G.; Yan, X.; et al. Dietary ellagic acid supplementation attenuates intestinal damage and oxidative stress by regulating gut microbiota in weanling piglets. *Anim. Nutr.* **2022**, *11*, 322–333. [CrossRef]
6. Wijtten, P.J.A.; van der Meulen, J.; Verstegen, M.W.A. Intestinal barrier function and absorption in pigs after weaning: A review. *Brit. J. Nutr.* **2011**, *105*, 967–981. [CrossRef]
7. Kang, H.J.; Lee, H.N.; Hong, S.J.; Park, B.R.; Ameer, K.; Cho, J.Y.; Kim, Y.M. Synthesis and characteristics of a rebaudioside-A like compound as a potential non-caloric natural sweetener by *Leuconostoc kimchii* dextransucrase. *Food Chem.* **2022**, *366*, 130623. [CrossRef]
8. Chatsudthipong, V.; Muanprasat, C. Stevioside and related compounds: Therapeutic benefits beyond sweetness. *Pharmacol. Therapeut.* **2009**, *121*, 41–54. [CrossRef]
9. Ragone, M.I.; Bonazzola, P.; Colareda, G.A.; Lazarte, M.L.; Bruno, F.; Consolini, A.E. Cardioprotection of stevioside on stunned rat hearts: A mechano-energetical study. *Phytomedicine* **2017**, *35*, 18–26. [CrossRef]
10. Casas-Grajales, S.; Ramos-Tovar, E.; Chavez-Estrada, E.; Alvarez-Suarez, D.; Hernandez-Aquino, E.; Reyes-Gordillo, K.; Cerda-Garcia-Roja, C.M.; Camacho, J.; Tsutsumi, V.; Lakshman, M.R.; et al. Antioxidant and immunomodulatory activity induced by stevioside in liver damage: In vivo, in vitro and in silico assays. *Life Sci.* **2019**, *224*, 187–196. [CrossRef]
11. Wei, F.Y.; Zhu, H.; Li, N.; Yu, C.L.; Song, Z.B.; Wang, S.Y.; Sun, Y.; Zheng, L.H.; Wang, G.N.; Huang, Y.X.; et al. Stevioside Activates AMPK to Suppress Inflammation in Macrophages and Protects Mice from LPS-Induced Lethal Shock. *Molecules* **2021**, *26*, 858. [CrossRef] [PubMed]
12. Sehar, I.; Kaul, A.; Bani, S.; Pal, H.C.; Saxena, A.K. Immune up regulatory response of a non-caloric natural sweetener, stevioside. *Chem. Biol. Interact.* **2008**, *173*, 115–121. [CrossRef] [PubMed]
13. Robinson, K.; Ma, X.; Liu, Y.L.; Qiao, S.Y.; Hou, Y.Q.; Zhang, G.L. Dietary modulation of endogenous host defense peptide synthesis as an alternative approach to in-feed antibiotics. *Anim. Nutr.* **2018**, *4*, 160–169. [CrossRef]
14. Han, X.F.; Chen, C.X.; Zhang, X.L.; Wei, Y.Q.; Tang, S.X.; Wang, J.R.; Tan, Z.L.; Xu, L.W. Effects of Dietary Stevioside Supplementation on Feed Intake, Digestion, Ruminal Fermentation, and Blood Metabolites of Goats. *Animals* **2019**, *9*, 32. [CrossRef] [PubMed]
15. Jiang, J.; Qi, L.; Wei, Q.; Shi, F. Maternal stevioside supplementation improves intestinal immune function of chicken offspring potentially via modulating gut microbiota and down-regulating the promoter methylation level of suppressor of cytokine signaling 1 (SOCS1). *Anim. Nutr.* **2022**, *10*, 329–346. [CrossRef]

16. Jiang, J.L.; Qi, L.N.; Lv, Z.P.; Jin, S.; Wei, X.H.; Shi, F.X. Dietary Stevioside Supplementation Alleviates Lipopolysaccharide-Induced Intestinal Mucosal Damage through Anti-Inflammatory and Antioxidant Effects in Broiler Chickens. *Antioxidants* **2019**, *8*, 575. [CrossRef]
17. Shin, Y.G.; Rathnayake, D.; Mun, H.S.; Dilawar, M.A.; Pov, S.; Yang, C.J. Sensory Attributes, Microbial Activity, Fatty Acid Composition and Meat Quality Traits of Hanwoo Cattle Fed a Diet Supplemented with Stevioside and Organic Selenium. *Foods* **2021**, *10*, 129. [CrossRef]
18. Jiang, J.; Liu, S.; Jamal, T.; Ding, T.; Qi, L.; Lv, Z.; Yu, D.; Shi, F. Effects of dietary sweeteners supplementation on growth performance, serum biochemicals, and jejunal physiological functions of broiler chickens. *Poult. Sci.* **2020**, *99*, 3948–3958. [CrossRef]
19. Xu, Q.L.; Liu, C.; Mo, X.J.; Chen, M.; Zhao, X.L.; Liu, M.Z.; Wang, S.B.; Zhou, B.; Zhao, C.X. Drinking Water Supplemented with Acidifiers Improves the Growth Performance of Weaned Pigs and Potentially Regulates Antioxidant Capacity, Immunity, and Gastrointestinal Microbiota Diversity. *Antioxidants* **2022**, *11*, 809. [CrossRef]
20. Chen, J.L.; Chen, D.W.; Yu, B.; Luo, Y.H.; Zheng, P.; Mao, X.B.; Yu, J.; Luo, J.Q.; Huang, Z.Q.; Yan, H.; et al. Chlorogenic Acid Attenuates Oxidative Stress-Induced Intestinal Mucosa Disruption in Weaned Pigs. *Front. Vet. Sci.* **2022**, *9*, 806253. [CrossRef]
21. Belle, N.M.; Ji, Y.B.; Herbine, K.; Wei, Y.; Park, J.; Zullo, K.; Hung, L.Y.; Srivatsa, S.; Young, T.; Oniskey, T.; et al. TFF3 interacts with LINGO2 to regulate EGFR activation for protection against colitis and gastrointestinal helminths. *Nat. Commun.* **2019**, *10*, 4408. [CrossRef] [PubMed]
22. Brosnahan, A.J.; Brown, D.R. Porcine IPEC-J2 intestinal epithelial cells in microbiological investigations. *Vet. Microbiol.* **2012**, *156*, 229–237. [CrossRef] [PubMed]
23. Livak, K.J.; Schmittgen, T.D. Analysis of relative gene expression data using real-time quantitative PCR and the 2(T)(-Delta Delta C) method. *Methods* **2001**, *25*, 402–408. [CrossRef]
24. Huang, S.B.; Wu, Z.F.; Huang, Z.H.; Hao, X.Y.; Zhang, L.M.; Hu, C.J.; Wei, J.F.; Deng, J.P.; Tan, C.Q. Maternal supply of cysteamine alleviates oxidative stress and enhances angiogenesis in porcine placenta. *J. Anim. Sci. Biotechnol.* **2021**, *12*, 91. [CrossRef]
25. Novais, A.K.; Martel-Kennes, Y.; Roy, C.; Deschene, K.; Beaulieu, S.; Bergeron, N.; Laforest, J.P.; Lessard, M.; Matte, J.J.; Lapointe, J. Tissue-specific profiling reveals modulation of cellular and mitochondrial oxidative stress in normal- and low-birthweight piglets throughout the peri-weaning period. *Animal* **2020**, *14*, 1014–1024. [CrossRef] [PubMed]
26. Ma, K.D.; Bai, Y.S.; Li, J.B.; Ren, Z.S.; Li, J.P.; Zhang, J.; Shan, A.S. *Lactobacillus rhamnosus* GG ameliorates deoxynivalenol-induced kidney oxidative damage and mitochondrial injury in weaned piglets. *Food Funct.* **2022**, *13*, 3905–3916. [CrossRef] [PubMed]
27. Tang, Q.; Huang, K.; Liu, J.Z.; Wu, S.; Shen, D.; Dai, P.Y.; Li, C.M. Fine particulate matter from pig house induced immune response by activating TLR4/MAPK/NF-kappa B pathway and NLRP3 inflammasome in alveolar macrophages. *Chemosphere* **2019**, *236*, 124373. [CrossRef]
28. He, Y.L.; Sang, Z.; Zhuo, Y.S.; Wang, X.Y.; Guo, Z.H.; He, L.H.; Zeng, C.P.; Dai, H.C. Transport stress induces pig jejunum tissue oxidative damage and results in autophagy/mitophagy activation. *J. Anim. Physiol. Anim. Nutr.* **2019**, *103*, 1521–1529. [CrossRef]
29. Frame, C.A.; Johnson, E.; Kilburn, L.; Huff-Lonergan, E.; Kerr, B.J.; Serao, M.R. Impact of dietary oxidized protein on oxidative status and performance in growing pigs. *J. Anim. Sci.* **2020**, *98*, skaa097. [CrossRef]
30. Suez, J.; Cohen, Y.; Valdes-Mas, R.; Mor, U.; Dori-Bachash, M.; Federici, S.; Zmora, N.; Leshem, A.; Heinemann, M.; Linevsky, R.; et al. Personalized microbiome-driven effects of non-nutritive sweeteners on human glucose tolerance. *Cell* **2022**, *185*, 3307–3328.e19. [CrossRef]
31. Alavala, S.; Sangaraju, R.; Nalban, N.; Sahu, B.D.; Jerald, M.K.; Kilari, E.K.; Sistla, R. Stevioside, a diterpenoid glycoside, shows anti-inflammatory property against Dextran Sulphate Sodium-induced ulcerative colitis in mice. *Eur. J. Pharmacol.* **2019**, *855*, 192–201. [CrossRef] [PubMed]
32. Zhong, K.L.; Lu, M.Y.; Liu, F.; Mei, Y.; Zhang, X.J.; Zhang, H.; Zan, J.; Sun, X.O.; Tan, W. Isosteviol Sodium Protects Neural Cells Against Hypoxia-Induced Apoptosis Through Inhibiting MAPK and NF-kappa B Pathways. *J. Stroke Cerebrovasc. Dis.* **2019**, *28*, 175–184. [CrossRef] [PubMed]
33. Velesiotis, C.; Kanellakis, M.; Vynios, D.H. Steviol glycosides affect functional properties and macromolecular expression of breast cancer cells. *IUBMB Life* **2022**, *74*, 1012–1028. [CrossRef] [PubMed]
34. Tang, X.P.; Liu, B.; Wang, X.R.; Yu, Q.F.; Fang, R.J. Epidermal Growth Factor, through Alleviating Oxidative Stress, Protect IPEC-J2 Cells from Lipopolysaccharides-Induced Apoptosis. *Int. J. Mol. Sci.* **2018**, *19*, 848. [CrossRef] [PubMed]
35. Lim, S.W.; Shin, Y.J.; Luo, K.; Quan, Y.; Cui, S.; Ko, E.J.; Chung, B.H.; Yang, C.W. Ginseng increases Klotho expression by FoxO3-mediated manganese superoxide dismutase in a mouse model of tacrolimus-induced renal injury. *Aging* **2019**, *11*, 5548–5569. [CrossRef]
36. Jin, Y.; Zhai, Z.; Jia, H.; Lai, J.; Si, X.; Wu, Z. Kaempferol attenuates diquat-induced oxidative damage and apoptosis in intestinal porcine epithelial cells. *Food Funct.* **2021**, *12*, 6889–6899. [CrossRef]
37. Detterich, J.A.; Liu, H.L.; Suriany, S.; Kato, R.M.; Chalacheva, P.; Tedla, B.; Shah, P.M.; Khoo, M.C.; Wood, J.C.; Coates, T.D.; et al. Erythrocyte and plasma oxidative stress appears to be compensated in patients with sickle cell disease during a period of relative health, despite the presence of known oxidative agents. *Free Radic. Biol. Med.* **2019**, *141*, 408–415. [CrossRef]
38. Vergauwen, H.; Prims, S.; Degroote, J.; Wang, W.; Casteleyn, C.; van Cruchten, S.; de Smet, S.; Michiels, J.; van Ginneken, C. In Vitro Investigation of Six Antioxidants for Pig Diets. *Antioxidants* **2016**, *5*, 41. [CrossRef]

39. Gori, M.; Altomare, A.; Cocca, S.; Solida, E.; Ribolsi, M.; Carotti, S.; Rainer, A.; Francesconi, M.; Morini, S.; Cicala, M.; et al. Palmitic Acid Affects Intestinal Epithelial Barrier Integrity and Permeability In Vitro. *Antioxidants* **2020**, *9*, 417. [CrossRef]
40. Kim, M.; Chung, K.S.; Hwang, S.J.; Yoon, Y.S.; Jang, Y.P.; Lee, J.K.; Lee, K.T. Protective Effect of *Cicer arietinum* L. (Chickpea) Ethanol Extract in the Dextran Sulfate Sodium-Induced Mouse Model of Ulcerative Colitis. *Nutrients* **2020**, *12*, 456. [CrossRef]
41. Feng, R.; Ma, L.J.; Wang, M.; Liu, C.; Yang, R.; Su, H.; Yang, Y.; Wan, J.B. Oxidation of fish oil exacerbates alcoholic liver disease by enhancing intestinal dysbiosis in mice. *Commun. Biol.* **2020**, *3*, 481. [CrossRef]
42. Zhang, X.J.; Gu, J.X.; Zhao, C.Y.; Hu, Y.Z.; Zhang, B.W.; Wang, J.; Lv, H.; Ji, X.M.; Wang, S. Sweeteners Maintain Epithelial Barrier Function through the miR-15b/RECK/MMP-9 Axis, Remodel Microbial Homeostasis, and Attenuate Dextran Sodium Sulfate-Induced Colitis in Mice. *J. Agric. Food Chem.* **2022**, *70*, 171–183. [CrossRef] [PubMed]
43. Kurtoglu, Y.E.; Navath, R.S.; Wang, B.; Kannan, S.; Romero, R.; Kannan, R.M. Poly(amidoamine) dendrimer-drug conjugates with disulfide linkages for intracellular drug delivery. *Biomaterials* **2009**, *30*, 2112–2121. [CrossRef] [PubMed]
44. Tang, L.; Zeng, Z.H.; Zhou, Y.H.; Wang, B.K.; Zou, P.; Wang, Q.; Ying, J.F.; Wang, F.; Li, X.; Xu, S.J.; et al. *Bacillus amyloliquefaciens* SC06 Induced AKT-FOXO Signaling Pathway-Mediated Autophagy to Alleviate Oxidative Stress in IPEC-J2 Cells. *Antioxidants* **2021**, *10*, 1545. [CrossRef]
45. Lopez, V.; Perez, S.; Vinuesa, A.; Zorzetto, C.; Abian, O. *Stevia rebaudiana* ethanolic extract exerts better antioxidant properties and antiproliferative effects in tumour cells than its diterpene glycoside stevioside. *Food Funct.* **2016**, *7*, 2107–2113. [CrossRef]
46. Jung, H.I.; Kong, M.S.; Lee, B.R.; Kim, T.H.; Chae, M.J.; Lee, E.J.; Jung, G.B.; Lee, C.H.; Sung, J.K.; Kim, Y.H. Exogenous Glutathione Increases Arsenic Translocation into Shoots and Alleviates Arsenic-Induced Oxidative Stress by Sustaining Ascorbate-Glutathione Homeostasis in Rice Seedlings. *Front. Plant Sci.* **2019**, *10*, 1089. [CrossRef]
47. Sun, K.Y.; Sun, Y.T.; Li, H.Y.; Han, D.Y.; Bai, Y.T.; Zhao, R.; Guo, Z.J. Anti-Ageing Effect of Physalis alkekengi Ethyl Acetate Layer on a d-galactose-Induced Mouse Model through the Reduction of Cellular Senescence and Oxidative Stress. *Int. J. Mol. Sci.* **2020**, *21*, 1836. [CrossRef]
48. Salcher, S.; Hagenbuchner, J.; Geiger, K.; Seiter, M.A.; Rainer, J.; Kofler, R.; Hermann, M.; Kiechl-Kohlendorfer, U.; Ausserlechner, M.J.; Obexer, P. C10ORF10/DEPP, a transcriptional target of FOXO3, regulates ROS-sensitivity in human neuroblastoma. *Mol. Cancer* **2014**, *13*, 224. [CrossRef]
49. Zou, Y.; Wang, J.; Peng, J.; Wei, H.K. Oregano Essential Oil Induces SOD1 and GSH Expression through Nrf2 Activation and Alleviates Hydrogen Peroxide-Induced Oxidative Damage in IPEC-J2 Cells. *Oxid. Med. Cell. Longev.* **2016**, *2016*, 5987183. [CrossRef] [PubMed]
50. Li, F.M.; Han, X.K.; Li, F.; Wang, R.; Wang, H.; Gao, Y.J.; Wang, X.J.; Fang, Z.Y.; Zhang, W.J.; Yao, S.; et al. LKB1 Inactivation Elicits a Redox Imbalance to Modulate Non-small Cell Lung Cancer Plasticity and Therapeutic Response. *Cancer Cell* **2015**, *27*, 698–711. [CrossRef]
51. Liu, S.; Xiong, Y.; Cao, S.; Wen, X.; Xiao, H.; Li, Y.; Chi, L.; He, D.; Jiang, Z.; Wang, L. Dietary Stevia Residue Extract Supplementation Improves Antioxidant Capacity and Intestinal Microbial Composition of Weaned Piglets. *Antioxidants* **2022**, *11*, 2016. [CrossRef] [PubMed]
52. Seyedsadjadi, N.; Berg, J.; Bilgin, A.A.; Grant, R. A Pilot Study Providing Evidence for a Relationship between a Composite Lifestyle Score and Risk of Higher Carotid Intima-Media Thickness: Is There a Link to Oxidative Stress? *Oxid. Med. Cell. Longev.* **2018**, *2018*, 4504079. [CrossRef] [PubMed]
53. Xu, X.X.; Wei, Y.J.; Shah, M.K.; Wang, X.Y.; Lin, J.T.; Wan, P.; Cui, L.; Yin, Q.Q. Effects of Compound Active Peptides on Protecting Liver and Intestinal Epithelial Cells from Damages and Preventing Hyperglycemia. *Oxid. Med. Cell. Longev.* **2020**, *2020*, 3183104. [CrossRef]
54. Boonkaewwan, C.; Burodom, A. Anti-inflammatory and immunomodulatory activities of stevioside and steviol on colonic epithelial cells. *J. Sci. Food Agric.* **2013**, *93*, 3820–3825. [CrossRef] [PubMed]
55. Wu, X.; Yang, P.; Sifa, D.; Wen, Z. Effect of dietary stevioside supplementation on growth performance, nutrient digestibility, serum parameters, and intestinal microflora in broilers. *Food Funct.* **2019**, *10*, 2340–2346. [CrossRef] [PubMed]
56. Zhou, J.; Jiang, L.Y.; Long, X.; Fu, C.P.; Wang, X.D.; Wu, X.D.; Liu, Z.L.; Zhu, F.; Shi, J.D.; Li, S.Q. Bone-marrow-derived mesenchymal stem cells inhibit gastric aspiration lung injury and inflammation in rats. *J. Cell. Mol. Med.* **2016**, *20*, 1706–1717. [CrossRef] [PubMed]
57. Chang, Y.; Al-Alwan, L.; Alshakfa, S.; Audusseau, S.; Mogas, A.K.; Chouiali, F.; Nair, P.; Baglole, C.J.; Hamid, Q.; Eidelman, D.H. Upregulation of IL-17A/F from human lung tissue explants with cigarette smoke exposure: Implications for COPD. *Respir. Res.* **2014**, *15*, 145. [CrossRef]
58. Shi, C.; Zhan, L.; Wu, Y.; Li, Z.; Li, J.; Li, Y.; Wei, J.; Zhang, Y.; Li, L. Kaji-Ichigoside F1 and Rosamultin Protect Vascular Endothelial Cells against Hypoxia-Induced Apoptosis via the PI3K/AKT or ERK1/2 Signaling Pathway. *Oxid. Med. Cell. Longev.* **2020**, *2020*, 6837982. [CrossRef]
59. Fengyang, L.; Yunhe, F.; Bo, L.; Zhicheng, L.; Depeng, L.; Dejie, L.; Wen, Z.; Yongguo, C.; Naisheng, Z.; Xichen, Z.; et al. Stevioside suppressed inflammatory cytokine secretion by downregulation of NF-kappaB and MAPK signaling pathways in LPS-stimulated RAW264.7 cells. *Inflammation* **2012**, *35*, 1669–1675. [CrossRef]
60. Kim, S.M.; Park, E.J.; Lee, H.J. Nuciferine attenuates lipopolysaccharide-stimulated inflammatory responses by inhibiting p38 MAPK/ATF2 signaling pathways. *Inflammopharmacology* **2022**, *30*, 2373–2383. [CrossRef]

61. Wang, C.; Shi, C.; Su, W.; Jin, M.; Xu, B.; Hao, L.; Zhang, Y.; Lu, Z.; Wang, F.; Wang, Y.; et al. Dynamics of the Physicochemical Characteristics, Microbiota, and Metabolic Functions of Soybean Meal and Corn Mixed Substrates during Two-Stage Solid-State Fermentation. *mSystems* **2020**, *5*, e00501-19. [CrossRef] [PubMed]
62. Cui, H.; Li, X.; An, X.R.; Liu, W.; Yuan, T. Masticadienonic acid from Chios mastic gum mitigates colitis in mice via modulating inflammatory response, gut barrier integrity and microbiota. *Phytomedicine* **2023**, *108*, 154518. [CrossRef]
63. Plaza-Diaz, J.; Pastor-Villaescusa, B.; Rueda-Robles, A.; Abadia-Molina, F.; Ruiz-Ojeda, F.J. Plausible Biological Interactions of Low- and Non-Calorie Sweeteners with the Intestinal Microbiota: An Update of Recent Studies. *Nutrients* **2020**, *12*, 1153. [CrossRef] [PubMed]

antioxidants

MDPI

Article

Effect of Thyme (*Thymus vulgaris* L.) Used in Diets with Extruded Flaxseed on the Antioxidant and Lipid Profile of the Blood and Tissues of Fattening Pigs

Kamila Klimiuk, Iwona Sembratowicz *, Krzysztof Tutaj and Anna Czech *

Department of Biochemistry and Toxicology, Faculty of Animal Sciences and Bioeconomy, University of Life Sciences in Lublin, Akademicka 13, 20-950 Lublin, Poland
* Correspondence: iwona.sembratowicz@up.lublin.pl (I.S.); anna.czech@up.lublin.pl (A.C.)

Abstract: Thyme has strong antioxidant properties and, therefore, can reduce the intensity of oxidative processes taking place in the body. The study aimed to assess whether the addition of thyme to diets for fattening pigs containing extruded flaxseeds, a source of n-3 PUFAs, which are particularly susceptible to oxidation, would have a positive effect on redox status and lipid metabolism. The experiment was conducted using 120 weaners (WBP × Neckar crosses) of about 30 kg BW, which were kept until the end of fattening (about 110 kg BW) and divided into three groups of 40 pigs. The control group received a diet with 4% extruded flaxseed. In groups T1 and T3, 1% or 3% of thyme was added to the basal diet. The introduction of 3% thyme resulted in a decrease in the total cholesterol level in the blood and the loin muscle. Moreover, an increase in SOD and CAT activity and a decrease in FRAP and LOOH was noted. Following supplementation with 3% thyme, the n-3 PUFA content and n-3/n-6 ratio increased, while the SFA content was significantly reduced. The results of the studies indicate that thyme has a positive effect on the redox status and lipid profile of the blood and muscles.

Keywords: thyme additive; flaxseed; pig tissue; antioxidant parameters; fatty acids

Citation: Klimiuk, K.; Sembratowicz, I.; Tutaj, K.; Czech, A. Effect of Thyme (*Thymus vulgaris* L.) Used in Diets with Extruded Flaxseed on the Antioxidant and Lipid Profile of the Blood and Tissues of Fattening Pigs. *Antioxidants* **2023**, *12*, 1045. https://doi.org/10.3390/antiox12051045

Academic Editors: Luciana Rossi and Matteo Dell'Anno

Received: 24 March 2023
Revised: 24 April 2023
Accepted: 2 May 2023
Published: 4 May 2023

1. Introduction

Due to health concerns, consumers are increasingly paying attention not only to the nutritional value of food products but to their health benefits as well. For this reason, components are added to diets for livestock in order to obtain products with health-promoting properties, e.g., with high content of polyunsaturated fatty acids (PUFA), especially n-3 acids. One ingredient used for this purpose is flaxseed, which improves meat quality due to increased content of PUFAs, such as α-linolenic (ALA), eicosatrienoic, eicosapentaenoic acid (EPA), and docosapentaenoic (DPA) acids [1]. This is a positive phenomenon, but PUFAs are highly susceptible to oxidation, which leads to redox imbalances in the body and reduces the nutritional value and quality of food [2]. This process leads not only to the degradation of valuable unsaturated fatty acids but also to losses of some vitamins [3] and reduced utilization of protein from feed [4]. Furthermore, lipid oxidation results in the formation of harmful metabolites for consumers, such as aldehydes, ketones, or peroxides [5]. Therefore, antioxidants such as tocopherols or selenium must be added in order to protect fat against oxidation. Although flaxseed contains substances with antioxidant properties (e.g., lignans and vitamin E), their quantity may not appear sufficient to effectively limit oxidation processes resulting from the presence of polyunsaturated fatty acids. An effective solution applied in practice is to supplement animal diets with plant preparations known as phytobiotics. These include herbs with antioxidant properties, rich in phenols and other antioxidants. Martini et al. [6] analysed the effect of the combined use of polyphenol-rich extracts of red grape skin, oregano, and extruded flaxseed on oxidation processes in raw, grilled, and in vitro-digested pork. The addition of natural antioxidants was shown to reduce oxidation processes during cooking and digestion and to be more effective than the use of synthetic vitamin E and sodium selenite. A significant observation was that polyphenols are accumulated in the muscle tissue, which most likely determines their beneficial effects in it.

Herbs of the family Labiatae are particularly rich in antioxidants. These include many medicinal and seasoning herbs, such as rosemary, thyme, oregano, sage, and mint [7]. Among herbs of the family Labiatae, thyme is worthy of attention. Extracts of thyme exhibit particularly high antioxidant activity [8] due to its rich and highly diverse chemical composition. The antioxidant properties of thyme are exhibited not only by oil components such as thymol and carvacrol but by other phenolic substances as well [9]. By reducing the oxidative degradation of PUFAs, phytobiotics (e.g., thymol and carvacrol) help to increase the content of these valuable acids in the muscles [10].

Phenolic compounds and other active plant substances can have beneficial effects on numerous biochemical processes in the body, including lipid metabolism, which leads to a decrease in the level of cholesterol and its atherogenic LDL fraction. Herbs with this type of activity include basil, ginseng, dandelion, and thyme [11,12]. The possibility of including natural antioxidant sources such as thyme in the diet of pigs in order to improve their health condition and increase the nutritional and health-promoting value of their meat is very interesting, especially as no research has yet been conducted on the effect of thyme herb on the redox and lipid profile of the blood and tissues of fattening pigs.

The aim of the study was to determine the effect of the administration of compound feed with 4% inclusion of extruded flaxseed in combination with different levels of thyme herb (1% or 3%) on lipid parameters and redox processes in fattening pigs, as well as on the oxidative stability and fatty acid profile of the tissues.

2. Materials and Methods

The material for the research was WBP × Neckar crossbred weaners, from a body weight of about 30.32 ± 0.24 kg (about 84 days of age) until the end of the fattening period (about 112.2 ± 3.79 kg BW; 168 days of age). The animals were individually tagged, weighed, and assigned to three feeding groups matched for sex and body weight. Each group comprised 40 animals, kept in 5 pens with 8 pigs in each (4 gilts and 4 barrows). The animals received complete Grower and Finisher diets ad libitum. The content of nutrients, amino acids, phosphorus, and calcium in all groups met the nutritional requirements laid out in NCR for pigs [13] (Table 1).

Table 1. Composition (g/kg) and nutritive value of growing and finishing pig diets [14].

Item	Grower			Finisher		
	C	T1	T3	C	T1	T3
Wheat	230	230	230	100	100	100
Rapeseed meal	80	80	80	120	122	125
Soybean meal	100	101	103	0	0	0
Thyme herb	0	10	30	0	10	30
Barley	483	472	450	647	635	612
Extruded flaxseed	40	40	40	40	40	40
Extruded soybean	20	20	20	20	20	20
Soya oil	20	20	20	0	0	0
Wheat bran	0	0	0	50	50	50
Dicalcium phosphate	3	3	3	0	0	0
Limestone	3	3	3	6	6	6
L-lysine chloride	1	1	1	2	2	2
Mineral–vitamin premix (1)	20	20	20	15	15	15
Total	1000	1000	1000	1000	1000	1000
Analysed (g/kg):						
Dry matter	893.5	893.6	893.6	893.8	893.7	893.9
Crude protein	169.4	169.3	169.2	150.2	150.1	150.1
Ether extract	56.1	56.1	56.1	38.2	38.1	38.3
Crude fibre	49.3	49.4	49.6	57.4	57.5	57.7

Table 1. *Cont.*

Item	Grower			Finisher		
	C	T1	T3	C	T1	T3
Total lysine	10.3	10.3	10.3	9.06	9.06	9.06
Methionine + cysteine	6.52	6.52	6.52	5.89	5.89	5.89
Total phosphorus	5.57	5.57	5.58	5.17	5.17	5.17
Calcium	7.12	7.12	7.12	6.49	6.49	6.49
ME, MJ (2)	13.07	13.07	13.06	12.56	12.56	12.55

(1) 1 kg of mineral–vitamin premix contained vitamins: A 600,000 IU, D_3 60,000 IU, E 3000 mg, K_3 120 mg, B_1 120 mg, B_2 240 mg, B_6 240 mg, nicotinic acid 1600 mg, pantothenic acid 800 mg, folic acid 160 mg, biotin 10 mg, and B_{12} 1.6 mg; choline chloride 12 g, Mg 0.8 g, Fe 6 g, Zn 5.6 g, Mn 2.4 g, Cu 6.4 g, I 40 mg, Se 16 mg, Co 16 mg. (2) Metabolizable energy was calculated according to the equation proposed by Kirchgessner and Roth [15].

All experimental procedures used in the study were approved by the Second Local Ethics Committee on Animal Experimentation of the University of Life Sciences in Lublin, Poland.

The pigs in the control group received a diet containing extruded flaxseed in the amount of 4%. This composition was chosen on the basis of previous research [14,16], which showed the beneficial effects of extruded flaxseed on the lipid profile and redox processes in pigs, as well as on the oxidative stability and fatty acid profile of raw pork. Despite these benefits, to additionally enhance antioxidant defense and protect PUFAs against oxidation, as well as to improve gastrointestinal function, thyme was added to the basal diet in the amount of 1% (group T1) or 3% (group T3).

The choice of thyme was influenced by a pilot experiment involving observations of the food preferences of piglets, which were given diets containing herbs with antioxidant properties: thyme, caraway, coriander, and purple coneflower. The observations showed that the diet containing thyme was most preferred by the piglets. Dried thyme herb was purchased from the herb shop Zakład Zielarski KAWON-HURT Nowak (Gostyń, Poland) and added to the diets when they were being prepared in the feed mixer.

2.1. Experimental Procedures

The feed was sampled three times during the experiment and analysed for the content of basic nutrients (crude protein, crude fibre, crude fat, lysine, methionine, and cysteine) and minerals (phosphorus and calcium) according to AOAC [17].

The thyme herb was analysed as well.

The samples were homogenized using a BUCHI mixer B-400 with ceramic blades.

The homogenates were analysed for total phenolic content, total antiradical activity using DPPH and ABTS radicals, and flavonoid content.

Thyme was extracted by a modification of the method described by Bakowska-Barczak and Kolodziejczyk [18]. Dry thyme (0.1 g) was extracted twice with 1 mL of 80% aqueous methanol containing 0.1% formic acid for analysis of total phenolic content and evaluation of antioxidant capacity. Total phenolic content (TPC) was determined according to Song et al. [19].

The ABTS•+ assay was carried out according to Re et al. [20]. DPPH• free radical scavenging activity was measured according to Bocco et al. [21]. Total flavonoid content was measured according to Dewanto et al. [22]. The content of individual substances was determined according to Czech et al. [23].

2.1.1. Analysis of Animal Material

Blood was sampled for analysis from 6 barrows from each group at a body weight of about 70 and 100 kg. The animals had no access to feed for 12 h before blood sampling. Blood was drawn from the jugular vein into 10 mL heparinized tubes. The pigs chosen for analysis were matched for body weight. The same animals from which blood was sampled were slaughtered at about 110 kg BW, and the material collected from them was used for further analysis.

2.1.2. Blood Analysis

Plasma was obtained by centrifuging whole blood at 3500 rpm for 15 min in a laboratory centrifuge (MPW-260R, Warsaw, Poland) and stored in Eppendorf tubes at −80 °C until analysis.

2.1.3. Biochemical Parameters

Tests from Cormay (Lublin, Poland) were used to analyse the plasma for selected biochemical parameters, i.e., total protein (TP; Cor-TOTAL PROTEIN 60, catalogue No. 2-236, Poland), uric acid (UA; Liquick Cor-URIC ACID 120, catalogue No. 2-208, Poland), urea (UREA; Liquick Cor-UREA 60, catalogue No. 2-206, Poland), creatinine (CREAT; Liquick Cor- CREATININE 60, catalogue No. 2-233, Poland), phosphorus (P; Liquick Cor-PHOSPHORUS 120, catalogue No. 2-338, Poland); total cholesterol (CHOL; Liquick Cor-CHOL 60, catalogue No. 2-204, Poland), triacylglycerols (TG; Liquick Cor-TG 30, catalogue No. 2-262, Poland), and HDL cholesterol (HDL-C; Liquick Cor-HDL, catalogue No. 2-053, Poland). LDL cholesterol (LDL-C) was calculated using the formula given by Friedewald [24]:

$$\text{LDL-C (mmol/l)} = \text{CHOL-HDL-C-TG}/2.2$$

Tests from Cormay (Lublin, Poland) were used to determine the activity of alkaline phosphatase (ALP; Liquick Cor-ALP 60, catalogue No. 1-212, Poland), alanine aminotransferase (ALT; Liquick Cor-ALAT 60, catalogue No. 1-216, Poland), and aspartate aminotransferase (AST; Liquick Cor-ASAT 60, catalogue No. 1-214, Poland).

Ca, Mg, Zn, Cu, and Fe content was determined by atomic absorption spectrometry (AAS) with a Varian model 720-ES ICP-OES spectrophotometer (Varian, Palo Alto, Santa Clara, CA, USA).

2.1.4. Redox Parameters

The activity of antioxidant enzymes superoxide dismutase (SOD) and catalase (CAT) and the concentrations of lipid peroxidation products, i.e., peroxides (LOOH) and malondialdehyde (MDA), were determined in the plasma and tissues and parameters of the antioxidant system were measured in the plasma, i.e., ferric reducing antioxidant power (FRAP) and content of vitamin C, according to methods described by Czech et al. [25].

The contents of fatty acids and cholesterol were analysed following fat extraction with a chloroform/methanol mixture, according to Folch et al. [26]. The analysis was conducted according to standards [27,28].

The percentages of fatty acid methyl esters were estimated by gas chromatography on a Varian CP-3800 chromatograph. The operating conditions for fatty acid separation were as follows: CP WAX 52CB DF 0.25 mm capillary column 60 m in length, gas carrier–helium, flow rate 1.4 mL/min, column temperature 120 °C gradually increased by 2 °C/min to 210 °C, determination time 127 min, feeder temperature 160 °C, detector temperature 160 °C, other gases—hydrogen and oxygen.

The content of fatty acids in the longissimus dorsi muscle was calculated according to Weihrauch et al. [29]. Lipid quality indicators, i.e., the atherogenicity index (AI) and thrombogenicity index (TI), were calculated using the following equations [30]:

$$AI = [(4 \times C14{:}0) + C16{:}0]/[\text{n-6 PUFA} + \text{n-3 PUFA} + \text{MUFA}]$$

$$TI = [C14{:}0 + C16{:}0 + C18{:}0]/[(0.5 \times \text{MUFA}) + (0.5 \times \text{n-6PUFA}) + (3 \times \text{n-3PUFA}) + \text{n-3/n-6 PUFA}]$$

2.2. Statistical Analysis

All data are expressed as means and SEM (standard error of the mean).

Data were analysed by ANOVA with treatment as the fixed effect and pig (tissues and blood analysis) as the experimental unit.

The normality of the data distribution was tested using the Shapiro–Wilk test and equality of variance was tested by Levene's test.

Treatment means were compared using Tukey's HSD (honest significant difference) test in Statistica 13 [31].

For all tests, a criterion α level of $p < 0.05$ was used to determine statistical significance.

3. Results

The content of chemical components, including the antioxidant parameters of thyme herb, is presented in Table 2.

Table 2. Chemical composition of thyme herb (dry weight).

Item	Content
Nutrient g/100 g	
Dry matter	99.43 ± 0.54
Total protein	5.87 ± 0.143
Crude fat	3.92 ± 0.076
Crude ash	7.51 ± 0.109
Crude fibre	21.22 ± 1.65
Total polyphenols, mg GAE/g	12.32 ± 2.101
Total flavonoids, mg CE/g	1.50 ± 0.15
ABTS•+, mmol Trolox/g	2.11 ± 0.034
DPPH•, mmol Trolox/g	1.76 ± 0.066

DPPH• 2,2-diphenyl-1-picrylhydrazyl radical. ABTS•+ 2,2′-azinobis-(3-ethylbenzthiazolin-6-sulfonic acid) radical.

The 3% inclusion of thyme in the diets in the grower period caused a significant decrease in the total cholesterol level compared to the control group ($p = 0.003$) and in the finisher period compared to the control group and T1 group ($p = 0.001$). The plasma of fatteners in the grower period receiving a diet with 1% thyme had a significantly lower level of TG than in the control ($p = 0.015$). In the grower period, the plasma of pigs receiving a diet with 1% and 3% thyme had a significantly higher content of HDL-C than in the control ($p = 0.046$). In the finisher period, the HDL-C content in the plasma of pigs receiving a diet with 3% thyme (T3) was significantly higher than in the control ($p = 0.012$). The plasma of group T3 fatteners had a significantly lower content of LDL-C ($p = 0.044$ grower and $p = 0.003$ finisher) compared to the other groups. The level of thyme significantly influenced the content of CHOL ($p = 0.006$ finisher), HDL-C ($p = 0.013$ finisher), TG ($p = 0.013$ finisher), and LDL-C ($p = 0.016$ grower and $p = 0.005$ finisher) (Table 3).

Table 3. Lipid profile parameters in the plasma of fattening pigs.

Item [1]	Treatment	Feeding Groups [2]			SEM [3]	*p* Value		
		C	T1	T3		TRT [4]	T [5]	D [6]
CHOL, mmol/L	Grower	2.10 ^{a}	1.96 ^{ab}	1.84 ^{b}	0.031	0.003	0.066	0.093
	Finisher	2.50 ^{a}	2.41 ^{a}	2.05 ^{b}	0.061	0.001	0.094	0.006
HDL-C, mmol/L	Grower	1.36 ^{a}	1.27 ^{b}	1.28 ^{b}	0.015	0.046	0.049	0.955
	Finisher	1.58 ^{a}	1.53 ^{ab}	1.39 ^{b}	0.029	0.012	0.074	0.013
%HDL	Grower	64.68	64.9	69.44	0.035	0.289	0.094	0.166
	Finisher	63.35	63.82	68.17	1.06	0.122	0.404	0.130
TG, mmol/L	Grower	0.681 ^{a}	0.576 ^{b}	0.633 ^{ab}	0.033	0.015	0.074	0.062
	Finisher	0.686 ^{ab}	0.575 ^{b}	0.718 ^{a}	0.011	0.039	0.088	0.013

Table 3. *Cont.*

Item [1]	Treatment	Feeding Groups [2]			SEM [3]	*p* Value		
		C	T1	T3		TRT [4]	T [5]	D [6]
LDL-C, mmol/L	Grower	0.433 [a]	0.426 [a]	0.272 [b]	0.012	0.044	0.018	0.016
	Finisher	0.605 [a]	0.615 [a]	0.332 [b]	0.044	0.003	0.234	0.005
CHOL/HDL	Grower	1.62	1.62	1.62	0.033	0.997	0.234	0.965
	Finisher	1.64	1.65	1.60	0.034	0.864	0.945	0.629

[a–b] Means with the same superscript are statistically the same across all 3 treatments ($p > 0.05$) according to Tukey's post hoc test. [1] CHOL–total cholesterol; HDL-C–high-density lipoprotein cholesterol; LDL-C–low-density lipoprotein cholesterol; TG–triacylglycerols. [2] There were 3 dietary treatments: a control diet (C) and diets T1 and T3, with 1% or 3% thyme, respectively. [3] SEM–standard error of the means. [4] TRT–p value for overall effect of dietary treatment (diets C vs. T1 vs. T3). [5] T–p value for thyme effect vs. control (diet C vs. T1 and T3). [6] D–p value for thyme dose effect (T1 vs. T3).

The levels of uric acid and urea in the plasma of finisher pigs receiving a diet with 3% inclusion of thyme were significantly higher than in the other experimental groups ($p = 0.044$ and $p = 0.009$, respectively). The level of inclusion of thyme also significantly influenced UA and UREA in the plasma of finisher pigs ($p = 0.048$ and $p = 0.11$, respectively) (Table 4). Alkaline phosphatase activity in the plasma was significantly increased by the addition of thyme ($p < 0.001$ grower and $p = 0.005$ finisher) relative to the control and was highest in the group receiving a diet with 3% thyme ($p = 0.039$ grower and $p = 0.009$ finisher) (Table 4).

Table 4. Biochemical parameters and enzyme activity in the plasma of fattening pigs.

Item [1]	Treatment	Feeding Groups [2]			SEM [3]	*p* Value		
		C	T1	T3		TRT [4]	T [5]	D [6]
TP, g/L	Grower	65.31	66.53	64.57	6.99	0.664	0.533	0.361
	Finisher	64.08	63.05	61.33	1.05	0.587	0.901	0.531
UA, mmol/L	Grower	0.077	0.073	0.082	0.003	0.555	0.927	0.200
	Finisher	0.067 [b]	0.066 [b]	0.088 [a]	0.001	0.044	0.927	0.048
UREA, mmol/L	Grower	4.62	4.96	5.08	0.163	0.524	0.263	0.775
	Finisher	4.62 [b]	4.53 [b]	5.71 [a]	0.190	0.009	0.263	0.011
CREAT, µmol/L	Grower	137.3	140.9	145.7	4.15	0.730	0.51	0.657
	Finisher	153.9	156.3	159.2	4.50	0.902	0.51	0.777
ALP, U/L	Grower	171.0 [b]	184.6 [ab]	204.8 [a]	2.17	<0.001	0.045	0.039
	Finisher	127.3 [b]	125.3 [b]	139.2 [a]	2.07	0.005	0.004	0.009
ALT, U/L	Grower	35.08	33.99	33.77	1.05	0.876	0.404	0.944
	Finisher	28.03	27.58	29.03	1.23	0.898	0.606	0.628
AST, U/L	Grower	33.89	32.31	28.03	1.14	0.187	0.099	0.133
	Finisher	24.26	23.87	29.96	1.16	0.054	0.127	0.060

[a–b] Means with the same superscript are statistically the same across all 3 treatments ($p > 0.05$) according to Tukey's post hoc test. [1] TP–total protein; UA–uric acid; UREA–urea; CREAT–creatinine; ALP-alkaline phosphatase, ALT-alanine aminotransferase; AST-aspartate aminotransferase. [2] There were 3 dietary treatments: a control diet (C) and diets T1 and T3, with 1% and 3% thyme, respectively. [3] SEM–standard error of the means [4] TRT–p value for overall effect of dietary treatment (diets C vs. T1 vs. T3) [5] T–p value for thyme effect vs. control (diet C vs. T1 and T3) [6] D–p value for thyme dose effect (T1 vs. T3).

Superoxide dismutase activity and vitamin C content were significantly higher in the plasma of pigs in group T3 than in groups C and T1 (grower period: SOD-$p < 0.001$, Vitamin C-$p = 0.001$; finisher period: SOD–$p = 0.047$, Vitamin C-$p = 0.033$). Ferric-reducing antioxidant power (FRAP) in the plasma of grower pigs from groups C and T3 was higher than in group T1 ($p = 0.001$), while in the finisher period, the FRAP value was significantly higher in the plasma of pigs receiving a diet with thyme (T1 and T3) than in the control ($p = 0.004$). LOOH content in the plasma of finisher pigs ($p = 0.020$) and MDA in grower ($p = 0.001$) and finisher ($p = 0.009$) pigs were significantly lower than in the control (Table 5).

Table 5. Redox parameters in the plasma of fattening pigs.

Item [1]	Treatment	Feeding Groups [2]			SEM [3]	*p* Value		
		C	T1	T3		TRT [4]	T [5]	D [6]
SOD, U/mL	Grower	16.71 [b]	16.38 [b]	23.93 [a]	0.957	<0.001	0.678	<0.001
	Finisher	35.67 [b]	35.54 [b]	37.98 [a]	3.19	0.047	0.090	0.050
CAT, U/mL	Grower	4.71	4.61	5.06	0.014	0.360	0.090	0.214
	Finisher	5.06 [b]	4.19 [c]	6.19 [a]	0.263	<0.001	0.678	<0.001
Vitamin C, mg/L	Grower	0.287 [b]	0.293 [b]	0.385 [a]	0.316	0.001	0.042	0.005
	Finisher	0.224 [b]	0.247 [b]	0.293 [a]	0.001	0.033	0.082	0.045
FRAP, µmol/L	Grower	12.38 [a]	10.39 [b]	12.62 [a]	0.021	0.001	0.204	0.004
	Finisher	9.53 [b]	13.37 [a]	13.69 [a]	0.338	0.038	0.004	0.222
LOOH, µmol/L	Grower	0.622	0.634	0.641	0.040	0.937	0.732	0.893
	Finisher	1.36 [a]	1.04 [ab]	1.19 [b]	0.029	0.020	0.732	0.065
MDA, µmol/L	Grower	1.56 [a]	1.59 [a]	1.31 [b]	0.012	0.001	0.204	0.004
	Finisher	1.36 [a]	1.30 [ab]	1.14 [b]	0.033	0.009	0.204	0.058

[a–c] Means with the same superscript are statistically the same across all 3 treatments ($p > 0.05$) according to Tukey's post hoc test. [1] SOD–superoxide dismutase; CAT–catalase; FRAP–ferric reducing antioxidant power; LOOH–lipid peroxides; MDA–malondialdehyde. [2] There were 3 dietary treatments: a control diet (C) and diets T1 and T3, with 1% and 3% thyme, respectively. [3] SEM–standard error of the means. [4] TRT–*p* value for overall effect of dietary treatment (diets C vs. T1 vs. T3). [5] T–*p* value for thyme effect vs. control (diet C vs. T1 and T3). [6] D–*p* value for thyme dose effect (T1 vs. T3).

SOD and CAT activity was significantly higher in the longissimus dorsi muscle of pigs from the group receiving a diet with 3% thyme compared to the control ($p = 0.016$ and $p = 0.039$, respectively). The level of inclusion of thyme significantly influenced CAT activity ($p = 0.026$). The content of LOOH and MDA in the longissimus dorsi muscle of pigs from group T3 was significantly lower than in the control ($p < 0.001$ and $p = 0.038$, respectively). The level of thyme significantly influenced the content of both LOOH ($p = 0.041$) and MDA ($p = 0.049$) (Table 6).

Table 6. Redox parameters in the longissimus dorsi muscle.

Item [1]	Feeding Groups [2]			SEM [3]	*p* Value		
	C	T1	T3		TRT [4]	T [5]	D [6]
SOD, U/g	237.4 [b]	244.8 [ab]	267.6 [a]	2.39	0.016	0.158	0.178
CAT, U/g	25.67 [b]	26.18 [b]	32.35 [a]	2.4	0.039	0.507	0.026
Vitamin C, mg/g	0.252	0.257	0.251	0.01	0.711	0.825	0.817
LOOH, µmol/mg	3.16 [a]	2.65 [b]	2.10 [c]	0.146	<0.001	0.031	0.041
MDA, nmol/mg	0.039 [a]	0.032 [b]	0.025 [c]	0.003	0.038	0.115	0.049

[a–c] Means with the same superscript are statistically the same across all 3 treatments ($p > 0.05$) according to Tukey's post hoc test. [1] SOD–superoxide dismutase; CAT–catalase; LOOH–peroxides; MDA–malondialdehyde [2] There were 3 dietary treatments: a control diet (C) and diets T1 and T3, with 1% and 3 % thyme, respectively. [3] SEM–standard error of the means [4] TRT–*p* value for overall effect of dietary treatment (diets C vs. T1 vs. T3) [5] T–*p* value for thyme effect vs. control (diet C vs. T1 and T3). [6] D–*p* value for thyme dose effect (T1 vs. T3).

In the case of fatty acids, the inclusion of thyme in diets for fattening pigs significantly reduced the concentration of 14:0 ($p = 0.016$) in the longissimus dorsi muscle relative to the control. The longissimus dorsi muscle of group T3 fatteners had significantly lower content of 18:0 ($p = 0.050$) and higher content of 20:1 ($p = 0.049$), 18:2 ($p = 0.023$), 18:3 ($p = 0.039$), and 20:3 ($p = 0.045$) compared to the other groups. The level of thyme significantly influenced the levels of 18:0 (T1 > T3, $p = 0.046$), 20:1 (T1 < T3, $p = 0.011$), 18:2 (T1 < T3, $p = 0.022$), 18:3 (T1 < T3, $p = 0.048$), and 20:3 (T1 > T3, $p = 0.022$) (Table 7).

Table 7. Fatty acid profile of total lipids in the longissimus dorsi muscle.

Item [1]	Feeding Groups [2]			SEM [3]	p Value		
	C	T1	T3		TRT [4]	T [5]	D [6]
	Individual SFA (g/100 g of TFA)						
14:00	3.11 [a]	2.67 [b]	2.69 [b]	0.090	0.044	0.016	0.825
16:00	22.62	22.74	22.65	0.088	0.878	0.714	0.768
18:00	12.79 [a]	12.51 [a]	11.43 [b]	0.090	0.050	0.098	0.046
20:00	0.375	0.341	0.346	0.012	0.485	0.221	0.841
	Individual MUFA (g/100 g of TFA)						
16:1 n-9	3.60	3.57	3.64	0.020	0.343	0.992	0.166
18:1 n-9	43.22	43.53	43.17	0.068	0.052	0.405	0.071
18:1 n-7	4.68	4.98	4.89	0.086	0.369	0.168	0.571
20:1 n-9	0.047 [b]	0.040 [b]	0.177 [a]	0.013	0.049	0.423	0.011
	Individual PUFA (g/100 g of TFA)						
18:2 n-6	4.87 [b]	4.87 [b]	5.09 [a]	0.041	0.023	0.222	0.022
22:2	0.301	0.300	0.311	0.004	0.514	0.635	0.392
18:3 n-3	0.846 [c]	0.978 [b]	1.51 [a]	0.014	0.039	0.024	0.048
20:3 n-3	0.782 [b]	0.735 [b]	1.34 [a]	0.007	0.045	0.607	0.022
20:4 n-6	0.411	0.403	0.368	0.016	0.574	0.498	0.136
Other	2.34	2.33	2.26	0.156	0.977	0.907	0.875
SFA	38.88 [a]	38.25 [a]	37.11 [b]	0.187	0.021	0.072	0.041
MUFA	51.56	52.12	51.88	0.125	0.189	0.093	0.331
PUFA	6.43 [b]	6.55 [ab]	7.24 [a]	0.042	0.029	0.051	0.137
n-3	1.62 [b]	1.71 [b]	2.85 [a]	0.023	0.028	0.275	0.027
n-6	5.29	5.28	5.46	0.045	0.194	0.409	0.056
n-3/n-6	0.310 [b]	0.325 [b]	0.523 [a]	0.016	0.046	0.359	0.037
AL	0.600	0.565	0.555	0.007	0.059	0.015	0.813
TI	1.15 [a]	1.10 [ab]	0.975 [b]	0.013	0.043	0.072	0.125
CHOL	111.8 [a]	102.3 [b]	99.25 [b]	2.05	0.013	0.004	0.338

[a–c] Means with the same superscript are statistically the same across all 3 treatments ($p > 0.05$) according to Tukey's post hoc test. [1] TFA–total fatty acids; SFA–saturated fatty acids; MUFA–monounsaturated fatty acids; PUFA–polyunsaturated fatty acids; AI–atherogenicity index; TI–thrombogenicity index. [2] There were 3 dietary treatments: a control diet (C) and diets T1 and T3, with 1% or 3% thyme, respectively. [3] SEM–standard error of the means. [4] TRT–p value for overall effect of dietary treatment (diets C vs. T1 vs. T3). [5] T–p value for thyme effect vs. control (diet C vs. T1 and T3). [6] D–p value for thyme dose effect (T1 vs. T3).

The content of SFAs in the longissimus dorsi muscle of pigs from group T3 was significantly lower than in the other experimental groups ($p = 0.021$), while the PUFA content in the longissimus dorsi muscle of pigs from group T3 was significantly higher than in group C ($p = 0.029$). Significantly higher content of n-3 acids ($p = 0.028$) and a higher n-3/n-6 ratio ($p = 0.046$) were noted in the muscle of T3 pigs compared to the other experimental groups (C and T1 groups). The 3% inclusion of thyme in diets for pigs caused a significant reduction in the TI index ($p = 0.043$) and the cholesterol level ($p = 0.013$) compared to the control. The level of thyme inclusion significantly influenced the content of SFAs (T1 > T3, $p = 0.041$), the content of n-3 acids (T1 < T3, $p = 0.027$), and the n-3/n-6 ratio (T1 < T3, $p = 0.037$) (Table 7).

4. Discussion

Formulation of diets for pigs presents many challenges associated with the choice of feed ingredients that will provide adequate amounts of essential nutrients and, at the same time, ensure that the animal products will meet consumer expectations. Consumers prefer products with the highest possible levels of substances with health-promoting effects, such as valuable protein, minerals, and polyunsaturated fatty acids (PUFAs). A valuable source of PUFAs for animals, especially n-3 PUFAs, is flaxseed, which is increasingly used as a feed ingredient for various livestock animals. This strategy results in beneficial modifications of the fatty acid composition of the meat or other products, such as milk or eggs. Our earlier research [16] and studies by other researchers [32] have shown that the inclusion of flaxseed in the diet of pigs increases the content of beneficial omega-3 fatty acids in animal tissues and also has a positive effect on the lipid profile of the blood. Due to the high susceptibility of PUFAs to oxidation, whose rate increases with the number of unsaturated bonds in the molecule [33], the use of antioxidant additives is recommended. Popular additives that can minimize lipid oxidation and positively affect the lipid profile of the blood include various phytobiotics, including herbs and their extracts, such as oregano (100, 200 mg/kg) or thyme essential oils (300, 450 mg/kg) [34,35].

In the present study, the diet containing thyme herb in the amount of 3% had a beneficial effect on the lipid metabolism of fattening pigs, resulting in a reduction in the level of cholesterol and its LDL fraction, as well as that of TG even in the case of 1% inclusion of the phytobiotic. Shao et al. [36] used a blend containing carvacrol, thymol, and cinnamaldehyde in the diet of early-weaned piglets and found that it caused a decrease in TG, total cholesterol, and its LDL-C and HDL-C fractions. A reduction in the concentrations of total cholesterol and triacylglycerols was also noted in calves receiving a diet supplemented with thyme or oregano essential oils [37]. Similar results have been reported in poultry, e.g., quail [38] and chicken broilers [39].

Lipid metabolism is regulated mainly by the active substances in thyme oil, i.e., thymol and carvacrol. These substances have been shown to exert an inhibitory effect on the activity of hydroxymethylglutaryl–coenzyme A reductase, a key enzyme regulating endogenous biosynthesis of cholesterol [40]. The capacity of the active components of thyme to reduce the TG level in the blood may be linked to the stimulation of lipoprotein lipase activity, as shown in research using other herbs [41].

There are indications that thyme has hepatoprotective effects in the case of damage to the liver caused by various factors [42]. The beneficial effect of thyme on liver function was manifested as a reduction in the activity of indicator enzymes AST and ALT. However, long-term intake of high doses of formulations containing essential oil components thymol and carvacrol has been shown to have the opposite hepatotoxic effect [43]. Therefore, caution and monitoring of liver markers are recommended during the administration of thymol or carvacrol. Analysis of the effect of the inclusion of thyme in the diet of fattening pigs in the present study did not reveal changes in the values of these liver indicators (ALT or AST) or of kidney markers (CREAT, UREA, UA), which suggests that this level of the phytobiotic does not disturb liver or kidney function. Improvement of kidney and liver function in rabbits fed diets with aqueous extracts of thyme (50 mg/kg BW) was reported by Abdel-Gabbar et al. [44]. The researchers noted a decrease in kidney markers, i.e., creatinine, uric acid, and urea, as well as in the activity of ALT and AST. These effects are mainly attributed to the antioxidant properties of thyme, which, like other plants of the family Labiatae, is a rich source of phenolic substances: phenolic acids (primarily rosmarinic acid) and flavonoids. Apart from their effect on redox status, the active compounds in thyme may have anti-inflammatory properties and also increase the activity of phase I and II xenobiotic detoxification enzymes, which also favours hepatoprotection [45].

The results of the present study, in which the effect of thyme herb on redox parameters of the blood was evaluated, confirmed the antioxidant effects of the plant. The plasma of pigs receiving a diet with 3% thyme herb showed increased activity of antioxidant enzymes SOD and CAT, which protect tissues against oxidation, as well as an increased FRAP value (ferric reducing antioxidant power). The value of this parameter is determined by the presence of vitamin C (15%), uric acid (60%), bilirubin (5%), and protein (10%) [46]. The level of low-molecular-weight antioxidants (uric acid, urea, and bilirubin) in the blood of pigs receiving a diet with 3% thyme was higher than in the control group but was still within reference values [47]. The content of vitamin C was also significantly higher in the group receiving 3% thyme. A study using rats with hyperlipidaemia [11] showed that polyphenol-rich extracts of various varieties of thyme exhibit significant antioxidant activity, as indicated by FRAP and RSA (Radical Scavenging Activity) values, owing to which they can protect organs from oxidative stress. Polyphenols have been shown to exhibit multi-faceted antioxidant activity, including scavenging of free radicals, inhibition of enzymes taking part in free radical production, and chelation of metal ions [48]. Furthermore, these compounds can protect natural antioxidants such as tocopherols or vitamin C against oxidation [49]. A study by Placha et al. [50] using rabbits confirmed that the addition of exogenous antioxidants, such as antioxidant-rich thyme oil, can improve the body's redox status, increasing the total antioxidant status of the plasma and GPx activity in the liver. In the present study, enhancement of endogenous antioxidant defense in pigs receiving a diet supplemented with thyme herb resulted in a reduction in indicators of lipid peroxidation in the blood, i.e., LOOH and MDA. The results pertaining to the levels of antioxidants and activity of antioxidant enzymes in the muscle tissue of the pigs indicate similar tendencies to those noted in the blood. The content of MDA and LOOH at both levels of thyme supplementation (1% and 3%) was significantly reduced, with a much more pronounced decrease in the case of 3% inclusion. Therefore, the higher inclusion of thyme inhibited lipid oxidation to a greater extent and also caused a significant increase in the activity of enzymes involved in antioxidant processes, i.e., SOD and CAT. Martini et al. [6] added natural extracts of grape skin and oregano to the diet of pigs receiving feed with extruded flaxseed. This resulted in a reduction in the amount of lipid oxidation products (hydroperoxides and TBA-RS) in the grilled pork, as well as in meat digested in vitro. In the group of pigs receiving synthetic antioxidants, this effect was observed only in the cooked meat but not in the digested meat. Studies by various authors indicate that supplementation of animal diets with antioxidant substances such as thymol, carvacrol, or oregano oil can have a beneficial impact by reducing oxidation phenomena in the meat [51–53]. In the case of poultry, a reduction in lipid oxidation products during meat storage was obtained by using thymol or carvacrol as feed additives (150 mg/kg), which proved to be equally as effective as BHT [51]. Ranucci et al. [52] used a combination of oregano essential oil and sweet chestnut wood extract (0.2%) to supplement feed for pigs and observed a beneficial increase in glutathione peroxidase and glutathione reductase activity in the longissimus lumborum muscle, accompanied by a decrease in lipid oxidation indicators. A trend of decreasing markers of lipid oxidation in pork following feed supplementation with oregano oleoresin (0.05%) was reported by Janz et al. [53].

The inclusion of thyme herb in the diet of pigs also modified the fatty acid profile of the loin meat. Both levels of the additive reduced the content of acids 14:0 and 18:0 and the total amount of saturated fatty acids. The use of 3% thyme herb increased the content of acids 20:1, 18:2, 18:3, and 20:3, the total PUFA content, and the n-3/n-6 ratio. In addition, there was a decrease in the thrombogenicity index TI and the cholesterol content. These changes should be regarded as highly favourable for consumers of pork, as a lower intake of saturated fatty acids and cholesterol can minimize the risk of atherosclerosis and other cardiovascular diseases [54]. The inclusion of thyme in the diet (group T3) inhibited the oxidation of unsaturated fatty acids, especially PUFAs, which are more susceptible to oxidation in the loin muscle of pigs. This resulted in an increase in the content of PUFAs in the muscle tissue.

This is also evidenced by the lower levels of lipid peroxidation markers in this group of pigs. The literature, unfortunately, lacks data on the fatty acid composition of pork from pigs fed with thyme or thyme oil. Chickens receiving an aqueous extract of thyme (50 and 100 mg/kg of feed) and thyme powder (150 and 250 mg/kg of feed) had reduced levels of some saturated fatty acids in the breast muscle, accompanied by an increase in health-promoting unsaturated fatty acids [55]. Results obtained in pigs receiving extracts from other plants, such as rosemary (1 g/kg), indicate an increase in PUFA content in the longissimus lumborum muscle, especially linoleic, arachidonic, and docosahexaenoic acids [56].

5. Conclusions

The presence of thyme herb in diets with extruded flaxseed has a beneficial effect by influencing the antioxidant status of the blood and muscle and improving the lipid profile of tissues.

The study showed that a 3% inclusion of thyme in the diet significantly slows down oxidant processes, which may be a successful means of protecting meat against spoilage during storage. Furthermore, the 3% inclusion of thyme herb had a clear beneficial effect on lipid metabolism, so the levels of cholesterol and HDL cholesterol were reduced in both the blood and the longissimus dorsi muscle. In addition, the level of n-3 PUFAs in the muscles increased while the level of SFAs significantly decreased. The results of this study must be confirmed by further research.

Author Contributions: Conceptualization, K.K. and A.C.; methodology, A.C.; software, K.T.; validation, K.T., I.S. and K.K.; formal analysis, K.K. and I.S.; investigation, A.C.; resources, K.T.; data control, A.C.; writing—original preparation of the project, K.K. and I.S.; writing—review and editing, A.C.; visualization, K.T.; supervision, A.C. All authors have read and agreed to the published version of the manuscript.

Funding: This research received no external funding.

Institutional Review Board Statement: All experimental procedures used during this study was approved by the Second Local Ethics Committee on Animal Experimentation of the University of Life Sciences in Lublin, Poland (Resolution No. 64/2011). The experiment was conducted according to the guidelines of the Declaration of Helsinki and in compliance with the European Union law (Directive 2010/63/UE, received in Poland by Legislative Decree 266/2015) of the European Parliament and of the Council on the protection of animals used for scientific or educational purposes.

Informed Consent Statement: Not applicable. None of the data were deposited in an official repository. The model was not deposited in an official repository.

Data Availability Statement: Not applicable.

Conflicts of Interest: The authors declare no conflict of interest.

References

1. Corino, C.; Rossi, R.; Cannata, S.; Ratti, S. Effect of dietary linseed on the nutritional value and quality of pork and pork products: Systematic review and metaanalysis. *Meat Sci.* **2014**, *98*, 679–688. [CrossRef] [PubMed]
2. Tao, L. Oxidation of polyunsaturated fatty acids and its impact on food quality and human health. *Adv. Food Technol. Nutr. Sci.-Open J.* **2015**, *1*, 135–142. [CrossRef]
3. Wąsowicz, E.; Gramza, A.; Hęś, M.; Jeleń, H.H.; Korczak, J.; Małecka, M.; Mildner-Szkudlarz, S.; Rudzińska, M.; Samotyja, U.; Zawirska-Wojtasiak, R. Oxidation of lipids in food. *Pol. J. Food Nutr. Sci.* **2004**, *13*, 87–100.
4. Korczak, J.; Hęś, M.; Gramza, A.; Jędrusek-Golińska, A. Influence of fat oxidation on the stability of lysine and protein digestibility in frozen meat products. *Electron. J. Pol. Agric. Univ. Ser. Food Sci. Technol.* **2004**, *7*, 1–13.
5. Papuc, C.; Goran, G.V.; Predescu, C.N.; Nicorescu, V. Mechanisms of oxidative processes in meat and toxicity induced by postprandial degradation products: A Review. *Compr. Rev. Food Sci. Food Saf.* **2017**, *16*, 96–123. [CrossRef]
6. Martini, S.; Tagliazucchi, D.; Minelli, G.; Lo Fiego, D.P. Influence of linseed and antioxidant-rich diets in pig nutrition on lipid oxidation during cooking and in vitro digestion of pork. *Food Res. Int.* **2020**, *137*, 109528. [CrossRef]
7. Khojasteh, A.; Mirjalili, M.H.; Alcalde, M.A.; Cusido, R.M.; Eibl, R.; Palazon, J. Powerful plant antioxidants: A new biosustainable approach to the production of rosmarinic acid. *Antioxidants* **2020**, *9*, 1273. [CrossRef]
8. Economou, K.D.; Oreopoulou, V.; Thomopoulos, C.D. Antioxidant activity of some plant extracts of the family labiatae. *J. Am. Oil Chem. Soc.* **1991**, *68*, 109–113. [CrossRef]
9. Youdim, K.A.; Deans, S.G.; Finlayson, H.J. The antioxidant properties of thyme (*Thymus zygis* L.) essential oil: An inhibitor of lipid peroxidation and a free radical scavenger. *J. Essent. Oil Res.* **2002**, *14*, 210–215. [CrossRef]
10. Hashemipour, H.; Kermanshahi, H.; Golian, A.; Veldkamp, T. Effect of thymol and carvacrol feed supplementation on performance, antioxidant enzyme activities, fatty acid composition, digestive enzyme activities, and immune response in broiler chickens. *Poult. Sci.* **2013**, *92*, 2059–2069. [CrossRef]
11. Ramchoun, M.; Harnafi, H.; Alem, C.; Büchele, B.; Simmet, T.; Rouis, M.; Atmani, F.; Amrani, S. Hypolipidemic and antioxidant effect of polyphenol-rich extracts from Moroccan thyme varieties. *e-SPEN J.* **2012**, *7*, e119–e124. [CrossRef]

12. Rouhi-Boroujeni, H.; Rouhi-Boroujeni, H.; Heidarian, E.; Mohammadizadeh, F.; Rafieian-Kopaei, M. Herbs with anti-lipid effects and their interactions with statins as a chemical anti- hyperlipidemia group drugs: A systematic review. *ARYA Atheroscler.* **2015**, *11*, 244–251.
13. National Research Council (NRC). *Nutrient Requirements of Swine*, 11th ed.; National Academies Press: Washington, DC, USA, 2012.
14. Klimiuk, K.; Sembratowicz, I.; Czech, A. Effect of the inclusion of extruded flaxseed in the diet of fattening pigs on performance parameters and blood parameters. *Ann. Anim. Sci.* **2023**. *ahead of print*. [CrossRef]
15. Kirchgessner, M.; Roth, F.X. Equation for prediction of the energy value in mixed feeds for pigs. *J. Anim. Physiol. Anim. Nutr.* **1983**, *50*, 270–275.
16. Czech, A.; Klimiuk, K.; Sembratowicz, I. Effect of the inclusion of extruded flaxseed in the diet of fattening pigs on lipid metabolism and tissue redox status. *Sci. Rep.* **2023**. *in the review*.
17. AOAC. *Official Methods of Analysis of AOAC International*, 20th ed.; Association of Official Analytical Chemists International: Gaithersburg, MD, USA, 2016.
18. Bakowska-Barczak, A.M.; Kolodziejczyk, P.P. Black currant polyphenols: Their storage stability and microencapsulation. *Ind. Crops Prod.* **2011**, *34*, 1301–1309. [CrossRef]
19. Song, F.L.; Gan, R.Y.; Zhang, Y.; Xiao, Q.; Kuang, L.; Li, H.B. Total phenolic contents and antioxidant capacities of selected Chinese medicinal plants. *Int. J. Mol. Sci.* **2010**, *11*, 2362–2372. [CrossRef]
20. Re, R.; Pellegrini, N.; Proteggente, A.; Pannala, A.; Yang, M.; Rice-Evans, C. Antioxidant activity applying an improved ABTS radical cation decolorization assay. *Free Radic. Biol. Med.* **1999**, *26*, 1231–1237. [CrossRef]
21. Bocco, A.; Cuvelier, M.-E.; Richard, H.; Berset, C. Antioxidant activity and phenolic composition of citrus peel and seed extracts. *J. Agric. Food Chem.* **1998**, *46*, 2123–2129. [CrossRef]
22. Dewanto, V.; Wu, X.; Adom, K.K.; Liu, R.H. Thermal processing enhances the nutritional value of tomatoes by increasing total antioxidant activity. *J. Agric. Food Chem.* **2002**, *50*, 3010–3014. [CrossRef]
23. Czech, A.; Malik, A.; Sosnowska, B.; Domaradzki, P. Bioactive substances, heavy metals, and antioxidant activity in whole fruit, peel, and pulp of citrus fruits. *Int. J. Food Sci.* **2021**, *2021*, 6662259. [CrossRef] [PubMed]
24. Friedewald, W.T.; Levy, R.I.; Fredrickson, D.S. Estimation of the concentration of low-density lipoprotein cholesterol in plasma, without use of the preparative ultracentrifuge. *Clin. Chem.* **1972**, *18*, 499–502. [CrossRef] [PubMed]
25. Czech, A.; Sembratowicz, I.; Kiesz, M. The effects of a fermented rapeseed or/and soybean meal additive on antioxidant parameters in the blood and tissues of piglets. *Animals* **2021**, *11*, 1646. [CrossRef] [PubMed]
26. Folch, J.; Lees, M.; Stanley, G.H.S. A simple method for the isolation and purification of total lipids from animal tissue. *J. Biol. Chem.* **1957**, *226*, 497–509. [CrossRef] [PubMed]
27. *PN-EN ISO 5508*; Animal and Vegetable Fats and Oils—Analysis by Gas Chromatography of Methyl Esters of Fatty Acids. ISO: Geneva, Switzerland, 1996.
28. *PN-EN ISO 5509*; Animal and Vegetable Fats and Oils—Preparation of Methyl Esters of Fatty Acids. ISO: Geneva, Switzerland, 2001.
29. Weihrauch, J.L.; Posati, L.P.; Anderson, B.A.; Exler, J. Lipid conversion factors for calculating fatty acids contents in foods. *J. Am. Oil Chem. Soc.* **1977**, *54*, 36–40. [CrossRef]
30. Ulbricht, T.L.V.; Southgate, D.A.T. Coronary disease seven dietary factors. *Lancet* **1991**, *338*, 985–992. [CrossRef]
31. Statistica 10. Available online: https://www.statsoft.pl/statistica-10/ (accessed on 20 September 2021).
32. Liu, W.C.; Kim, I.H. Effects of different dietary n-6:n-3 PUFA ratios on growth performance, blood lipid profiles, fatty acid composition of pork, carcass traits and meat quality in finishing pigs. *Ann. Anim. Sci.* **2018**, *18*, 143–154. [CrossRef]
33. Ahmed, M.; Pickova, J.; Ahmad, T.; Liaquat, M.; Farid, A.; Jahangir, M. Oxidation of lipids in foods. *Sarhad J. Agric.* **2016**, *32*, 230–238. [CrossRef]
34. Caceda-Gallardo, L.; Mendoza-Ordoñez, G.; Loyaga-Cortéz, B.; Ybañez-Julca, R.; Gonzales-Nonato, D.; Asunción-Alvarez, D. Oregano essential oil supplementation improves productive performance, oxidative stability, and lipid parameters in turkeys. *Sci. Agropecu.* **2020**, *11*, 187–193. [CrossRef]
35. Gumus, R.; Ercan, N.; Imik, H. The effect of thyme essential oil (*Thymus vulgaris*) added to quail diets on performance, some blood parameters, and the antioxidative metabolism of the serum and liver tissues. *Braz. J. Poult. Sci.* **2017**, *19*, 297–304. [CrossRef]
36. Shao, Y.; Peng, Q.; Wu, Y.; Peng, C.; Wang, S.; Zou, L.; Qi, M.; Peng, C.; Liu, H.; Li, R.; et al. The effect of an essential oil blend on growth performance, intestinal health, and microbiota in early-weaned piglets. *Nutrients* **2023**, *15*, 450. [CrossRef]
37. Seirafy, H.; Sobhanirad, S. Effects of oregano (*Origanum vulgare*) and thyme (*Thymus vulgaris*) oils on growth performance and blood parameters in holstein suckling calves. *Iran. J. Appl. Anim. Sci.* **2017**, *7*, 585–593.
38. Khaksar, V.; Krimpen, M.V.; Hashemipour, H.; Pilevar, M. Effects of thyme essential oil on performance, some blood parameters and ileal microflora of Japanese quail. *J. Poult. Sci.* **2012**, *25*, 201–210. [CrossRef]
39. El-Ghousein, S.S.; Al-Beitawi, N.A. The effect of feeding of crushed thyme (*Thymus valgaris* L) on growth, blood constituents, gastrointestinal tract and carcass characteristics of broiler chickens. *J. Poult. Sci.* **2009**, *46*, 100–104. [CrossRef]
40. Elson, C.E. Suppression of mevalonate pathway activities by dietary isoprenoids: Protective roles in cancer and cardiovascular disease. *J. Nutr.* **1995**, *125*, 1666–1672.

41. Mohammed, H.A.; Sallau, A.B.; Nzelibe, H.C.; Muhammad, A.; Erukainure, O.L.; Musa, B.; Anigo, M.K. Bioactive fraction of *Mentha piperita* leaves extract stimulates lipoprotein lipase at cellular and molecular levels via augmentation of antioxidant defense system in hyperlipidemic rats. *J. Herbs Spices Med. Plants* **2021**, *27*, 267–288. [CrossRef]
42. Rostami, R.; Eslamifar, Z.; Nazemi, S.; Hosseini, S.Z.; Behvandi, M.M.; Jafaripour, L. The effect of thyme essential oil on liver injuries caused by renal ischemia-reperfusion in rats. *BioMed Res. Int.* **2022**, *2022*, 2988334. [CrossRef]
43. Kraft, K.; Hobbs, C. *Pocket Guide to Herbal Medicine*; Thieme: Stuttgart, Germany, 2004.
44. Abdel-Gabbar, M.; Ahmed, R.R.; Kandeil, M.A.; Mohamed, A.E.H.; Ali, S.M. Administration of ginger and/or thyme has ameliorative effects on liver and kidney functions of V-line rabbits: Histological and biochemical studies. *J. Anim. Physiol. Anim. Nutr.* **2019**, *103*, 1758–1767. [CrossRef]
45. Sasaki, K.; Wada, K.; Tanaka, Y.; Yoshimura, T.; Matuoka, K.; Anno, T. Thyme (*Thymus vulgaris* L) leaves and its constituents increase the activities of xenobiotic-metabolizing enzymes in mouse liver. *J. Med. Food* **2005**, *8*, 184–189. [CrossRef]
46. Benzie, I.F.F.; Strain, J.J. The ferric reducing ability of plasma (FRAP) as a measure of "Antioxidant Power": The FRAP assay. *Anal. Biochem.* **1996**, *239*, 70–76. [CrossRef]
47. Klem, T.B.; Bleken, E.; Morberg, H.; Thoresen, S.I.; Framstad, T. Hematologic and biochemical reference intervals for Norwegian crossbreed grower pigs. *Vet. Clin. Pathol.* **2010**, *39*, 221–226. [CrossRef] [PubMed]
48. Leopoldini, M.; Russo, N.; Toscano, M. The molecular basis of working mechanism of natural polyphenolic antioxidants. *Food Chem.* **2011**, *125*, 288–306. [CrossRef]
49. Majewska, M.; Czeczot, H. Flavonoids in prevention and therapy. *Farm. Pol.* **2009**, *65*, 369–377. (In Polish)
50. Placha, I.; Chrastinova, L.; Laukova, A.; Cobanova, K.; Takacova, J.; Strompfova, V. Effect of thyme oil on small intestine integrity and antioxidant status, phagocytic activity and gastrointestinal microbiota in rabbits. *Acta Vet. Hung.* **2013**, *61*, 197–208. [CrossRef] [PubMed]
51. Luna, A.; Lábaque, M.C.; Zygadlo, J.A.; Marin, R.H. Effects of thymol and carvacrol feed supplementation on lipid oxidation in broiler meat. *Poult. Sci.* **2010**, *89*, 366–370. [CrossRef]
52. Ranucci, D.; Beghelli, D.; Trabalza-Marinucci, M.; Branciari, R.; Forte, C.; Olivieri, O.; Badillo Pazmay, G.V.; Cavallucci, C.; Acuti, G. Dietary effects of a mix derived from oregano (*Origanum vulgare* L.) essential oil and sweet chestnut (*Castanea sativa* Mill.) wood extract on pig performance, oxidative status and pork quality traits. *Meat Sci.* **2015**, *100*, 319–326. [CrossRef]
53. Janz, J.; Morel, P.; Wilkinson, B.; Purchas, R. Preliminary investigation of the effects of low-level dietary inclusion of fragrant essential oils and oleoresins on pig performance and pork quality. *Meat Sci.* **2007**, *75*, 350–355. [CrossRef]
54. Berek, K.; Bobiński, R. Atherosclerosis—The multifactors disease. *Nurs. Probl.* **2009**, *17*, 257–262. (In Polish)
55. Belali, M.; Seidavi, A.; Bouyeh, M. Effects of thyme powder and aqueous extract on fatty acid profile of breast meat in broiler chicks. *MESO Prvi Hrvat. Časopis O Mesu* **2021**, *23*, 38–43. [CrossRef]
56. Liotta, L.; Chiofalo, V.; D'Alessandro, E.; Lo Presti, V.; Chiofalo, B. Supplementation of Rosemary extract in the diet of Nero Siciliano pigs: Evaluation of the antioxidant properties on meat quality. *Animal* **2015**, *9*, 1065–1072. [CrossRef]

antioxidants

MDPI

Article

Dietary Vitamin E and/or Hydroxytyrosol Supplementation to Sows during Late Pregnancy and Lactation Modifies the Lipid Composition of Colostrum and Milk

Hernan D. Laviano [1], Gerardo Gómez [2], María Muñoz [3], Juan M. García-Casco [3], Yolanda Nuñez [3], Rosa Escudero [1], Ana Heras Molina [1], Antonio González-Bulnes [4], Cristina Óvilo [3], Clemente López-Bote [1] and Ana I. Rey [1,*]

[1] Departamento Producción Animal, Facultad de Veterinaria, Universidad Complutense de Madrid, Avda. Puerta de Hierro s/n., 28040 Madrid, Spain; hernanlavianomvz03@gmail.com (H.D.L.); rmescude@ucm.es (R.E.); andelash@ucm.es (A.H.M.); clemente@ucm.es (C.L.-B.)

[2] Instituto Regional de Investigación y Desarrollo Agroalimentario y Forestal de Castilla-La Mancha (IRIAF), 13700 Toledo, Spain; g.gomez.mat@gmail.com

[3] Departamento de Mejora Genética Animal, Instituto Nacional de Investigación y Tecnología Agraria y Alimentaria (INIA), Consejo Superior de Investigaciones Científicas (CSIC), Ctra Coruña km 7.5, 28040 Madrid, Spain; mariamm@inia.csic.es (M.M.); garcia.juan@inia.csic.es (J.M.G.-C.); nunez.yolanda@inia.csic.es (Y.N.); ovilo@inia.csic.es (C.Ó.)

[4] Departamento de Producción y Sanidad Animal, Facultad de Veterinaria, Universidad Cardenal Herrera-CEU, CEU Universities, C/ Tirant lo Blanc, 7, Alfara del Patriarca, 46115 Valencia, Spain; antonio.gonzalezbulnes@uchceu.es

* Correspondence: anarey@ucm.es; Tel.: +34-913-943-889

Citation: Laviano, H.D.; Gómez, G.; Muñoz, M.; García-Casco, J.M.; Nuñez, Y.; Escudero, R.; Molina, A.H.; González-Bulnes, A.; Óvilo, C.; López-Bote, C.; et al. Dietary Vitamin E and/or Hydroxytyrosol Supplementation to Sows during Late Pregnancy and Lactation Modifies the Lipid Composition of Colostrum and Milk. *Antioxidants* **2023**, *12*, 1039. https://doi.org/10.3390/antiox12051039

Academic Editor: Alessandra Napolitano

Received: 17 March 2023
Revised: 14 April 2023
Accepted: 28 April 2023
Published: 3 May 2023

Abstract: Modifying the composition of a sow's milk could be a strategy to improve the intestinal health and growth of her piglet during the first weeks of life. This study evaluated how dietary supplementation of vitamin E (VE), hydroxytyrosol (HXT) or VE+HXT given to Iberian sows from late gestation affected the colostrum and milk composition, lipid stability and their relationship with the piglet's oxidative status. Colostrum from VE-supplemented sows had greater C18:1n−7 than non-supplemented sows, whereas HXT increased polyunsaturated (∑PUFAs), ∑n−6 and ∑n−3 fatty acids. In 7-day milk, the main effects were induced by VE supplementation that decreased ∑PUFAs, ∑n−6 and ∑n−3 and increased the Δ-6-desaturase activity. The VE+HXT supplementation resulted in lower desaturase capacity in 20-day milk. Positive correlations were observed between the estimated mean milk energy output and the desaturation capacity of sows. The lowest concentration of malondialdehyde (MDA) in milk was observed in VE-supplemented groups, whereas HXT supplementation increased oxidation. Milk lipid oxidation was negatively correlated with the sow's plasma oxidative status and to a great extent with the oxidative status of piglets after weaning. Maternal VE supplementation produced a more beneficial milk composition to improve the oxidative status of piglets, which could promote gut health and piglet growth during the first weeks, but more research is needed to clarify this.

Keywords: sow's lactation; colostrum composition; milk composition; vitamin E; hydroxytyrosol

1. Introduction

Milk is a vital food to guarantee the survival of mammalian animals during the first weeks of life. In the case of a piglet that is born without hair and with scarce energy reserves, milk is the main energy vehicle to support thermogenesis [1]. Thus, sow milk is characterized by containing, compared with other non-ruminant and ruminant species, a greater proportion of protein, fat and lactose [2,3]. However, milk is not only a nutrient-rich fluid suitable for growth but its composition can enhance intestinal mucosa development and promote the growth of certain bacteria in the intestine or limit the growth of others, thus affecting intestinal health [4].

Milk components, mainly those that are not synthetized in the mammary gland, can be modified by factors such as feeding [5]. Hence, some previous research on the effects of liposoluble micronutrients, such as vitamin E (VE), on milk composition found that when administered at high doses during the gestation and lactation periods, it increased its concentration in colostrum and milk and the fat component, and thus, affected the health status and growth of piglets [6]. This could be a relevant aspect in the case of the Iberian pig, which was shown to have lower growth rates than those of improved genotypes, partly due to the limited use of energy and milk protein [3]. The health and productivity effect of VE-supplementation in piglets [6] was attributed to their potent chain-breaking antioxidant effect on cells [7], which could also protect the epithelial barrier function [8]. It is widely documented that supplementation with high doses of VE is an effective strategy to avoid the important decline of VE plasma concentration post-weaning [9–11] and may help to counteract the oxidative state of piglets in this critical period [6,9–11]. Moreover, a study that investigated the effects of supplementation with VE (150 mg/kg) in its different chemical forms on the fatty acid profile of sow's milk showed a positive effect on the proportion of some specific fatty acids [12]. Fatty acids may participate in the oxidative balance of the body to some extent by promoting or inhibiting lipid oxidation [13], and they can also be used with different efficiencies for energy production [14] or affect gut health [8]. Therefore, the study of how fatty acids or other milk components can be modified by antioxidant micronutrient supplementation deserves further attention.

The effect of supplementing diets with phenolic hydrophilic antioxidants on milk composition was also investigated in other species. In lactating buffaloes, dried stoned olive pomace supplementation was reported to improve milk tocopherols and retinol [15]. Other antioxidants derived from olive leaves, such as oleuropein or hydroxytyrosol, were shown to not only have powerful antioxidant effects but also to produce changes in the metabolic use of fatty acids [16], as well as hypolipidemic and hypoglycemic effects [17]. However, the effect of dietary supplementation on the composition of milk has not been investigated. There is also a lack of information on the possible combined effects of different antioxidants (lipophilic and non-lipophilic, such as polyphenols) on sow's milk and colostrum composition. Considering that the growth and gut health of nursery piglets depend largely on the quantity and quality of milk production [18,19], extending knowledge on how different antioxidants supplementation in diets or their combined administration affects colostrum and milk composition is a matter of interest.

Furthermore, the post-weaning period is a major challenge for a piglet when trying to maintain homeostasis by having to deal with numerous pro-oxidant agents [9–11]. This control of the oxidative stress to which the animal is subjected during the first days of life plays an important role in the metabolism and health of the adult due to its effects on the development of inflammation and the immune response [20]. On many occasions, the extra contribution of antioxidants in order to achieve antioxidant/pro-oxidant balance must be provided by food, including breast milk [9–11]. In this sense, it is unknown how HXT can modify the oxidative stability of milk compared with the use of VE.

It is hypothesized that VE, HXT or their combined administration to sows from gestation could modify the composition and lipid stability of milk in different ways and then affect the oxidative status.

Thus, the objective of the present research was to study the effects of dietary supplementation of VE (100 mg/kg), hydroxytyrosol (HXT) (1.5 mg/kg) or a combined administration provided to Iberian sows during late gestation and lactation on the colostrum and milk composition, lipid stability and their possible relationships with piglets' oxidative status.

2. Materials and Methods

2.1. Chemicals

All chemicals used were analytical grade and were supplied by Sigma-Aldrich (Alcobendas, Madrid, Spain), Panreac (Castellar del Vallès, Barcelona, Spain) or Scharlau (Sentmenat, Barcelona, Spain).

2.2. Animals, Experimental Diets and Sample Collection

The experimental procedures used in this study were in compliance with the Spanish guidelines (RD53/2013) [21] and European Union Directive 2010/63/UE for the care and use of animals in research [22]. The experimental procedures (report ORCEEA 2019-10) were approved by the INIA Committee of Ethics in Animal Research.

The study was carried out at the facilities of Dehesón del Encinar (Oropesa, Toledo, Spain). Fifty pregnant Iberian sows (half primiparous and half multiparous with between 4 and 5 parity) were divided into four experimental groups (each group with equal distribution of primiparous and multiparous), according to the dietary supplementation from day 85 of gestation (average weight of 126.2 ± 29.3 kg) until weaning (28 days): (1) the control group received a 30 mg supplementation of α-tocopheryl acetate/kg feed, (2) the VE group received 100 mg of α-tocopheryl acetate/kg, (3) the HXT group received 30 mg α-tocopheryl acetate/kg and 1.5 mg hydroxytyrosol/kg, and (4) the VE+HXT group received 100 mg of α-tocopheryl acetate/kg + 1.5 mg hydroxytyrosol/kg feed. The basal level of α-tocopherol in the feed ingredients was 12–14 mg/kg; thus, the total vitamin E concentration in all groups was in compliance with values recommended for reproductive sows [23]. The highest VE dose of 100 mg/kg was used after taking into account the effective antioxidant effect observed in a previous study [16]. During this experimental period, the feed administration was adjusted to fulfill daily requirements according to the National Research Council [23] (Table A1, Appendix A), and water was provided ad libitum. In the pre-experimental period (from the end of natural service to day 85 of pregnancy), sows were given a standard grain-based diet (g/kg: 888 dry matter, 124.6 protein, 29.9 fat, 49.3 fiber, 62.1 ash; 3050 kcal/kg metabolic energy) (Sanchez Romero Carvajal, Jabugo, Spain). The α-tocopheryl acetate used in the diets was purchased from DSM Nutritional Products (Alcalá de Henares, Madrid, Spain) and the hydroxytyrosol extract (*Olea europaea* L. dry extract, N20130102) was obtained from Natac (Alcorcón, Madrid, Spain).

Milk and colostrum (5 mL) samples were taken from a representative number of multiparous sows (n = 7 per treatment) with a homogeneous litter size on the day of delivery and 7 and 20 days post-partum. Piglets were separated from their mother and milk was collected from the functional glands by hand-milking after disinfecting the operator's hands and nipples with soap and water, and put into plastic tubes of 20 mL. Samples were kept under refrigeration while handling, and then immediately frozen at −20 °C until analysis (within 2 months after collection). The mean milk energy output (kcal/day) was estimated according to the following equation: 4.92× mean litter gain (g/day) − (90× litter size) [22].

Blood samples from sows for oxidative status analysis were obtained coinciding with the taking of milk samples (at the peak of lactation on day 20 after farrowing), which is the moment when the production is at a maximum and composition could be affected [5,24]. In addition, blood was taken from piglets (1 male piglet was selected from each litter at weaning). Blood was collected in sterile EDTA vacuum tubes (Vacutainer, BD, Franklin Lakes, NJ, USA), immediately centrifuged at 2500 rpm for 10 min and plasma samples were kept at −80 °C until analysis (less than 1 month).

2.3. Laboratory Analysis

2.3.1. Lactose Concentrations of the Colostrum and Milk

Lactose was measured according to Beutler [25] with few modifications. For the sample preparation, milk or colostrum (100 μL) was placed in a 1.5 mL Eppendorf tube in the presence of distilled water. The mixture was heated in a stirring bath for 15 min at 50 °C. For sample clarification, potassium hexacyanoferrate (Carrez I solution), zinc sulfate (Carrez II solution) and NaOH (100 mM) were then added in a total volume of 1 mL, mixed and centrifuged at 10,000 r.p.m. (Hermle Z383-K, Wehingen, Germany). The intermediate phase was collected for lactose determination. Lactose was determined via the addition of β-galactosidase after shaking and incubating the sample extract at 25 °C for 10 min in the presence of a tamponed solution (pH 8.6), NAD+ and β-galactose dehydrogenase. The lactose concentration was measured via the increase in absorbance at 340 nm according to the commercial kit procedure (K-Lacgar 05/17, Megazyme, Scotland, UK)

2.3.2. Fat Quantifications of the Colostrum and Milk

Fat was quantified according to a modification of the Rose–Gottlieb method (905.02) [26]. Milk was placed in the presence of NAOH (2.5%), and fat was extracted via the addition of hexane after centrifugation (Hermle Z383-K, Wehingen, Germany). The superior phase was then collected and evaporated under a nitrogen stream and the residue was weighted for fat quantification.

2.3.3. Tocopherol and Retinol Quantifications of the Colostrum and Milk

The α-tocopherol and retinol concentrations in colostrum and milk samples were quantified as described by Rey et al. [27]. Tocopherols were extracted in duplicate samples via saponification in the presence of KOH (50%) and analyzed via reverse phase HPLC (HP 1200, equipped with a diode array detector) (Agilent Technologies, Waldbronn, Germany). The mobile phase was methanol:water (97:3 v/v) at a flow rate of 2 mL/min and the diode array detector was set at 292 nm. Separation was performed on a LiChrospher 100 RP-18 column (250-4 column size, 5 µm particle size) (Agilent Technologies GmbH, Waldbronn, Germany). Identification and quantification were carried out by means of a standard curve (R^2 = 0.999) built using the pure compound (Sigma, Alcobendas, Madrid, Spain).

2.3.4. Fatty Acid Compositions of the Colostrum and Milk

Lipids from colostrum and milk were determined via the one-step procedure proposed by Sukhija and Palmquist [28] with modifications. A lyophilized sample (200 mg) was placed in a test tube and toluene (1 mL) containing internal standard (10 mg C15:0/mL) (Sigma-Aldrich, Tres Cantos, Madrid, Spain) and 3 mL of freshly made 10% methanolic-acetyl chloride solution were added. The tubes were vortexed and heated in a shaking water bath at 70 °C for 2 h. Then, potassium carbonate (5%) and toluene were added, mixed and centrifuged at 600× g for 5 min. The superior layer containing the fatty acid methyl esters was collected for analysis. Fatty acids were identified and quantified via gas chromatography as described by Rey et al. [12] using a 6890 Hewlett Packard (Avondale, PA, USA) gas chromatograph equipped with an automatic injector maintained at 250 °C and a flame ionization detector. The fatty acid methyl esters were separated in a capillary column (HP-Innowax, 30 m × 0.32 mm id and 0.25 µm cross-linked polyethylene glycol) (Agilent Technologies GmbH, Waldbronn, Germany) using a temperature program of 170 to 245 °C. A split ratio of 1:50 was used. Identification of each fatty acid was carried out via the use of the mixtures of known standards (Sigma-Aldrich, Tres Cantos, Madrid, Spain). Fatty acids were expressed as a percentage of the total fatty acids.

Different indices were measured to estimate desaturase or elongase activities [29–31]:

The Δ-9 − desaturase index was calculated as the ratio of the monounsaturated fatty acid to the sum of the monounsaturated fatty acid plus the saturated fatty acid of the same number of carbons using the following equation [29,30]: Δ-9 − desaturase index = (C14:1 n−5 + C16:1 n−7 + C18:1n−9)/(C14:0 + C14:1 n−5+ C16:0 + C16:1 n−7 + C18:0 + C18:1n−9).

The Δ-5 and Δ-6−desaturase indices are meant for the evaluation of the enzymes that participate in the desaturation of C18:2 n−6 and C18:3 n−3 to their long-chain fatty acids and were calculated with the following equations [30,31]:

$$\Delta\text{-5-desaturase} = (\text{C20:4 n}-6)/(\text{C20:3 n}-6 + \text{C20:4 n}-6)$$

$$\Delta\text{-6-desaturase} = (\text{C18:3 n}-6 + \text{C18:4 n}-3)/(\text{C18:2 n}-6 + \text{C18:3 n}-3 + \text{C18:3 n}-6 + \text{C18:4 n}-3)$$

The elongase indices were calculated as the ratios of C18:0 to C16:0 and C20:0 to C18:0, whereas the thioesterase index was calculated as the ratio of C16:0 to C14:0 [29,31].

2.3.5. Oxidative Statuses of the Milk Samples, Sows and Piglets

The susceptibility of milk homogenates to iron-induced lipid oxidation was determined as described elsewhere [32] with few modifications. Briefly, samples were incubated at 37 °C in the presence of a Tris–malate buffer (pH 7.4) and ascorbic acid in a total volume of 5 mL. To start the lipid oxidation, 1 mM $FeSO_4$ was added to homogenates. At fixed time intervals (0, 30, 60 and 90 min), 1 mL aliquots were removed, mixed with TBA-TCA-HCl reagent in 1:2 proportion and after heating, extract containing thiobarbituric acid-reactive substances (TBARs) were measured spectrophotometrically (532 nm) (ScanGo, Thermofisher Scientific, Alcobendas, Spain). TBARs were expressed as nM malondialdehyde (MDA)/mg protein. The protein content was measured via the procedure of Bradford [33].

The oxidative status of plasma samples from sows (catalase enzyme activity and α-tocopherol) and piglets (MDA) was evaluated spectrophotometrically (Multiscan ScanGo, Thermo-Fisher Scientific, Alcobendas, Spain) as described elsewhere [10,11,16]. Analysis of catalase was carried out according to the kit's instructions (Arbor Assays, Ann Arbor, MI, USA). Sow's plasma (50 μL) was diluted with an assay buffer (1:10) and these dilutions (25 μL) were mixed with 25 μL of the hydrogen peroxide reagent (H_2O_2), 25 μL of the substrate solution and 25 μL of horseradish peroxidase (HRP), and incubated for 15 min at room temperature. The HRP reacted in the presence of H_2O_2 to convert the colorless substrate into a pink-colored product that was read spectrophotometrically at 560 nm. The catalase activity was expressed as U/mL.

2.4. Statistical Analysis

The experimental unit for analysis of all data was the individual sow. Data were analyzed following a completely randomized design using the general linear model (GLM) procedure contained in SAS (version 9.4; SAS Inst. Inc., Cary, NC, USA) that included the fixed effects of VE and HXT supplementation and their interaction in a factorial model. A comparative analysis between means was conducted using the Duncan test. Data are presented as the mean of each group and the root-mean-square error (RMSE), together with the significance levels (p-value) of the main effects and interaction. The relationship between the mean milk energy output and the C16:1/C16:0 or C18:1/C18: ratios, milk MDA concentration and milk fatty acid proportions, milk MDA and sow's plasma oxidative stability or piglet's oxidative status were quantified with regression equations using Statgraphics-19. Differences between means were considered statistically significant at $p < 0.05$ and p-values between 0.05 and 0.10 were considered a trend.

3. Results

3.1. General Compositions of the Colostrum and Milk

The composition of colostrum and milk as affected by VE or HXT supplementation is shown in Table 1. Neither the overall composition of milk nor colostrum (dry matter, lactose, protein, fat) was affected by the dietary antioxidant supplementation of sows. However, dietary VE supplementation tended ($p = 0.090$) to increase the lactose percentage of colostrum and decrease the protein percentage of milk on day 7 ($p = 0.090$) and day 20 ($p = 0.081$). Furthermore, the lactose percentage of milk on day 7 ($p = 0.056$) tended to increase by sow's HXT supplementation. Moreover, the sows that received VE at 100 mg/kg during late gestation and lactation increased α-tocopherol concentrations mainly in colostrum ($p = 0.006$), whereas accumulation of this compound was not as marked in day 7 milk ($p = 0.062$) and no changes were found on day 20. Dietary supplementation of HXT to sows from day 85 of pregnancy to day 28 of lactation also tended to increase ($p = 0.093$) the α-tocopherol concentration in colostrum and produced a marked increase in the day 7 milk retinol (vitamin A) concentration ($p = 0.010$).

No changes were observed by the combined supplementation of both antioxidants, except for the α-tocopherol concentration of colostrum that tended ($p = 0.058$) to be the greatest in the VE+HXT group than when VE was administered individually without HXT.

Table 1. General composition of colostrum and milk from sows given different levels of α-tocopherol (VE) or hydroxytyroxol (HXT) from day 85 of gestation.

	Control	VE	HXT	VE+HXT	RMSE [1]	*p* VE [2]	*p* HXT	*p* VE × HXT
Colostrum								
Dry matter, %	21.60	20.64	22.40	22.39	3.981	0.760	0.466	0.793
Lactose, %	3.64	5.04	3.91	4.32	1.437	0.090	0.728	0.399
Fat, %	5.55	5.39	5.98	6.97	2.504	0.938	0.498	0.810
Tocopherol, μg/mL	13.44 ^{b}	15.65 ^{ab}	12.90 ^{b}	25.62 ^{a}	5.972	0.006	0.093	0.058
Retinol, μg/mL	0.83	0.80	0.68	1.46	0.589	0.200	0.439	0.169
Milk day 7								
Dry matter, %	18.06	16.05	17.33	17.16	3.957	0.445	0.893	0.518
Lactose, %	5.36 ^{ab}	5.23 ^{b}	6.51 ^{ab}	6.48 ^{a}	1.696	0.896	0.056	0.942
Fat, %	7.37	6.40	7.49	9.08	2.359	0.721	0.113	0.144
Protein, %	5.59 ^{a}	4.89 ^{b}	5.00 ^{ab}	4.91 ^{b}	0.632	0.091	0.226	0.196
Tocopherol, μg/mL	3.49 ^{b}	3.95 ^{ab}	3.50^{b}	4.25 ^{a}	0.888	0.062	0.641	0.641
Retinol, μg/mL	0.25 ^{b}	0.33 ^{ab}	0.46 ^{ab}	0.52 ^{a}	0.202	0.346	0.010	0.911
Milk day 20								
Dry matter, %	17.37	16.91	17.94	17.10	1.879	0.418	0.637	0.811
Lactose, %	6.05	5.21	7.99	6.06	2.106	0.105	0.101	0.515
Fat, %	7.47	7.57	7.97	7.55	2.257	0.859	0.784	0.773
Protein, mg/mL	5.35 ^{ab}	5.11 ^{b}	5.87 ^{a}	5.30 ^{ab}	0.563	0.081	0.125	0.449
Tocopherol, μg/mL	2.90	3.32	2.44	2.84	0.737	0.167	0.117	0.966
Retinol, μg/mL	0.39	0.48	0.48	0.55	0.122	0.415	0.421	0.891

[1] RMSE—root-mean-square error (pooled SD) (n = 14 replicates for each main effect, n = 7 replicates for the interaction); [2] *p*—differences were statistically different when $p < 0.05$. ^{a,b} Different superscript letters signify a statistically significant difference.

The estimated mean milk energy output (EMe) increased in the VE-supplemented groups ($p = 0.0005$), whereas HXT supplementation did not modify this parameter (Figure 1).

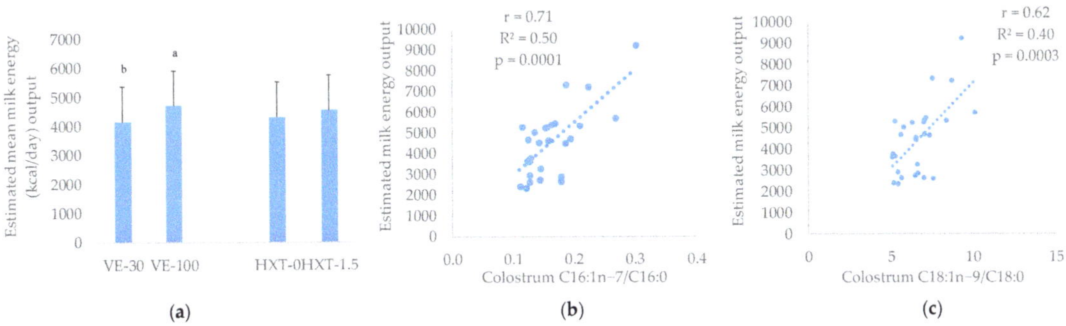

Figure 1. Estimated mean milk energy output (Kcal/day) (EMe) during lactation from sows given α-tocopherol (VE: 30 vs. 100 mg/kg) or hydroxytyroxol (HXT: 0 vs. 1.5 mg/kg) from day 85 of gestation (**a**); correlations between EMe and desaturase capacity of sows measured as C16:1n−7/C16:0 (**b**) and C18:1n−9/C18:0 (**c**) in colostrum. ^{a,b} Different superscript letters signify a statistically significant difference; *p*—differences were statistically different when $p < 0.05$.

3.2. Fatty Acid Composition of Colostrum and Milk

The fatty acid composition of colostrum and milk on days 7 and 20 of lactation are presented in Tables 2 and 3 and Figure 2.

The colostrum from sows supplemented with 100 mg/kg VE showed a greater proportion of C18:1 n−7 ($p = 0.046$) and tended to have a greater ($p = 0.056$) elongase C16 to C14 index than those groups that received 30 mg/kg VE (Table 2). The colostrum was also modified to a greater extent by the supplementation of 1.5 mg/kg HXT in the sow's diets. Therefore, the HXT-supplemented groups had greater proportions of C18:2 n−6 ($p = 0.002$), C18:3 n−3 ($p = 0.002$), sum of total polyunsaturated fatty acids (∑PUFAs) ($p = 0.002$),

∑n−6 PUFAs (p = 0.002) and ∑n−3 PUFAs (p = 0.008) and lesser proportions of C18:1 n−9 (p = 0.027) and C20:1 n−9 (p = 0.044) than those groups without HXT supplementation.

Table 2. Fatty acid composition (%) of colostrum from sows given different levels of α-tocopherol (VE) or hydroxytyroxol (HXT) from day 85 of gestation.

% Fatty Acids	Control	VE	HXT	VE+HXT	RMSE [12]	p VE [13]	p HXT	p VE × HXT
C14:0	1.25	1.15	1.27	1.22	0.139	0.180	0.418	0.678
C14:1	0.04	0.03	0.04	0.02	0.021	0.146	0.240	0.796
C16:0	22.01	22.27	22.12	23.37	1.449	0.328	0.470	0.601
C16:1n−9	1.26	1.20	1.30	1.21	0.169	0.446	0.395	0.847
C16:1n−7	3.70	3.89	3.48	3.11	0.873	0.916	0.275	0.636
C17:0	1.10	1.07	1.04	0.96	0.306	0.652	0.495	0.851
C17:1	0.46	0.47	0.49	0.50	0.068	0.982	0.433	0.807
C18:0	6.27	6.43	6.06	6.75	0.972	0.580	0.676	0.889
C18:1n−9	41.44	41.84	39.76	38.70	2.521	0.836	0.027	0.542
C18:1n−7	3.05	3.58	3.14	3.15	0.608	0.046	0.829	0.848
C18:2n−6	15.76 [ab]	14.58 [b]	17.52 [a]	17.53 [a]	1.643	0.225	0.002	0.529
C18:3n−6	0.23	0.22	0.21	0.17	0.059	0.659	0.481	0.995
C18:3n−3	0.94 [ab]	0.82 [b]	1.10 [a]	1.09 [a]	0.142	0.100	0.002	0.583
C18:4n−3	0.20	0.18	0.18	0.17	0.063	0.671	0.830	0.732
C20:0	0.09	0.08	0.07	0.05	0.041	0.540	0.494	0.790
C20:1n−9	0.34	0.39	0.29	0.25	0.092	0.466	0.044	0.511
C20:2	0.48	0.50	0.46	0.45	0.079	0.388	0.558	0.938
C20:3n−6	0.24	0.21	0.23	0.23	0.044	0.773	0.490	0.126
C20:4n−6	1.13	1.09	1.22	1.06	0.198	0.611	0.239	0.979
∑SAT [1]	30.72	30.99	30.57	32.36	2.276	0.479	0.819	0.691
∑MUFAs [2]	50.30	51.41	48.51	46.94	3.410	0.828	0.057	0.535
∑PUFAs [3]	18.99 [ab]	17.60 [b]	20.92 [a]	20.70 [a]	1.783	0.193	0.002	0.496
∑n−6 [4]	17.37 [ab]	16.10 [b]	19.18 [a]	18.99 [a]	1.649	0.198	0.002	0.502
∑n−3 [5]	1.14 [ab]	1.00 [b]	1.28 [a]	1.25 [a]	0.162	0.109	0.008	0.540
Δ-9-desaturase [6]	0.61	0.61	0.60	0.57	0.033	0.621	0.199	0.571
Δ-5-desaturase [7]	0.83	0.84	0.84	0.82	0.019	0.828	0.819	0.056
Δ-6-desaturase [8]	0.03	0.03	0.02	0.02	0.006	0.923	0.065	0.830
Thioesterase (16-14) [9]	0.95	0.95	0.95	0.95	0.006	0.067	0.759	0.943
Elongase (18-16) [10]	0.22	0.22	0.21	0.22	0.021	0.816	0.385	0.922
Elongase (20-18) [11]	0.01	0.01	0.01	0.01	0.007	0.529	0.757	0.765

[1] ∑SAT—sum of total saturated fatty acids; [2] ∑MUFAs—sum of total monounsaturated fatty acids; [3] ∑PUFAs—sum of total polyunsaturated fatty acids; [4] ∑n−6—sum of total n−6 fatty acids; [5] ∑n−3—sum of total n−3 fatty acids; [6] Δ-9−desaturase index = (C14:1 + C16:1 + C18:1)/C14:0 + C14:1 + C16:0 + C16:1 + C18:0 + C18:1); [7] Δ-5-desaturase = (C20:4n−6)/(C20:4n−6 + C20:3n−6); [8] Δ-6-desaturase = (C18:3n−6 + C18:4n−3)/(C18:2n−6 + C18:3n−3 + C18:3n−6 + C18:4n−3); [9] thioesterase index = C16:0/C14:0; [10] elongase (18/16) index = C18:0/C16:0; [11] elongase (20/18) index = C20:0/C18:0; [12] RMSE—root-mean-square error (pooled SD) (n = 14 replicates for each main effect, n = 7 replicates for the interaction); [13] p—differences were statistically different when $p < 0.05$; [a,b] Different superscript letters signify a statistically significant difference.

EMe was also positively and linearly correlated with the ratio C18:1n−9/C18:0 of colostrum (r = 0.62, p = 0.0003), C16:1n−7/C16:0 (r = 0.71, p = 0.0001) (Figure 1), and in general with the Δ-9 and Δ-6 desaturase capacities (r = 0.52, p = 0.005 and r = 0.48, p = 0.012, respectively) of the sow.

Changes in the fatty acid proportion of milk on day 7 of lactation were also observed by dietary antioxidants supplementation, mainly in the VE-supplemented groups (Table 3). Hence, the administration of 100 mg/kg VE to sows from day 85 of pregnancy (VE effect) decreased C18:2 n−6 (p = 0.015), C18:3 n−3 (p = 0.011), ∑PUFAs (p = 0.028), ∑n−6 PUFAs (p = 0.027) and ∑n−3 PUFAs (p = 0.018) but increased C17:1 (p = 0.008), Δ-6-desaturase (p = 0.016) and Δ-5 + Δ-6-desaturase (p = 0.032) indices when compared with the groups that received 30 mg/kg VE. Moreover, C20:3 n−6 (p = 0.070), ∑PUFAs (p = 0.093) and ∑n−3 PUFAs (p = 0.062) tended to be lower in milk on day 7 when the sows were given the HXT supplementation. The combined administration of both antioxidants also modified the proportion of fatty acids in milk on day 7 of lactation in a different way (interaction effect). Thus, when VE was supplemented with HXT in the sow's diets, this produced a significant decrease in C18:2 n−6 (p = 0.034), C18:3 n−6 (p = 0.018), C20:3 n−6 (p = 0.011), C20:4 n−6

($p = 0.030$), ∑PUFAs ($p = 0.013$), ∑n−6 ($p = 0.018$), ∑n−3 ($p = 0.014$) and Δ-6 desaturase ($p = 0.023$) of milk than when VE was administered individually. The desaturase indices were not statistically affected on day 7 ($p > 0.05$) when both antioxidants were combined at high doses in the diets.

Table 3. Fatty acid composition (%) of milk on day 7 of lactation from sows given different levels of α-tocopherol or hydroxytyroxol from day 85 of gestation.

% Fatty Acids	Control	VE	HXT	VE+HXT	RMSE [12]	*p* VE [13]	*p* HXT	*p* VE × HXT
C14:0	3.08	2.77	2.96	2.90	0.544	0.370	0.981	0.534
C14:1	0.19	0.15	0.19	0.17	0.058	0.168	0.777	0.763
C16:0	29.81	28.29	28.07	28.55	3.049	0.639	0.507	0.372
C16:1n−9	0.37	0.54	0.42	0.40	0.189	0.294	0.530	0.177
C16:1n−7	7.82	7.01	7.57	7.13	1.653	0.305	0.916	0.756
C17:0	0.71	0.90	0.78	0.72	0.195	0.358	0.417	0.076
C17:1	0.41 [b]	0.52 [ab]	0.46 [ab]	0.56 [a]	0.096	0.008	0.205	0.896
C18:0	4.62	4.92	4.45	5.35	0.877	0.071	0.694	0.351
C18:1n−9	34.83	35.86	35.78	38.02	3.545	0.212	0.235	0.636
C18:1n−7	1.85	2.44	2.42	2.39	0.515	0.149	0.170	0.111
C18:2n−6	13.55 [a]	13.35 [a]	13.97 [a]	11.19 [b]	1.594	0.015	0.143	0.034
C18:3n−6	0.10 [b]	0.15 [a]	0.13 [b]	0.12 [b]	0.028	0.094	0.700	0.018
C18:3n−3	0.97 [a]	0.94 [a]	0.99 [a]	0.78 [b]	0.120	0.011	0.135	0.054
C18:4n−3	0.18	0.22	0.19	0.18	0.042	0.390	0.343	0.149
C20:0	0.10	0.12	0.10	0.09	0.044	0.573	0.285	0.302
C20:1n−9	0.30	0.38	0.34	0.37	0.108	0.160	0.762	0.576
C20:2	0.35	0.44	0.40	0.40	0.091	0.239	0.916	0.192
C20:3n−6	0.14 [b]	0.22 [a]	0.16 [b]	0.13 [b]	0.052	0.285	0.071	0.012
C20:4n−6	0.60 [b]	0.80 [a]	0.64 [ab]	0.56 [b]	0.166	0.288	0.101	0.029
∑SAT [1]	38.31	37.01	36.35	37.60	3.300	0.980	0.570	0.293
∑MUFAs [2]	45.78	46.89	47.17	49.04	3.128	0.199	0.128	0.740
∑PUFAs [3]	15.91 [a]	16.11 [a]	16.48 [a]	13.36 [b]	1.731	0.028	0.093	0.013
∑n−6 [4]	14.40 [a]	14.52 [a]	14.90 [a]	12.00 [b]	1.650	0.027	0.102	0.018
∑n−3 [5]	1.15 [a]	1.16 [a]	1.18 [a]	0.96 [b]	0.116	0.018	0.062	0.014
Δ-9-desaturase [6]	0.53	0.55	0.55	0.55	0.035	0.625	0.326	0.683
Δ-5-desaturase [7]	0.80	0.79	0.80	0.81	0.048	0.945	0.675	0.498
Δ-6-desaturase [8]	0.02 [b]	0.03 [a]	0.02 [ab]	0.02 [ab]	0.005	0.016	0.883	0.427
Δ-5+Δ-6-desaturase	0.05 [b]	0.07 [a]	0.05 [b]	0.05 [ab]	0.013	0.032	0.279	0.135
Thioesterase (16-14) [9]	0.91	0.91	0.90	0.91	0.011	0.255	0.597	0.959
Elongase (18-16) [10]	0.14	0.15	0.14	0.16	0.031	0.096	0.579	0.662
Elongase (20-18) [11]	0.02	0.02	0.02	0.02	0.008	0.917	0.173	0.157

[1] ∑SAT—sum of total saturated fatty acids; [2] ∑MUFAs—sum of total monounsaturated fatty acids; [3] ∑PUFAs—sum of total polyunsaturated fatty acids; [4] ∑n−6—sum of total n−6 fatty acids; [5] ∑n−3—sum of total n−3 fatty acids; [6] Δ-9−desaturase index = (C14:1 + C16:1 + C18:1)/C14:0 + C14:1 + C16:0 + C16:1 + C18:0 + C18:1); [7] Δ-5-desaturase = (C20:4n−6)/(C20:4n−6 + C20:3n−6); [8] Δ-6-desaturase = (C18:3n−6 + C18:4n−3)/(C18:2n−6 + C18:3n−3 + C18:3n−6 + C18:4n−3); [9] thioesterase index = C16:0/C14:0; [10] elongase (18/16) index = C18:0/C16:0; [11] elongase (20/18) index = C20:0/C18:0; [12] RMSE—root-mean-square error (pooled SD) (n = 14 replicates for each main effect, n = 7 replicates for the interaction); [13] *p*—differences were statistically different when $p < 0.05$; [a,b] Different superscript letters signify a statistically significant difference.

On day 20 of lactation, changes in the fatty acid profile of the milk were not as marked as in the milk on day 7. The milk from sows supplemented with HXT tended ($p = 0.096$) to have a greater C18:2 n−6 proportion when compared with the milk from non-HXT-supplemented sows but no significant changes were observed by the individual administration of VE ($p > 0.05$).

The combined administration of both antioxidants modified the composition of day 20 milk in a different way than that of day 7 milk. Hence, sows given VE+HXT produced milk with a greater proportion of C18:2 n−6 ($p = 0.024$), C18:3 n−3 ($p = 0.047$), ∑PUFAs ($p = 0.029$) and ∑n−6 PUFAs ($p = 0.015$) than when VE was supplemented individually in the diet (Figure 2). Moreover, the Δ-5 and Δ-6 desaturase indices were lower in the VE+HXT group than when VE was supplemented without HXT ($p < 0.05$).

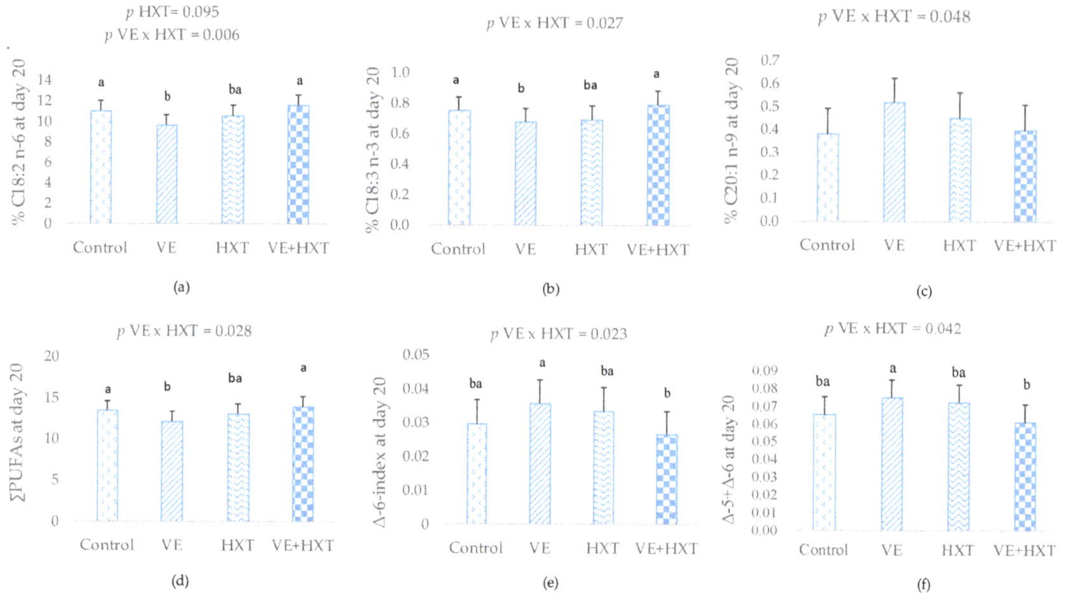

Figure 2. Interaction effect of dietary antioxidant α-tocopherol (VE: 30 vs. 100 mg/kg), hydroxytyroxol (HXT: 0 vs. 1.5 mg/kg) or their combination provided to sows from day 85 of gestation to weaning on the fatty acid composition (%) of milk on day 20 of lactation: (**a**) % C18:2 n−6; (**b**) % C18:3 n−3; (**c**) % C20:1 n−9; (**d**) ∑PUFAs—sum of total polyunsaturated fatty acids; (**e**,**f**) Δ-5 = (C20:4n−6)/(C20:4n−6 + C20:3n−6); Δ-6 = (C18:3n−6 + C18:4n−3)/(C18:2n−6 + C18:3n−3 + C18:3n−6 + C18:4n−3); [a,b] Different superscript letters signify a statistically significant difference; *p*—differences were statistically different when $p < 0.05$.

3.3. Oxidative Stability of Milk and Oxidative Status of Sows and Piglets

The iron-induced oxidation of milk on days 7 and 20 of lactation is presented in Table 4. Sows given a VE-supplemented diet at 100 mg/kg produced milk that was less susceptible to oxidation on days 7 and 20 of lactation. Hence, the milk collected on day 7 of lactation from the VE-supplemented groups had lower malondialdehyde (MDA) concentrations after 30 min of incubation in the presence of ferrous sulfate and total MDA ($p = 0.001$) than those groups that received 30 mg/kg of VE. However, HXT supplementation did not statistically modify the lipid stability of milk on day 7 of lactation ($p > 0.05$) but decreased the stability of milk collected on day 20 ($p = 0.028$). Thus, milk from these HXT-supplemented groups had greater MDA concentrations than groups without HXT supplementation.

Moderate linear and positive correlations (Table 5) were found between the total MDA concentration of milk on day 7 and their composition in unsaturated fatty acids, mainly C18:2 n−6 (r = 0.39, $p = 0.034$), C18:3 n−3 (r = 0.42, $p = 0.021$) and ∑PUFAs (r = 0.36, $p = 0.050$). In addition, high linear and positive correlations were also observed between the MDA of milk on day 20 of lactation and the C18:2 n−6 proportion (r = 0.65, $p = 0.001$) and ∑PUFAs (r = 0.59, $p = 0.002$).

The plasma catalase activity of the sows was not statistically affected ($p > 0.05$), whereas the plasma α-tocopherol concentration was increased by the dietary VE supplementation ($p = 0.0001$). Moreover, lipid oxidation of the milk on day 20 was negatively correlated with the sow's plasma catalase (r = −0.58, $p = 0.017$) and sow's plasma α-tocopherol concentration on day 20 (r = −0.67, $p = 0.006$) (Table 5). Finally, the MDA concentration of milk on days 7 and 20 correlated to a great extent with the oxidative status of piglets (MDA) after weaning (r = 0.72, $p = 0.003$ and r = 0.73, $p = 0.005$, respectively) (Table 5).

Table 4. Iron-induced lipid oxidation (nmols MDA/mg protein) of milk on days 7 and 20 of lactation from sows given α-tocopherol (VE: 30 vs. 100 mg/kg) or hydroxytyroxol (HXT: 0 vs. 1.5 mg/kg) from day 85 or gestation.

	Control	VE	HXT	VE+HXT	RMSE [1]	*p* VE [2]	*p* HXT	*p* VE × HXT
Milk day 7, nmols MDA/mg protein								
MDA 0 min	0.15 [a,b]	0.14 [b]	0.17 [a]	0.15 [ab]	0.024	0.134	0.075	0.263
MDA 30 min	0.14	0.10	0.14	0.12	0.063	0.183	0.713	0.664
MDA 90 min	0.61 [a]	0.28 [b]	0.42 [ab]	0.25 [b]	0.188	0.001	0.126	0.240
MDA 120 min	0.30 [a]	0.11 [b]	0.31	0.16 [ab]	0.164	0.010	0.592	0.770
Total ∑MDA [3]	1.20 [a]	0.56 [c]	1.05 [ab]	0.69 [bc]	0.368	0.001	0.932	0.298
Milk day 20, nmols MDA/mg protein								
MDA 0 min	0.15	0.12	0.14	0.14	0.037	0.317	0.760	0.253
MDA 30 min	0.11 [b]	0.10 [b]	0.24 [a]	0.13 [b]	0.091	0.097	0.039	0.182
MDA 90 min	0.22 [b]	0.15 [b]	0.48 [a]	0.23 [b]	0.128	0.005	0.003	0.079
MDA 120 min	0.24	0.10	0.23	0.21	0.167	0.259	0.456	0.394
Total ∑MDA	0.72 [ab]	0.48 [b]	1.09 [a]	0.72 [ab]	0.328	0.026	0.028	0.604

[1] RMSE—root-mean-square error (pooled SD) (n = 14 replicates for each main effect, n = 7 replicates for the interaction); [2] *p*—differences were statistically different when $p < 0.05$. [3] MDA—malondialdehyde concentration; [a,b,c] Different superscript letters signify a statistically significant difference.

Table 5. Linear equations between the total MDA concentration (nmols/mg protein) of milk at 7 or 20 days of lactation and milk fatty acid proportion, sow's plasma oxidative status (catalase and α-tocopherol concentrations) or piglet's oxidative status (MDA).

	Intercept ±s.d. [1]	Slope ±s.d.	Variable x	r	R^2	*p* (Linear) [2]
Milk on day 7						
Total MDA [3]	−0.424 ± 0.62	0.107 ± 0.05	% C18:2 n−6 in milk	0.39	0.16	0.034
Total MDA	−0.486 ± 0.59	1.593 ± 0.65	% C18:3 n−3 in milk	0.42	0.18	0.021
Total MDA	−0.391 ± 0.68	0.088 ± 0.04	∑PUFAs in milk [4]	0.36	0.13	0.050
Total MDA	0.525 ± 0.13	9.133 ± 2.44	MDA in piglets after weaning	0.72	0.52	0.003
Milk on day 20						
Total MDA	−1.298 ± 0.51	0.189 ± 0.05	% C18:2 n−6 in milk	0.65	0.43	0.001
Total MDA	−1.510 ± 0.66	0.173 ± 0.05	∑PUFAs in milk	0.59	0.35	0.002
Total MDA	2.683 ± 0.73	−0.097 ±0.04	Sow's plasma catalase on day 20	−0.58	0.34	0.017
Total MDA	1.063 ± 0.13	−0.183 ± 0.06	Sow's plasma α-tocopherol (μg/mL) on day 20	−0.67	0.45	0.006
Total MDA	0.344 ± 0.12	7.238 ± 2.06	MDA in piglets after weaning	0.73	0.53	0.005

[1] s.d.—standard deviation of mean; [2] *p*—differences were statistically different when $p < 0.05$; [3] MDA—malondialdehyde concentration; [4] ∑PUFAs—sum of total polyunsaturated fatty acids.

4. Discussion

Milk is the main source of nutrients for a piglet before weaning and can affect its gut health and growth rate [4]. Therefore, in the present study, the use of antioxidants in sows' diets (vitamin E and/or hydroxytyrosol) that could produce changes at the metabolic level [12,16] was evaluated in order to know whether their individual or combined dietary administration could modify the composition and lipid stability of the milk through lactation.

The composition of lactose, fat, and protein of colostrum and milk at different times of lactation were within the ranges described in the literature for Iberian sows [3] and pigs with improved genotypes [5,34]. As described by the latter authors, colostrum had a lower content of lactose and fat than milk, in agreement with the results of the present study. The evolution in protein composition could not be evaluated because the protein was not quantified in colostrum due to an insufficient sample. Dietary VE or HXT given to sows from day 85 of gestation to day 28 of lactation did not hardly modify the general composition of colostrum or milk on day 7 or 20. Rosales et al. [35] found that VE supplementation to ewes during late gestation and early lactation increased the lactose concentration of colostrum and fat concentration of milk, although the protein and lactose in milk did not differ between treatments. Moreover, Wang et al. [6] observed that milk from sows supplemented with 250 IU VE/kg feed during the last week of gestation and lactation had a greater content of fat. In contrast, the fat content of milk was not affected in the present study, which could have been due to the VE supplementation dose that was half of that used by Wang et al. [6], or in part to a smaller litter size in the case of the Iberian sow than in the white genotype used in Wang study (11 piglets born alive). Hence, after supplementing 90 IU/kg of VE to five-parity Landrace × Yorkshire sows, Chen et al. [36] did not find any effect on the 11-day milk composition. On the other hand, a concomitant fat increase was observed when milk protein decreased [34]; however, despite the decreasing trends in protein observed in the present study, no significant increases in fat content were observed.

Concerning the dietary HXT supplementation, either alone or in combination with VE, there is a lack of literature reporting the effects on general colostrum and milk composition. According to our findings, the general composition was not modified by the HXT in any case.

The concentrations of α-tocopherol and retinol in colostrum or milk were within the expected values according to dietary intervention [10,11,36–38] and decreased with lactation length, in agreement with previous research in the literature [11]. Dietary treatments modified the vitamin concentration of milk to different extents. Hence, dietary VE at 100 mg/kg feed increased the α-tocopherol in colostrum but did not modify the concentration in milk. In the present study, differences in supplementation dose (30 mg/kg vs. 100 mg/kg) and consequently VE accumulation were not as marked as in other investigations in the literature [39], in which the VE content was 2–3-fold above in milk samples from supplemented sows (200 mg/kg and 400 mg/kg VE) when compared with those that were non-supplemented (36 mg/kg VE). According to Mahan et al. [40] and Pinelly-Saavedra et al. [39], colostrum and milk α-tocopherol increased as dietary VE increased. In contrast, dietary HXT supplementation to sows did not modify the concentration of α-tocopherol of the colostrum or milk but increased retinol concentration on day 7 of lactation. To our knowledge, there were no previous studies on the possible effects of HXT on the vitamin composition of sow's milk. In lactating ruminants, Terramoccia et al. [15] reported increased tocopherols and retinol concentrations in milk from buffaloes supplemented with dried stoned olive pomace. In addition, it was reported in rats that olive polyphenols may modify the metabolism of retinol and other lipid components and consequently have positive effects on some diseases [41].

The supplementation of both antioxidants did not modify the composition of the colostrum and milk. Bars-Cortina et al. [42] reported that a diet supplemented with some phenolic compounds of olive and thyme increased α-tocopherol in the tissues of rats. The combined administration of both antioxidants could produce a protective effect from their use or provide other antioxidant compounds [42]. However, other studies on pigs did not find any effect of the combination of VE and an olive-derived extract [16]. Discrepancies found in the literature can be explained by differences in the antioxidant source and/or supplementation dose.

The fatty acid proportions of colostrum and milk were within the values reported by other authors in sows [12,38,43]. Supplementation with VE in a sow's diet only produced limited effects on the fatty acid composition of colostrum, as was observed by other authors in the literature [12]. Moreover, it should be noted that sows supplemented with VE during pregnancy and lactation allocate high proportions of C18:1n−7 to colostrum. This fatty acid and those of its n−7 group, such as C16:1 n−7, are easily β-oxidized to obtain energy [14], which may be an especially interesting aspect during the first hours of the piglet's life. It is important to highlight that VE supplementation to sow's diets produced more marked effects on milk at day 7 of lactation with a decrease in the C18-polyunsaturated fatty acids and ∑PUFAs, as well as a marked increase in desaturase indices of the n−6 and n−3 series that could, in part, explain the observed decrease in polyunsaturated C18 proportions.

Then, when VE was given to sows independently, milk had the highest proportion of long PUFAs (C20:3 n−6 and C20:4 n−6) on day 7. A direct relationship between VE levels and the activity of the desaturase and elongase enzymes was described in the literature [44,45]. Thus, the lower the VE levels, the lower the activity of these enzymes. This fact was explained by the possible antioxidant effect of VE on desaturase and elongase enzymes [12,44,46], although a possible protective antioxidant effect on long-chain fatty acids was also postulated [46]. The changes in the desaturation index were maintained throughout lactation and were the highest when VE was given without HXT, although long PUFAs were not statistically modified at day 20 of lactation. These changes in the desaturation and, consequently, the different fatty acid classes imply that a piglet can better use them for metabolic purposes or other functions. Therefore, it was described that fatty acid is better metabolized when it has a higher number of unsaturations [47]. This is a relevant aspect in the particular case of the Iberian pig in which a reduced energy efficiency for growth and higher energy cost of body fat deposition during suckling was found [3,43]. In addition, a different transfer of long-chain fatty acids through milk could affect the composition of the tissue membrane or intestinal epithelium of the piglet [8] and affect the membrane fluidity or cell signaling [48] and, consequently, the animal's health.

HXT supplementation of the sows' diets also produced interesting changes in the fatty acid profile of colostrum and, to a lesser extent, in milk. The colostrum from HXT-supplemented groups had the highest content of ∑n−6 and ∑n−3 PUFAs that resulted in lower proportions of ∑MUFAs, mainly C18:1 n−9 and C20:1 n−9, and lower desaturase indices of the n−6 series. According to the results of the present study, the sows supplemented with HXT allocated a higher proportion of unsaturated fatty acids on the day of farrowing than the rest of the groups; however, after farrowing this potent ability to derive PUFAs to milk would be reduced. It was reported that olive-derived extracts increase glucose absorption [17] and help to produce faster fatty acid mobilization for different purposes [16]; in the specific case of a lactating sow, this could be used to address the energy needs for lactation and colostrum formation. However, it is of interest to observe the fact that the sows supplemented with HXT allocated many PUFAs to the colostrum, which may have resulted in a decrease in the desaturation indices and possibly a lower capacity of the sow to increase long PUFAs. Furthermore, a high proportion of C18-PUFAs in colostrum might result in a reduced capacity of the piglet to desaturate their long-chain derivatives, which, as reported before, are better utilized for energy purposes [43,47]; however, more research is needed. The fact of having less Δ-6 and Δ-5 desaturase activity should be taken into consideration since alterations of the Δ-5/Δ-6 activity have been associated with several diseases, from inflammation to tumorigenesis [49], and the modulation of Δ-5/Δ-6 activity could be considered as a possible therapeutic application [49].

In addition, although no significant change in the general milk composition was observed, it is interesting to point out that the transfer of nutrients to milk estimated as EMe output, that is, as a function of the litter size and litter weight, was not affected in the HXT groups but increased with VE supplementation. The higher desaturase capacity found in sows supplemented with VE could have preserved the utilization of PUFAs by the mother. Hence, Lauridsen and Danielsen [38] reported that a higher metabolic use and mobilization of C18:2 during lactation was related to lower fat and energy in milk, whereas milk from sows receiving MUFAs or SAT-enriched diets and lower PUFAs contained higher fat and energy. This fact was also observed in the present study in which analysis of data by multiple regression procedures indicated that 28% of the variation in EMe was explained by the proportion of C18:0 in colostrum and, consequently, with the desaturase capacity of the sow at the initial stage of lactation. Similarly, this would explain the positive correlation found between the desaturase capacity (C18:1/C18:0 and C16:1/C16:0) and the EMe output observed in the present research.

The supplementation of these different antioxidants in the diets of sows causes the mother to use different strategies to be able to promote the development of the piglet by diverting different fatty acids to the lactating gland. This could imply that the use of one or the other may affect gut health or favor the growth of the piglet at different times during lactation, depending on the specific fatty acid profile or milk composition. However, the use of the combination of both antioxidants during the first week of lactation did not seem to have beneficial effects on the fatty acid profile of the colostrum or milk compared with the independent use of each of them. In particular, the results of the supplementation with VE+HXT antioxidants implied a decrease in the derivation of unsaturated fatty acids from the n−6, n−3 or n−9 series toward milk on day 7, as well as a lower activity on the desaturation or elongation capacity of the sow. However, as lactation progressed and close to weaning, the combination of both antioxidants in the sows' diet increased the milk PUFA content, especially in the C18 n−6 series. It would be expected that PUFA transfer decreased with the advancing of lactation [50] since C18:2 n−6 followed by the n−7 fatty acid series could be one of the most easily used for energy utilization by the mother [14]. This increase in PUFA enrichment in milk with the use of both antioxidants in the sow's diet could, in part, be related to the decrease in the ability of the sow to desaturate fatty acids and obtain their long-chain derivatives close to the weaning time, which could also result in lower efficiency of milk energy for piglet growth [43]. However, more research is needed to clarify the possible effects on piglet growth or gut health.

The lipid stability of milk was measured in order to evaluate the oxidative capacity of the tested antioxidants and, consequently, the possible transfer to piglets of derivatives of the oxidation or other substances that may contribute to the control of oxidative stress. The groups supplemented with vitamin E had the lowest production of MDA in milk when compared with the other groups, whereas HXT supplementation resulted in an increase in oxidative-derived products. The antioxidant effects of vitamin E as a radical scavenger [7] and its capacity to protect against lipid oxidation were widely documented [7,10,11,32]. However, the higher range of oxidation values in milk from HXT groups could have been due to the high proportion of PUFAs, mainly on day 20 of lactation. Hence, in the present study, the total MDA production of milk was mainly explained by their PUFA content, as confirmed by the significant linear and positive correlation found between these components. It was reported that the more unsaturated the fatty acid chain, the higher the degree of oxidation [32]. In addition, it is of interest to point out that milk stability was also correlated with the oxidative status of the sow and their plasma tocopherol concentration. Data are only presented for day 20 because the plasma oxidative status of sows was not measured on day 7 of lactation, but similar results would be expected for the whole period. This is because, in the present research, a high and positive correlation between the milk MDA concentration and the piglet's oxidative status was also detected at different lactation stages. The fact that the piglet's plasma MDA concentration (as a derivative of oxidation and measurement of oxidative stress) was positive and highly correlated with the total MDA concentration of milk could indicate the high importance of milk as a main vehicle for different components, not only the antioxidant substances from the mother but also those resulting from the different metabolic state of the sow that determines a different proportion of fatty acids during the process of milk synthesis in the sow's mammary gland. This is of relevance for those nutrients with limited placental transfer, such as vitamin E [9], which are very important for not only maintaining the oxidative status but also those that provide energy during the first weeks of life.

5. Conclusions

In conclusion, the supplementation of sows with VE or HXT slightly modified the general composition of milk in different ways. The changes were more marked in the fatty acid profile. VE produced an increase in the n−7 of the colostrum and long PUFAs and Δ-5 and Δ-6 indices of the milk during lactation, as well as the lower production of TBARs products. Nevertheless, HXT supplementation markedly increased the C18-PUFAs of the colostrum and moderately so as lactation progressed with the decrease in the sow's ability to desaturate. Maternal VE supplementation produced a more beneficial milk composition to improve the oxidative status of piglets, which could promote gut health and the piglet's growth during the first weeks of life; however, further studies are needed to clarify this.

Author Contributions: Conceptualization, C.Ó., A.G-.B., C.L.-B. and A.I.R.; methodology, G.G., H.D.L., Y.N., M.M., J.M.G.-C., A.H.M., R.E., C.Ó., A.G.-B., C.L.-B. and A.I.R.; software, A.I.R.; investigation, H.D.L., G.G., Y.N., M.M., J.M.G.-C., C.Ó., A.G.-B., C.L.-B. and A.I.R.; resources, G.G., H.D.L., Y.N., M.M., J.M.G.-C., C.Ó., A.G.-B., C.L.-B. and A.I.R.; data curation, A.I.R.; writing—original draft preparation, A.I.R. and H.D.L.; writing—review and editing, H.D.L., G.G., Y.N., M.M., J.M.G.-C., A.H.M., R.E., C.Ó., A.G.-B., C.L.-B. and A.I.R.; project administration, C.Ó., A.G.-B., C.L.-B. and A.I.R.; funding acquisition, C.Ó., A.G.-B., C.L.-B. and A.I.R. All authors read and agreed to the published version of the manuscript.

Funding: This research was funded by the Spanish Agencia Estatal de Investigación (PID2019-108695RB-C32/AEI/10.13039/501100011033) co-funded by Fondo Europeo de Desarrollo Regional (FEDER). H.D.L. and G.G. received grants from the Spanish Government (FPI National Program grant: PRE2020-092891) and IRIAF-Sánchez Romero Carvajal Jabugo, S.A., agreement, respectively.

Institutional Review Board Statement: The animal study protocol was approved by the Ethics Committee for Animal Research of INIA (protocol code ORCEEA 2019-10).

Informed Consent Statement: Not applicable.

Data Availability Statement: The data are contained within this article.

Acknowledgments: Authors thank the personnel at "Deheson del Encinar" and "Sánchez-Romero Carvajal" for their technical support and assistance.

Conflicts of Interest: The authors declare no conflict of interest. The funders had no role in the design of the study; in the collection, analyses, or interpretation of data; in the writing of the manuscript; or in the decision to publish the results.

Appendix A

Table A1. Major nutrients and fatty acid compositions of the experimental diets.

Major Nutrients [1]	Control	VE	HXT [2]	HXT+VE
Dry matter, %	90.8	89.8	91.1	91.2
Crude protein, %	13.1	14.0	15.0	14.5
Fat, %	4.3	3.6	3.8	3.7
Ash, %	6.5	6.4	6.9	6.0
Fiber, %	4.3	4.4	4.3	4.4
Starch,%	49.0	46.1	41.1	43.0
Energy (kcal GE/kg)	4339.6	4251.7	4264.3	4313.3
Vitamin E, mg/kg	70.5	103.6	95.4	114.1
Fatty Acid Composition				
C14:0	0.60	0.66	0.56	0.60
C16:0	19.80	21.48	19.10	20.11
C16:1n−9	0.12	0.15	0.12	0.12
C16:1n−7	0.79	0.87	0.73	0.78
C18:0	4.77	4.80	4.45	4.54
C18:1n−9	28.97	25.26	30.33	26.94
C18:1n−7	1.92	1.49	1.57	1.60
C18:2n−6	38.97	40.81	38.97	40.89
C18:3n−3	3.04	3.44	3.15	3.36
C20:0	0.30	0.26	0.30	0.31
C20:1n−9	0.52	0.58	0.55	0.58
∑SAT	25.47	27.20	24.41	25.55
∑MUFAs	32.33	28.35	33.29	30.02
∑PUFAs	42.01	44.25	42.11	44.25

[1] Ingredients (%)—corn: 16.03; wheat: 10; barley: 50; soya meal 47: 6.73; wheat bran: 6.54; pork lard: 1; beetroot pulp: 5; calcium carbonate: 1.98; bicalcium phosphate: 0.54; salt: 0.5; L-lysine: 0.50; methionine: 0.04; threonine: 0.18; choline chloride: 0.03; premix: 0.3 (per kg/diet—vitamin A: 12,000 IU; vitamin D_3: 1400 IU; vitamin B1: 1.1 mg; vitamin B_2: 6 mg; vitamin B_{12}: 0.08 mg; vitamin B_6: 12 mg; nicotinic acid: 21 mg; biotin: 0.12 mg; pantothenic acid: 12 mg; vitamin K_3: 1.1 mg; choline chloride: 225 mg; Fe (ferrous carbonate): 60 mg; Cu (pentahydrate sulfate): 14.2 mg; Zn (oxide): 100 mg; Mn (monohydrate sulfate): 30 mg; I (potasium iodide): 0.8 mg; Se (sodium selenite): 0.3 mg; [2] *Olea europea* extract containing hydroxytyrosol (%): 0.40.

References

1. Noblet, J.; Le Dividich, J. Energy metabolism of the newborn pig during the first 24 hrs of life. *Biol. Neonate* **1981**, *40*, 175–182. [CrossRef] [PubMed]
2. Roy, D.; Ye, A.; Moughan, P.J.; Singh, H. Composition, structure, and digestive dynamics of milk from different species—A Review. *Front. Nutr.* **2020**, *7*, 577759. [CrossRef] [PubMed]
3. Aguinaga, M.A.; Gomez-Carballar, F.; Nieto, R.; Aguilera, J.F. Production and composition of Iberian sow's milk and use of milk nutrients by the suckling Iberian piglet. *Animal* **2011**, *5*, 1390–1397. [CrossRef] [PubMed]
4. Boudry, G.; Charton, E.; Le Huerou-Luron, I.; Ferret-Bernard, S.; Le Gall, S.; Even, S.; Blat, S. The relationship between breast milk components and the infant gut microbiota. *Front. Nutr.* **2021**, *8*, 629740. [CrossRef]
5. Hurley, W.L. Composition of sow colostrum and milk. In *The Gestating and Lactating Sow*; Farmer, C., Ed.; Wageningen Academic Publishers: Wageningen, The Netherlands, 2015. [CrossRef]
6. Wang, L.; Xiaodong, X.; Ge, S.; Baoming, S.; Anshan, S. High concentration of vitamin E supplementation in sow diet during the last week of gestation and lactation affects the immunological variables and antioxidative parameters in piglets. *J. Dairy Res.* **2017**, *84*, 8–13. [CrossRef] [PubMed]
7. Brigelius-Flohé, R.; Traber, M.G. Vitamin E: Function and metabolism. *FASEB J.* **1999**, *13*, 1145–1155. [CrossRef] [PubMed]
8. Lauridsen, C. Effects of dietary fatty acids on gut health and function of pigs pre- and post-weaning. *J. Anim. Sci.* **2020**, *98*, skaa086. [CrossRef]
9. Lauridsen, C.; Engel, H.; Jensen, S.K.; Craig, A.M.; Traber, M.G. Lactating sows and suckling piglets preferentially incorporate RRR- over all-rac-α-tocopherol into milk, plasma and tissues. *J. Nutr.* **2002**, *132*, 1258–1264. [CrossRef]
10. Amazan, D.; Rey, A.I.; Fernandez, E.; López-Bote, C.J. Natural vitamin E (d-α-tocopherol) supplementation in drinking water prevents oxidative stress in weaned piglets. *Livest. Sci.* **2002**, *145*, 55–62. [CrossRef]
11. Amazan, D.; Cordero, G.; López-Bote, C.J.; Lauridsen, C.; Rey, A.I. Effects of oral micellized natural vitamin E (d-α-tocopherol) vs. synthetic vitamin E (dl-α-tocopherol) in feed on α-tocopherol levels, stereoisomer distribution, oxidative stress and the immune response in piglets. *Animal* **2014**, *8*, 410–419. [CrossRef]
12. Rey, A.I.; Amazan, D.; Cordero, G.; Olivares, A.; López-Bote, C.J. Lower Oral Doses of Micellized α-Tocopherol Compared to α-Tocopheryl Acetate in Feed Modify Fatty Acid Profiles and Improve Oxidative Status in Pigs. *Int. J. Vitam. Nutr. Res.* **2014**, *84*, 229–243. [CrossRef] [PubMed]
13. Balasubramanian, K.A.; Nalini, S.; Cheesman, K.H.; Slater, T.F. Non esterified fatty acids inhibit iron dependent lipid peroxidation. *Biochim. Biophys. Acta* **1989**, *1003*, 232–237. [CrossRef] [PubMed]
14. Power, G.W.; Cake, M.H.; Newsholme, E.A. Influence of diet on the kinetic behavior of hepatic carnitine palmitoyltransferase I toward different acyl CoA esters. *Lipids* **1997**, *32*, 31–37. [CrossRef] [PubMed]
15. Terramoccia, S.; Bartocci, S.; Taticchi, A.; Di Giovanni, S.; Pauselli, M.; Mourvaki, E.; Urbani, S.; Servili, M. Use of dried stoned olive pomace in the feeding of lactating buffaloes: Effect on the quantity and quality of the milk produced. Asian Australas. *J. Anim. Sci.* **2013**, *26*, 971–980. [CrossRef]
16. Rey, A.I.; de-Cara, A.; Calvo, L.; Puig, P.; Hechavarría, T. Changes in Plasma Fatty Acids, Free Amino Acids, Antioxidant Defense, and Physiological Stress by Oleuropein Supplementation in Pigs Prior to Slaughter. *Antioxidants* **2020**, *9*, 56–74. [CrossRef]
17. Hadrich, F.; Garcia, M.; Maalej, A.; Modes, M.; Isoda, H.; Feve, B.; Sayadi, S. Oleuropein activated AMPK and induced insulin sensitivity in C2C12 muscle cells. *Life Sci.* **2016**, *151*, 167–173. [CrossRef] [PubMed]
18. Theil, P.K.; Hansen, A.V.; Sørensen, M. Effects of gestation and transition diets on colostrum intake and piglet survival. In *Book of Abstracts of the 61st Annual Meeting of the European Association for Animal Production, Heraklion, Greece, 23–27 August 2010*; Wageningen Academic Publishers: Wageningen, The Netherlands, 2010; p. 14.
19. Kim, S.W. Sow milk. In *Milk and Dairy Products in Human Nutrition: Production, Composition and Health*; Park, Y.W., Haenlein, F.W., Eds.; John Wiley & Sons Ltd.: New York, NY, USA, 2013. [CrossRef]
20. Lugrin, J.; Rosenblatt-Velin, N.; Parapanov, R.; Liaudet, L. The role of oxidative stress during inflammatory processes. *Biol. Chem.* **2014**, *395*, 203–230. [CrossRef]
21. BOE. RD 53/2013, de 21 de octubre por la que se establecen las normas básicas aplicables para la protección de los animales utilizados en experimentación y otros fines científicos, incluyendo la docencia, Spain. *Boletín Of. Estado* **2013**, *252*, 34367–34391.
22. EC. Council Regulation (EC) No 2010/63/CE of 22 September 2010 on the protection of animals used for scientific purposes. *Off. J. Eur. Union.* **2010**, *276*, 33–79.
23. *Nutrient Requirement of Swine*, 11th ed.; National Research Council: Ottawa, ON, Canada; National Academy Press: Washington, DC, USA, 2012.
24. Hansen, A.V.; Strathe, A.B.; Kebreab, E.; France, J.; Theil, P.K. Predicting milk yield and composition in lactating sows: A Bayesian approach. *J. Anim. Sci.* **2012**, *90*, 2285–2298. [CrossRef]
25. Beutler, H.O. Lactose and D-Galactose. In *Methods of Enzymatic Analysis*, 3rd ed.; Bergmeyer, H.U., Ed.; VCH Publishers, Ltd.: Cambridge, UK, 1988; Volume VI, pp. 104–112.
26. Fat in milk. In *Methods of Test for Dairy Industry*, 17th ed.; Official method 905.02; A.O.A.C.: Gaithersburg, MD, USA, 2000.
27. Rey, A.I.; Daza, A.; López-Carrasco, C.; López-Bote, C.J. Quantitative study of the alpha- and gamma-tocopherols accumulation in muscle and backfat from Iberian pigs kept free-range as affected by time of free-range feeding or weight gain. *Anim. Sci.* **2006**, *82*, 901–908. [CrossRef]

28. Sukhija, P.S.; Palmquist, D.L. Rapid method for determination of total fatty-acid content and composition of feedstuffs and feces. *J. Agric. Food Chem.* **1988**, *36*, 1202–1206. [CrossRef]
29. Zhang, S.; Knight, T.J.; Stalder, K.J.; Goodwing, R.N.; Lonergan, S.M.; Beitz, D.C. Effects of breed, sex, and halothane genotype on fatty acid composition of pork longissimus muscle. *J. Anim. Sci.* **2007**, *85*, 583–591. [CrossRef] [PubMed]
30. Korniluk, K.; Gabryszuk, M.; Kowalczyk, I.; Czauderna, M. Effect of diet supplementation with selenium, zinc and α-tocopherol on fatty acid composition in the liver and loin muscle of lambs. *Anim. Sci. Pap. Rep.* **2008**, *26*, 59–70.
31. Dal Bosco, A.; Mugnai, C.; Ruggeri, S.; Mattioli, S.; Castellini, C. Fatty acid composition of meat and estimated indices of lipid metabolism in different poultry genotypes reared under organic system. *Poult. Sci.* **2012**, *91*, 2039–2045. [CrossRef]
32. Rey, A.I.; López-Bote, C.J.; Kerry, J.P.; Lynch, P.B.; Buckley, D.J.; Morrissey, P.A. Modification of lipid composition and oxidation in porcine muscle and muscle microsomes as affected by dietary supplementation of n−3 with either n−9 or n−6 fatty acids and α-tocopheryl acetate. *Anim. Feed Sci. Technol.* **2004**, *113*, 223–238. [CrossRef]
33. Bradford, M. A rapid and sensitive method for the quantification of microgram quantities of protein utilizing the principle of protein-dye binding. *Anal. Biochem.* **1976**, *72*, 248–254. [CrossRef]
34. Klobasa, F.; Werhahn, E.; Butler, J.E. Composition of sow milk during lactation. *J. Anim. Sci.* **1987**, *64*, 1458–1466. [CrossRef]
35. Rosales, C.A.; Meza-Herrera, C.A.; Morón, F.J.; Flores, M.J.; Gámez, H.G.; Cuevas, V.; Liu, S.M. Effects of vitamin E supply during late gestation and early lactation upon colostrum composition, milk production and quality in nutritional restricted ewes. *Small Rum. Res.* **2015**, *133*, 77–81. [CrossRef]
36. Chen, J.; Han, J.H.; Guan, W.T.; Chen, F.; Wang, C.X.; Zhang, Y.Z.; Lv, Y.T.; Lin, G. Selenium and vitamin E in sow diets: I. Effect on antioxidant status and reproductive performance in multiparous sows. *Anim. Feed Sci. Technol.* **2016**, *221*, 111–123. [CrossRef]
37. Mahan, D.C.; Kim, Y.Y.; Stuart, R.I. Effect of vitamin E sources (RRR- or all-rac-α-tocopheryl acetate) and levels on sow reproductive performance, serum, tissue, and milk α-tocopherol contents over a fiveparity period, and the effects on the progeny. *J. Anim. Sci.* **2000**, *78*, 110–119. [CrossRef] [PubMed]
38. Lauridsen, C.; Danielsen, V. Lactational dietary fat levels and sources influence milk composition and performance of sows and their progeny. *Livest. Prod. Sci.* **2004**, *91*, 95–105. [CrossRef]
39. Pinelli-Saavedraa, A.; Scaife, J.R. Pre- and postnatal transfer of vitamins E and C to piglets in sows supplemented with vitamin E and vitamin C. *Livest. Prod. Sci.* **2005**, *97*, 231–240. [CrossRef]
40. Mahan, D.C. Effects of dietary vitamin E on sow reproductive performance over a five-parity period. *J. Anim. Sci.* **1994**, *72*, 2870–2879. [CrossRef] [PubMed]
41. Lemonakis, N.; Poudyal, H.; Halabalaki, M.; Brown, L.; Tsarbopoulos, A.; Skaltsounis, A.; Gikas, E. The LC–MS-based metabolomics of hydroxytyrosol administration in rats reveals amelioration of the metabolic syndrome. *J. Chromatogr. B* **2017**, *1041*, 45–59. [CrossRef] [PubMed]
42. Bars-Cortina, D.; Lopez de las Hazas, M.C.; Benavent-Vallés, A.; Motilva, M.J. Impact of dietary supplementation with olive and thyme phenols on alpha-tocopherol concentration in the muscle and liver of adult Wistar rats. *Food Funct.* **2018**, *9*, 1433–1443. [CrossRef] [PubMed]
43. Aguinaga, M.A.; Haro, A.; Lara, L.; Gómez-Carballar, F.; Nieto, R.; Aguilera, J.F. Utilization of milk fatty acids by the suck ling Iberian piglets. *Animal* **2016**, *10*, 1786–1795. [CrossRef]
44. Okayasu, T.; Kameda, K.; Ono, T.; Imai, Y. Effect of dietary vitamin B2 and vitamin E on the D-9-desaturase and catalase activities in rat liver microsomes. *Biochim. Biophys. Acta* **1977**, *489*, 389–402.
45. Fuhrmann, H.; Sallman, H.P. Phospholipid fatty acid of brain and liver are modified by -tocopherol and dietary fat in growing chicks. *Br. J. Nutr.* **1996**, *76*, 109–122. [CrossRef]
46. Lebold, K.M.; Jump, D.B.; Miller, G.W.; Wright, C.L.; Labut, E.M.; Barton, C.L.; Tanguay, R.L.; Traber, M.G. Vitamin E Deficiency Decreases Long-Chain PUFA in Zebrafish (Danio rerio). *J. Nutr.* **2011**, *141*, 2113–2118. [CrossRef]
47. Raclot, T.; Groscolas, R. Differential mobilization of white adipose tissue fatty acids according to chain length, unsaturation, and positional isomerism. *J. Lipid Res.* **1993**, *34*, 1515–1526. [CrossRef] [PubMed]
48. Nakamura, M.T.; Nara, T.Y. Structure, function, and dietary regulation of Δ6, Δ5, and Δ9 desaturases. *Ann. Rev. Nutr.* **2004**, *24*, 345–376. [CrossRef] [PubMed]
49. Tosi, F.; Sartori, F.; Guarini, P.; Olivieri, O.; Martinelli, N. Delta-5 and Delta-6 desaturases: Crucial enzymes in polyunsaturated fatty acid-related pathways with pleiotropic influences in health and disease. *Adv. Exp. Med. Biol.* **2014**, *824*, 61–81. [PubMed]
50. Hu, P.; Yang, H.; Bo, L.; Zhao, D.; Wang, J.; Zhu, W. Dynamic changes of fatty acids and minerals in sow milk during lactation. *J. Anim. Physiol. Anim. Nutr.* **2019**, *103*, 603–611. [CrossRef] [PubMed]

antioxidants

MDPI

Article

Intestinal Damages by F18+ *Escherichia coli* and Its Amelioration with an Antibacterial Bacitracin Fed to Nursery Pigs

Marcos Elias Duarte [1], Chad H. Stahl [2] and Sung Woo Kim [1,*]

[1] Department of Animal Science, North Carolina State University, Raleigh, NC 27695, USA
[2] Department of Animal and Avian Sciences, University of Maryland, College Park, MD 20742, USA
* Correspondence: sungwoo_kim@ncsu.edu

Abstract: This study investigated intestinal oxidative damage caused by F18+ *Escherichia coli* and its amelioration with antibacterial bacitracin fed to nursery pigs. Thirty-six weaned pigs (6.31 ± 0.08 kg BW) were allotted in a randomized complete block design. Treatments were: NC, not challenged/not treated; PC, challenged (F18+ *E. coli* at 5.2 × 10^9 CFU)/not treated; AGP challenged (F18+ *E. coli* at 5.2 × 10^9 CFU)/treated with bacitracin (30 g/t). Overall, PC reduced ($p < 0.05$) average daily gain (ADG), gain to feed ratio (G:F), villus height, and villus height to crypt depth ratio (VH:CD), whereas AGP increased ($p < 0.05$) ADG, and G:F. PC increased ($p < 0.05$) fecal score, F18+ *E. coli* in feces, and protein carbonyl in jejunal mucosa. AGP reduced ($p < 0.05$) fecal score and F18+ *E. coli* in jejunal mucosa. PC reduced ($p < 0.05$) *Prevotella stercorea* populations in jejunal mucosa, whereas AGP increased ($p < 0.05$) *Phascolarctobacterium succinatutens* and reduced ($p < 0.05$) *Mitsuokella jalaludinii* populations in feces. Collectively, F18+ *E. coli* challenge increased fecal score and disrupted the microbiota composition, harming intestinal health by increasing oxidative stress, and damaging the intestinal epithelium, ultimately impairing growth performance. Dietary bacitracin reduced reduced F18+ *E. coli* populations and the oxidative damages they cause, thereby improving intestinal health and the growth performance of nursery pigs.

Keywords: F18+ *E. coli*; growth performance; intestinal health; oxidative damages; pigs

Citation: Duarte, M.E.; Stahl, C.H.; Kim, S.W. Intestinal Damages by F18+ *Escherichia coli* and Its Amelioration with an Antibacterial Bacitracin Fed to Nursery Pigs. *Antioxidants* **2023**, *12*, 1040. https://doi.org/10.3390/antiox12051040

Academic Editor: Stanley Omaye

Received: 24 February 2023
Revised: 18 April 2023
Accepted: 1 May 2023
Published: 3 May 2023

1. Introduction

In swine production, the post-weaning period is associated with immunological, physiological, psychological, and nutritional challenges that can impair the intestinal immune system and growth performance of pigs [1–3]. The impaired intestinal immune system increases pigs susceptibility to pathogen invasion [3,4]. Enterotoxigenic *Escherichia coli*, which causes post-weaning diarrhea (PWD), is a pathogen of concern for producers around the world. As a consequence of *E. coli* infection, changes in intestinal microbiota can led to increased inflammation and oxidative damage in the intestine, ultimately resulting in growth retardation [5–7]. According to Duarte and Kim [8], the changes in the intestinal microbiota in pigs challenged with F18+ *E. coli* are positively correlated with oxidative damages in the jejunal mucosa.

Different strategies have been utilized to reduce the susceptibility of pigs to potential pathogens [9,10]. Since the 1950s, antibiotics have been used in swine feed to promote growth by improving intestinal health [11,12]. Bacitracin is an antibiotic commonly used in animal feeds as a growth promoter and to treat and control infections [13]. In the US, bacitracin use as a growth promoter is not subjected to the veterinary feed directive rule and, therefore, does not require veterinary prescription [14]. Although the use of bacitracin has been primarily thought of as effective against Gram-positive pathogens, its use has also been reported to modulate the intestinal microbiota in nursery pigs [7], rabbits [15], and poultry [16,17]. This modulation of microbiota may explain the ability of bacitracin ability to prevent the deleterious effects of *E. coli* infection [7].

If the damage caused by F18$^+$ *E. coli* infection is partially due to alterations in intestinal microbiota, which led to increased oxidative damage and increased intestinal inflammatory responses, understanding ways to mediate this is important for improving the efficiency of swine production. Bacitracin may be a useful tool to minimize the disruption of the intestinal microbiota due to F18$^+$ *E. coli* infection, consequently promoting the growth of challenged pigs. To test this hypothesis, this study evaluated the intestinal oxidative damages caused by F18$^+$ *E. coli* and its protection with the antibacterial bacitracin fed to nursery pigs.

2. Materials and Methods

The Institutional Animal Care and Use Committee at North Carolina State University approved the experimental protocol used in this study, as stated in the North Carolina State Animal Care and Use Procedures (REG 10.10.01).

2.1. Animals, Experimental Design, Diets, and Inoculation

An amount of 36 newly weaned pigs (18 barrows and 18 gilts) with 6.31 ± 0.08 kg body weight (BW) and 21 d of age were allotted to 3 treatments using a randomized complete block design (RCBD). Sex and initial BW were considered as blocks. The treatments were: NC, not challenged/not treated; PC, challenged (F18$^+$ *E. coli* at 5.2×10^9 CFU)/not treated; AGP, challenged (F18$^+$ *E. coli* at 5.2×10^9 CFU)/treated with bacitracin (30 g/t). Pigs were fed diets for 28 d divided into 2 phases (P1 for 14 d, and P2 for 14 d). Basal diets were formulated to meet the nutrient requirements suggested by NRC [18] (Table 1).

Table 1. Composition of basal diets (Exp. 1; as-fed basis).

Item	Phase 1	Phase 2
Ingredient, %		
Corn, yellow	40.45	54.47
Soybean meal, 48% CP	22.00	23.50
Whey permeate	20.00	10.00
Blood plasma	6.00	3.00
Poultry meal	5.00	4.00
Poultry fat	3.50	1.80
L-Lys HCl	0.48	0.47
DL-Met	0.22	0.18
L-Thr	0.15	0.13
L-Trp	0.00	0.00
Dicalcium phosphate	0.60	0.85
Limestone	0.95	0.95
Vitamin premix [1]	0.03	0.03
Mineral premix [2]	0.15	0.15
Salt	0.22	0.22
Zinc oxide	0.25	0.25
Calculated composition:		
Dry matter, %	90.8	90.1
ME, kcal/kg	3481	3388
CP, %	23.00	21.60
SID [3] Lys, %	1.50	1.35
SID Met + Cys, %	0.82	0.74
SID Trp, %	0.25	0.22
SID Thr, %	0.88	0.79
Ca, %	0.86	0.81
STTD [4] *p*, %	0.45	0.40
Total *p*, %	0.67	0.64

[1] The vitamin premix provided the following per kilogram of complete diet: 6613.8 IU of vitamin A as vitamin A acetate, 992.0 IU of vitamin D3, 19.8 IU of vitamin E, 2.64 mg of vitamin K as menadione sodium bisulfate, 0.03 mg of vitamin B12, 4.63 mg of riboflavin, 18.52 mg of D-pantothenic acid as calcium pantothenate, 24.96 mg of niacin, and 0.07 mg of biotin. [2] The trace mineral premix provided the following per kilogram of complete diet: 4.0 mg of Mn as manganous oxide, 165 mg of Fe as ferrous sulfate, 165 mg of Zn as zinc sulfate, 16.5 mg of Cu as copper sulfate, 0.30 mg of I as ethylenediamine di-hydroiodide, and 0.30 mg of Se as sodium selenite. [3] SID, standardized ileal digestible. [4] STTD, standardized total tract digestible.

Bacitracin methylene disalicylate (BMD) was added to the diets as a source of bacitracin. After 7 d of feeding (pre-challenge period), all pigs on PC and AGP received an oral dose of F18$^+$ *E. coli* (5.2×10^9 CFU), and pigs on NC received an oral dose of sterile saline solution. The F18$^+$ *E. coli* culture was prepared and inoculated to the challenged pigs, as previously reported by Duarte and Kim [8] and Xu et al. [7]. The inoculum was produced by utilizing the F18ac (O147) strain that generates heat-stable toxins A (STa) and B (STb). The strain stock was tested to confirm the expression of F18ac, STa, and STb.

2.2. Growth Performance and Fecal Score

Body weight and feed intake were measured weekly to calculate the average daily gain (ADG), average feed intake (ADFI), and the gain to feed ratio (G:F) in order to evaluate the growth performance of pigs. The fecal scores were recorded every other day using a scoring system where 1 = very hard and dry stool, 2 = firm stool; 3 = normal stool; 4 = loose stool; and 5 = watery stool, as previously reported by [19,20]

2.3. Sample Collection and Processing

Fecal and blood samples were collected from all pigs at d 14 and 28. Fecal samples were freshly collected to evaluate the microbiota composition in the post-challenge period. Blood (10 mL) was collected from the jugular vein into vacutainer tubes without anticoagulant to obtain serum to determine the concentration of tumor necrosis-alpha (TNF-α), as an indicator of inflammatory status [21] and protein carbonyl, as an indicator of oxidative stress status [22]. Sera were stored at −80 °C until analysis.

After 28 d feeding, all pigs were euthanized by penetrating captive bolt followed by exsanguination. Jejunal tissue and mucosa were collected 3 m distal to the pyloric-duodenal junction. Jejunal tissue (5 cm) was collected in 10% buffered formalin, and mucosa was obtained from the next 20 cm of jejunum and snap frozen in liquid nitrogen. The mucosa samples were used to evaluate the microbiota composition, the inflammatory and the oxidative stress status. Protein extracts from the mucosa were obtained by homogenization homogenizer (Tissuemiser; Fisher Scientific Inc., Waltham, MA, USA) in phosphate-buffered saline (PBS). The homogenate was then centrifuged at $10{,}000\times\ g$ at 4 °C for 15 min, and the supernatant stored at −80 °C for further analysis.

2.4. Immune and Oxidative Stress Status

Protein concentration of samples were determined using the Protein Assay Kit (23225#, Thermo Fisher Scientific Inc., Wilmington, DE, USA). Prior to analysis, the samples were diluted in PBS at 1:80 and 1:40 for serum and mucosa samples, respectively. Concentrations of TNF-α in mucosa and protein carbonyl in mucosa and sera were normalized to total protein content, as previously reported by Cheng et al. [23]. The concentration of TNF-α was measured in serum and mucosa samples using the Porcine TNF-α Immunoassay Kit (#PTA00; R&D Systems, Minneapolis, MN, USA) as previously described by Holanda and Kim [24]. The concentration of protein carbonyl was measured using the OxiSelect Protein carbonyl ELISA Kit (Cell Biolabs, Inc., San Diego, CA, USA) as previously described by Jang et al. [25].

2.5. Intestinal Morphology and Crypt Cell Proliferation

Jejunal tissue samples were sent to the North Carolina State University Histology Laboratory (College of Veterinary Medicine, Raleigh, NC, USA) for Ki-67 staining [21]. Fifteen fields of view at 40× magnification of villi and their respective crypts per pig were used to measure villi height and width and crypt depth. The villi height to crypt depth ratio (VH:CD) was then calculated. Fifteen fields of view at 100× magnification were used to determine the proportion of Ki-67$^+$ to total cells in the crypt as an estimator of cell proliferation rate in crypts, as previously described by Duarte and Kim [22].

2.6. Intestinal Microbiota

DNA was extracted (DNA Stool Mini Kit,#51604, Qiagen; Germantown, MD, USA) from fecal and mucosa samples for 16S rRNA analysis and for quantification of F18+ *E. coli* by qPCR. The DNA samples were sent to MAKO laboratories (Raleigh, NC, USA) for 16S rRNA and qPCR analysis according to their protocol, as reported by Duarte et al. [26]. The relative abundance of microbiota was calculated, and the OTU (operational taxonomic unit) with <0.5% relative abundance was combined and reported as "Others".

The F18+ *E. coli* in the mucosa and fecal samples was quantified by qPCR following the protocol used by MAKO laboratories. Briefly, the plasmid containing the F18 fimbriae genes fedA (NCBI GeneBank, accession no. M61713) was constructed using the GeneArt (Thermo Fisher Scientific). The synthetic F18 gene was assembled from synthetic oligonucleotides. The fragment was inserted into the pMK-RQ-Bs vector GeneArt (Thermo Fisher Scientific). The concentration of plasmid DNA was measured by UV spectroscopy after the purification from the transformed bacteria. The similarity of sequence within the insertion sites was 100%. A TaqMan probe specific to the fedA gene was provided by Thermo Fisher. For quantification of F18+ plasmid in the samples, the assembled vector was used as standard.

The standard vector was linearized using the SmaI digestion (#FD0664, Thermo Fisher Scientific) prior to sequencing using qPCR. The count of the stock standard was calculated based on the vector size (914 bp). Then, the standard was diluted to 2.86×10^7, 2.86×10^6, 2.86×10^5, 2.86×10^4, and 2.86×10^3. The Taqpath qPCR Master Mix CG (#A15297, Thermo Fisher Scientific) and the QuantStudio 12K Flex (Thermo Fisher Scientific) were used for the qPCR of samples and standards following the instructions of the manufacturer. Based on the count of the plasmid on the standard, linear regression was used to calculate the concentration of the F18+ plasmid in the samples. Before statistical analysis, the concentration of F18+ plasmid was Log transformed.

2.7. Statistical Analysis

The Mixed procedure of SAS 9.4 Software (Cary, NC, USA) was used to analyze all data based on a randomized complete block design. The main effect was the treatments, and the random effects were sex and initial BW. Pre-planned contrasts were used to test the effect of the F18+ *E. coli* challenge (NC vs. PC) and the effect of AGP on challenged pigs (PC vs. AGP). Statistical differences were considered significant with $p < 0.05$, and the tendency was considered when $0.05 \leq p < 0.10$.

3. Results

3.1. Growth Performance and Fecal Score

Prior to challenge (d 0 to 7), the treatments did not affect BW, ADG, ADFI, or G:F (Table 2). After the *E. coli* challenge, the PC had lower ($p < 0.05$) BW at d 14, 21, and 28 when compared with the NC. The AGP-treated pigs had higher ($p < 0.05$) BW at d 14 and tended to have higher BW ($p = 0.066$) at d 28 when compared with PC. The PC reduced ($p < 0.05$) the ADG of pigs post-challenge (d7 to 14, d 14 to 21, and d 7 to 28) and over the entire experiment (d 0 to 28) when compared to the NC. The AGP increased ($p < 0.05$) the ADG of pigs post-challenge (d 14 to 21, and d 7 to 28) and over the entire experiment (d 0 to 28) when compared with PC. During the last week of the experiment, d 21 to 28, the treatments did not affect the ADG, ADFI, nor the G:F. The PC did not affect the ADFI during the entire experiment, whereas AGP tended to increase ADFI ($p = 0.073$) from d 7 to 14. The PC reduced ($p < 0.05$) the G:F of pigs during the post-challenge (d7 to 14, and d 7 to 28) and the overall experiment (d 0 to 28) when compared with NC. The AGP increased ($p < 0.05$) the G:F of pigs, compared to the PC, from d 14 to 21.

Before the *E. coli* challenge (d 0 to 7), the treatments did not affect the fecal score of pigs (Figure 1). After the challenge, the PC pigs had higher ($p < 0.05$) fecal scores during the first- and second-week post-challenge when compared with NC. The AGP pigs had fecal scores that were intermediate to the PC and the NC during the first week post-challenge ($p < 0.05$), and they were not significantly different than those of the NC during the second week post-challenge. There were no significant differences in fecal score among the treatments during the final week of the experiment.

Table 2. Growth performance of pigs challenged with F18+ *Escherichia coli* and fed diets with bacitracin as a growth promoter.

Item	Treatment [1]			SEM	*p* Value	
	NC	PC	AGP		NC vs. PC	PC vs. AGP
BW, kg						
Initial	6.31	6.31	6.30	0.08	0.985	0.912
d 7	6.91	6.90	6.93	0.15	0.958	0.852
d 14	8.64	7.72	8.20	0.28	0.036	0.231
d 21	11.93	10.21	11.88	0.46	0.019	0.018
d 28	16.25	14.19	15.94	0.64	0.040	0.066
ADG, kg						
Pre-challenge (d 0 to 7)	0.080	0.084	0.091	0.021	0.980	0.804
Post-challenge (d 7 to 28)	0.445	0.348	0.429	0.026	0.020	0.039
d 7 to 14	0.247	0.118	0.181	0.028	0.005	0.119
d 14 to 21	0.470	0.356	0.526	0.035	0.038	0.003
d 21 to 28	0.617	0.569	0.580	0.042	0.430	0.857
Overall	0.353	0.282	0.344	0.022	0.022	0.046
ADFI, kg						
Pre-challenge (d 0 to 7)	0.120	0.149	0.142	0.024	0.415	0.844
Post-challenge (d 7 to 28)	0.643	0.582	0.663	0.048	0.403	0.254
d 7 to 14	0.340	0.303	0.386	0.034	0.436	0.073
d 14 to 21	0.627	0.572	0.681	0.064	0.467	0.149
d 21 to 28	0.972	0.870	0.921	0.071	0.300	0.589
Overall	0.512	0.474	0.532	0.038	0.497	0.285
G:F						
Pre-challenge (d 0 to 7)	0.61	0.55	0.57	0.10	0.679	0.845
Post-challenge (d 7 to 28)	0.72	0.56	0.62	0.03	0.001	0.121
d 7 to 14	0.76	0.36	0.46	0.07	0.001	0.276
d 14 to 21	0.77	0.65	0.78	0.04	0.060	0.036
d 21 to 28	0.64	0.67	0.63	0.03	0.634	0.378
Overall	0.71	0.55	0.61	0.03	0.009	0.236

[1] NC, not challenged/not treated; PC, challenged (*E. coli* F18+ at 5.2 × 10^9 CFU)/not treated; AGP, challenged (*E. coli* F18+ at 5.2 × 10^9 CFU)/treated with bacitracin (30 g/t).

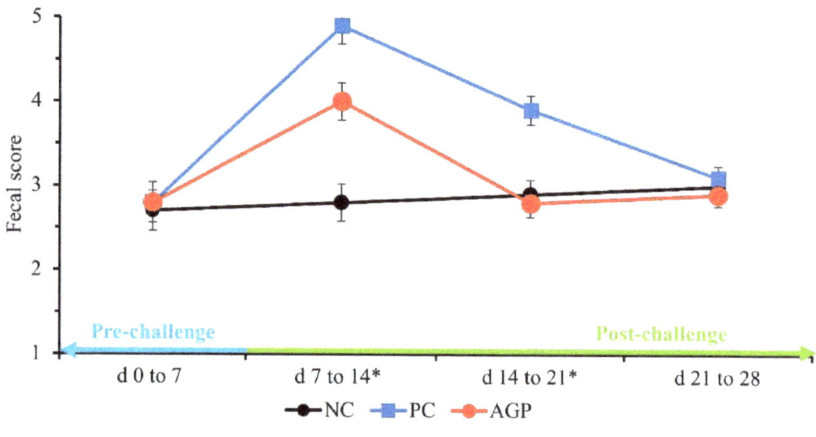

Figure 1. The fecal score of pigs challenged with F18+ *Escherichia coli* and fed diets with bacitracin as a growth promoter. NC, not challenged/not treated; PC, challenged (*E. coli* F18+ at 5.2 × 10^9 CFU)/not treated; AGP, challenged (*E. coli* F18+ at 5.2 × 10^9 CFU)/treated with bacitracin (30 g/t). * d 7 to 14: NC vs. PC: (p = 0.001), PC vs. AGP: (p = 0.004); d 14 to 21: NC vs. PC: (p = 0.001), PC vs. AGP: (p = 0.001).

3.2. F18⁺ *E. coli* Counting

The PC had increased ($p < 0.05$) copies of fedA, indicating higher populations of F18⁺ *E. coli* in the feces of pigs at d 14 when compared with the NC. There are no significant differences in feces at d 28 (Figure 2). The AGP did not significantly impact the copies of fedA in the feces on d 14. The PC tended to have greater ($p = 0.056$) concentrations of fedA in samples from jejunal mucosa, compared to the NC at d 28, and AGP appeared to have concentrations that were significantly lower than those of the PC group.

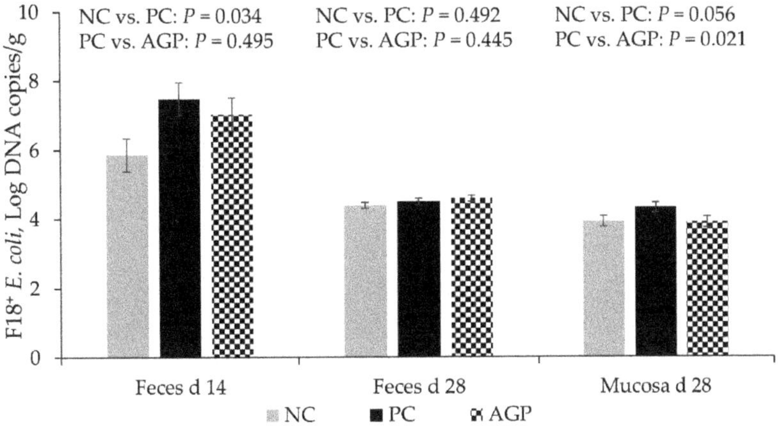

Figure 2. F18⁺ *E. coli* counting in feces and jejunal mucosa of pigs challenged with F18⁺ *Escherichia coli* and fed diets with bacitracin as a growth promoter. NC, not challenged/not treated; PC, challenged (*E. coli* F18⁺ at 5.2 × 10^9 CFU)/not treated; AGP, challenged (*E. coli* F18⁺ at 5.2 × 10^9 CFU)/treated with bacitracin (30 g/t).

3.3. Immune and Oxidative Stress Status

The concentration of TNF-α in jejunal mucosa was not affected by the treatments (Table 3). The PC tended to increase the concentration of TNF-α in sera at d 14 when compared with NC. The PC increased ($p < 0.05$) the concentration of protein carbonyl in serum and jejunal mucosa at d 28 when compared with NC. The AGP tended to reduce ($p < 0.05$) the concentration of protein carbonyl in the jejunal mucosa of pigs at d 28 when compared with PC.

Table 3. Immune and oxidative stress status of pigs challenged with F18⁺ *Escherichia coli* and fed diets with bacitracin as a growth promoter.

Item	Treatment [1]			SEM	*p* Value	
	NC	PC	AGP		NC vs. PC	PC vs. AGP
Tumor necrosis factor-alpha						
d 14 serum, pg/mL	110.4	128.7	119.4	6.3	0.054	0.365
d 28 serum, pg/mL	114.7	107.9	104.8	8.6	0.586	0.796
Jejunal mucosa, pg/mg protein	1.31	1.66	1.76	0.26	0.361	0.773
Protein carbonyl, nmol/mg protein						
d 14 serum	2.20	2.08	2.16	0.11	0.419	0.605
d 28 serum	1.63	2.37	2.32	0.19	0.010	0.851
Jejunal mucosa	2.15	3.61	2.56	0.35	0.012	0.059

[1] NC, not challenged/not treated; PC, challenged (*E. coli* F18⁺ at 5.2 × 10^9 CFU)/not treated; AGP, challenged (*E. coli* F18⁺ at 5.2 × 10^9 CFU)/treated with bacitracin (30 g/t).

3.4. Intestinal Morphology and Cell Proliferation in Crypt

The PC reduced ($p < 0.05$) the villus height and the VH:CD in the jejunum of pigs when compared with NC (Table 4). The villus width, crypt depth, and cell proliferation in jejunal crypts were not affected by the treatments.

Table 4. Intestinal morphology and cell proliferation in crypts of pigs challenged with F18$^+$ *Escherichia coli* and fed diets with bacitracin as a growth promoter.

Item	Treatment [1]			SEM	*p* Value	
	NC	PC	AGP		NC vs. PC	PC vs. AGP
Villus height, μm	527	394	436	32	0.003	0.296
Villus width, μm	85	91	85	7	0.378	0.419
Crypt depth, μm	245	253	253	12	0.591	0.975
VH:CD [2]	2.22	1.58	1.73	0.14	0.002	0.398
Ki-67^{+},[2] (%)	52.0	47.3	51.6	5.3	0.473	0.511

[1] NC, not challenged/not treated; PC, challenged (*E. coli* F18$^+$ at 5.2 × 10^9 CFU)/not treated; AGP, challenged (*E. coli* F18$^+$ at 5.2 × 10^9 CFU)/treated with bacitracin (30 g/t). [2] Cell proliferation rate.

3.5. Relative Abundance and Diversity of the Fecal and Mucosa-Associated Microbiota

The PC reduced ($p < 0.05$) the relative abundance of Tenericutes and tended to reduce ($p = 0.095$) the relative abundance of Deferribacteres in the feces of pigs at d 28 when compared with NC (Table 5). The PC tended to increase ($p = 0.072$) the relative abundance of Firmicutes in the feces of pigs at d 28 when compared with NC. The AGP tended to increase ($p = 0.088$) the relative abundance of Actinobacteria in the feces of pigs at d 28 when compared with PC. The PC tended to reduce ($p = 0.055$) the relative abundance of Bacteroidetes in the jejunal mucosa of pigs when compared with NC.

Table 5. Relative abundance of fecal and mucosa-associated microbiota at the phylum level in pigs challenged with F18$^+$ *Escherichia coli* and fed diets with bacitracin as a growth promoter.

Item	Treatment [1]			SEM	*p* Value	
	NC	PC	AGP		NC vs. PC	PC vs. AGP
d 14 (Feces)						
Bacteroidetes	41.3	37.2	39.3	3.3	0.380	0.651
Firmicutes	36.2	42.9	45.6	5.1	0.375	0.705
Proteobacteria	19.6	15.4	11.6	5.9	0.629	0.653
Spirochaetes	2.1	3.2	2.3	1.3	0.571	0.619
Tenericutes	0.4	1.0	0.4	0.3	0.170	0.146
Actinobacteria	0.1	0.1	0.6	0.2	0.878	0.121
Deferribacteres	0.1	0.2	0.1	0.1	0.628	0.337
d 28 (Feces)						
Firmicutes	46.1	53.9	55.7	2.8	0.072	0.649
Bacteroidetes	41.2	38.8	37.1	2.2	0.458	0.595
Proteobacteria	10.1	5.8	5.2	2.5	0.234	0.855
Tenericutes	1.4	0.1	0.2	0.4	0.043	0.906
Deferribacteres	0.5	0.0	0.0	0.2	0.095	0.912
Spirochaetes	0.4	0.7	0.6	0.4	0.653	0.865
Actinobacteria	0.1	0.5	1.2	0.3	0.438	0.088
d 28 (Mucosa)						
Firmicutes	50.5	46.2	51.0	14.6	0.839	0.821
Proteobacteria	25.9	43.1	39.3	15.6	0.452	0.867
Bacteroidetes	22.0	9.6	8.9	4.3	0.055	0.904
Actinobacteria	0.8	0.8	0.4	0.5	0.946	0.571
Spirochaetes	0.4	0.2	0.3	0.2	0.367	0.463

[1] NC, not challenged/not treated; PC, challenged (*E. coli* F18$^+$ at 5.2 × 10^9 CFU)/not treated; AGP, challenged (*E. coli* F18$^+$ at 5.2 × 10^9 CFU)/treated with bacitracin (30 g/t).

The PC increased ($p < 0.05$) the relative abundance of *Lachnospiraceae* and tended to increase ($p = 0.072$) the relative abundance of *Campylobacteraceae* in the feces of pigs at d 14 when compared with NC (Table 6). The AGP tended to reduce ($p = 0.094$) the relative abundance of Others in the feces of pigs at d 14 when compared with PC. The PC tended to increase ($p = 0.099$) the relative abundance of *Acidaminococcaceae* in the jejunal mucosa of pigs when compared with NC (Table 7). The AGP did not affect the relative abundance of mucosa-associated microbiota at the family level in the jejunum of pigs.

Table 6. Relative abundance of fecal microbiota at the family level in pigs challenged with F18$^+$ *Escherichia coli* and fed diets with bacitracin as a growth promoter.

	Treatment [1]				***p* Value**	
Item	**NC**	**PC**	**AGP**	**SEM**	**NC vs. PC**	**PC vs. AGP**
d 14						
Prevotellaceae	29.6	29.0	26.6	3.4	0.874	0.571
Veillonellaceae	10.8	10.9	14.2	4.3	0.989	0.599
Enterobacteriaceae	8.9	6.5	5.8	5.5	0.764	0.926
Ruminococcaceae	7.0	6.1	6.1	2.2	0.767	0.977
Acidaminococcaceae	2.5	4.5	6.2	1.6	0.402	0.480
Porphyromonadaceae	4.4	4.4	7.8	2.1	0.978	0.266
Lactobacillaceae	1.7	4.2	7.3	2.2	0.449	0.339
Lachnospiraceae	3.0	5.1	4.4	0.7	0.049	0.469
Succinivibrionaceae	6.6	3.6	2.9	2.0	0.309	0.825
Eubacteriaceae	2.0	2.8	3.0	0.5	0.301	0.813
Clostridiaceae	2.3	3.3	2.2	1.2	0.576	0.544
Cytophagaceae	3.3	1.3	0.9	1.3	0.302	0.813
Campylobacteraceae	0.6	3.7	1.9	1.2	0.072	0.284
Erysipelotrichaceae	1.2	1.3	1.3	0.6	0.919	0.984
Spirochaetaceae	1.4	1.3	1.4	1.1	0.983	0.967
Rikenellaceae	1.9	0.3	1.5	1.1	0.310	0.453
Peptococcaceae	2.6	0.9	0.1	1.4	0.414	0.709
Bacteroidaceae	0.6	1.4	1.8	0.6	0.368	0.627
Others	9.6	9.6	5.0	1.8	0.995	0.094
d 28						
Prevotellaceae	36.6	33.5	30.8	2.5	0.407	0.456
Veillonellaceae	21.0	27.1	21.1	2.8	0.144	0.150
Lactobacillaceae	7.7	4.6	3.6	2.7	0.438	0.793
Ruminococcaceae	4.5	5.0	7.2	0.9	0.737	0.121
Lachnospiraceae	3.5	4.9	6.4	1.4	0.404	0.544
Acidaminococcaceae	3.4	5.1	6.3	1.4	0.390	0.533
Porphyromonadaceae	2.5	2.6	4.5	0.9	0.922	0.189
Succinivibrionaceae	5.2	2.7	2.2	2.3	0.450	0.882
Eubacteriaceae	1.8	2.3	3.3	0.4	0.423	0.132
Streptococcaceae	0.9	1.3	4.2	1.6	0.864	0.226
Clostridiaceae	1.4	1.5	1.5	0.4	0.772	0.949
Campylobacteraceae	2.3	1.1	0.4	0.8	0.312	0.567
Others	9.2	8.2	8.5	1.9	0.703	0.904

[1] NC, not challenged/not treated; PC, challenged (*E. coli* F18$^+$ at 5.2×10^9 CFU)/not treated; AGP, challenged (*E. coli* F18$^+$ at 5.2×10^9 CFU)/treated with bacitracin (30 g/t).

The PC tended to reduce ($p = 0.079$) the relative abundance of *Succinivibrio dextrinosolvens* in the feces of pigs at d 14 when compared with NC (Table 8). The PC tended to reduce ($p = 0.065$) the relative abundance of *Prevotella stercorea* and increased the relative abundance of *Mitsuokella jalaludinii* in the feces of pigs at d 28 when compared with NC. The AGP increased ($p < 0.05$) the relative abundance of *Phascolarctobacterium succinatutens* whereas reduced ($p < 0.05$) the relative abundance of *Mitsuokella jalaludinii* in feces of pigs at d 28 when compared with PC.

Table 7. Relative abundance of mucosa-associated microbiota at the family level in pigs challenged with F18⁺ *Escherichia coli* and fed diets with bacitracin as a growth promoter.

Item	Treatment [1]			SEM	*p* Value	
	NC	PC	AGP		NC vs. PC	PC vs. AGP
Lactobacillaceae	25.6	23.8	21.4	12.7	0.922	0.893
Helicobacteraceae	14.6	35.6	34.7	15.7	0.365	0.968
Prevotellaceae	19.6	8.9	8.4	4.5	0.117	0.938
Veillonellaceae	7.6	5.38	19.2	5.7	0.763	0.115
Streptococcaceae	9.4	12.1	6.3	6.7	0.778	0.551
Campylobacteraceae	6.5	6.0	2.6	4.7	0.948	0.615
Acidaminococcaceae	2.6	1.0	0.8	0.7	0.099	0.862
Enterobacteriaceae	4.2	1.7	2.2	1.1	0.128	0.766
Lachnospiraceae	2.1	0.8	1.2	0.6	0.151	0.675
Ruminococcaceae	1.2	0.6	0.8	0.3	0.233	0.691
Erysipelotrichaceae	0.4	1.3	0.2	0.6	0.379	0.271
Clostridiaceae	0.4	0.6	0.2	0.2	0.556	0.268
Bifidobacteriaceae	0.6	0.7	0.3	0.4	0.897	0.539
Porphyromonadaceae	1.3	0.4	0.4	0.4	0.129	0.929
Succinivibrionaceae	0.7	0.5	1.0	0.5	0.749	0.479
Others	4.2	1.7	2.2	1.1	0.128	0.766

[1] NC, not challenged/not treated; PC, challenged (*E. coli* F18⁺ at 5.2 × 10^9 CFU)/not treated; AGP, challenged (*E. coli* F18⁺ at 5.2 × 10^9 CFU)/treated with bacitracin (30 g/t).

Table 8. Relative abundance of fecal microbiota at the specie level in pigs challenged with F18⁺ *Escherichia coli* and fed diets with bacitracin as a growth promoter.

Item	Treatment [1]			SEM	*p* Value	
	NC	PC	AGP		NC vs. PC	PC vs. AGP
d 14						
Prevotella copri	36.8	29.7	20.1	5.6	0.376	0.238
Phascolarctobacterium succinatutens	4.8	8.8	13.2	3.8	0.468	0.427
Prevotella stercorea	10.2	6.8	9.4	3.4	0.482	0.582
Faecalibacterium prausnitzii	10.1	8.0	9.1	3.2	0.648	0.811
Succinivibrio dextrinosolvens	6.8	3.0	0.7	1.5	0.079	0.288
Dialister succinatiphilus	2.5	1.6	6.3	3.0	0.845	0.290
Roseburia faecis	3.7	3.1	2.1	1.3	0.723	0.598
Campylobacter coli	1.1	4.8	6.0	2.4	0.275	0.720
Prevotella sp.	3.0	1.4	2.7	1.3	0.388	0.495
Mitsuokella jalaludinii	2.8	2.3	3.9	1.2	0.783	0.364
Brachyspira hampsonii	1.8	3.0	2.6	1.5	0.605	0.877
Treponema porcinum	0.9	1.8	3.7	1.2	0.594	0.262
Campylobacter lanienae	0.6	3.0	3.2	1.3	0.214	0.924
Dorea longicatena	0.4	0.9	1.2	0.4	0.410	0.641
Lactobacillus mucosae	0.5	0.4	0.4	1.2	0.939	0.985
Mitsuokella multacida	0.3	1.5	1.8	0.6	0.159	0.675
Lactobacillus kitasatonis	0.3	1.1	0.6	0.5	0.290	0.542
Gemmiger formicilis	1.7	1.2	0.6	0.6	0.585	0.472
Others	12.0	17.9	12.5	3.6	0.261	0.308
d 28						
Prevotella copri	39.0	35.2	30.3	6.0	0.656	0.568
Prevotella stercorea	9.3	5.0	8.6	1.6	0.065	0.119
Dialister succinatiphilus	4.2	8.0	3.2	3.2	0.409	0.294
Phascolarctobacterium succinatutens	4.3	3.0	10.0	2.0	0.658	0.024
Faecalibacterium prausnitzii	5.2	4.0	7.3	1.5	0.551	0.121
Mitsuokella jalaludinii	3.3	8.7	2.0	1.6	0.025	0.007
Roseburia faecis	3.1	3.6	3.6	1.1	0.754	0.992
Prevotella sp.	2.2	4.5	2.2	1.4	0.262	0.259
Streptococcus alactolyticus	1.0	1.6	4.5	2.6	0.867	0.435
Succinivibrio dextrinosolvens	4.4	2.3	1.8	2.2	0.502	0.876
Lactobacillus delbrueckii	4.4	1.1	0.5	2.6	0.373	0.873
Lactobacillus kitasatonis	1.7	0.7	0.8	1.5	0.654	0.958
Acidaminococcus fermentans	1.4	3.9	1.4	1.1	0.110	0.108
Gemmiger formicilis	1.1	1.2	1.6	0.8	0.901	0.699
Mitsuokella multacida	0.7	2.3	1.5	0.8	0.194	0.523
Campylobacter coli	2.7	0.8	0.5	0.8	0.127	0.837
Selenomonas bovis	1.6	1.3	2.2	0.9	0.797	0.489
Lactobacillus mucosae	1.4	1.3	0.6	0.7	0.911	0.496
Dorea longicatena	0.2	1.3	4.4	1.5	0.595	0.168
Megasphaera hominis	1.5	0.9	1.3	0.4	0.367	0.568
Selenomonas lipolytica	0.9	1.4	2.2	0.8	0.637	0.476
Others	6.4	7.9	9.5	1.9	0.578	0.574

[1] NC, not challenged/not treated; PC, challenged (*E. coli* F18⁺ at 5.2 × 10^9 CFU)/not treated; AGP, challenged (*E. coli* F18⁺ at 5.2 × 10^9 CFU)/treated with bacitracin (30 g/t).

The PC tended to reduce the relative abundance of *Prevotella copri* (p = 0.090), *Phascolarctobacterium succinatutens* (p = 0.053), and *Lactobacillus delbrueckii* (p = 0.050) in jejunal mucosa of pigs when compared with NC (Table 9). The AGP did not affect the relative abundance of mucosa-associated microbiota in pigs challenged with F18$^+$ *E. coli*.

Table 9. Relative abundance of mucosa-associated microbiota at the specie level in pigs challenged with F18$^+$ *Escherichia coli* and fed diets with bacitacin as growth promoter.

Item	Treatment [1]			SEM	*p* Value	
	NC	PC	AGP		NC vs. PC	PC vs. AGP
Helicobacter rappini	7.3	26.3	14.2	11.1	0.240	0.449
Helicobacter mastomyrinus	7.5	11.2	22.4	7.6	0.732	0.311
Lactobacillus kitasatonis	9.9	10.4	5.4	7.2	0.959	0.631
Lactobacillus mucosae	9.7	4.5	11.3	7.4	0.627	0.526
Prevotella copri	17.8	9.3	12.1	3.4	0.090	0.560
Streptococcus alactolyticus	10.4	10.1	4.7	4.5	0.967	0.408
Campylobacter upsaliensis	2.8	3.8	2.6	5.7	0.905	0.879
Streptococcus infantarius	3.2	5.4	3.9	2.6	0.549	0.679
Dialister succinatiphilus	2.9	1.9	2.9	1.2	0.555	0.568
Prevotella stercorea	5.1	1.5	1.2	1.1	0.036	0.853
Phascolarctobacterium succinatutens	4.8	1.0	0.8	1.3	0.053	0.937
Lactobacillus salivarius	0.7	0.9	1.2	1.7	0.945	0.906
Faecalibacterium prausnitzii	1.7	0.8	1.4	0.7	0.346	0.524
Lactobacillus delbrueckii	2.2	0.6	0.9	0.6	0.050	0.751
Helicobacter sp.	1.3	0.0	0.0	1.2	0.473	0.992
Mitsuokella jalaludinii	0.7	1.4	1.8	0.9	0.600	0.788
Others	12.0	11.0	13.5	4.6	0.875	0.705

[1] NC, not challenged/not treated; PC, challenged (*E. coli* F18$^+$ at 5.2 × 10^9 CFU)/not treated; AGP, challenged (*E. coli* F18$^+$ at 5.2 × 10^9 CFU)/treated with bacitracin (30 g/t).

The alpha diversity of fecal microbiota was not affected by the treatments at d 14 and d 28 (Figures 3 and 4). However, AGP tended to increase (p = 0.052) the alpha diversity of mucosa-associated microbiota estimated with Chao1 (Figure 5).

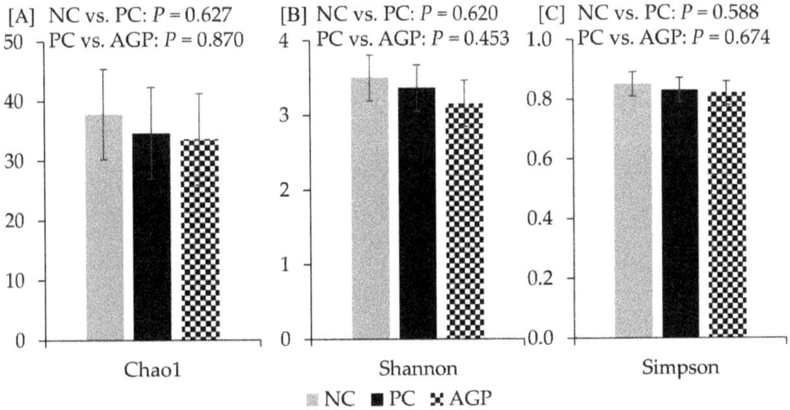

Figure 3. Alpha diversity of fecal microbiota at d 14 estimated with Chao1 richness (**A**), Shannon diversity (**B**), and Simpson diversity (**C**) in pigs challenged with F18$^+$ *Escherichia coli* and fed diets with bacitracin as a growth promoter. NC, not challenged/not treated; PC, challenged (*E. coli* F18$^+$ at 5.2 × 10^9 CFU)/not treated; AGP, challenged (*E. coli* F18$^+$ at 5.2 × 10^9 CFU)/treated with bacitracin (30 g/t).

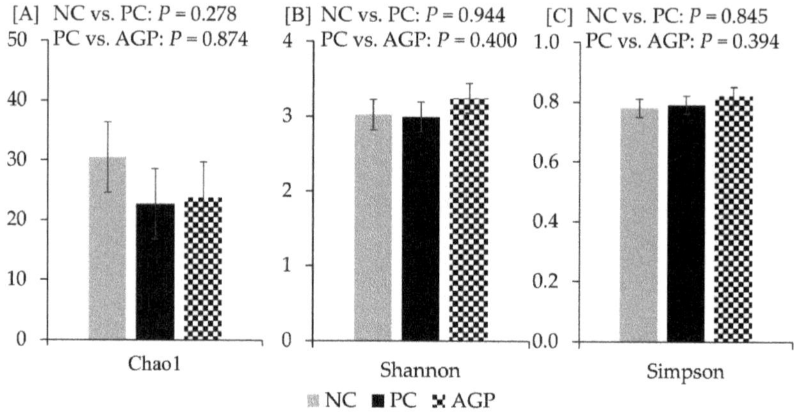

Figure 4. Alpha diversity of fecal microbiota at d 28 estimated with Chao1 richness (**A**), Shannon diversity (**B**), and Simpson diversity (**C**) in pigs challenged with F18$^+$ *Escherichia coli* and fed diets with bacitracin as a growth promoter. NC, not challenged/not treated; PC, challenged (*E. coli* F18$^+$ at 5.2×10^9 CFU)/not treated; AGP, challenged (*E. coli* F18$^+$ at 5.2×10^9 CFU)/treated with bacitracin (30 g/t).

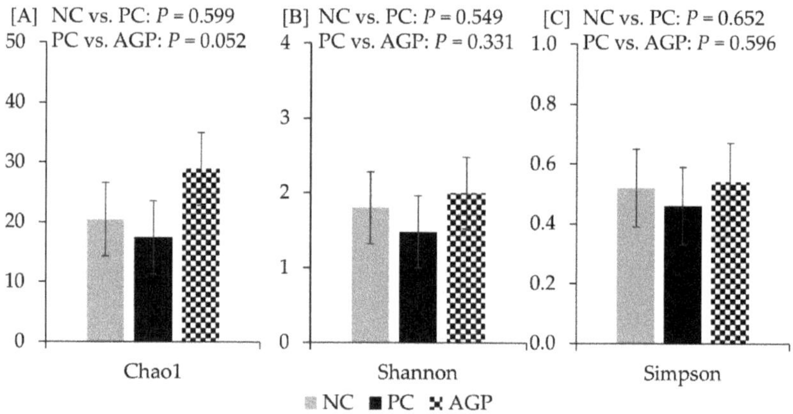

Figure 5. Alpha diversity of mucosa-associated microbiota at d 28 estimated with Chao1 richness (**A**), Shannon diversity (**B**), and Simpson diversity (**C**) in pigs challenged with F18$^+$ *Escherichia coli* and fed diets with bacitracin as a growth promoter. NC, not challenged/not treated; PC, challenged (*E. coli* F18$^+$ at 5.2×10^9 CFU)/not treated; AGP, challenged (*E. coli* F18$^+$ at 5.2×10^9 CFU)/treated with bacitracin (30 g/t).

4. Discussion

In this study, direct oral challenge with F18$^+$ *E. coli* caused PWD, increased oxidative damage, and reduced the growth performance of weaned pigs, which is in agreement with previous reports [7,27]. The reduced feed efficiency seen among the *E. coli* challenged pigs can be attributed to the impaired intestinal health of challenged pigs as observed by increased fecal score, increased inflammation and oxidative stress, and the damaged villi and disrupted microbial community. The health challenged pigs may have had reduced nutrient absorption and/or altered partitioning of nutrients for immune response and growth, resulting in reduced feed efficiency [5,28]. However, bacitracin ameliorated many of the effects of the *E. coli* challenge, as evidenced by improved fecal scores, reduced oxidative damage and improved the feed efficiency. These benefits were seen without a significant reduction in fecal shedding of *E. coli* post-challenge.

Changes in diet, environment, social interaction, and the removal of the passive immunity from sow's milk during a period where the immune system is not fully mature increase the vulnerability of newly weaned pigs to opportunistic pathogens [1,10,29]. The F18$^+$ *E. coli* attaches to glycoproteins on the brush border in the intestine mediating resistance to flushing and promoting colonization [30–32]. In the current study, the increased F18$^+$ *E. coli* counting in feces at d 14 matches with the increased fecal score in the period of 7 to 14 d of the experiment and may be an indicator of proliferation on the intestinal epithelium. At d 28, 21 d after challenge, the F18$^+$ *E. coli* counting in feces did not differ among treatments, and fecal scores returned to normal, indicating pigs had controlled the *E. coli* infection to a less harmful level. The trend toward increased F18$^+$ *E. coli* in the jejunal mucosa of challenged pigs indicates that F18$^+$ *E. coli* can persist in the gastrointestinal tract for up to 21 d post-challenge. Duarte and Kim [8] reported that F18$^+$ *E. coli* has a long-lasting effect in jejunal mucosa when compared with feces.

Interestingly, bacitracin reduced the F18$^+$ *E. coli* population in jejunal mucosa. Antibiotics have been used to overcome or mitigate the challenges associated with health and nutrition, mainly by impairing the growth of pathogens [11,12]. Bacitracin, produced by *Bacillus licheniformis*, is an antibiotic with a narrow spectrum against primarily Gram-positive bacteria [33]. Bacitracin inhibits the synthesis of peptidoglycan and teichoic acids in the cell wall of bacteria inhibiting their proliferation [34,35]. Gram-positive bacteria are the main target for bacitracin due to the thicker peptidoglycan layer [35,36]. However, Gram-negative bacteria also contain peptidoglycan on the cell wall [37]. Xu et al. [7] demonstrated that bacitracin can mitigate the effects of PWD caused by F18$^+$ *E. coli* in nursery pigs.

During proliferation, *E. coli* can produce enterotoxins, including STa and STb, that induce the secretion of fluid in the lumen of the small intestine, causing diarrhea [7,30]. Pigs challenged with F18$^+$ *E. coli* in this study had increased fecal scores until d 21 of the experiment. Challenged pigs that received bacitracin showed improved fecal scores at d 14, although they remained higher than those of the unchallenged pigs. By d 21, the fecal scores of the F18$^+$ *E. coli* challenged pigs treated with bacitracin were similar to those of the unchallenged controls. These results demonstrate the efficacy of bacitracin in mitigating PWD in pigs, as previously reported by Xu et al. [7]. Duarte and Kim [8] reported that, although the diarrhea symptoms ceased 14 d after an F18$^+$ *E. coli*-challenge, the effects of F18$^+$ *E. coli* on intestinal health lasted for at least 21 d. In addition to diarrhea, F18$^+$ *E. coli* infections can also result in an inflammatory response [7,8]. A systemic inflammatory response was seen 14 d after challenge in this study, with a trend toward increased concentration of TNF-α in the serum of challenged pigs. However, there were no differences at d 28 and in TNF-α concentrations in jejunal mucosa. Other studies have reported increased expression of IL-6 and IL-8 in the jejunal mucosa of F18$^+$ *E. coli*-challenged pigs without a significant change in TNF-α expression [7,38]. Due to the complex timing of cytokine cascades during an immune response, it is not necessarily surprising that TNF-α concentrations in the intestinal mucosa were not elevated at the end of the study. At the completion of the study, sera and mucosal concentrations of protein carbonyl were increased in the challenged pigs. These findings are in agreement with previous works that have reported that a F18$^+$ *E. coli* challenge increases oxidative stress in nursery pigs [5,7,27]. During infection, ROS, including nitrite, are produced by immune cells to fight the infection [39–41]. The antioxidant enzymes scavenge the ROS maintaining homeostasis [40]. When the production of ROS exceeds the antioxidant capacity, products from oxidative stress, including protein carbonyls, are generated [42]. Protein carbonyl has been reported as an important biomarker of oxidative stress because it can be produced by all ROS, and it has higher stability compared with other products of oxidative damage [43]. Protein carbonyls lead to the dysfunction of cellular proteins, which can induce apoptosis [41,44]. In this study, bacitracin treatment tended to reduce protein carbonyl concentrations in challenged pigs, possibly by altering gut microbiota and reducing the intestinal mucosa's immunoreaction in response to the F18$^+$ *E. coli* or by altering the production of toxins and other antigens by the *E. coli* [45].

The altered fluid secretion induced by enterotoxins from *E. coli* can reduce water absorption and increase the flux of water from the enterocyte into the lumen of the intestine, causing dehydration and cell apoptosis [46–48]. Previous studies have shown that the apoptosis induced by cell dehydration and oxidative damage in challenged pigs is associated with the reduction in villus length [46,49]. In this study, pigs challenged with F18$^+$ *E. coli* had the lower villus height in jejunum, confirming the deleterious effects of the *E. coli* on the epithelium. Enterocyte damage in the villi can induce cell proliferation in crypts to provide new enterocytes [49]. Increased cell proliferation can increase crypt depth, therefore reducing the villus height to crypt depth ratio [50–52], which was seen with *E. coli* challenge in this study. According to Pluske et al. [50], the atrophy of villi and the hyperplasia of crypts can reduce the digestion and absorption of nutrients, thereby reducing the feed efficiency of pigs. Additionally, undigested nutrients can further contribute to PWD due to the increased amount of substrate available for microbial fermentation [6,53].

Increased fluid secretion, products from an immune response, and undigested nutrients can all modulate the microbiome toward a more inflammatory microbiota, such as increasing the abundance of Proteobacteria [5–8]. This change in the microbiota composition is associated with the increased production of ROS, including nitrite, released during the immune response. The nitrite is transformed into nitrate in the lumen favoring the growth of bacteria expressing nitrate reductase, such as Proteobacteria [5,54]. However, 7 d after the challenge, there was a trend of increasing Firmicutes on the feces of pigs mainly by increasing *Lachnospiraceae* while reducing Tenericutes and Deferribacteries. The environmental changes near the mucosa may have exerted pressure on the microbiota, moving *Lachnospiraceae* toward the luminal content, consequently modulating the luminal and the mucosa-associated microbiota [6,8]. Additionally, it was observed a trend towards increasing *Campylobacteraceae* in the feces of challenged pigs. Interestingly, at 21 d after the challenge, the abundance of was increased in the feces of challenged pigs. According to Duarte and Kim [17], *Mitsuokella* spp. and *Campylobacter* spp. are highly correlated to inflammatory and oxidative stress in pigs challenged with F18$^+$ *E. coli*.

According to Belkaid et al. [55], the immune system plays a pivotal role in modulating the mucosa-associated microbiota, which in turn modulate the luminal microbiota. The relative abundance of *Prevotella* spp. and *Phascolarctobacterium succinatutens* in jejunal mucosa-associated microbiota was reduced in challenged pigs, possibly due to the oxidative environment promoted by the immune response against *E. coli*. *Prevotella* is a Bacteroidetes that is associated with health conditions, and its relative abundance increases in pigs after weaning due to the fiber content in the diet [6,56]. The unbalance in the microbiota composition by reducing the abundance of fiber-degrading bacteria, in turn, can increase the immune response in the intestine [5,6]. Interestingly, bacitracin tended to increase the alpha diversity of mucosa-associated microbiota in the jejunum. Previous studies have demonstrated that bacitracin can increase microbial diversity [7,17,57], although, in general, antibiotics are associated with reduced diversity [58]. According to Proctor and Phillips [17], the bacitracin may have inhibited the proliferation of certain bacteria allowing the growth of others. These effects were observed in fecal samples at d 28, where the abundance of *P. succinatutens* was increased, and *M. jalaludinii* populations were reduced. *M. jalaludinii* are Gram-negative bacteria confirming that bacitracin can also affect bacteria other than Gram-positive. *Phascolarctobacterium succinatutens*, a strict anaerobic bacteria belonging to Firmicutes, are associated with propionate production through the succinate scavenge [59,60]. Succinate is normally produced by different bacteria within the intestine, especially from carbohydrate fermentation [60]. It has been reported that succinate exerts inflammatory [61] and oxidative [62] roles. Therefore, these findings suggest that the reduction in protein carbonyl reported in the current study can also be associated with the increased abundance of bacteria associated with fiber utilization, including *P. succinatutens*.

5. Conclusions

The F18$^+$ *E. coli* challenge resulted in increased fecal scores, altered intestinal histology, increased oxidative damage, all demonstrating reduced intestinal health. This resulted in impaired growth performance of pigs challenged with F18$^+$ *E. coli*. Dietary supplementation with bacitracin ameliorates many of the intestinal health challenges caused by F18$^+$ *E. coli*, resulting in improved growth performance. Whereas further studies are needed to elucidate the protective mechanisms of bacitracin on a F18$^+$ *E. coli* infection, alterations in the microbiota towards a less harmful milieu may underlay this effect and ultimately provide greater insight into the role of microbiota on improving growth performance.

Author Contributions: Conceptualization, S.W.K.; methodology, S.W.K. and C.H.S.; formal analysis, S.W.K. and M.E.D.; investigation, S.W.K.; resources, S.W.K.; data curation, M.E.D.; and S.W.K.; writing—original draft preparation, M.E.D. and S.W.K.; writing—review and editing, M.E.D., C.H.S., and S.W.K.; supervision, S.W.K.; project administration, S.W.K.; funding acquisition, S.W.K. All authors have read and agreed to the published version of the manuscript.

Funding: USDA-NIFA (Hatch #02893) and North Carolina Agricultural Foundation (#660101).

Institutional Review Board Statement: The experimental protocols used in this study were reviewed and approved by the Institutional Animal Care and Use Committee at North Carolina State University (IACUC #: 19-834). The experiments were performed by trained scientists in full compliance with the North Carolina State Animal Care and Use Procedures (REG 10.10.01).

Informed Consent Statement: Not applicable.

Data Availability Statement: The data presented in this study are available in the article.

Acknowledgments: K. L. Brooks, H. Chen, J. Guo, K. B. Jang, Y. I. Kim; J. K. Lee, I. Park, W. Parnsen, and L. Zheng of the Kim Lab for their assistance in animal management, sampling, and sample analysis.

Conflicts of Interest: The authors declare no conflict of interest.

References

1. McLamb, B.L.; Gibson, A.J.; Overman, E.L.; Stahl, C.; Moeser, A.J. Early weaning stress in pigs impairs innate mucosal immune responses to enterotoxigenic *E. coli* challenge and exacerbates intestinal injury and clinical disease. *PLoS ONE* **2013**, *8*, e59838. [CrossRef]
2. Campbell, J.M.; Crenshaw, J.D.; Polo, J. The biological stress of early weaned piglets. *J. Anim. Sci. Biotechnol.* **2013**, *4*, 2–5. [CrossRef]
3. Moeser, A.J.; Pohl, C.S.; Rajput, M. Weaning stress and gastrointestinal barrier development: Implications for lifelong gut health in pigs. *Anim. Nutr.* **2017**, *3*, 313–321. [CrossRef]
4. Sun, Y.; Kim, S.W. Intestinal challenge with enterotoxigenic *Escherichia coli* in pigs, and nutritional intervention to prevent postweaning diarrhea. *Anim. Nutr.* **2017**, *3*, 322–330. [CrossRef]
5. Duarte, M.E.; Tyus, J.; Kim, S.W. Synbiotic effects of enzyme and probiotics on intestinal health and growth of newly weaned pigs challenged with enterotoxigenic F18$^+$ *Escherichia coli*. *Front. Vet. Sci.* **2020**, *7*, 573. [CrossRef] [PubMed]
6. Duarte, M.E.; Kim, S.W. Intestinal microbiota and its interaction to intestinal health in nursery pigs. *Anim. Nutr.* **2022**, *8*, 169–184. [CrossRef] [PubMed]
7. Xu, X.; Duarte, M.E.; Kim, S.W. Postbiotic effects of Lactobacillus fermentate on intestinal health, mucosa-associated microbiota, and growth efficiency of nursery pigs challenged with F18$^+$ *Escherichia coli*. *J. Anim. Sci.* **2022**, *100*, skac210. [CrossRef] [PubMed]
8. Duarte, M.E.; Kim, S.W. Significance of mucosa-associated microbiota and its impacts on intestinal health of pigs challenged with F18$^+$ *E. coli*. *Pathogens* **2022**, *11*, 589. [CrossRef]
9. Jang, K.B.; Purvis, J.M.; Kim, S.W. Dose–response and functional role of whey permeate as a source of lactose and milk oligosaccharides on intestinal health and growth of nursery pigs. *J. Anim. Sci.* **2021**, *99*, skab008. [CrossRef]
10. Zheng, L.; Duarte, M.E.; Sevarolli Loftus, A.; Kim, S.W. Intestinal health of pigs upon weaning: Challenges and nutritional intervention. *Front. Vet. Sci.* **2021**, *8*, 628258. [CrossRef]
11. Carpenter, L.E. The effect of antibiotics and vitamin B12 on the growth of swine. *Arch. Biochem. Biophys.* **1951**, *32*, 187–191. [CrossRef] [PubMed]
12. Luecke, R.W.; Thorp, F.; Newland, H.W.; Mcmillen, W.N. The growth promoting effects of various antibiotics on pigs. *J. Anim. Sci.* **1951**, *10*, 538–542. [CrossRef] [PubMed]
13. Bridges, J.H.; Hale, F.; Kunkel, H.O.; Lyman, C.M. The effects of bacitracin, penicillin and arsanilic acid on growth rate and feed efficiency in swine. *J. Anim. Sci.* **1954**, *13*, 912–917. [CrossRef]

14. FDA Food and Drug Administration. *Guidance for Industry #213: New Animal Drugs and New Animal Drug Combination Products Administered in or on Medicated Feed or Drinking Water of Food- Producing Animals: Recommendations for Drug Sponsors for Voluntarily Aligning Product Use Conditions with*; Center for Veterinary Medicine: Rockville, MD, USA, 2013; ISBN 2402769084.
15. Chen, Y.; Hu, S.; Li, J.; Zhao, B.; Yang, N.; Zhou, T.; Liang, S.; Bai, S.; Wu, X. Bacitracin methylene disalicylate improves intestinal health by modulating its development and microbiota in weaned rabbits. *Front. Microbiol.* **2021**, *12*, 579006. [CrossRef]
16. Johnson, T.A.; Sylte, M.J.; Looft, T. In-feed bacitracin methylene disalicylate modulates the turkey microbiota and metabolome in a dose-dependent manner. *Sci. Rep.* **2019**, *9*, 8212. [CrossRef] [PubMed]
17. Proctor, A.; Phillips, G.J. Differential effects of bacitracin methylene disalicylate (BMD) on the distal colon and cecal microbiota of young broiler chickens. *Front. Vet. Sci.* **2019**, *6*, 114. [CrossRef] [PubMed]
18. NRC. *Nutrient Requirements of Swine*, 11th ed.; National Academies Press: Washington, DC, USA, 2012; ISBN 978-0-309-22423-9.
19. Deng, Z.; Duarte, M.E.; Jang, K.B.; Kim, S.W. Soy protein concentrate replacing animal protein supplements and its impacts on intestinal immune status, intestinal oxidative stress status, nutrient digestibility, mucosa-associated microbiota, and growth performance of nursery pigs. *J. Anim. Sci.* **2022**, *100*, skac255. [CrossRef]
20. Jang, K.B.; Duarte, M.E.; Purvis, J.M.; Kim, S.W. Impacts of weaning age on dietary needs of whey permeate for pigs at 7 to 11 kg body weight. *J. Anim. Sci. Biotechnol.* **2021**, *12*, 111. [CrossRef]
21. Moita, V.H.C.; Duarte, M.E.; Kim, S.W. Functional roles of xylanase enhancing intestinal health and growth performance of nursery pigs by reducing the digesta viscosity and modulating the mucosa-associated microbiota in the jejunum. *J. Anim. Sci.* **2022**, *100*, skac116. [CrossRef]
22. Duarte, M.E.; Kim, S.W. Phytobiotics from oregano extracts enhance the intestinal health and growth performance of pigs. *Antioxidants* **2022**, *11*, 2066. [CrossRef]
23. Cheng, Y.-C.; Duarte, M.E.; Kim, S.W. Effects of Yarrowia lipolytica supplementation on growth performance, intestinal health and apparent ileal digestibility of diets fed to nursery pigs. *Anim. Biosci.* **2022**, *35*, 605–613. [CrossRef]
24. Holanda, D.M.; Kim, S.W. Investigation of the efficacy of mycotoxin-detoxifying additive on health and growth of newly-weaned pigs under deoxynivalenol challenges. *Anim. Biosci.* **2021**, *34*, 405–416. [CrossRef] [PubMed]
25. Jang, K.B.; Kim, J.H.; Purvis, J.M.; Chen, J.; Ren, P.; Vazquez-Anon, M.; Kim, S.W. Effects of mineral methionine hydroxy analog chelate in sow diets on epigenetic modification and growth of progeny. *J. Anim. Sci.* **2020**, *98*, skaa271. [CrossRef] [PubMed]
26. Duarte, M.E.; Sparks, C.; Kim, S.W. Modulation of jejunal mucosa-associated microbiota in relation to intestinal health and nutrient digestibility in pigs by supplementation of β-glucanase to corn–soybean meal-based diets with xylanase. *J. Anim. Sci.* **2021**, *99*, skab190. [CrossRef] [PubMed]
27. Jang, K.B.; Moita, V.H.C.; Martinez, N.; Sokale, A.; Kim, S.W. Efficacy of zinc glycinate reducing zinc oxide on intestinal health and growth of nursery pigs challenged with F18$^+$ *Escherichia coli*. *J. Anim. Sci.* **2023**, *101*, skad035. [CrossRef]
28. Huntley, N.F.; Nyachoti, C.M.; Patience, J.F. Lipopolysaccharide immune stimulation but not β-mannanase supplementation affects maintenance energy requirements in young weaned pigs. *J. Anim. Sci. Biotechnol.* **2018**, *9*, 47. [CrossRef]
29. Kim, S.W.; Duarte, M.E. Understanding intestinal health in nursery pigs and the relevant nutritional strategies. *Anim. Biosci.* **2021**, *34*, 338–344. [CrossRef]
30. Dubreuil, J.D.; Isaacson, R.E.; Schifferli, D.M. Animal enterotoxigenic *Escherichia coli*. *EcoSal Plus* **2016**, *7*, 1–47. [CrossRef]
31. Peterson, J.W.; Whipp, S.C. Comparison of the mechanisms of action of cholera toxin and the heat-stable enterotoxins of *Escherichia coli*. *Infect. Immun.* **1995**, *63*, 1452–1461. [CrossRef]
32. Nagy, B.; Whipp, S.C.; Imberechts, H.; Bertschinger, H.U.; Dean-Nystrom, E.A.; Casey, T.A.; Salajka, E. Biological relationship between F18ab and F18ac fimbriae of enterotoxigenic and verotoxigenic *Escherichia coli* from weaned pigs with oedema disease or diarrhoea. *Microb. Pathog.* **1997**, *22*, 1–11. [CrossRef]
33. Huyghebaert, G.; De Groote, G. The bioefficacy of zinc bacitracin in practical diets for broilers and laying hens. *Poult. Sci.* **1997**, *76*, 849–856. [CrossRef]
34. Tay, W.M.; Epperson, J.D.; da Silva, G.F.Z.; Ming, L.-J. H NMR, Mechanism, and Mononuclear oxidative activity of the antibiotic metallopeptide bacitracin: The Role of d -Glu-4, Interaction with Pyrophosphate Moiety, DNA Binding and Cleavage, and Bioactivity. *J. Am. Chem. Soc.* **2010**, *132*, 5652–5661. [CrossRef] [PubMed]
35. Mascher, T.; Margulis, N.G.; Wang, T.; Ye, R.W.; Helmann, J.D. Cell wall stress responses in Bacillus subtilis: The regulatory network of the bacitracin stimulon. *Mol. Microbiol.* **2003**, *50*, 1591–1604. [CrossRef] [PubMed]
36. Rajagopal, M.; Walker, S. Envelope structures of Gram-positive bacteria. In *Current Topics in Microbiology and Immunology*; Springer: Cham, Switzerland, 2017; Volume 404, pp. 1–44. [CrossRef]
37. Cheng, Y.-C.; Kim, S.W. Use of microorganisms as nutritional and functional feedstuffs for nursery pigs and broilers. *Animals* **2022**, *12*, 3141. [CrossRef] [PubMed]
38. Wong, B.T.; Park, S.; Kovanda, L.; He, Y.; Kim, K.; Xu, S.; Lingga, C.; Hejna, M.; Wall, E.; Sripathy, R.; et al. Dietary supplementation of botanical blends enhanced performance and disease resistance of weaned pigs experimentally infected with enterotoxigenic *Escherichia coli* F18. *J. Anim. Sci.* **2022**, *100*, skac353. [CrossRef]
39. Yang, Y.; Bazhin, A.V.; Werner, J.; Karakhanova, S. Reactive oxygen species in the immune system. *Int. Rev. Immunol.* **2013**, *32*, 249–270. [CrossRef]
40. Morris, G.; Gevezova, M.; Sarafian, V.; Maes, M. Redox regulation of the immune response. *Cell. Mol. Immunol.* **2022**, *19*, 1079–1101. [CrossRef]

41. Celi, P.; Gabai, G. Oxidant/antioxidant balance in animal nutrition and health: The role of protein oxidation. *Front. Vet. Sci.* **2015**, *2*, 48. [CrossRef]
42. Schieber, M.; Chandel, N.S. ROS function in redox signaling and oxidative stress. *Curr. Biol.* **2014**, *24*, R453–R462. [CrossRef]
43. Dalle-Donne, I.; Rossi, R.; Giustarini, D.; Milzani, A.; Colombo, R. Protein carbonyl groups as biomarkers of oxidative stress. *Clin. Chim. Acta* **2003**, *329*, 23–38. [CrossRef]
44. DalleDonne, I.; Milzani, A.; Colombo, R. The tert -butyl hydroperoxide-induced oxidation of actin cys-374 is coupled with structural changes in distant regions of the protein. *Biochemistry* **1999**, *38*, 12471–12480. [CrossRef] [PubMed]
45. Settle, T.; Leonard, S.S.; Falkenstei, E.; Fix, N.; Van Dyke, K.; Klandorf, H. Effects of a phytogenic feed additive versus an antibiotic feed additive on oxidative stress in broiler chicks and a possible mechanism determined by electron spin resonance. *Int. J. Poult. Sci.* **2014**, *13*, 62–69. [CrossRef] [PubMed]
46. Johnson, A.M.; Kaushik, R.S.; Rotella, N.J.; Hardwidge, P.R. Enterotoxigenic *Escherichia coli* modulates host intestinal cell membrane asymmetry and metabolic activity. *Infect. Immun.* **2009**, *77*, 341–347. [CrossRef]
47. Dubreuil, J.D. Enterotoxigenic *Escherichia coli* and probiotics in swine: What the bleep do we know? *Biosci. Microbiota Food Health* **2017**, *36*, 75–90. [CrossRef] [PubMed]
48. Kennedy, D.J.; Greenberg, R.N.; Dunn, J.A.; Abernathy, R.; Ryerse, J.S.; Guerrant, R.L. Effects of *Escherichia coli* heat-stable enterotoxin STb on intestines of mice, rats, rabbits, and piglets. *Infect. Immun.* **1984**, *46*, 639–643. [CrossRef]
49. Assimakopoulos, S.F.; Tsamandas, A.C.; Louvros, E.; Vagianos, C.E.; Nikolopoulou, V.N.; Thomopoulos, K.C.; Charonis, A.; Scopa, C.D. Intestinal epithelial cell proliferation, apoptosis and expression of tight junction proteins in patients with obstructive jaundice. *Eur. J. Clin. Investig.* **2011**, *41*, 117–125. [CrossRef] [PubMed]
50. Pluske, J.R.; Hampson, D.J.; Williams, I.H. Factors influencing the structure and function of the small intestine in the weaned pig: A review. *Livest. Prod. Sci.* **1997**, *51*, 215–236. [CrossRef]
51. Shaw, D. Intestinal mucosal atrophy and adaptation. *World J. Gastroenterol.* **2012**, *18*, 6357. [CrossRef]
52. Pluske, J.R.; Williams, I.H.; Aherne, F.X. Maintenance of villous height and crypt depth in piglets by providing continuous nutrition after weaning. *Anim. Sci.* **1996**, *62*, 131–144. [CrossRef]
53. Kim, K.; He, Y.; Xiong, X.; Ehrlich, A.; Li, X.; Raybould, H.; Atwill, E.R.; Maga, E.A.; Jørgensen, J.; Liu, Y. Dietary supplementation of Bacillus subtilis influenced intestinal health of weaned pigs experimentally infected with a pathogenic *E. coli*. *J. Anim. Sci. Biotechnol.* **2019**, *10*, 52. [CrossRef]
54. Bäumler, A.J.; Sperandio, V. Interactions between the microbiota and pathogenic bacteria in the gut. *Nature* **2016**, *535*, 85–93. [CrossRef]
55. Belkaid, Y.; Hand, T.W. Role of the microbiota in immunity and inflammation. *Cell* **2014**, *157*, 121–141. [CrossRef] [PubMed]
56. Kovatcheva-Datchary, P.; Nilsson, A.; Akrami, R.; Lee, Y.S.; De Vadder, F.; Arora, T.; Hallen, A.; Martens, E.; Björck, I.; Bäckhed, F. Dietary Fiber-induced improvement in glucose metabolism is associated with increased abundance of Prevotella. *Cell Metab.* **2015**, *22*, 971–982. [CrossRef] [PubMed]
57. Díaz Carrasco, J.M.; Redondo, E.A.; Pin Viso, N.D.; Redondo, L.M.; Farber, M.D.; Fernández Miyakawa, M.E. Tannins and bacitracin differentially modulate gut microbiota of broiler chickens. *Biomed Res. Int.* **2018**, *2018*, 1879168. [CrossRef] [PubMed]
58. Grazul, H.; Kanda, L.L.; Gondek, D. Impact of probiotic supplements on microbiome diversity following antibiotic treatment of mice. *Gut Microbes* **2016**, *7*, 101–114. [CrossRef]
59. Ikeyama, N.; Murakami, T.; Toyoda, A.; Mori, H.; Iino, T.; Ohkuma, M.; Sakamoto, M. Microbial interaction between the succinate-utilizing bacterium Phascolarctobacterium faecium and the gut commensal Bacteroides thetaiotaomicron. *Microbiologyopen* **2020**, *9*, e1111. [CrossRef]
60. Fernández-Veledo, S.; Vendrell, J. Gut microbiota-derived succinate: Friend or foe in human metabolic diseases? *Rev. Endocr. Metab. Disord.* **2019**, *20*, 439–447. [CrossRef]
61. Tannahill, G.M.; Curtis, A.M.; Adamik, J.; Palsson-McDermott, E.M.; McGettrick, A.F.; Goel, G.; Frezza, C.; Bernard, N.J.; Kelly, B.; Foley, N.H.; et al. Succinate is an inflammatory signal that induces IL-1β through HIF-1α. *Nature* **2013**, *496*, 238–242. [CrossRef]
62. Chouchani, E.T.; Pell, V.R.; Gaude, E.; Aksentijević, D.; Sundier, S.Y.; Robb, E.L.; Logan, A.; Nadtochiy, S.M.; Ord, E.N.J.; Smith, A.C.; et al. Ischaemic accumulation of succinate controls reperfusion injury through mitochondrial ROS. *Nature* **2014**, *515*, 431–435. [CrossRef]

antioxidants

MDPI

Article

Novel Antioxidant Insights of Myricetin on the Performance of Broiler Chickens and Alleviating Experimental Infection with *Eimeria* spp.: Crosstalk between Oxidative Stress and Inflammation

Waleed Rizk El-Ghareeb [1,2,*], Asmaa T. Y. Kishawy [3], Reham G. A. Anter [4], Asmaa Aboelabbas Gouda [4], Walaa S. Abdelaziz [5], Bassam Alhawas [1], Ahmed M. A. Meligy [6,7], Sherief M. Abdel-Raheem [1,8], Hesham Ismail [1,9] and Doaa Ibrahim [3,*]

1 Department of Public Health, College of Veterinary Medicine, King Faisal University, P.O. Box 400, Hofuf 31982, Al-Ahsa, Saudi Arabia
2 Food Control Department, Faculty of Veterinary Medicine, Zagazig University, Zagazig 44519, Egypt
3 Department of Nutrition and Clinical Nutrition, Faculty of Veterinary Medicine, Zagazig University, Zagazig 44511, Egypt
4 Department of Parasitology, Faculty of Veterinary Medicine, Zagazig University, Zagazig 44511, Egypt
5 Avian and Rabbit Medicine Department, Faculty of Veterinary Medicine, Zagazig University, Zagazig 44511, Egypt
6 Department of Clinical Science, Central Lab, College of Veterinary Medicine, King Faisal University, P.O. Box 400, Hofuf 31982, Al-Ahsa, Saudi Arabia
7 Department of Physiology, Agricultural Research Center (ARC), Giza 12511, Egypt
8 Department of Animal Nutrition and Clinical Nutrition, Faculty of Veterinary Medicine, Assiut University, Assiut 71526, Egypt
9 Food Hygiene Department, Faculty of Veterinary Medicine, Assiut University, Assiut 71526, Egypt
* Correspondence: welsaid@kfu.edu.sa (W.R.E.-G.); doibrahim@vet.zu.edu.eg (D.I.)

Citation: El-Ghareeb, W.R.; Kishawy, A.T.Y.; Anter, R.G.A.; Aboelabbas Gouda, A.; Abdelaziz, W.S.; Alhawas, B.; Meligy, A.M.A.; Abdel-Raheem, S.M.; Ismail, H.; Ibrahim, D. Novel Antioxidant Insights of Myricetin on the Performance of Broiler Chickens and Alleviating Experimental Infection with *Eimeria* spp.: Crosstalk between Oxidative Stress and Inflammation. *Antioxidants* **2023**, *12*, 1026. https://doi.org/10.3390/antiox12051026

Academic Editors: Luciana Rossi and Matteo Dell'Anno

Received: 6 March 2023
Revised: 6 April 2023
Accepted: 24 April 2023
Published: 28 April 2023

Abstract: In the modern poultry industry, the application of novel phytogenic bioactive compounds with antioxidant potential aims to enhance productivity and quality and to minimize the stress of associated diseases. Herein, myricetin, a natural flavonoid, was evaluated for the first time on broiler chickens' performance, antioxidants and immune modulating functions, and tackling avian coccidiosis. A total of 500 one-day-old chicks were divided into five groups. The negative (NC) and infected control (IC) groups were fed a control diet without additives, and the latter was infected with *Eimeria* spp. Groups supplemented with myricetin (Myc) were fed a control diet of Myc (200, 400 and 600 mg/kg diet each). On d 14, all chicks except those in NC were challenged with oocysts of mixed *Eimeria* spp. Significant improvements in the overall growth rate and feed conversion ratio were detected in the group that was fed 600 mg/kg, unlike the IC group. Notably, groups that were fed 400 and 600 mg/kg showed higher total meat antioxidant capacity with an inverse reduction in oxidative and lipid peroxidation biomarkers (hydrogen peroxide: H_2O_2; reactive oxygen species: ROS; Malondialdehyde: MDA). Of note, the upregulation of glutathione peroxidase; *GSH-Px*, catalase; *CAT*, superoxide dismutase; *SOD*, heme oxygenase-1; *HO-1* and NAD(P)H dehydrogenase quinone 1 *NQO1* genes in jejunum and muscle were prominently observed with increasing levels of supplemental Myc. At 21 dpi, the severity of coccoidal lesions ($p < 0.05$) induced by mixed *Eimeria* spp. and oocyst excretion were greatly reduced in the group that was fed 600 mg/kg of Myc. In the IC group, higher serum levels of C-reactive protein; CRP and nitric oxide; and NO and the upregulated expression of inflammatory biomarkers (interleukin-1β; IL-1β, interleukin-6; IL-6, tumor necrosis factor-α; TNF-α, chemotactic cytokines; CCL20, stromal cell-derived factor-1; CXCL13, and avian defensins; AvBD612) were subsided in higher levels in the Myc-fed groups. Taken together, these findings indicate the promising antioxidant role of Myc in modulating immune responses and reducing growth depression associated with coccidia challenges.

Keywords: myricetin; flavonoid; performance; antioxidant; inflammation; *Eimeria* spp.

1. Introduction

The poultry gastrointestinal tract (GIT) offers a biological environment for nutrient digestion and absorption, as well as protection from pathogens and toxins. Oxidative stress in birds' GIT is derived from nutritional factors, environmental heat stress, and pathological factors, which alter the overall performance as well as meat quality [1,2]. Biological damage associated with oxidative stress can cause many degenerative health issues, which have a great impact on the overall performance and productivity of livestock [3]. Studies have suggested that the interaction of mucosa with microbes or their toxins triggers oxidative stress [4–6]. The supplementation of antioxidant-rich diets and plant extracts having antioxidant properties that scavenge reactive oxygen species (ROS) are beneficial in mitigating oxidative stress in the GITs of animals [7–9]. Avian coccidiosis is among the most common parasitic diseases caused by genus *Eimeria* spp. in the poultry industry that is responsible for great economic losses, increasing mortality and lowering growth rates [10]. *Eimeria* spp., primarily producing proinflammatory mediators together with oxidative stress, contributes to lipid peroxidation, antioxidant insults, damage of the intestinal epithelial barrier, inflammatory injury and diarrhea [11]. Following infection with parasites, particularly with *Eimeria* spp., the antioxidant systems of chickens are significantly disrupted. The use of antioxidants is critical in combating the oxidative stressors caused by the production of ROS and for the maintenance of homeostasis [12]. Under the conditions of ROS overproduction, supplementation with compounds of high antioxidant potential is immensely valuable [13,14]. Concomitantly, regarding the prolonged use of common traditional approaches, such as coccidacidal pharmaceuticals that are used for controlling avian coccidiosis, these chemotherapeutics have been banned due to decreased *Eimeria* spp. sensitivity and developed drug resistance [15,16]. Therefore, the dietary inclusion of bioactive compounds that have new antioxidant and immunological prophylactic properties can solve these previous issues by exerting specific coccidiastat effects [17,18]. Prior studies have recommended that natural phytogenic-derived agents, such as resveratrol, oregano essential oil and aloe vera, have profound impacts on ameliorating oxidative stress for animals [19,20]. Moreover, phytogenic substances, rich in isoflavones, have been recognized as alternative additives to replace antibiotic usage in poultry farming via enhancing intestinal integrity and controlling inflammatory signaling pathways [21]. The anticoccidial properties of several natural herbal products (or their extracts) is mainly attributed to their ability to lower the impact on the output of oocyst via the inhibition or suppression of the invasion, replication and expansion of *Eimeria* ssp. in chickens' gastrointestinal tissues [22].

Moreover, the therapeutic potential and defensive role of these phytogenics against artificially induced coccidiosis in chickens is mainly due to their phenolic compounds that interact with cytoplasmic membranes, triggering coccidial sporozoite death, attenuating intestinal lipid peroxidation, facilitating epithelial injury repair and lessening the intestinal permeability triggered by *Eimeria* spp. [23]. Phenolic compounds, such as flavones (apigenin and luteolin), hydroxycinnamic, caffeic and sinapic acids, have been tested for their inhibitory effective role on the sporulation of coccidian oocysts [24–26]. Thus, searching for a new bioactive phenolics candidate with anticoccidial efficacy is an important prerequisite. Myricetin is a natural flavonoid with known strong antioxidant, anti-inflammatory and anticancer properties [27,28]. Additionally, the antiparasitic effect of myricetin was recently documented against Schistosomiasis [29]. The potential antioxidant activity of myricetin could be related to the presence of three hydroxyl groups on its B ring like other flavonoids [30]. Furthermore, myricetin has also proved to modify inflammatory diseases by suppressing pro-inflammatory mediators (inhibition of inflammatory mediators such as TNF-α, IL-6, IL-12 and iNOS) [31,32]. Myricetin can effectively reduce oxidative stress via down-stream expression of HO-1 and NQ1 [33]. At present, the impact of myricetin on combating oxidative stress and avian coccidiosis has not been reported yet in the poultry industry. Hence, this study was designed to assess the effective role of in-feed myricetin on broiler chickens' performance indices, antioxidant and immune modulation, fecal oocyst excretion and intestinal lesion score following *Eimeria* ssp. challenges.

2. Materials and Methods

Husbandry, ethics and guidelines for animal management practices and techniques were obtained from the Institutional Animal Care and Use Committee of the Faculty of Veterinary Medicine at Zagazig University (ZU-IACUC/2/F/320/2020).

2.1. Birds, Experimental Design and Growth Monitoring

Five hundred one-day-old male chicks with an initial body weight of 44.2 ± 0.2 g were delivered by a local hatchery (commercial Ross 308 broiler chicks). Chicks were weighed and randomly divided into five equal experimental groups, with 5 replicate pens of 10 birds each. All birds were housed in floor pens with wood shavings (bird density: 10 broilers/m^2) in similar environmental and sanitary circumstances throughout the experimental time. The room temperature throughout the 1st week was adjusted to be 33 °C and then progressively declined until reaching 23 °C. Three diets were offered during the trial in a crumble form: starter (1 to 10 d), grower (11 to 20 d) and finisher (21 to 42 d) diets, which were formulated to furnish the Ross broilers' nutrient requirements in accordance with the nutritional specification of ROSS [34]. All birds were fed and watered with ad libitum access. Experimental groups were established as follows. There were two control groups. NC birds received a basal diet without additives and were not challenged. IC birds received a basal diet without additives and were challenged at d 14 of age with *Eimeria* spp. Birds in groups 2, 3 and 4 received a basal diet supplemented with myricetin (Myc, sigma Aldrich, Product No, 529-44-2) at concentrations of 200 (Myc 200), 400 (Myc 400) and 600 (Myc 600) mg/kg diet, respectively, and were challenged at d 14 of age with *Eimeria* spp. The NC group was isolated in a separate area and was checked to be free from coccidian infection via fecal examination at d 14 and 21 of the experiment.

The levels of feed ingredient and chemical composition of the control diet are listed in Table 1, formulated according to the Ross broiler guidelines [35]. The proximate analysis of the feed components was made along with the standard procedures of the Association of Official Agricultural Chemists [36].

2.2. Experimental Challenge by Eimeria spp.

A potassium dichromate solution (2%) was used for the induction of oocysts sporulation, followed by washing numerous times with tap water for potassium dichromate removal. At 14 days of age, experimental birds were challenged with *Eimeria* spp. via gavaging a 2 mL suspension of sporulated oocysts from *E. tenella* (5.0×10^3), *E. maxima* (7.0×10^3) and *E. acervulina* (3.5×10^4) into the crop via a plastic syringe that fitted with a plastic cannula. Before gavaging the birds with sporulated coccidian oocysts, their number per mL was checked via microscope utilizing a McMaster counting chamber.

2.3. Growth Performance Parameters

The initial individual weights of the chicks were determined on the day of arrival, and then body weight (BW) and the average daily feed intake (FI) in each replicate pen were recorded to calculate the body weight gain (BWG) and feed conversion ratio (FCR) for the entire experimental period (d 1–42), as previously reported by Kishawy et al. [37]. Mortality was noted daily and then calculated for each at the end of the experiment as a percentage of the total bird number.

Table 1. Feed ingredients and chemical analysis of the basal diet (as dry matter).

Ingredients g/kg	Starter (0–10 Day)	Grower (11–24 Day)	Finisher (25–42 Day)
Yellow corn grain	57.00	60.60	62.00
Soybean meal, 47.5%	35.00	29.00	25.00
Corn gluten, 60%	2.80	4.50	4.00
Wheat bran	–	–	1.90
Soybean oil	1.20	2.00	3.66
Calcium carbonate	1.00	1.00	0.90
Dicalciumphosphate	1.96	1.90	1.60
Common salt	0.30	0.30	0.30
Premix *	0.30	0.30	0.30
DL-Methionine, 98%	0.18	0.14	0.11
Lysine, Hcl, 78%	0.16	0.16	0.13
Anti-mycotoxin	0.10	0.10	0.10
		Analyzed chemical composition	
Metabolic energy, Kcal/Kg	3000.40	3104.52	3202.02
Crude protein, %	23.02	21.44	19.57
Ether extract, %	3.72	4.60	6.24
Crude fiber, %	2.66	2.55	2.63
Calcium, %	1.01	0.98	0.86
Available phosphorus, %	0.50	0.48	0.41
Lysine, %	1.38	1.22	1.10
Methionine, %	0.56	0.52	0.46

* Vitamin and mineral premix supplied per kg of diet, as follows: Vitamin A, 12,500 IU; Vitamin D3, 2300 IU; Vitamin E, 30 IU; Vitamin K3, 6.00 mg; Vitamin B1, 3.85 mg; Vitamin B2, 6.62 mg; Vitamin B6, 1.6 g; Pantothenic acid, 20 mg; Vitamin B12, 0.5 mg; Niacin, 40 mg; Folic acid, 1.5 mg; Biotin, 0.7 mg; Fe, 55 mg; Mn, 65 mg; Cu, 7 mg; I, 0.9 mg; Co, 1.2 mg; Se, 0.30 mg; Zn, 55 mg; and Choline chloride, 600 mg.

2.4. *Fecal Oocytes Shedding of Eimeria spp.*

For oocyst output determination, at 7-, 14- and 21-days post-infection (dpi), the total fecal output of each replicate pen was weighed, and fecal samples were collected daily, homogenized thoroughly, and directly examined for *Eimeria* spp. oocytes. One gram of a homogenized fecal sample was diluted 10-fold first with tap water and was then diluted with a saturated saline solution (1:10), and finally, the oocyst counts were determined according to [38] and expressed as the number of oocytes for each g of feces.

2.5. *Intestinal Lesion Score*

Five birds from each group were euthanized and slaughtered via cervical dislocation at 7 dpi for the determination of the intestinal lesion score. The intestines were directly removed and segmented into the duodenum, jejunum, ileum and cecum, and then each segment was opened. The lesion scores were performed according to [39] and ranged from 0 (no gross lesion) to 4 (most severe gross lesion), as follows: 0 refers to a normal intestinal sigment with no observed lesions, 1 refers to a small scattered petechiae, 2 refers to numerous petechiae, 3 refers to extensive hemorrhaging, and 4 refers to extensive hemorrhaging that causes dark colors in the intestinal sigment.

2.6. *Biochemical Measurements*

Blood samples were aseptically collected from birds' wing veins (one bird/replicate) for immunological analysis. Blood samples were centrifuged for 10 min at 3000 rpm for serum separation. At 4 days pre-infection and at 14 and 21 dpi post *Eimeria* spp. challenge, a clear serum was used for the assessment of myeloperoxidase (MPO), nitric oxide (NO) and C-reactive protein (CRP) activities utilizing commercial kits (Jiancheng Biotechnology Institute, Nanjing, China). Serum immunoglobulin G (IgG) levels were quantified via an enzyme-linked immunosorbent assay (*ELISA*) as previously described [40].

2.7. Oxidative and Antioxidant Evaluation

At the end of the experimental period (42 days), intestinal tissue samples and breast meat were collected and thoroughly homogenized for evaluating the oxidative and antioxidant biomarkers. Lipid oxidation measurements were assessed via a thiobarbituric acid-reactive assay (TBARS) value, as defined by Ahn, Olson [41], and TBARS values were expressed as nmol/g of tissue. The total antioxidant capacity (T-AOC) was determined using a commercial assay kit (Sigma-Aldrich, MAK187). Hydrogen peroxide (H_2O_2) levels were assessed according to Loreto and Velikova [42], and their levels were expressed as μmoL/g of tissue. Moreover, reactive oxygen species (ROS) were assessed in accordance with the method of LeBel, Ischiropoulos [43].

2.8. Gene Expression by Reverse Transcription Quantitative Real-Time PCR (RT-qPCR)

Intestinal tissues and breast meat samples were collected at the end of the experiment (42 days of age) for examining the expression levels of genes encoding glutathione peroxidase (*GSH-Px)*, super oxide dismutase (*SOD*), catalase, NAD(P)H dehydrogenase quinone 1 (*NQO1*) and heme oxygenase-1 (*HO-1*), cyclooxygenase-2 (*COX-2*),. interleukin (IL)-6, *IL-1β* and *IL-10*, tumor necrosis factor-α (*TNF-α*), (chemokine C–C motif ligand 4, also known as macrophage inflammatory proteins-1β), (*CCL4*), chemokine C–C motif ligand 20, also known as macrophage inflammatory proteins-3 α (*CCL20*) and stromal cell-derived factor-1 (*CXCL13*) Avian β-defensin 6 and 12 (*AvBD6* and *AvBD612*). The isolation of total RNA was performed utilizing QIAamp RNeasy Mini kit (Qiagen, Hilden, Germany), and the RNA concentration was quantified at 260 nm via a spectrophotometer. The assay of one-step RT-qPCR was achieved on the Stratagene MX3005P real-time PCR using a QuantiTect SYBR Green RT-PCR Kit (Qiagen, Hilden, Germany). All measurements of PCR were employed in triplicate. The distinction of each PCR amplification assay was validated via an analysis of the final melting curve. Various transcript levels were then standardized by using glyceraldehyde 3-phosphate dehydrogenase (GAPDH) as an endogenous control. All gene-distinct primer sequences employed in RT-qPCR assay are listed in Table 2. The results of the relative mRNA expression of the investigated genes were evaluated using the $2^{-\Delta\Delta Ct}$ method [44].

Table 2. Primer sequences employed for reverse transcription quantitative polymerase chain reaction assay.

Specificity/Target Gene	Primer Sequence (5′-3′)	Accession No.
CAT	F-GGGGAGCTGTTTACTGCAAG R-GGGGAGCTGTTTACTGCAAG	NM_001031215.2
SOD	F-GGCAATGTGACTGCAAAGGG R-CCCCTCTACCCAGGTCATCA	NM_205064.1
GSH-Px	F-AACCAATTCGGGCACCAG R-CCGTTCACCTCGCACTTCTC	HM590226
HO-1	F-AAGAGCCAGGAGAACGGTCA R-AAGAGCCAGGAGAACGGTCA	NM_205344
NQO1	F-TCGCCGAGCAGAAGAAGATTGAAG R-CGGTGGTGAGTGACAGCATGG	NM_001277620.1
COX-2	F-TGTCCTTTCACTGCTTTCCAT R-TTCCATTGCTGTGTTTGAGGT	NM_0,011,67718.1
IL-6	F-AGGACGAGATGTGCAAGAAGTTC R-TTGGGCAGGTTGAGGTTGTT	NM_204,628
IL-1β	GCTCTACATGTCGTGTGTGATGAG TGTCGATGTCCCGCATGA	NM_204,524
TNF-α	F-CCCCTACCCTGTCCCACAA R-ACTGCGGAGGGTTCATTCC	XM_046900549.1

Table 2. *Cont.*

Specificity/Target Gene	Primer Sequence (5′-3′)	Accession No.
CCL4	F: GCAGTTGTTCTCGCTCTTC R: GCGCTCCTTCTTTGTGAT	NM_204720.1
CCL20	F: AGGCAGCGAAGGAGCAC R: GCAGAGAAGCCAAAATCAAAC	NM_204438
CXCL13	F: GCCTGTGCCTGGTGCTC R: TGCCCCCTTCCCCTAAC	NM_001348657.1
AVBD6	F:GCCCTACTTTTCCAGCCCTATT R: GGCCCAGGAATGCAGACA	NM 001001193.1
AVBD12	F:TGTAACCACGACAGGGGATTG R: GGGAGTTGGTGACAGAGGTTT	NM 001001607.2
GAPDH	F: GGTGGTGCTAAGCGTGTTA R: CCCTCCACAATGCCAA	NM205518

Catalase (*CAT*), superoxide dismutase (*SOD*), glutathione peroxidase (*GSH-Px*), heme oxygenase-1 (*HO-1*), NAD(P)H dehydrogenase quinone 1 (*NQO1*), cyclooх-ygenase-2 (*COX-2*) and interleukin (*IL*)-*6*, *IL-1β* and *IL-10*, tumor necrosis factor-α (*TNF-α*), (chemokine C–C motif ligand 4, also known as macrophage inflammatory proteins-1β), (*CCL4*), chemokine C–C motif ligand 20, also known as macrophage inflammatory proteins-3 α (*CCL20*) and stromal cell-derived factor-1 (*CXCL13*) Avian β-defensin 6 and 12 (*AvBD6* and *AvBD12*), glyceraldahyde-3-phosphate dehydrogenase (GAPDH).

2.9. Statistical Analysis

Statistical data analysis was achieved via the general linear method (GLM) of SPSS. The homogeneity among experimental groups was estimated using Levene's test, and normality was assessed using the Shapiro–Wilk test. Tukey's test was performed to distinguish the mean values and variations that were significant. Data variation was conveyed as the standard error of the mean (SEM), and the statistical significance was adjusted at a p value of less than 0.05. All graphs were created by GraphPad Prism software Version 8.

3. Results

3.1. Growth Performance

The impact of supplementing diets with varying levels of Myc on the growth performance parameters in response to *Eimeria* spp. infection is shown in Table 3. At the end of the starter period, the BW and BWG were significantly ($p < 0.05$) improved in groups supplemented with Myc in a dose-dependent manner compared to the control groups. Feed intake was also significantly ($p < 0.05$) increased in the Myc-supplemented groups compared to the control groups. A superior feed conversion ratio was found ($p > 0.05$) in the group that was fed the 600 mg/kg diet of Myc. Regarding the grower period, the most lowered BW and BWG was ($p < 0.05$) detected in the IC group. Moreover, Myc supplementation at higher levels restored the BW and BWG nearly the same as that in the NC group. Higher feed intake was recorded for the IC group, and then it decreased ($p < 0.05$) gradually with Myc inclusion in a dose-dependent manner. The best FCR was noted for the NC group, and the worst one was recorded for the IC group. Feeding with Myc improved ($p < 0.05$) FCR when compared with the IC group in a dose-dependent manner. The performance parameters at the end of the finisher period were nearly the same as that of the grower period, as BW and BWG were significantly ($p > 0.05$) decreased in the IC group compared with the NC group and then significantly ($p > 0.05$) increased with higher levels of Myc supplementation (400 and 600 mg/kg diet). The lowest feed intake during the finisher period ($p < 0.05$)) was found in the IC and Myc200 groups, and the NC group exhibited the highest significant ($p < 0.05$) feed intake. Moreover, FCR was significantly ($p > 0.05$) increased in the IC group, and the NC, Myc 400 and Myc 600 supplemented groups did not significantly ($p > 0.05$) differ. Regarding the overall performance parameters, inferior ($p < 0.05$) performance parameters were noticed for the IC group, and Myc supplementation improved all these parameters. Notably, no significant difference ($p > 0.05$) in FCR was detected between the infected group fed with Myc 600 and the non-infected group fed with the basal diet.

Table 3. Effect of supplementing diets with varying levels of myricetin on growth performance parameters of broiler Ross chickens.

	Myricetin (mg/kg Diet)						
Parameters	**NC**	**IC**	**Myc 200**	**Myc 400**	**Myc 600**	***p*-Value**	**SEM**
			Starter period (0–10 day)				
Initial BW (g/bird)	44.40	44.40	44.40	44.20	44.40	0.979	0.11
BW (g/bird)	269.20 ^{d}	268.80 ^{d}	280.60 ^{c}	286.60 ^{b}	291.00 ^{a}	<0.001	1.87
BWG (g/bird)	224.80 ^{d}	224.40 ^{d}	236.20 ^{c}	242.40 ^{b}	246.60 ^{a}	<0.001	1.88
FI (g/bird)	311.60 ^{b}	310.80 ^{b}	318.40 ^{a}	324.00 ^{a}	320.40 ^{a}	<0.001	1.24
FCR	1.39 ^{a}	1.39 ^{a}	1.35 ^{b}	1.34 ^{b}	1.30 ^{c}	<0.001	0.01
			Grower period (11–24 day)				
BW (g/bird)	1130.40 ^{a}	1065.80 ^{d}	1083.40 ^{c}	1097.00 ^{b}	1127.40 ^{a}	<0.001	5.15
BWG (g/bird)	861.20 ^{a}	797.00 ^{d}	802.80 ^{d}	810.40 ^{c}	836.40 ^{b}	<0.001	4.96
FI (g/bird)	1281.20 ^{cd}	1356.40 ^{a}	1329.20 ^{b}	1286.40 ^{c}	1266.40 ^{d}	<0.001	7.09
FCR	1.49 ^{e}	1.70 ^{a}	1.66 ^{b}	1.59 ^{c}	1.51 ^{d}	<0.001	0.02
			Finisher period (25–42 day)				
BW (g/bird)	2602.33 ^{a}	2145.67 ^{d}	2267.33 ^{c}	2459.67 ^{b}	2496.67 ^{b}	<0.001	34.00
BWG (g/bird)	1471.93 ^{a}	1079.87 ^{d}	1183.93 ^{c}	1362.67 ^{b}	1369.27 ^{b}	<0.001	29.24
FI (g/bird)	2595.73 ^{a}	2360.00 ^{c}	2342.67 ^{c}	2483.67 ^{b}	2416.67 ^{bc}	<0.001	22.23
FCR	1.76 ^{c}	2.19 ^{a}	1.98 ^{b}	1.82 ^{c}	1.76 ^{c}	<0.001	0.03
			Overall performance (0–42 day)				
Final BW (g/bird)	2602.33 ^{a}	2145.67 ^{d}	2267.33 ^{c}	2459.67 ^{b}	2496.67 ^{b}	<0.001	34.00
Total BWG (g/bird)	2557.93 ^{a}	2101.27 ^{d}	2222.93 ^{c}	2415.47 ^{b}	2452.27 ^{b}	<0.001	34.01
Total FI (g/bird)	4188.53 ^{a}	4027.20 ^{b}	3990.27 ^{b}	4093.73 ^{ab}	4003.47 ^{b}	<0.001	19.29
Overall FCR	1.64 ^{d}	1.92 ^{a}	1.80 ^{b}	1.69 ^{c}	1.63 ^{d}	<0.001	0.02

BW = body weight; BWG = body weight gain; FI = feed intake; FCR = feed conversion ratio. ^{a,b,c,d,e} Means within a row carrying different superscript letters denote significant differences ($p < 0.05$). NC (negative control): birds fed with basal diet; IC (positive control): birds fed with basal diet and challenged with *Eimeria* spp. at d 14 of age; Myricetin (Myc) 200, 400 and 600: birds fed with basal diet supplemented with Myc at the levels of 200, 400 and 600 mg/kg.

3.2. Oxidative and Antioxidant Status in Muscle and Intestinal Tissues

As displayed in Table 4, significant higher T-AOC levels ($p < 0.05$) were detected with increasing the supplemental level of Myc in both intestinal and muscle tissues. A noticeable reduction ($p < 0.05$) in lipid peroxidation biomarkers (MDA) was found in groups fed with Myc-supplemented diets in a dose-dependent manner in intestinal and muscle tissues. A remarkable decline ($p < 0.05$) in ROS production was detected in the muscle tissues of the group fed with Myc at the level of 600 (mg/kg diet). ROS contents in jejunal tissues were significantly decreased dose dependently. Groups fed with higher Myc levels exhibited a remarkable decline in H_2O_2 in both muscle and intestinal tissues.

3.3. Fecal Oocytes Count, Intestinal Lesion Score and Mortality Percent

The impact of supplementing diets with different Myc levels on oocyte fecal shedding, intestinal lesion score and mortality percent is demonstrated in Table 5. *Eimeria* spp. oocytes were not detected in the feces of the NC group. At 7 dpi, the fecal oocyte count was elevated ($p < 0.05$) in the IC and Myc 200 groups compared with the Myc 400 and 600 groups, and at 14 and 21 dpi, the oocyte count decreased ($p > 0.05$) in the feces of the Myc-supplemented groups in a dose-dependent manner compared with the IC group. Regarding the intestinal lesion score, the IC group displayed ($p < 0.05$) the most severe lesion score in the duodenum, jejunum, ileum and cecum, and the Myc-600-fed group had ($p < 0.05$) a reduced lesion score in all intestinal segments. The highest mortality percent was detected in ($p < 0.05$) the IC group. Notably, the highest ($p < 0.05$) mortality rate was detected at 7 dpi in all infected groups. The maximum ($p < 0.05$) mortality rate was noticed in the infected control group compared with the other groups; moreover, the mortality rate was significantly ($p < 0.05$) decreased with the increase in myricetin level.

Table 4. Effect of supplementing diets with varying levels of myricetin on oxidative/antioxidant biomarkers in muscle and intestinal tissues.

	Myricetin (mg/kg Diet)						
Parameters	**NC**	**IC**	**Myc 200**	**Myc 400**	**Myc 600**	***p*-Value**	**SEM**
Muscle Tissues							
T-AOC (U/mg of protein)	1.79 [c]	1.62 [c]	2.46 [b]	2.98 [ab]	3.35 [a]	<0.001	0.69
MDA (nmol/g tissue)	19.96 [b]	21.69 [a]	19.10 [b]	18.32 [c]	16.20 [d]	<0.001	1.29
ROS	63.47 [b]	86.14 [a]	64.12 [b]	63.52 [b]	57.66 [c]	<0.001	3.03
H_2O_2 (μmoL/g tissue)	2.69 [b]	3.85 [a]	2.31 [c]	2.21 [cd]	2.01 [d]	<0.001	0.53
Intestinal Tissues (Jejunum)							
T-AOC (U/mg of protein)	1.13 [c]	0.63 [d]	1.26 [b]	1.29 [b]	1.39 [a]	<0.001	0.022
MDA (nmol/g tissue)	15.32 [b]	21.69 [a]	13.55 [c]	12.90 [d]	12.11 [e]	<0.001	1.36
ROS (μL/g tissue)	58.17 [b]	66.10 [a]	58.18 [b]	55.12 [c]	51.60 [d]	<0.001	5.15
H_2O_2 (μmoL/g tissue)	2.23 [b]	3.99 [a]	1.96 [c]	1.78 [cd]	1.61 [d]	<0.001	0.50

T-AOC: total antioxidant capacity; MDA: malondialdehyde, ROS: reactive oxygen species, H_2O_2: hydrogen peroxide. [a,b,c,d,e] Mean values with various letters in the same row are significantly different at $p < 0.05$. Data are described as means ± SE. NC (negative control): birds fed with basal diet; IC (positive control): birds fed with basal diet and challenged with *Eimeria* spp. at d 14 of age; Myricetin (Myc) 200, 400 and 600: birds fed with basal diet supplemented with Myc at the levels of 200, 400 and 600 mg/kg.

3.4. Serum Inflammatory and Immune-Related Biomarkers

The impact of supplementing diets with varying levels of Myc on the serum immune-related biomarkers of Ross broiler chickens challenged with *Eimeria* spp. is presented in Table 6. At d 4 pre-infection, the supplementation of the broilers' diets with Myc improved the birds' immune-associated parameters via increasing ($p < 0.05$) the levels of IgG and decreasing CRP values, especially at higher doses. At 14 dpi, post-coccidial infection, NO, CRP and MPO were significantly ($p < 0.05$) elevated in the IC group compared to the NC group, and supplementation with Myc decreased ($p < 0.05$) their serum levels, especially at higher doses. Moreover, the highest serum IgG level was detected in the Myc-600-fed group ($p < 0.05$). At 21 dpi, NO, CRP and MPO levels were significantly ($p < 0.05$) decreased in groups supplemented with Myc when compared with the IC group. Moreover, serum IgG levels were significantly ($p < 0.05$) elevated by Myc supplementation, especially at high doses compared to the IC group.

3.5. Intestinal and Muscles Antioxidants Gene Expression

The impact of supplementing diets with varying levels of Myc on the expression of antioxidant-related genes is shown in Figure 1. Intestinal and muscle *CAT, SOD, GSH-Px, HO-1* and *NQO1* gene expression were significantly downregulated in the IC group when compared with the NC group. Moreover, Myc supplementation significantly upregulated ($p < 0.05$) their expression with increasing their dose. In contrast, the expression of COX-2 genes was significantly upregulated in the IC group, unlike the NC group. Interestingly, groups fed with Myc significantly downregulated COX-2 expression ($p < 0.05$) in a dose-dependent manner in both muscles and intestines.

Table 5. Effect of supplementing diets with varying levels of myricetin on fecal oocyte count, intestinal lesion score and mortality % of broiler Ross chickens.

			Myricetin (mg/kg Diet)				
Parameters	**NC**	**IC**	**Myc 200**	**Myc 400**	**Myc 600**	**p-Value**	**SEM**
			Fecal oocytes count ($\times 10^3$/g feces)				
7 dpi	ND	355.40 ^{a}	352.20 ^{a}	340.00 ^{b}	339.60 ^{b}	<0.001	28.36
14 dpi	ND	227.60 ^{a}	212.80 ^{b}	169.60 ^{c}	135.20 ^{d}	<0.001	16.62
21 dpi	ND	74.20 ^{a}	54.40 ^{b}	44.00 ^{c}	34.80 ^{d}	<0.001	5.03
			Intestinal lesion score 7 dpi				
Duodenal lesion score	ND	3.40 ^{a}	3.00 ^{b}	2.80 ^{b}	2.60 ^{c}	<0.001	0.27
Jujenal lesion score	ND	3.80 ^{a}	3.60 ^{a}	3.20 ^{a}	2.20 ^{b}	<0.001	0.29
Ileal lesion score	ND	3.80 ^{a}	3.40 ^{a}	3.40 ^{a}	2.00 ^{b}	<0.001	0.30
Cecal lesion score	ND	3.60 ^{a}	3.20 ^{a}	3.20 ^{a}	2.20 ^{b}	<0.001	0.28
			Mortality % overall the experimental period				
Mortality % 7 dpi	– ^{e}	18.00 ^{a}	10.00 ^{b}	8.80 ^{c}	7.40 ^{d}	<0.001	1.20
Mortality % 14 dpi	0.80 ^{d}	11.00 ^{a}	8.00 ^{b}	5.00 ^{c}	5.00 ^{c}	<0.001	0.76
Mortality % 21 dpi	– ^{e}	5.00 ^{a}	4.00 ^{b}	3.00 ^{c}	2.00 ^{d}	<0.001	0.45

^{a,b,c,d,e} Means within a row carrying different superscript letters denote significant differences ($p < 0.05$). ND: not detected; dpi: days post infection; NC (negative control): birds fed with basal diet; IC (positive control): birds fed with basal diet and challenged with *Eimeria* spp. at d 14 of age; Myricetin (Myc) 200, 400 and 600: birds fed with basal diet supplemented with Myc at the levels of 200, 400 and 600 mg/kg.

Table 6. Effect of supplementing diets with varying levels of myricetin on serum inflammatory and immune-related biomarkers of Ross broiler chickens.

			Myricetin (mg/kg Diet)				
Parameters	**NC**	**IC**	**Myc 200**	**Myc 400**	**Myc 600**	**p-Value**	**SEM**
			Pre infection (4 days pre coccidian infection)				
NO (µmol/L)	0.64	0.64	0.63	0.60	0.59	<0.07	0.01
CRP (mg/L)	3.51 ^{a}	3.51 ^{a}	3.32 ^{b}	3.01 ^{c}	2.73 ^{d}	<0.001	0.06
MPO (µmol/L, OD 450 nm)	0.45	0.45	0.42	0.44	0.47	0.013	0.06
IgG (mg/dL)	1.67 ^{c}	1.67 ^{c}	1.81 ^{b}	1.86 ^{b}	2.38 ^{a}	<0.001	0.06
			14 days post infection				
NO (µmol/L)	0.65 ^{d}	1.28 ^{a}	0.99 ^{b}	0.99 ^{b}	0.87 ^{c}	<0.001	0.04
CRP (mg/L)	3.51 ^{d}	10.58 ^{a}	8.60 ^{b}	8.67 ^{b}	6.92 ^{c}	<0.001	0.49
MPO (µmol/L, OD 450 nm)	0.46 ^{d}	1.67 ^{a}	1.50 ^{b}	1.49 ^{b}	1.25 ^{c}	<0.001	0.09
IgG (mg/dL)	1.70 ^{d}	5.50 ^{c}	6.67 ^{b}	6.78 ^{b}	7.46 ^{a}	<0.001	0.42
			21 days post infection				
NO (µmol/L)	0.65 ^{e}	1.02 ^{a}	0.85 ^{b}	0.79 ^{c}	0.72 ^{d}	<0.001	0.03
CRP (mg/L)	3.49 ^{d}	7.16 ^{a}	5.46 ^{b}	5.45 ^{b}	4.58 ^{c}	<0.001	0.25
MPO (µmol/L, OD 450 nm)	0.44 ^{d}	1.25 ^{a}	0.90 ^{b}	0.80 ^{b}	0.66 ^{c}	<0.001	0.06
IgG (mg/dL)	1.66 ^{d}	6.40 ^{c}	7.80 ^{b}	7.71 ^{b}	8.64 ^{a}	<0.001	0.51

NO: nitric oxide; CRP: c-reactive protein; MPO: myeloperoxidase; IgG: immunoglobulin-G. ^{a,b,c,d,e} Mean values with various letters in the same row are different significantly at $p < 0.05$. Data are described as means $\pm$ SE. NC (negative control): birds fed with basal diet; IC (positive control): birds fed with basal diet and challenged with *Eimeria* spp. at d 14 of age; Myricetin (Myc) 200, 400 and 600: birds fed with basal diet supplemented with Myc at the levels of 200, 400 and 600 mg/kg.

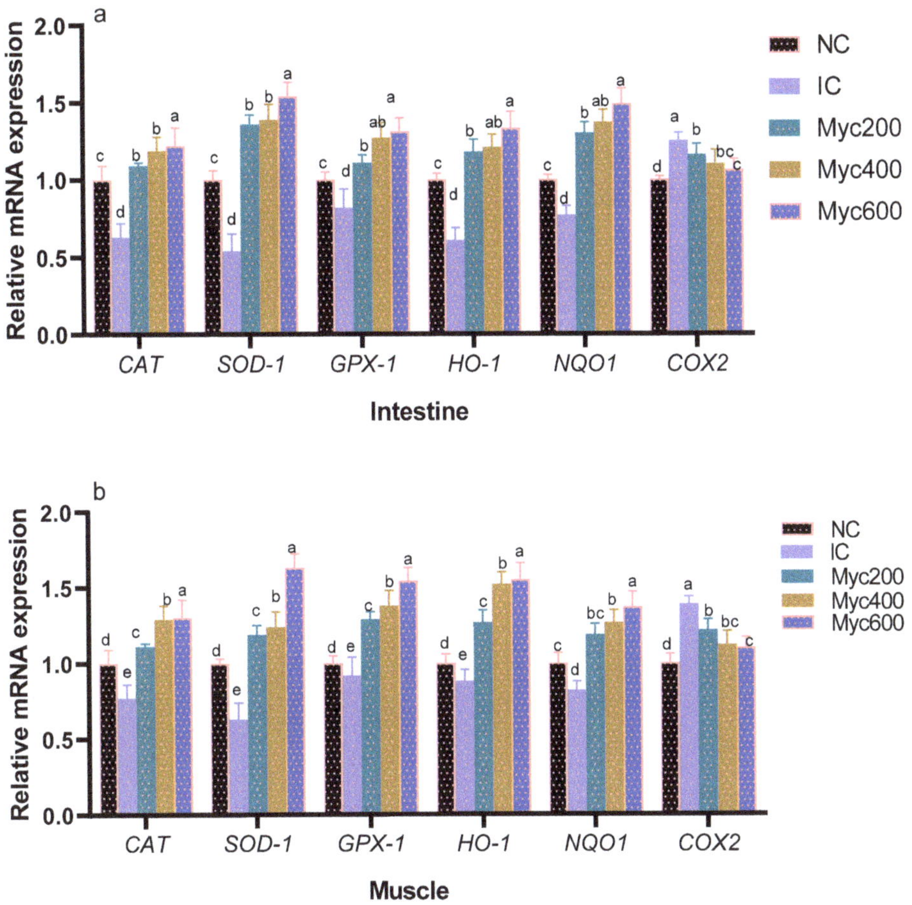

Figure 1. Effect of supplementing diets with varying levels of myricetin on the relative expression of glutathione peroxidase (*GSH-Px*), superoxide dismutase (*SOD*), catalase (*CAT*), NAD(P)H dehydrogenase quinone 1 (*NQO1*), heme oxygenase-1 (*HO-1*), and cyclooxygenase-2 (*COX-2*) genes in intestinal (**a**) and muscle tissues (**b**) of Ross broiler chickens post infection with *Eimeria* spp. Various letters in columns point to statistical significance ($p < 0.05$). Data are described as means ± SE. NC (negative control): birds fed with basal diet; IC (positive control): birds fed with basal diet and challenged with *Eimeria* spp. at d 14 of age; Myricetin (Myc) 200, 400 and 600: birds fed with basal diet supplemented with Myc at the levels of 200, 400 and 600 mg/kg.

3.6. Cytokines and Chemokines Gene Expression

The impact of supplementing diets with varying levels of Myc on the expression of cytokine- and chemokine-related genes in intestinal tissues is shown in Figure 2. The highest expression of proinflammatory cytokine-related genes as *IL-1β*, *IL-6* and *TNF-α* was detected ($p > 0.05$) in the IC group, and Myc supplementation significantly ($p < 0.05$) downregulated them in a dose-dependent manner. In contrast, the expression of anti-inflammatory cytokine IL-10 and the AvBD6 and AvBD612 genes were significantly ($p > 0.05$) downregulated in the IC group. Moreover, Myc supplementation prominently ($p > 0.05$) upregulated their levels. Regarding the expression of chemokines involving CCL4, CCL20 and CXCL13, their expression was significantly ($p > 0.05$) upregulated in the IC group in comparison; moreover, higher levels of Myc supplementation significantly ($p > 0.05$) downregulated their expression.

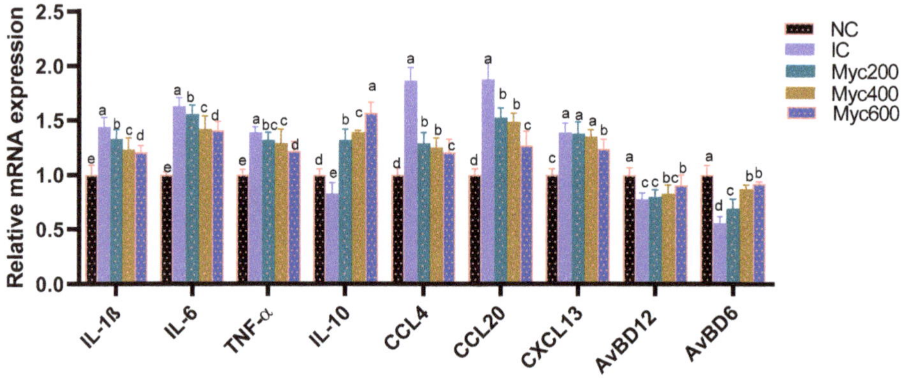

Figure 2. Effect of supplementing diets with varying levels of myricetin on relative expression of interleukin (*IL*)-*6*, *IL-1β* and *IL-10*, tumor necrosis factor-α (*TNF-α*), (chemokine C–C motif ligand 4, also known as macrophage inflammatory proteins -1β), (*CCL4*), chemokine C–C motif ligand 20, also known as macrophage inflammatory proteins -3 α (*CCL20*) and stromal cell-derived factor-1 (*CXCL13*) Avian β-defensin 6 and 12 (*AvBD6* and *AvBD612*), post infection of Ross broiler chickens with *Eimeria* spp. Various letters in columns point to statistical significance ($p < 0.05$). Data are described as means ± SE. NC (negative control): birds fed with basal diet; IC (positive control): birds fed with basal diet and challenged with *Eimeria* spp. at d 14 of age; Myricetin (Myc) 200, 400 and 600: birds fed with basal diet supplemented with Myc at the levels of 200, 400 and 600 mg/kg.

4. Discussion

The distribution of the oxidative balance that emerges from oxidative stress, owing to higher free radical production, can trigger cell chain reactions, which in turn results in cell damage or even death. Invasion by the *Eimeria* species can evoke the over production of ROS and free radicals in the host's cellular immune response and goes beyond the protection capability of the natural antioxidant defense system, which contributes to tissue damage and pathological lesions [45]. Oxidative stress can trigger inflammation that modifies gene expression related to antioxidant status, playing a critical role in the physiological function of the gastro intestinal tract [46]. Antioxidant-enriched diets are among the most substantial dietary factors for poultry, which have unique consequences not only for maintaining their better growth and preventing them from various diseases but also affecting the quality of their products offered for consumers. Plants or their bioactive principles enriched with flavonoids are considered an alternative approach to treat coccidiosis [47]. Herein, dietary myricetin harbors strong antioxidants properties that attenuate the impaired host oxidative equilibrium resulting from coccidian experimental infection. During the starter stage, the dietary inclusion of myricetin in broilers' diets improved body weight gain and the efficiency of the feed conversion ratio. In accordance, feeding with dietary polyphenol-enriched grape seeds improves the body weight gain and antioxidant status of broiler chicks [48]. In many studies, plant-derived flavonoids have been reported to have an effective role in reducing lipid oxidation, decreasing the pathogenic microbial loads in birds' intestines and intestinal pH and improving the histomorphology of the intestine, leading to maximizing nutrient absorption and promoting growth performance [49,50]. Myricetin is a plant-derived flavonoid that exhibits many activities in animals' bodies as a growth promoter with anti-inflammatory, antioxidant and anticancer properties [51]. Before infection, dietary myricetin enhances the immune response and antioxidant status of birds and consequently enhances their performance, including improving their growth rate and FCR. Coccidiosis can induce intestinal oxidative stress that greatly impairs the growth rate and feed efficiency of birds [11]. Remarkably, after *Eimeria* spp. challenges, the body gain of birds was greatly impaired in the IC group that was fed with no additives, and the Myc-supplemented groups, especially at 600 mg/kg, restored these impaired growth performance parameters. Similar results were reported by Bozkurt et al. [52], as the growth performance of broilers and feed conversion ratio worsened after coccidial infection. Moreover, the main clinical signs of coccidial infection are reductions in body weight gain and feed intake with high feed conversion and high economic losses. The alleviated growth performance due to flavonoid supplementation is in agreement with the findings of Wang et al. [47], who reported improved growth performance of experimentally coccidial-infected birds fed with diets supplemented by grape seed extract. The improvement of the birds' performance may be attributed to the anti-inflammatory effect of flavonoids that reduces the effect of the coccidial destructive effect on the intestine, ameliorating intestinal health status and decreasing diarrhea [53]. Moreover, the anticipated mechanism of action of flavonoids also resulted from their hydroxyl groups, which acted as pro-oxidants that oxidized via ROS inside cell membranes, in turn delaying the bad consequences (lipid oxidation and DNA damage) [54].

In the current study, we assessed the efficacy of myricetin against the severity of coccidian infection via evaluating fecal oocyst counts after infection at 7, 14 and 21 dpi. Coccidian infection exaggerated oocyst excretion per gram of feces after 7 dpi, especially in the infected non supplemented control. The rate of oocyst excretion in feces was observed to decrease as the number of days post infection increased in all groups. Interestingly, the rate of oocyst excretion was much lower in groups supplemented with higher doses of dietary myricetin compared to that of the infected control. In correlation, the intestinal lesion score in all segments of the intestine was reduced with increasing the supplementation levels of myricetin. Phytochemicals as herbal plant extracts or their active substances with antioxidant functions have been developed to be used as anti-parasitic agents, especially coccidiosis such as garlic extract [55], cinnamaldhyde [56], Chinese herbs [57] and a mixture of thyme, oregano and garlic [58]. The effective role of flavonoids and polyphenols in reducing oocyst shedding in the feces of broilers was in accordance with the findings of [16,59], in which it was found that the addition of herbal extract enriched with flavonoids reduces coccidial oocyst shedding in feces simultaneously with decreasing the severity of infection. Moreover, our findings are in accordance with Wang et al. [47], who observed decreased fecal oocysts and intestinal lesion scores with increased levels of grape seed extract in broiler chickens. In line with our findings, Liu et al. [53] reported that chlorogenic acid, which is an antioxidant and anti-inflammatory substance that reduces oocyst count in feces, intestinal lesion score and bloody diarrhea, indicates the inhibition of coccidial infection in broiler chickens. The coccidiocidal or coccidiostatic role of flavonoids could be attributed to interrupting the parasitic life cycle via inhibiting its sporulation [60]. Additionally, the main role of flavonoids and polyphenols as anti-coccidials may be due to its mode of action as anti-inflammatory and antioxidant substances that improve gut health through maintaining mucus secretion, increasing gut epithelial integrity, reducing the colonization of pathogenic microbes and improving local intestinal and body immune defense [61,62].

Dietary polyphenolic supplementation can improve the immune systems of the birds via several ways: binding to the immune cells' receptors and changing the signaling pathway of the cell, causing the regulation and modulation of the immune response of the host against invasive microorganisms; and enhancing the release of anti-inflammatory cytokines that improve the birds' resistance against infection [63]. The enhanced immunity that decreased fecal oocyst shedding, intestinal lesions and bloody diarrhea due to myricetin supplementation, from our point of view, is the main cause of the decreased mortality percent compared to the IC group. In accordance, reduction in mortality due to coccidial infection was also observed in broiler chickens supplemented with grape seed extract, which is rich in flavonoids and polyphenol compounds [47]. Moreover, Ageratum conyzoides, enriched with flavonoids at the level of 500–1000 mg/kg, revealed a considerable decrease in the oxidative stress produced by *E. tenella*, improving broiler chickens' performance and reducing mortality [64].

Under normal management conditions of broiler chickens, the dietary supplementation of polyphenols and flavonoids has a crucial role in protecting birds from oxidative stressors and neutralizing the free radicals produced in body cells such as ROS and reactive nitrogen species (RNS) [65]. Plant-derived polyphenolics compounds have been proven to play an important role in the stimulation of immunity, either through cellular immunity through the modulation of the function of immune cells by binding to immune cells receptors, altering their signaling pathway and stimulating their proliferation [66]; or humoral immunity through the elevation of an antibody titer, increasing lysozyme activity and increasing serum immune globulins [67]. Herein, the role of myricetin (400 and 600 mg/kg) in stimulating the immune status of broiler chickens before infection was clear through boosting IgG and lowering CRP levels.

Parasitic infection such as by *Eimeria* spp. can induce inflammatory responses of the host [53]. In our results, after coccidial challenge on day 14, the NO, CRP and MPO levels were significantly elevated in the IC group compared to the non-infected one, and myricetin addition significantly decreased their levels. *Eimeria* spp. infection has been reported to induce plasma NO levels that may be involved in their pathogenesis, as follows; NO is considered to be a toxic substance to sporulated oocysts [68], and the ingestion of NOS inhibitors increases oocyst output [69]. Also, Yan et al. [68] proved that exogenous NO causes the egress of *E. tenella* sporozite from primary chicken kidney cell cultures before parasite replication. However, the high production of NO by the host cell above the host cell's tolerance due to coccidian infection can cause tissue damage and cell cytotoxicity, which can induce the inflammation and development of clinical signs such as diarrhea, mortality and intestinal lesions [70]. C-reactive protein, considered an acute inflammatory protein that lowers inflammation levels and is highly produced at the site of inflammation or infection by many cells, such as macrophages, lymphocytes and endothelial cells, is considered potential marker of decreased body inflammation and cells damage [71]. Moreover, CRP plays a crucial role in response to the host's infection through NO release, phagocytosis, apoptosis and cytokine production, particularly IL-6 and TNF-α [72]. Myeloperoxidase is a pro-inflammatory enzyme generated from neutrophilic granulocytes, and it plays an important role in innate cellular immune responses through its potential effect to injure healthy tissue, thus contributing to disease initiation in poultry [73]. Flavonoid supplementation, such as with curcumin, resveratrol and thymol, have been proven to have an immunostimulatory effect through inhibiting the generation of ROS and NOS by supressing MPO and reducing MPO mRNA expression in neutrophiles [74,75]. Additionally, the hummoral immune response of the host has been reported to be activated through an increased antibody titer, espcially the protective IgG titer, after coccidial infection in laying hens [76]. Our results indicate a high titer of IgG in infected birds supplemented with myricetin, escially at higher dosese. Similarly, Liu et al. [53], who reported an improved antibody titer in *Eimeria* spp. challenged birds, fed birds a diet supplemented with antioxidant chlorogenic acid. Notably, at 21 dpi, the excessive inflammatory response in the IC group subsided in myricetin-supplemented groups, which indicates its potent role against coccidial challenges.

The protective consequence of dietary myricetin against coccidian infection in our study was also achieved through the downregulation of proinflammatory cytokines (*IL-1β*, *IL-6* and *TNF-α*) and chemokines (*CCL4*, *CCL20* and *CXCL13*) and the upregulation of anti-inflammatory cytokines (*IL-10*) and *AvBD6* and *AvBD612* mRNA gene expression. In agreement with our observations, flavonoids such as resveratrol have been proven to have a role in the generation and modulation of cytokines and chemokines in different immune cells [77]. The essential role of proinflammatory cytokines such as interleukin 1 (IL-1), IL-6 and TNF-α have a responsibility in the acute-phase inflammation that is associated with general and metabolic changes [78]. Moreover, these proinflammatory cytokines have a crucial role in modulating the host immune response during infection [79]. Moreover, the anti-inflammatory cytokine IL-10 has been reported to have a role in controlling the host's immune response by limiting the target cell damage during inflammation [80]. Parasite invasion can use IL-10 to downregulate host immunity and reduce pathogen-damaging inflammatory reactions [81].

Macrophage inflammatory proteins are also known as chemotactic cytokines that comprise CCL4 and CCL20, which play an important role in coordinating the host's immune responses against infection [82]. Furthermore, CCL4 acts as chemoattractant for important immune cells such as monocytes, macrophages, T-lymphocytes, dendritic cells and natural killer cells [83]. Moreover, CCL4 secretion from neutrophils participates in inflammation by attracting other leukocytic cells to the area of inflammation, resulting in resolving the inflammation by macrophage-mediated cells and developing chronic inflammation [84]. Regarding the CCL20 chemokine, it plays a vital role in the initiation of chronic intestinal inflammation in broiler chickens [85]. CXCL13, which is also recognized as stromal cell-derived factor-1, has a high chemotactic impact on lymphocytes that are involved in inflammatory responses of the host against infections [86]. Moreover, resveratrol was found to improve phagocytes' killing capability, and the inhibition of TNF-α and NF-kB was found to relieve inflammation in damaged livers with hydrogen peroxide [87]. In the same vein, the immune responses of tilapia have been reported to be improved by the upregulation of *IL-10* and *TGF-β* and the downregulation of *IL-1β*, *IL-8* and *TNF-α* mRNA levels after supplementation with quercetin nano particles [88].

Defensin is an indispensable peptide for the host's defense mechanism, giving it instant defense against microbial invasion. However, the particular role of defensin proteins in local resistance against the infection of *Eimeria* ssp. has not been well explored [89]. Avian β-defensin 6 and 12 exhibit a chemotactic effect and lipopolysaccharide-neutralizing effect for chicken macrophages. In addition, AvBD12 has been proven to be involved in the induction of murine immature dendritic cell migration to the site of inflammation [90]. Herein, increasing the expression of AvBD6 and 12, following dietary supplementation with Myc, indicates its protective role against coccidia infection.

Additionally, COX-2 is an enzyme whose intermediate, arachidonic acid, undergoes bioconversion to inflammatory prostaglandin with consequential cytokine release [91].

Invasion of host cells with *Eimeria* spp. is known to produce oxidative stress through releasing high amounts of free radicals that play a crucial role in the host's defense mechanism against parasite infection [92]. The concentration of those free radicals may increase cell tolerance and cause cell cytotoxicity and death, cascading the pathogenesis of the disease. Moreover, the production of this massive amount of ROS and NOS in parasitic diseases can exhaust both low molecular antioxidants such as vitamin A, E and C [70] and metal-dependent antioxidants such as GPX, SOD and CAT [93]. In the current study, it seemed that, after coccidia invasion, higher free radical release and high levels of NO and MDA production were the most important factors impairing the natural antioxidant defense system, which comes in agreement with Georgieva, Koinarski [14].

As evidenced in our study of post coccidia challenge, ROS production decreased, and the expression of the *COX-2* gene in the group that was fed with higher Myc levels was downregulated, which are the main messengers that modify the expression of numerous genes implicated in inflammation [94]. Furthermore, coccidia infection greatly downregulated the expression of antioxidant-related genes such as *GPX*, *SOD*, *CAT*, *HO-1* and *NQO1* in both intestinal and muscle tissues, and the supplementation of strong flavanol compounds such as myricetin alleviated oxidative stress and improved the expression of these antioxidants. Herein, the contents of MDA and ROS and H_2O_2 levels in intestinal and breast muscle tissues were significantly reduced after the inclusion of elevated levels of Myc. In contrast, higher T-AOC in intestinal and muscle tissues following supplementation with Myc indicated decreased free radical production and lipid peroxidation. These findings suggest that feeding with myricetin strengthens the oxidative stability of birds via activating antioxidant mediators. The role of plant extracts in protection against coccidiosis may be related to their ability to control the impact on lipid peroxidation in intestinal mucosa and decreasing ROS and NOS production and consequently their destructive effects [95]. The same results were obtained by Idris et al. [96], who described that the inclusion of antioxidant-enriched essential oils can alleviate the oxidative stress caused by the invasion of *Eimeria* spp. In agreement with our results, Tsiouris et al. [16] reported improved antioxidant markers and the alleviation of coccidian infection occurring after the addition of high-polyphenol herbal extract.

5. Conclusions

The great benefits of natural antioxidant-rich flavonoids encourage the application of novel ones to tackle the stressors that face the modern poultry industry. Considered together, our findings recommend that the dietary inclusion of Myc can reduce the intestinal sporulation and fecal oocyst shedding of *Eimeria* spp. of infected birds. These beneficial outcomes of Myc could be related to its unique antioxidant and immunomodulatory properties that invoke protection against avian coccidiosis with consequences of superior broiler chicken growth. After dietary supplementation with Myc, its ability to maintain good antioxidant capacity for both intestinal tissues and meat, even after coccidia experimental infection, proves its role as a powerful antioxidant additive. Finally, evaluating the proposed mechanisms beyond Myc's beneficial effects and modifying the immune and antioxidant responses of birds have prospective future applications in poultry farming.

Author Contributions: Conceptualization, W.R.E.-G., A.T.Y.K., W.S.A., R.G.A.A., A.A.G., B.A., A.M.A.M., H.I., S.M.A.-R., H.I. and D.I.; methodology, W.R.E.-G., A.T.Y.K., S.M.A.-R., B.A., A.A.G. and D.I.; software, A.T.Y.K. and D.I.; validation, A.T.Y.K., W.S.A., R.G.A.A., A.A.G., B.A. and D.I.; formal analysis, W.R.E.-G., A.T.Y.K. and D.I.; investigation, W.R.E.-G., A.T.Y.K., W.S.A., R.G.A.A., A.A.G., B.A., A.M.A.M., S.M.A.-R., H.I. and D.I.; resources, W.R.E.-G., W.S.A., R.G.A.A., A.A.G., A.M.A.M, S.M.A.-R., H.I. and D.I.; data curation, W.R.E.-G., A.T.Y.K., W.S.A., R.G.A.A., A.A.G., A.M.A.M., S.M.A.-R., H.I. and D.I.; writing—original draft preparation, D.I., A.T.Y.K. and S.M.A.-R.; writing—review and editing W.R.E.-G., A.T.Y.K., W.S.A., R.G.A.A., A.A.G., A.M.A.M., S.M.A.-R. and D.I.; visualization, W.R.E.-G., A.T.Y.K., W.S.A., R.G.A.A., B.A., A.M.A.M., S.M.A.-R. and D.I.; supervision, W.R.E.-G., A.T.Y.K., W.S.A., R.G.A.A., A.A.G., B.A., A.M.A.M., S.M.A.-R. and D.I.; project administration, W.R.E.-G., A.T.Y.K., W.S.A., R.G.A.A., A.A.G., B.A., A.M.A.M., S.M.A.-R. and D.I.; funding acquisition, W.R.E.-G., A.T.Y.K., R.G.A.A., A.A.G., A.M.A.M., S.M.A.-R., H.I. and D.I. All authors have read and agreed to the published version of the manuscript.

Funding: This research was funded by the Deputyship for Research and Innovation, Ministry of Education in Saudi Arabia through Project number [INST172].

Institutional Review Board Statement: All experimental procedures, bird rearing and management were conducted upon approval from animal resources at the Faculty of Veterinary Medicine, Zagazig University, Egypt, according to the rules of the Institutional Animal Care and Use Committee (ZU-IACUC/2/F/320/2022 approval number).

Informed Consent Statement: Not applicable.

Data Availability Statement: The data presented in this research will be offered upon request to the corresponding author.

Acknowledgments: The authors extend their appreciation to the Deputyship for Research and Innovation, Ministry of Education in Saudi Arabia, for funding this research work through project number [INST172].

Conflicts of Interest: The authors declare no conflict of interest.

References

1. Mishra, B.; Jha, R. Oxidative stress in the poultry gut: Potential challenges and interventions. *Front. Vet. Sci.* **2019**, *6*, 60. [CrossRef] [PubMed]
2. Xing, T.; Gao, F.; Tume, R.K.; Zhou, G.; Xu, X. Stress effects on meat quality: A mechanistic perspective. *Compr. Rev. Food Sci. Food Saf.* **2019**, *18*, 380–401. [PubMed]
3. Yang, T.; Sun, Z.; Li, X. Generation mechanism of oxidative stress during early weaning and Its Impacts. *Chin. J. Anim. Nutr.* **2013**, *25*, 705–714.

4. Ibrahim, D.; Ismail, T.A.; Khalifa, E.; El-Kader, A.; Shaimaa, A.; Mohamed, D.I.; Mohamed, D.T.; Shahin, S.E.; El-Hamid, A.; Marwa, I. Supplementing Garlic Nanohydrogel Optimized Growth, Gastrointestinal Integrity and Economics and Ameliorated Necrotic Enteritis in Broiler Chickens Using a *Clostridium perfringens* Challenge Model. *Animals* **2021**, *11*, 2027. [CrossRef] [PubMed]
5. Ibrahim, D.; Arisha, A.H.; Khater, S.I.; Gad, W.M.; Hassan, Z.; Abou-Khadra, S.H.; Mohamed, D.I.; Ahmed Ismail, T.; Gad, S.A.; Eid, S.A. Impact of Omega-3 Fatty Acids Nano-Formulation on Growth, Antioxidant Potential, Fillet Quality, Immunity, Autophagy-Related Genes and Aeromonas hydrophila Resistance in Nile Tilapia (*Oreochromis niloticus*). *Antioxidants* **2022**, *11*, 1523. [CrossRef]
6. Khater, S.I.; Lotfy, M.M.; Alandiyjany, M.N.; Alqahtani, L.S.; Zaglool, A.W.; Althobaiti, F.; Ismail, T.A.; Soliman, M.M.; Saad, S.; Ibrahim, D. Therapeutic Potential of Quercetin Loaded Nanoparticles: Novel Insights in Alleviating Colitis in an Experimental DSS Induced Colitis Model. *Biomedicines* **2022**, *10*, 1654. [CrossRef] [PubMed]
7. Surai, P.F. *Natural Antioxidants in Avian Nutrition and Reproduction*; Nottingham University Press: Nottingham, UK, 2002.
8. Ibrahim, D.; Nem, A.N.A.; Ibrahim, S.M.; Eissa, H.M.; Fawzey, M.; Mostafa, D.I.; Abd El-Kader, S.A.; Khater, S.; Khater, S.I. Dual effect of Selenium loaded Chitosan Nanoparticles on growth, antioxidant, immune related genes expression, transcriptomics modulation of caspase 1, cytochrome P450 and heat shock protein and Aeromonas hydrophila resistance of Nile Tilapia (*Oreochromis niloticus*). *Fish Shellfish Immunol.* **2021**, *110*, 91–99.
9. Ibrahim, D.; Moustafa, A.; Metwally, A.S.; Nassan, M.A.; Abdallah, K.; Eldemery, F.; Tufarelli, V.; Laudadio, V.; Kishawy, A.T. Potential Application of Cornelian Cherry Extract on Broiler Chickens: Growth, Expression of Antioxidant Biomarker and Glucose Transport Genes, and Oxidative Stability of Frozen Meat. *Animals* **2021**, *11*, 1038. [CrossRef]
10. Guo, F.; Suo, X.; Zhang, G.; Shen, J. Efficacy of decoquinate against drug sensitive laboratory strains of *Eimeria tenella* and field isolates of *Eimeria* spp. in broiler chickens in China. *Vet. Parasitol.* **2007**, *147*, 239–245. [CrossRef]
11. Naidoo, V.; McGaw, L.J.; Bisschop, S.; Duncan, N.; Eloff, J.N. The value of plant extracts with antioxidant activity in attenuating coccidiosis in broiler chickens. *Vet. Parasitol.* **2008**, *153*, 214–219. [CrossRef]
12. Ibrahim, D.; Moustafa, A.; Shahin, S.; Sherief, W.; Farag, M.; Nassan, M. Impact of fermented or enzymatically fermented dried olive pomace on growth, expression of digestive enzymes and glucose transporters genes, oxidative stability of frozen meat and economic efficiency of broiler chickens. *Front. Vet. Sci.* **2021**, *8*, 442. [CrossRef] [PubMed]
13. Yang, Q.; Li, Y.; Luo, L. Effect of myricetin on primary open-angle glaucoma. *Transl. Neurosci.* **2018**, *9*, 132–141. [CrossRef]
14. Georgieva, N.; Koinarski, V.; Gadjeva, V. Antioxidant status during the course of *Eimeria tenella* infection in broiler chickens. *Vet. J.* **2006**, *172*, 488–492. [CrossRef] [PubMed]
15. Bozkurt, M.; Giannenas, I.; Küçükyilmaz, K.; Christaki, E.; Florou-Paneri, P. An update on approaches to controlling coccidia in poultry using botanical extracts. *Br. Poult. Sci.* **2013**, *54*, 713–727. [CrossRef]
16. Tsiouris, V.; Giannenas, I.; Bonos, E.; Papadopoulos, E.; Stylianaki, I.; Sidiropoulou, E.; Lazari, D.; Tzora, A.; Ganguly, B.; Georgopoulou, I. Efficacy of a Dietary Polyherbal Formula on the Performance and Gut Health in Broiler Chicks after Experimental Infection with *Eimeria* spp. *Pathogens* **2021**, *10*, 524. [CrossRef]
17. Jang, S.I.; Jun, M.-H.; Lillehoj, H.S.; Dalloul, R.A.; Kong, I.-K.; Kim, S.; Min, W. Anticoccidial effect of green tea-based diets against *Eimeria maxima*. *Vet. Parasitol.* **2007**, *144*, 172–175. [CrossRef] [PubMed]
18. Ibrahim, D.; Sewid, A.H.; Arisha, A.H.; Abd El-Fattah, A.H.; Abdelaziz, A.M.; Al-Jabr, O.A.; Kishawy, A.T. Influence of Glycyrrhiza glabra Extract on Growth, Gene Expression of Gut Integrity, and Campylobacter jejuni Colonization in Broiler Chickens. *Front. Vet. Sci.* **2020**, *7*, 612063. [CrossRef] [PubMed]
19. Amber, K.; Nofel, R.; Ghanem, R.; Sayed, S.; Farag, S.A.; Shukry, M.; Dawood, M.A. Enhancing the growth rate, biochemical blood indices, and antioxidative capacity of broilers by including aloe vera gel in drinking water. *Front. Vet. Sci.* **2021**, *7*, 632666. [CrossRef]
20. Abou-Elkhair, R.; Gaafar, K.M.; Elbahy, N.; Helal, M.A.; Mahboub, H.; Sameh, G. Bioactive effect of dietary supplementation with essential oils blend of oregano, thyme and garlic oils on performance of broilers infected with *Eimeria* species. *Glob. Vet.* **2014**, *13*, 977–985.
21. Shehata, A.A.; Yalçın, S.; Latorre, J.D.; Basiouni, S.; Attia, Y.A.; Abd El-Wahab, A.; Visscher, C.; El-Seedi, H.R.; Huber, C.; Hafez, H.M. Probiotics, prebiotics, and phytogenic substances for optimizing gut health in poultry. *Microorganisms* **2022**, *10*, 395. [CrossRef]
22. Muthamilselvan, T.; Kuo, T.-F.; Wu, Y.-C.; Yang, W.-C. Herbal remedies for coccidiosis control: A review of plants, compounds, and anticoccidial actions. *Evid.-Based Complement. Altern. Med. Ecam* **2016**, *2016*, 2657981. [CrossRef] [PubMed]
23. Qureshi, N.A. In vitro Anticoccidial, Antioxidant Activities and Biochemical Screening of Methanolic and Aqueous Leaves Extracts of Selected Plants. *Pak. Vet. J.* **2021**, *41*, 57–63.
24. Igwe, E.O.; Charlton, K.E. A systematic review on the health effects of plums (*Prunus domestica* and *Prunus salicina*). *Phytother. Res.* **2016**, *30*, 701–731. [CrossRef] [PubMed]
25. Abbas, A.; Iqbal, Z.; Abbas, R.Z.; Khan, M.K.; Khan, J.A. In-vitro anticoccidial potential of *Saccharum officinarum* extract against *Eimeria* oocysts. *Bol. Latinoam. Caribe Plantas Med. Y Aromat.* **2015**, *14*, 456–461.
26. Ugwuoke, G.M.; Pewan, S. Effect of methanol extract of Parkia biglobosa root bark on organ and carcass weight and histopathological changes in *Eimeria tenella* infected broiler chickens. *Anim. Res. Int.* **2020**, *17*, 3587–3595.
27. Ong, K.C.; Khoo, H.-E. Biological effects of myricetin. *Gen. Pharmacol. Vasc. Syst.* **1997**, *29*, 121–126. [CrossRef]

28. Lin, H.-H.; Huang, C.-Y. Characterization of flavonol inhibition of DnaB helicase: Real-time monitoring, structural modeling, and proposed mechanism. *J. Biomed. Biotechnol.* **2012**, *2012*, 735368. [CrossRef]
29. Huang, P.; Zhou, M.; Cheng, S.; Hu, Y.; Gao, M.; Ma, Y.; Limpanont, Y.; Zhou, H.; Dekumyoy, P.; Cheng, Y. Myricetin possesses anthelmintic activity and attenuates hepatic fibrosis via modulating TGFβ1 and Akt signaling and shifting Th1/Th2 balance in *Schistosoma japonicum*-infected mice. *Front. Immunol.* **2020**, *11*, 593. [CrossRef]
30. Imran, M.; Saeed, F.; Hussain, G.; Imran, A.; Mehmood, Z.; Gondal, T.A.; El-Ghorab, A.; Ahmad, I.; Pezzani, R.; Arshad, M.U. Myricetin: A comprehensive review on its biological potentials. *Food Sci. Nutr.* **2021**, *9*, 5854–5868. [CrossRef]
31. Cho, B.O.; Yin, H.H.; Park, S.H.; Byun, E.B.; Ha, H.Y.; Jang, S.I. Anti-inflammatory activity of myricetin from Diospyros lotus through suppression of NF-κB and STAT1 activation and Nrf2-mediated HO-1 induction in lipopolysaccharide-stimulated RAW264. 7 macrophages. *Biosci. Biotechnol. Biochem.* **2016**, *80*, 1520–1530. [CrossRef]
32. Dörr, R.; Müller-Wieland, D.; Tschoepe, D. Diabetes and the Heart-a Never-ending Story. *Herz* **2010**, *35*, 129. [CrossRef]
33. Kan, X.; Liu, J.; Chen, Y.; Guo, W.; Xu, D.; Cheng, J.; Cao, Y.; Yang, Z.; Fu, S. Myricetin protects against H2O2-induced oxidative damage and apoptosis in bovine mammary epithelial cells. *J. Cell. Physiol.* **2021**, *236*, 2684–2695. [CrossRef]
34. Aviagen, W. *Ross 308: Broiler Nutrition Specification*; Aviagen Inc.: Huntsville, AL, USA, 2014.
35. Aviagen, W. *Ross 308: Broiler Nutrition Specification*; Aviagen Inc.: Huntsville, AL, USA, 2022.
36. AOAC. *Official Methods of Analysis of AOAC International, Association of Official Analytical Chemists*; AOAC: Rockville, MD, USA, 2012.
37. Kishawy, A.T.; Al-Khalaifah, H.S.; Nada, H.S.; Roushdy, E.M.; Zaglool, A.W.; Ahmed Ismail, T.; Ibrahim, S.M.; Ibrahim, D. Black Pepper or Radish Seed Oils in a New Combination of Essential Oils Modulated Broiler Chickens' Performance and Expression of Digestive Enzymes, Lipogenesis, Immunity, and Autophagy-Related Genes. *Vet. Sci.* **2022**, *9*, 43. [CrossRef] [PubMed]
38. Christaki, E.; Florou-Paneri, P.; Giannenas, I.; Papazahariadou, M.; Botsoglou, N.A.; Spais, A.B. Effect of a mixture of herbal extracts on broiler chickens infected with *Eimeria tenella*. *Anim. Res.* **2004**, *53*, 137–144. [CrossRef]
39. Johnson, J.; Reid, W.M. Anticoccidial drugs: Lesion scoring techniques in battery and floor-pen experiments with chickens. *Exp. Parasitol.* **1970**, *28*, 30–36. [CrossRef] [PubMed]
40. Islam, M.R.; Oomah, D.B.; Diarra, M.S. Potential immunomodulatory effects of non-dialyzable materials of cranberry extract in poultry production. *Poult. Sci.* **2017**, *96*, 341–350. [CrossRef]
41. Ahn, D.; Olson, D.; Jo, C.; Love, J.; Jin, S. Volatiles production and lipid oxidation in irradiated cooked sausage as related to packaging and storage. *J. Food Sci.* **1999**, *64*, 226–229. [CrossRef]
42. Loreto, F.; Velikova, V. Isoprene produced by leaves protects the photosynthetic apparatus against ozone damage, quenches ozone products, and reduces lipid peroxidation of cellular membranes. *Plant Physiol.* **2001**, *127*, 1781–1787. [CrossRef]
43. LeBel, C.P.; Ischiropoulos, H.; Bondy, S.C. Evaluation of the probe 2′, 7′-dichlorofluorescin as an indicator of reactive oxygen species formation and oxidative stress. *Chem. Res. Toxicol.* **1992**, *5*, 227–231. [CrossRef]
44. Livak, K.J.; Schmittgen, T.D. Analysis of relative gene expression data using real-time quantitative PCR and the $2-\Delta\Delta CT$ method. *Methods* **2001**, *25*, 402–408. [CrossRef]
45. Georgieva, N.; Gabrashanska, M.; Koinarski, V.; Ermidou-Pollet, S. Antioxidant status in *Eimeria acervulina* infected chickens after dietary selenium treatment. *Trace Elem. Electrolytes* **2011**, *28*, 42. [CrossRef]
46. Ruff, J.; Tellez, G., Jr.; Forga, A.J.; Señas-Cuesta, R.; Vuong, C.N.; Greene, E.S.; Hernandez-Velasco, X.; Uribe, Á.J.; Martínez, B.C.; Angel-Isaza, J.A. Evaluation of three formulations of essential oils in broiler chickens under cyclic heat stress. *Animals* **2021**, *11*, 1084. [CrossRef] [PubMed]
47. Wang, M.; Suo, X.; Gu, J.; Zhang, W.; Fang, Q.; Wang, X. Influence of grape seed proanthocyanidin extract in broiler chickens: Effect on chicken coccidiosis and antioxidant status. *Poult. Sci.* **2008**, *87*, 2273–2280. [CrossRef] [PubMed]
48. Abu Hafsa, S.; Ibrahim, S. Effect of dietary polyphenol-rich grape seed on growth performance, antioxidant capacity and ileal microflora in broiler chicks. *J. Anim. Physiol. Anim. Nutr.* **2018**, *102*, 268–275. [CrossRef] [PubMed]
49. Abolfathi, M.-E.; Tabeidian, S.A.; Foroozandeh Shahraki, A.D.; Tabatabaei, S.N.; Habibian, M. Comparative effects of n-hexane and methanol extracts of elecampane (*Inula helenium* L.) rhizome on growth performance, carcass traits, feed digestibility, intestinal antioxidant status and ileal microbiota in broiler chickens. *Arch. Anim. Nutr.* **2019**, *73*, 88–110. [CrossRef] [PubMed]
50. Nath, S.; Kumar.k, A. Role of Flavonoids in Poultry Nutrition. *Acta Sci. Vet. Sci.* **2021**, *3*, 88–91. [CrossRef]
51. Semwal, D.K.; Semwal, R.B.; Combrinck, S.; Viljoen, A. Myricetin: A Dietary Molecule with Diverse Biological Activities. *Nutrients* **2016**, *8*, 90. [CrossRef]
52. Bozkurt, M.; Aysul, N.; Küçükyilmaz, K.; Aypak, S.; Ege, G.; Catli, A.; Akşit, H.; Çöven, F.; Seyrek, K.; Çınar, M. Efficacy of in-feed preparations of an anticoccidial, multienzyme, prebiotic, probiotic, and herbal essential oil mixture in healthy and *Eimeria* spp.-infected broilers. *Poult. Sci.* **2014**, *93*, 389–399. [CrossRef]
53. Liu, H.; Chen, P.; Lv, X.; Zhou, Y.; Li, X.; Ma, S.; Zhao, J. Effects of chlorogenic acid on performance, anticoccidial indicators, immunity, antioxidant status, and intestinal barrier function in coccidia-infected broilers. *Animals* **2022**, *12*, 963. [CrossRef]
54. Bakkali, F.; Averbeck, S.; Averbeck, D.; Idaomar, M. Biological effects of essential oils—A review. *Food Chem. Toxicol.* **2008**, *46*, 446–475.
55. Kim, D.K.; Lillehoj, H.S.; Lee, S.H.; Lillehoj, E.P.; Bravo, D. Improved resistance to *Eimeria acervulina* infection in chickens due to dietary supplementation with garlic metabolites. *Br. J. Nutr.* **2013**, *109*, 76–88. [CrossRef] [PubMed]
56. Lee, S.H.; Lillehoj, H.S.; Jang, S.I.; Lee, K.W.; Park, M.S.; Bravo, D.; Lillehoj, E.P. Cinnamaldehyde enhances in vitro parameters of immunity and reduces in vivo infection against avian coccidiosis. *Br. J. Nutr.* **2011**, *106*, 862–869. [CrossRef] [PubMed]

57. Chang, C.L.; Chung, C.-Y.; Kuo, C.-H.; Kuo, T.-F.; Yang, C.-W.; Yang, W.-C. Beneficial effect of Bidens pilosa on body weight gain, food conversion ratio, gut bacteria and coccidiosis in chickens. *PLoS ONE* **2016**, *11*, e0146141. [CrossRef] [PubMed]
58. Pop, L.M.; Varga, E.; Coroian, M.; Nedișan, M.E.; Mircean, V.; Dumitrache, M.O.; Farczádi, L.; Fülöp, I.; Croitoru, M.D.; Fazakas, M. Efficacy of a commercial herbal formula in chicken experimental coccidiosis. *Parasites Vectors* **2019**, *12*, 343. [CrossRef]
59. Allen, P. Avian Dis.: Anticoccidial effects of xanthohumol. *J. Avian Med. Surg.* **2007**, *21*, 241–242.
60. Sharma, U.N.S.; Fernando, D.D.; Wijesundara, K.K.; Manawadu, A.; Pathirana, I.; Rajapakse, R.J. Anticoccidial effects of Phyllanthus emblica (*Indian gooseberry*) extracts: Potential for controlling avian coccidiosis. *Vet. Parasitol. Reg. Stud. Rep.* **2021**, *25*, 100592. [CrossRef]
61. Koutsos, E.; Arias, V. Intestinal ecology: Interactions among the gastrointestinal tract, nutrition, and the microflora. *J. Appl. Poult. Res.* **2006**, *15*, 161–173. [CrossRef]
62. Mountzouris, K.; Paraskevas, V.; Tsirtsikos, P.; Palamidi, I.; Steiner, T.; Schatzmayr, G.; Fegeros, K. Assessment of a phytogenic feed additive effect on broiler growth performance, nutrient digestibility and caecal microflora composition. *Anim. Feed. Sci. Technol.* **2011**, *168*, 223–231. [CrossRef]
63. Abdel-Moneim, A.-M.E.; Shehata, A.M.; Alzahrani, S.O.; Shafi, M.E.; Mesalam, N.M.; Taha, A.E.; Swelum, A.A.; Arif, M.; Fayyaz, M.; Abd El-Hack, M.E. The role of polyphenols in poultry nutrition. *J. Anim. Physiol. Anim. Nutr.* **2020**, *104*, 1851–1866. [CrossRef]
64. Nweze, N.; Obiwulu, I. Anticoccidial effects of Ageratum conyzoides. *J. Ethnopharmacol.* **2009**, *122*, 6–9. [CrossRef]
65. Lipiński, K.; Mazur, M.; Antoszkiewicz, Z.; Purwin, C. Polyphenols in monogastric nutrition—A review. *Ann. Anim. Sci.* **2017**, *17*, 41–58. [CrossRef]
66. Liang, Y.; Zhou, J.; Ji, K.; Liu, H.; Degen, A.; Zhai, M.; Jiao, D.; Guo, J.; Zhao, Z.; Yang, G. Protective effect of resveratrol improves systemic inflammation responses in LPS-injected lambs. *Animals* **2019**, *9*, 872. [CrossRef] [PubMed]
67. Badran, A.M. Effect of dietary curcumin and curcumin nanoparticles supplementation on growth performance, immune response and antioxidant of broilers chickens. *Egypt. Poult. Sci. J.* **2020**, *40*, 325–343. [CrossRef]
68. Yan, X.; Han, W.; Liu, X.; Suo, X. Exogenous nitric oxide stimulates early egress of *Eimeria tenella* sporozoites from primary chicken kidney cells in vitro. *Parasite* **2021**, *28*, 11. [CrossRef]
69. Shen, Z.-J.; Zhu, B.-L.; Jiang, J.-S. The effect of NO during *E. tenella* or *E. acervulina* infection of broilers. *Acta Vet. Zootech. Sin.* **2002**, *33*, 395–399.
70. Evans, P.; Halliwell, B. Micronutrients: Oxidant/antioxidant status. *Br. J. Nutr.* **2001**, *85*, S67–S74. [CrossRef]
71. O'Reilly, E.; Eckersall, D. Acute phase proteins: A review of their function, behaviour and measurement in chickens. *World's Poult. Sci. J.* **2014**, *70*, 27–43. [CrossRef]
72. Sproston, N.R.; Ashworth, J.J. Role of C-reactive protein at sites of inflammation and infection. *Front. Immunol.* **2018**, *9*, 754. [CrossRef] [PubMed]
73. Klebanoff, S.J. Myeloperoxidase: Friend and foe. *J. Leukoc. Biol.* **2005**, *77*, 598–625. [PubMed]
74. Castro, R.; Lamas, J.; Morais, P.; Sanmartín, M.L.; Orallo, F.; Leiro, J. Resveratrol modulates innate and inflammatory responses in fish leucocytes. *Vet. Immunol. Immunopathol.* **2008**, *126*, 9–19. [CrossRef] [PubMed]
75. Chang, C.Y.; Choi, D.-K.; Lee, D.K.; Hong, Y.J.; Park, E.J. Resveratrol confers protection against rotenone-induced neurotoxicity by modulating myeloperoxidase levels in glial cells. *PLoS ONE* **2013**, *8*, e60654. [CrossRef]
76. Sharman, P.A.; Smith, N.C.; Wallach, M.G.; Katrib, M. Chasing the golden egg: Vaccination against poultry coccidiosis. *Parasite Immunol.* **2010**, *32*, 590–598. [CrossRef]
77. Schwager, J.; Richard, N.; Widmer, F.; Raederstorff, D. Resveratrol distinctively modulates the inflammatory profiles of immune and endothelial cells. *BMC Complement. Altern. Med.* **2017**, *17*, 309. [CrossRef] [PubMed]
78. Moraes, P.O.; Andretta, I.; Cardinal, K.M.; Ceron, M.; Vilella, L.; Borille, R.; Frazzon, A.P.; Frazzon, J.; Santin, E.; Ribeiro, A.M.L. Effect of functional oils on the immune response of broilers challenged with *Eimeria* spp. *Animal* **2019**, *13*, 2190–2198. [CrossRef] [PubMed]
79. Allen, P.C.; Fetterer, R. Recent advances in biology and immunobiology of *Eimeria* species and in diagnosis and control of infection with these coccidian parasites of poultry. *Clin. Microbiol. Rev.* **2002**, *15*, 58–65. [CrossRef] [PubMed]
80. Arendt, M.; Elissa, J.; Schmidt, N.; Michael, E.; Potter, N.; Cook, M.; Knoll, L.J. Investigating the role of interleukin 10 on *Eimeria* intestinal pathogenesis in broiler chickens. *Vet. Immunol. Immunopathol.* **2019**, *218*, 109934. [CrossRef]
81. Cyktor, J.C.; Turner, J. Interleukin-10 and immunity against prokaryotic and eukaryotic intracellular pathogens. *Infect. Immun.* **2011**, *79*, 2964–2973. [CrossRef]
82. Ren, M.; Guo, Q.; Guo, L.; Lenz, M.; Qian, F.; Koenen, R.R.; Xu, H.; Schilling, A.B.; Weber, C.; Ye, R.D. Polymerization of MIP-1 chemokine (CCL3 and CCL4) and clearance of MIP-1 by insulin-degrading enzyme. *EMBO J.* **2010**, *29*, 3952–3966. [CrossRef]
83. Von Stebut, E.; Metz, M.; Milon, G.; Knop, J.R.; Maurer, M. Early macrophage influx to sites of cutaneous granuloma formation is dependent on MIP-1α/β released from neutrophils recruited by mast cell–derived TNFα. *Blood J. Am. Soc. Hematol.* **2003**, *101*, 210–215. [CrossRef]
84. Maurer, M.; Von Stebut, E. Macrophage inflammatory protein-1. *Int. J. Biochem. Cell Biol.* **2004**, *36*, 1882–1886. [CrossRef]
85. Dal Pont, G.C.; Lee, A.; Bortoluzzi, C.; Farnell, Y.; Gougoulias, C.; Kogut, M. Novel model for chronic intestinal inflammation in chickens: (2) Immunologic mechanism behind the inflammatory response. *Dev. Comp. Immunol.* **2023**, *138*, 104524. [CrossRef] [PubMed]

86. Razmkhah, M.; Talei, A.-R.; Doroudchi, M.; Khalili-Azad, T.; Ghaderi, A. Stromal cell-derived factor-1 (SDF-1) alleles and susceptibility to breast carcinoma. *Cancer Lett.* **2005**, *225*, 261–266. [CrossRef] [PubMed]
87. Martin, L.M.; Johnson, P.J.; Amorim, J.R.; Honaker, A.R.; Donaldson, R.S.; DeClue, A.E. Investigation of the potential immunomodulatory effects of resveratrol on equine whole blood: An in vitro investigation. *Res. Vet. Sci.* **2016**, *106*, 97–99. [CrossRef] [PubMed]
88. Ibrahim, D.; Kishawy, A.T.Y.; Khater, S.I.; Khalifa, E.; Ismail, T.A.; Mohammed, H.A.; Elnahriry, S.S.; Tolba, H.A.; Sherief, W.R.I.A.; Farag, M.F.M.; et al. Interactive effects of dietary quercetin nanoparticles on growth, flesh antioxidant capacity and transcription of cytokines and Aeromonas hydrophila quorum sensing orchestrating genes in Nile tilapia (*Oreochromis niloticus*). *Fish Shellfish Immunol.* **2021**, *119*, 478–489. [CrossRef]
89. Zhao, C.; Nguyen, T.; Liu, L.; Sacco, R.E.; Brogden, K.A.; Lehrer, R.I. Gallinacin-3, an inducible epithelial β-defensin in the chicken. *Infect. Immun.* **2001**, *69*, 2684–2691. [CrossRef]
90. Zhao, B.-C.; Lin, H.-C.; Yang, D.; Ye, X.; Li, Z.-G. Disulfide bridges in defensins. *Curr. Top. Med. Chem.* **2016**, *16*, 206–219. [CrossRef]
91. Yu, H.Y.; Kim, K.-S.; Lee, Y.-C.; Moon, H.-I.; Lee, J.-H. Oleifolioside A, a new active compound, attenuates LPS-stimulated iNOS and COX-2 expression through the downregulation of NF-κB and MAPK activities in RAW 264.7 macrophages. *Evid. Based Complement. Altern. Med.* **2012**, *2012*, 637512. [CrossRef]
92. Vladimirov, Y.A. Reactive oxygen and nitrogen species: Diagnostic, preventive and therapeutic values. *Biochemistry* **2004**, *69*, 1. [CrossRef]
93. Koinarski, V.; Georgieva, N.; Gadjeva, V.; Petkov, P. Antioxidant status of broiler chickens, infected with *Eimeria acervulina*. *Rev. Méd. Vét.* **2005**, *156*, 498.
94. Finkel, T.; Holbrook, N.J. Oxidants, oxidative stress and the biology of ageing. *Nature* **2000**, *408*, 239–247. [CrossRef]
95. Abbas, R.; Iqbal, Z.; Mansoor, M.; Sindhu, Z.; Zia, M.; Khan, J. Role of natural antioxidants for the control of coccidiosis in poultry. *Pak. Vet. J.* **2013**, *33*, 401–407.
96. Idris, M.; Abbas, R.; Masood, S.; Rehman, T.; Farooq, U.; Babar, W.; Hussain, R.; Raza, A.; Riaz, U. The potential of antioxidant rich essential oils against avian coccidiosis. *World's Poult. Sci. J.* **2017**, *73*, 89–104. [CrossRef]

antioxidants MDPI

Article

Nano-Sized Selenium Maintains Performance and Improves Health Status and Antioxidant Potential While Not Compromising Ultrastructure of Breast Muscle and Liver in Chickens

Damian Bień [1,*], Monika Michalczuk [1,*], Małgorzata Łysek-Gładysińska [2,3], Artur Jóźwik [3], Anna Wieczorek [2], Arkadiusz Matuszewski [4], Misza Kinsner [5] and Paweł Konieczka [5,6]

1 Department of Animal Breeding, Institute of Animal Sciences, Warsaw University of Life Sciences WULS-SGGW, 02-786 Warszawa, Poland
2 Division of Medical Biology, Institute of Biology, University of Jan Kochanowski, Uniwersytecka 7, 25-406 Kielce, Poland
3 Institute of Genetics and Animal Breeding PAS, Jastrzębiec, Postępu 36A, 05-552 Magdalenka, Poland
4 Department of Animal Environment Biology, Institute of Animal Sciences, Warsaw University of Life Sciences WULS-SGGW, 02-786 Warszawa, Poland
5 Department of Animal Nutrition, The Kielanowski Institute of Animal Physiology and Nutrition, Polish Academy of Sciences, Instytucka 3, 05-110 Jabłonna, Poland
6 Department of Poultry Science and Apiculture, University of Warmia and Mazury in Olsztyn, 10-719 Olsztyn, Poland
* Correspondence: damian_bien@sggw.edu.pl (D.B.); monika_michalczuk@sggw.edu.pl (M.M.)

Abstract: The poultry industry is looking for the most effective sources of selenium (Se) for commercial use. Over the past five years, nano-Se has attracted a great deal of attention in terms of its production, characterisation and possible application in poultry production. The objective of this study was to evaluate the effects of dietary levels of inorganic and organic Se, selenised yeast and nano forms of selenium on breast meat quality, liver and blood markers of antioxidants, the ultrastructure of tissue and the health status of chickens. A total of 300 one-day-old chicks Ross 308 were divided into 4 experimental groups, in 5 replications, with 15 birds per replication. Birds were fed the following treatments: a standard commercial diet containing inorganic Se in the form of inorganic Se at the level of 0.3 mg/kg diet and an experimental diet with an increased level of Se (0.5 mg/kg diet). The use of other forms of Se (nano-Se) versus sodium selenate significantly influences ($p \leq 0.05$) a higher collagen content and does not impair physico-chemical properties in the breast muscle or the growth performance of the chickens. In addition, the use of other forms of selenium at an increased dose versus sodium selenate affected ($p \leq 0.01$) the elongation of sarcomeres in the pectoral muscle while reducing ($p \leq 0.01$) mitochondrial damage in hepatocytes and improving ($p \leq 0.05$) oxidative indices. The use of nano-Se at a dose of 0.5 mg/kg feed has high bioavailability and low toxicity without negatively affecting the growth performance and while improving breast muscle quality parameters and the health status of the chickens.

Keywords: chickens; selenium; nanoparticles; quality of meat; liver; ultrastructure; antioxidant potential; health status

check for updates

Citation: Bień, D.; Michalczuk, M.; Łysek-Gładysińska, M.; Jóźwik, A.; Wieczorek, A.; Matuszewski, A.; Kinsner, M.; Konieczka, P. Nano-Sized Selenium Maintains Performance and Improves Health Status and Antioxidant Potential While Not Compromising Ultrastructure of Breast Muscle and Liver in Chickens. *Antioxidants* **2023**, *12*, 905. https://doi.org/10.3390/antiox12040905

Academic Editors: Luciana Rossi and Matteo Dell'Anno

Received: 6 March 2023
Revised: 22 March 2023
Accepted: 8 April 2023
Published: 10 April 2023

1. Introduction

Selenium is an essential mineral element with important biological functions for the whole body, through incorporation into at least 30 selenoproteins [1–4]. The effect of selenium on the health of birds depends on the form in which it is supplied to the diet. Therefore, the poultry industry is currently searching for the most effective sources of Se for commercial use [5]. In animal nutrition, Se is most commonly used in two forms: inorganic, as sodium selenite (Na_2SeO_3) or sodium selenate (Na_2SeO_4), and organic, as selenomethionine (SeMet) or selenocysteine (SeCys) [6–9]. Se is an essential bio-element that, together with other micronutrients such as zinc and iodine, plays a key role in the proper functioning, development and growth of various organisms [10]. The effect of selenium on the health of birds depends on the form in which it is supplied to the diet. Inorganic Se is passively absorbed from the intestine through a simple diffusion process and competes with many mineral elements for absorption pathways, and organic Se is actively absorbed through amino acid transport mechanism and has a higher bioavailability than the inorganic form [11]. The combination of SeMet + sodium selenite is more efficient than their individual treatments for Se deposition in egg and chicken embryo tissues. Although many studies have shown that organic Se is easier to absorb than inorganic Se, the reduction of competitive absorption leads to higher absorption efficiency and production performance of the combination of the two Se sources [12]. A thorough understanding of this mechanism requires further research and analysis. Se alleviates oxidative stress and peroxidative damage to unsaturated fatty acids and influences the efficiency of fatty acid biosynthesis in animal tissues [13,14]. Dietary selenium deficiency can adversely affect the fatty acid profile and the conversion of linolenic acid (ALA) to eicosapentaenoic acid (EPA) and docosahexaenoic acid (DHA), resulting in an unfavourable $n-6/n-3$ ratio in tissue lipids [15–19]. Se supplementation has been found to stimulate the immune response, improve growth and reproductive performance and increase disease resistance. Se deficiency, on the other hand, increases the risk of myopathy, muscular dystrophy, causes immune declines and reduces poultry performance [20,21]. European Union regulations (list of authorised feed additives published in application of Article 9t (b) of Council Directive 70/524/EEC concerning additives in feeding stuffs (2004/C 50/01)) specify that the maximum level of Se in chicken feed must not exceed 0.5 mg/kg to ensure safe use of the feed. The requirement for Se ranges from 0.1 to 0.15 mg/kg in various poultry diets. In contrast, other authors [22,23] recommend increasing this dose to 0.3–0.75 mg/kg feed.

One of the most rapidly developing fields of science today is nanobiotechnology. The last 10 years have seen a breakthrough in research into the use of nanoparticles in poultry husbandry. Current research is mainly focused on the use of elements in nanometric form as replacements for their traditional sources [24]. Over the past five years, nano-Se has attracted a great deal of attention in terms of its production, characterisation and possible application in poultry production and other livestock species and medical science [25–31]. Nano-Se supplementation has variable effects compared to classical Se sources on the production performance of chickens, probably due to the variable amount of Se in the feed. However, nano-Se has significant effects on Se retention in tissues, GPx activity, or modification of the fatty acid profile in poultry, with less toxic effects. The authors of this study [32] found that the addition of nano-Se at doses up to 30 mg/kg had a significant effect on ma-weight gain and lower FCR. Comparing nanoparticles with Na_2SeO_3 in broiler chickens, it was found that the addition of nanoparticles in amounts of 0.2 and 0.5 mg/kg and the addition of sodium selenite (0.2 mg/kg) relative to the control group improved body weight gain and FCR, but the form of Se used had no significant effect on improving the aforementioned parameters. In a similar study, where the same Se sources were compared at four doses (0.15, 0.30, 0.60 and 1.2 mg/kg) in broiler chickens, Hu et al. [33] found that chickens had better weight gain using a dose of up to 30 mg/kg, which, however, deteriorated in later stages of rearing. With nano-Se, average daily gains and FCR improved already at a dose of 0.15 mg/kg with no change over time as with the application of Na_2SeO_4. The application of nano-Se at 0.5 mg/kg of feed to broiler chickens has a significant effect on the higher PUFA (polyunsaturated fatty acid) content and the protection of lipids against the action of reactive oxygen species, with its high bioavailability and low toxicity for the chicken organism. The use of higher doses of Se in feed (than the recommended 0.3 mg/kg) is a response to the increasingly rapid growth rate of the chicken body. This will provide consumers with a high-quality product rich in good-quality fats [34]. Nano-Se at a dose higher than 2 mg/kg appears to be toxic to the organisms of birds, as when studying the effects of nano-Se at a dose of 0.30 mg/kg versus 2 mg/kg, deterioration of glutathione peroxidase (GPx) activity, concentration of immunoglobulin M (the main antibodies produced during first contact of the organism with an antigen), glutathione, malondialdehyde and thus deterioration of the free radical scavenging capacity in the serum, liver, and muscle of birds was found for a dose of 2 mg/kg. Studies conducted to date [22,33] suggest that nano-Se has a higher absorption rate and antioxidant capacity at a wide range between nutritional and toxic dose compared to sodium selenite. This allows for the conclusion that the toxicity of nano-Se is even lower than even selenomethionine.

Given the steadily increasing growth performance of broiler chickens, it seems appropriate to determine the effect of increased dietary supplementation with Se, especially Se in the form of nanoparticles. However, there is an urgent need to address issues related to the absorption, assimilation and metabolism of nano-Se in animals/poultry before it can find its way into animal/poultry production as a feed supplement [35]. The aim of this study was to compare the effects of different forms of Se, including nano-Se, on the health status, antioxidant potential and ultrastructural changes of breast muscle and liver in chickens.

2. Materials and Methods

2.1. Animals and Diets

The experiment was carried out with 300 Ross 308 chicken broilers randomly allocated to 4 experimental groups, in 5 replications, 15 birds per replication. Chickens were reared under standard conditions for 6 weeks (from hatching day until 42 days of life). They had free access to water and were kept under a controlled light cycle (according to the Aviagen Ross 308: Broiler Performance Objectives. Aviagen Inc., Huntsville, AL, USA. 2019, 1–15). For the first 10 days, all birds were fed the same starter diet balanced to meet their nutritional demands. On day 11 of life, birds started to receive respective diets. Chicken diets are presented in our previous manuscript [34]. Experimental groups differed in terms of Se form implemented in the diet, for example:

CON (control group)—diet meeting nutritional demands of Ross 308 broilers with the basic (recommended) dose of inorganic Se (SS, 0.3 mg/kg feed),

T1 (SS)—diet with upper dose of inorganic Se (sodium selenate, 0.5 mg/kg feed),

T2 (SY)—diet with upper dose of Se in the organic form (selenised yeast, commercial preparation) (0.5 mg/kg feed),

T3 (nano-Se)—diet with upper dose of Se in the form of nanoparticles (commercial preparation) (0.5 mg/kg feed).

2.2. Selenium Forms Used in the Diets

In the CON treatment, a sodium selenate form was used, which was provided in the diet with a vitamin–mineral premix. This form of Se is commonly used in the formulation of diets for broiler chicken [36]. In the treatment of T2, a premix-provided Se was substituted by the Se-enriched yeast provided in the form of commercial product (SelPlex 1000, Alltech, Nicholasville, KY, USA) according to manufacturer declaration. In the treatment T3, Se was provided with the premix in the form of nanoparticles. Briefly, nano-Se in the form of nano-powder was obtained by chemical synthesis. According to manufacturer declaration (American Elements, Los Angeles, CA, USA), average particle size was 10–45 nm, specific surface area was approximately 30–50 m^2/g, and purity was 99.9%.

2.3. Sampling Procedures

Forty cockerels were chosen (10 birds from each treatment; 2 birds for each replicate) for slaughter at the age of 42 days of life that had a body weight similar to the group mean. The pectoral muscle, liver, and blood samples were taken for analysis: chemical composition, physico-chemical properties, selenium content determination, indicators of health status and antioxidant potential and analysis of ultrastructure.

2.4. Assessment of Slaughter Efficiency, Chemical Composition, and Physico-Chemical Properties

Upon completion of the feeding experiment, 10 cockerels (n = 10) per experimental group were randomly selected and weighed before slaughter. After slaughter and cooling of the carcasses, the slaughter efficiency of the chickens was assessed based on a previous report by Michalczuk et al. [37], determining the percentages of the pectoral muscle, legs, and giblets. The basic chemical composition was determined for the collected samples of pectoral muscles: dry weight, crude fat, crude protein, and ash. The determinations were made using the NIR method [37]. The pH value, drip loss, WHC and colour parameter (L*, a*, b*) were analysed according to the method by Michalczuk et al. [38] and ΔE parameter by the method described by Bendowski et al. [39].

2.5. Selenium Content Determination

Determination of Se content in breast muscle and liver (n = 10) was performed according to the PB-28/LF method in an accredited laboratory (PCA Accreditation Certificate No. AB 1095, Issue No. 19, dated 1 January 2022). A full description of the method can be found in our previous manuscript [34].

2.6. Indicators of Health Status and Antioxidant Potential

On the day of slaughter, ten birds (n = 10) per experimental group were randomly chosen to collect blood post-mortem in the amount of 1.5 mL per bird. Laboratory analysis aimed to also determine the activity of selected enzymes and antioxidant compounds by analysing blood samples and 5 g fragments of breast muscle and liver tissue.

In order to determine the effect of the different sources and levels of Se on the health of the chickens, the activity of the following hydrolytic enzymes from the blood, liver and pectoral muscles of the chickens was determined: alanine aminopeptidase (AlaAP, EC 3.4.11.2), leucine aminopeptidase (LeuAP, EC 3.4.11.1), and arginine aminopeptidase (ArgAP, EC 3.4.11.6), which are all responsible for limiting harmful metabolism and accelerating protein circulation in the body, which translates into better weight gain. The following compounds were also determined: the activity of acid phosphatase (AcP, EC 3.1.3.2), beta-glucuronidase (BGRD, EC 3.2.1.31), beta-galactosidase (BGAL, EC 3.2.1.23), beta-glucosidase (BGLU, EC 3.2.1.21), alpha-glucosidase (aGlu, EC 3.2.1.20), mannosidase (MAN, EC 3.2.1.25), and N-acetyl-BD-hexosaminidase (HEX, EC 3.2.1.52). These compounds are responsible for breaking down complex sugars into simple sugars and for the removal of harmful metabolites formed inside the cell [39].

The activity of aminopeptidases was measured as Fast Blue BB salt derivatives at 540 nm by the method of [40], whereas the activity of AcP, BGRD, BGAL, BGLU, aGLU, aMAN, and HEX was measured as 4-nitrophenyl derivatives at 420 nm according to [41]. All determinations were performed with use of Varian Cary 50 Bio UV–VIS spectrophotometer (Santa Clara, CA, USA). The enzyme activity was expressed in nmol/mg protein/h.

The effect of Se type and dose on the oxidative status in analysed tissues of broiler chickens was tested using the following determinations: vitamin C, glutathione (GSH), and 2,2-diphenyl-1-picrylhydrazyl (DPPH). The concentration of vitamin C in the collected tissues was determined using a LambdaBio-20 spectrophotometer (Perkin Elmer, Waltham, MA, USA), whereas the level of GSH was determined by means of the OxisReasearch™ Bioxytech® GSH/GSSG—412™ test (Foster City, CA, USA) according to the methods described by [42]. Measuring the radical scavenging activity was carried out using a routine test procedure employing the synthetic radical DPPH [42,43].

2.7. Analysis of Ultrastructure

Immediately after chicken slaughter, small fragments of the pectoral muscle and liver were taken and cut into properly sized pieces (2 mm^3) and fixed by immersion in buffered 3% glutaraldehyde in cacodylate buffer (pH 7.2) for at least 2 h at 4 °C. The tissue specimens were then post-fixed in 2% osmium tetroxide in cacodylate buffer (pH 7.2) for 1 h at 4 °C. Dehydration of the fixed tissues was performed using an ascending series of ethanol and then transferred into epoxy resin via propylene oxide [44]. Finally, the liver samples were embedded in a mixture of DDSA/NMA/EMbed-812 (Agar Scientific Ltd., London, UK). Ultra-thin sections (40–60 nm) were cut on a Reichert–Jung ultramicrotome and double stained with uranyl acetate and lead citrate. Evaluation of ultrastructure was performed using a transmission electron microscope TESLA BS-500 with Frame Transfer-1K-CCD-Camera (TRS, Mannheim, Germany).

2.8. Statistical Evaluation

Mean values in the analysed samples were processed using the PS IMAGO PRO 8.0 statistical package employing one-way analysis of variance (ANOVA). The normality of the data was checked with the Shapiro–Wilk test. The homogeneity of variance was also checked with the help of Levene's test for homogeneity of variance. Tukey's test was used to determine the significance of differences between the examined groups. The results were considered statistically significant when associated with a probability lower than 5%. The results with a probability lower than 1% were considered highly significant.

3. Results

The use of different forms of Se in the chickens' diets at increased doses did not significantly ($p > 0.05$) affect the birds' growth performance (Table 1). The results obtained are representative and do not differ from the expected performance for chicken [45].

Table 1. Chicken growing results.

Indices	Group				SEM	p Value
	CON	T1	T2	T3		
Body weight (BW), g:						
1 day	37.32	37.87	38.24	38.18	0.167	0.415
42 day	3012.22	2999.30	3037.50	2989.70	22.728	0.554
FCR, kg kg^{-1}	1.63	1.55	1.57	1.53	0.023	0.130
Mortality, %	2.66	1.33	2.66	1.33	0.012	0.479

Data are given as mean ± SEM (n = 10); CON—control group; T1—diet with upper dose of inorganic Se (0.5 mg/kg feed); T2—diet with upper dose of Se in the organic form (0.5 mg/kg feed); T3—diet with upper dose of Se in the form of nanoparticles (0.5 mg/kg feed).

The form of Se at the increased dose had no significant effect on the results of the slaughter analysis of chickens (Table 2). Final weight and carcass weight, slaughter yield and proportion of individual internal organs relative to BW were not significantly different from the CON group ($p > 0.05$).

In the present study, neither the supplemental form of Se nor the different doses of Se were found to have a negative effect on selected parameters of the chemical composition of the pectoral muscles (BM) of cockerels ($p > 0.05$) (Table 3). Only the total collagen content of the BM differed significantly ($p \leq 0.01$). T2 pectoral muscles had the highest concentration of collagen in the muscle tissue, nearly 40% more collagen relative to CON. The addition of nano-Se did not significantly affect ($p > 0.05$) the collagen content of the BM with respect to the application of SS at 0.5 mg/kg feed.

The highest Se concentration (Figure 1A) in the pectoral muscles (CON = 0.27; T1 = 0.37; T2 = 0.42; T3 = 0.12 mg/kg) was determined in the chickens of the T2 group ($p \leq 0.01$). For the liver (Figure 1B), the highest Se concentration in these tissues was found ($p \leq 0.01$) in the CON group (CON = 3.79; T1 = 2.90; T2 = 0.86; T3 = 0.49 mg/kg).

The use of different forms of Se in the diet of chickens at a dose of 0.5 mg/kg feed had a significant effect ($p \leq 0.05$) on the physico-chemical properties of breast muscle (Table 4). The BM of cockerels fed with SS (0.5 mg/kg feed) had the lowest pH_{24} (5.53) and the highest drip loss (3.01%) and WHC (3.11 cm^2/g) relative to the analysed groups. Increasing the dose of Se in the form of other sources relative to CON significantly ($p \leq 0.05$) affected WHC and drip loss parameters. However, the use of a dose of 0.5 mg/kg SS in the chickens' diet had a significant effect on the deterioration of parameters related to the suitability of the meat for use in a specified product. The addition of SS at an increased dose had a significant effect ($p \leq 0.01$) on higher levels of L* and b* parameters relative to the other test groups. This was also confirmed by analysis of the ΔE parameter. The muscles from cockerels with the increased dose of SS supplementation in the diet had the highest and visually apparent colour deviation (ΔE = 4.53) relative to the muscles from the CON.

Table 2. Results of male broiler chicken slaughter analysis.

Indices	Group				SEM	*p* Value
	CON	T1	T2	T3		
Body weight, g	3343.75	3235.44	3251.76	3236.00	19.964	0.166
Carcass weight, g	2525.60	2446.22	2438.46	2417.74	17.626	0.143
Dressing percentage, g per 100 g BW	75.56	75.60	75.00	74.71	3.293	0.738
Breast muscles, g per 100 g BW	28.19	28.42	28.00	28.20	1.484	0.811
Leg muscles, g per 100 g BW	19.38	19.77	18.94	19.78	3.212	0.785
Gizzard, g per 100 g BW	0.82	0.83	0.83	0.83	0.023	0.710
Heart, g per 100 g BW	0.75	0.79	0.80	0.80	0.014	0.620
Liver, g per 100 g BW	2.16	2.29	2.29	2.33	0.042	0.380
Abdominal fat, g per 100 g BW	1.48	1.49	1.51	1.54	0.023	0.810

Data are given as mean ± SEM (n = 10); CON—control group; T1—diet with upper dose of inorganic Se (0.5 mg/kg feed); T2—diet with upper dose of Se in the organic form (0.5 mg/kg feed); T3—diet with upper dose of Se in the form of nanoparticles (0.5 mg/kg feed).

Table 3. Selected chemical components of breast muscles of male broiler chickens.

Indices, %	Group				SEM	*p* Value
	CON	T1	T2	T3		
Moisture	75.32	76.41	75.46	74.86	0.254	0.065
Total fat	0.87	0.89	1.48	1.10	0.243	0.121
Total protein	23.53	23.32	22.85	23.35	0.194	0.383
Total ash	1.40	1.41	1.44	1.50	0.198	0.175
Total collagen	0.37^{A}	0.44^{AB}	0.93^{C}	0.75^{BC}	0.129	<0.001

Data are given as mean ± SEM (n = 10); $^{A, B, C}$—values denoted with various letters differ significantly at $p \leq 0.01$; CON—control group; T1—diet with upper dose of inorganic Se (0.5 mg/kg feed); T2—diet with upper dose of Se in the organic form (0.5 mg/kg feed); T3—diet with upper dose of Se in the form of nanoparticles (0.5 mg/kg feed).

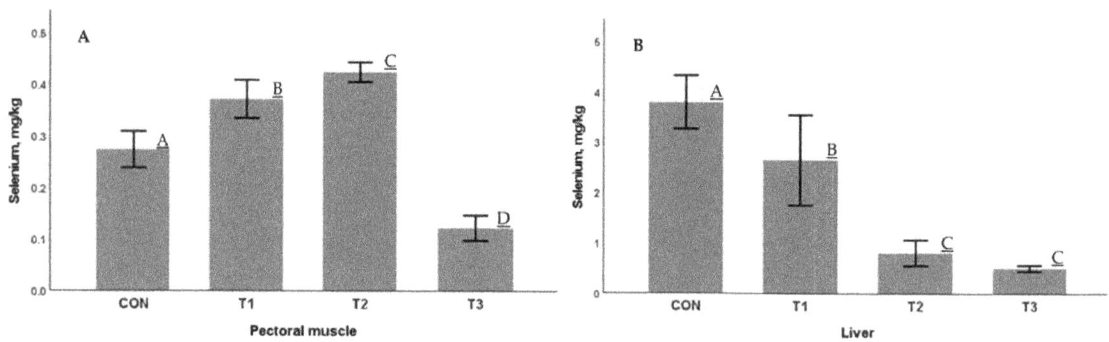

Figure 1. The effect of using increased doses of various forms of Se on Se content in tissues: (**A**) breast muscle; (**B**) liver. Data are given as mean ± SEM (n = 10). $^{A, B, C, D}$—values denoted with various letters differ significantly at $p \leq 0.01$; CON—control group; T1—diet with upper dose of inorganic Se (0.5 mg/kg feed); T2—diet with upper dose of Se in the organic form (0.5 mg/kg feed); T3—diet with upper dose of Se in the form of nanoparticles (0.5 mg/kg feed).

Table 4. Physico-chemical properties of breast muscles of male broiler chickens.

Indices	Group				EM	*p* Value
	CON	T1	T2	T3		
pH_{24}	5.64 ^{A}	5.53 ^{C}	5.73 ^{B}	5.72 ^{AB}	0.017	<0.001
Drip loss, %	2.81 ^{A}	3.01 ^{B}	2.57 ^{A}	2.51 ^{A}	0.212	0.023
WHC, cm^2/g	2.34 ^{A}	3.11 ^{B}	2.70 ^{A}	2.54 ^{A}	0.123	0.016
L* lightness	63.70 ^{A}	67.46 ^{B}	61.61 ^{A}	60.71 ^{A}	0.572	<0.001
a* redness	12.50	12.71	12.73	12.53	0.161	0.943
b* yellowness	9.69 ^{A}	12.20 ^{B}	9.16 ^{AC}	8.11 ^{C}	0.294	<0.001
ΔE CON: T1–T3	0.00	4.53	2.17	3.38	-	-

Data are given as mean ± SEM (n = 10). $^{A, B, C}$—values denoted with various letters differ significantly at $p \leq 0.01$; parameter L* (colour brightness) can have values from 0 to 100. Parameters a* (redness) and b*(yellowness) are tri-chromaticity coordinates and can have positive and negative values: +a* corresponds to red, +b* to yellow. ΔE—absolute colour difference; $^{A, B, C}$—values denoted with various letters differ significantly at $p \leq 0.01$; CON—control group; T1—diet with upper dose of inorganic Se (0.5 mg/kg feed); T2—diet with upper dose of Se in the organic form (0.5 mg/kg feed); T3—diet with upper dose of Se in the form of nanoparticles (0.5 mg/kg feed).

Additionally, tissue sections of chicken livers and pectoral muscles from all four experimental groups were examined by transmission electron microscopy to determine the ultrastructural changes that correlate with the effects of the Se diet. The comparison of ultrastructural changes in pectoral muscle fibres is shown in Figure 2.

Ultrastructural analysis of the muscle fibres of broiler chickens fed a standard commercial diet containing inorganic Se at a level of 0.3 mg/kg BW of diet (CON) showed myofibrils with an irregular arrangement, becoming constricted and disrupted in places. Myofibrils are separated by sarcoplasm with visible mitochondria with slightly damaged structure. Sarcomeres showed regular organization of actin and myosin fibrils (Figure 2A,B). More ultrastructural changes in muscle fibres were found after supplementation with inorganic Se at a dose of 0.5 mg/kg BW. Irregularly arranged, markedly constricted myofibrils with a loose structure, patchy in places, separated by puffy sarcoplasm were seen. Also shown are swollen mitochondria with damaged cristae. Of note are sarcomeres with a significantly narrowed profile and damaged structure (Figure 2C,D). Figure 2E,F shows the ultrastructure of myofibrils of broiler chickens remaining dieting with an upper dose of Se in the organic form at a dose of 0.5 mg/kg feed. Normal parallel-running myofibrils with a compact regular structure are visible. The sarcomere structure with regular organization of actin and myosin filaments is well preserved. Mitochondria have the correct structure. The ultrastructure of the pectoral muscle of broiler chickens from the T3 group shows normal, parallel-running myofibrils with a compact, regular structure. Sarcomeres showed a regular organization of actin and myosin filaments. A proper mitochondrial profile with a slightly translucent mitochondrial matrix is shown (Figure 2G,H).

The ultrastructural analysis also showed statistically significant differences in sarcomere lengths ($p \leq 0.01$). The shortest sarcomeres were found after supplementation with inorganic Se at a dose of 0.5 mg/kg BW, which was 1.64 µm compared to the control. The length of sarcomeres increased significantly in both groups of birds (T2 and T3) remaining on diets containing organic Se and nano-Se (Figure 3). In birds fed from the T2 group, the length of sarcomeres was 1.79 µm, and in birds from the T3 group, it was 1.81 µm.

The hepatocytes of broilers fed the diet containing inorganic Se at the level of 0.3 mg/kg feed (CON) show almost normal morphology (Figure 4A,B). Inside hepatocytes, a centrally located single nucleus of varying shape, usually spherical or oval, is visible. It is characterized by a conventional pattern of nuclear architecture: euchromatin is located mostly in the nuclear centre, whereas heterochromatin tends to be found in perinucleolar and perinuclear positions (the rim of heterochromatin typically lines the nuclear periphery) as well as forming random clumps throughout the nucleoplasm. The double membrane of the nuclear envelope (including perinuclear space and nuclear pores), is well visualized, and it is not affected (Figure 4A,B).

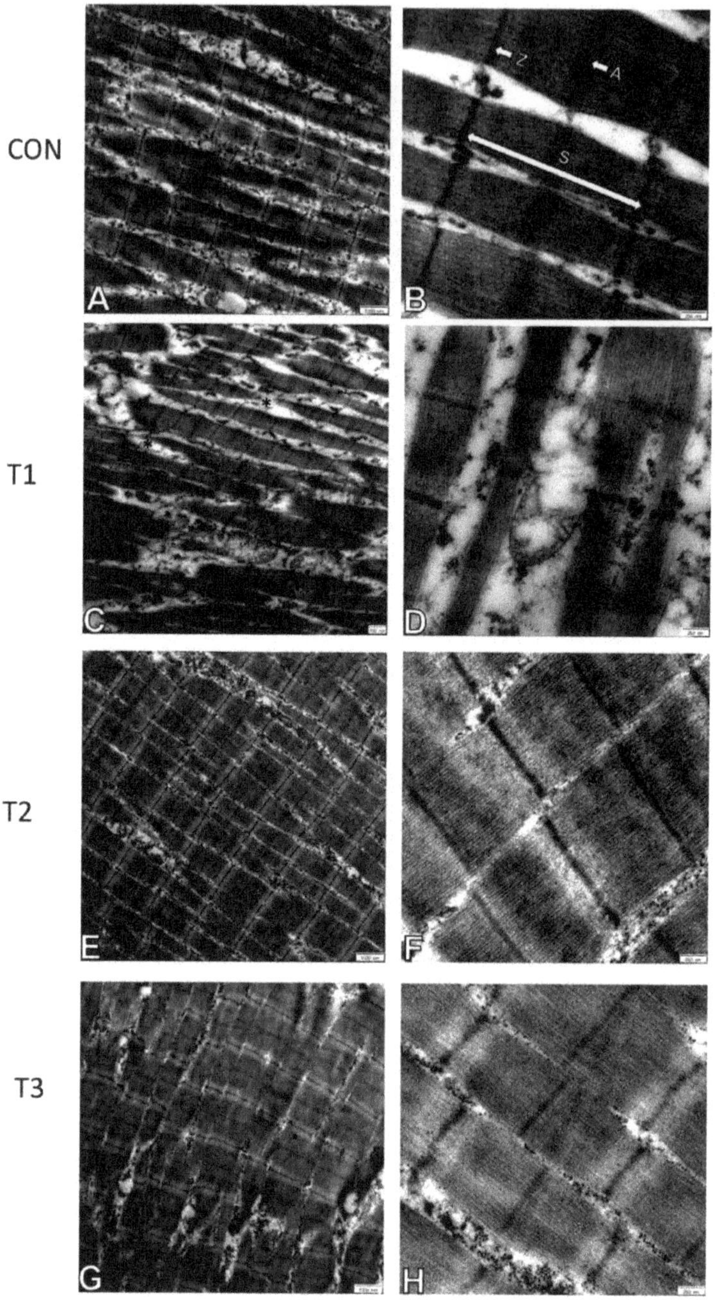

Figure 2. Ultrastructure of representative myofibers of the pectoral muscle of cockerels: CON—control group (**A**,**B**), T1—diet with upper dose of inorganic Se (**C**,**D**), T2—diet with upper dose of Se in the organic form (**E**,**F**), T3—diet with upper dose of Se in the form of nanoparticles (**G**,**H**). Scale bar 1000 nm (**A**,**C**,**E**,**G**) and 250 nm (**B**,**D**,**F**,**H**). S—sarcomere; A—A band; Z—Z band; *—indicates the disruption of myofibers.

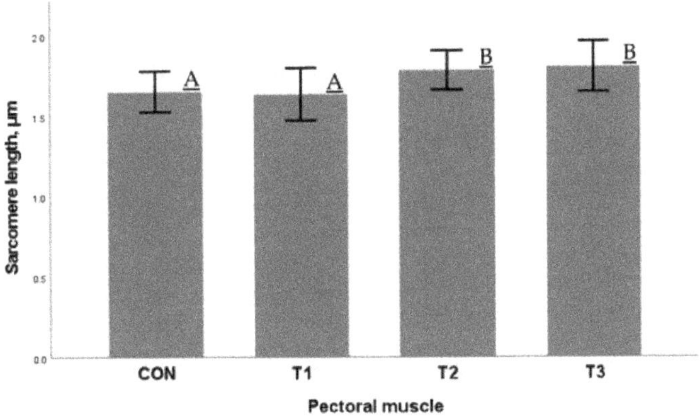

Figure 3. Length of sarcomeres in the pectoral muscle of cockerels. Data are given as mean ± SEM (n = 10); $^{A, B}$—values denoted with various letters differ significantly at $p \leq 0.01$; CON—control group; T1—diet with upper dose of inorganic Se (0.5 mg/kg feed); T2—diet with upper dose of Se in the organic form (0.5 mg/kg feed); T3—diet with upper dose of Se in the form of nanoparticle s (0.5 mg/kg feed).

The cytoplasm is rich in normal, double-membraned mitochondria, whose inner membrane periodically invaginates into laminar cristae and surrounds a mitochondrial matrix of normal density (Figure 4B). The vast majority of mitochondria appear normal or show only little changes. In a few mitochondria, the presence of spherical electron-dense inclusions in the mitochondrial matrix is noteworthy (Figure 4A). A moderate amount of rough endoplasmic reticulum of normal organization forms a membrane network that surrounds mitochondria and extends across the entire cytoplasm. Ribosomes are bound to the cytoplasmic surfaces of cisternae in a more or less regular pattern. The characteristic lamellar membranous structure of the Golgi apparatus (ribbon-like shape) and its typical location near the nucleus can be seen. In addition, few primary lysosomes are visible near the nucleus (Figure 4A). Figure 4C,D presents the hepatocytes of the chicken fed the diet in the T1 group. Their morphology differs significantly from that described above. Although the cell nuclei have the correct structure, cytoplasmic organelles, especially mitochondria, are significantly altered. As shown in Figure 4C, almost all mitochondria exhibit varying degrees of hydropic degeneration; they have a vacuolated matrix and damaged inner membrane, including a high degree of cristae disorganization. In addition, the mitochondrial matrix of numerous mitochondria contains clusters of osmiophilic spherical deposits/inclusions (Figure 4D). Damaged mitochondria ($p \leq 0.01$) represent 91.83% of the total amount in the cell (Figure 5).

There is a noticeable reduction in the number of cisternae of the rough endoplasmic reticulum while maintaining its unique structure. The appearance of autophagic vacuoles (Figure 4C) and the reduction of primary lysosome numbers are noteworthy.

The ultrastructure of hepatocytes of the chicken fed the diet with an upper dose of Se in the organic form (SY) at the level of 0.5 mg/kg feed (T2) appears normal (Figure 4E,F) and is similar to that observed in chickens fed the recommended diet. Mitochondria show no significant abnormalities in their morphology. However, some mitochondria are swollen and have damaged cristae. They account for 25.43% of the total amount in the cell (Figure 5). This is noteworthy that they do not contain any electron-dense material in the matrix. Instead, numerous primary lysosomes appeared (Figure 4F). They are distributed randomly within the cytoplasm. A nearly identical picture of hepatocytes was also observed in the chicken in group T3 (Figure 4G,H). Swollen mitochondria with damaged cristae also occur. They account for 26.96% of the total amount in the cell (Figure 5).

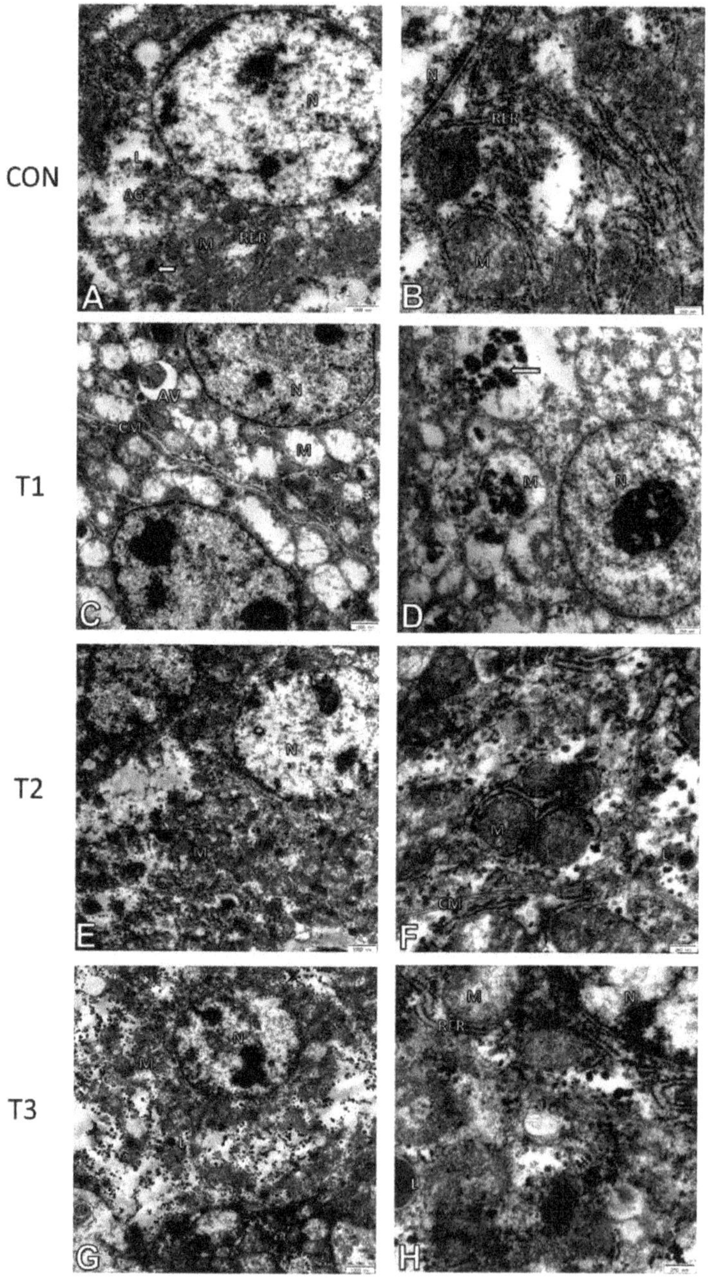

Figure 4. Ultrastructure of representative hepatocytes of cockerels: CON—control group (**A**,**B**), T1—diet with upper dose of inorganic Se (**C**,**D**), T2—diet with upper dose of Se in the organic form (**E**,**F**), T3—diet with upper dose of Se in the form of nanoparticles (**G**,**H**). Scale bar 1000 nm (**A**,**C**,**E**,**G**) and 250 nm (**B**,**D**,**F**,**H**). N—nucleus; M—mitochondria; RER—rough endoplasmic reticulum; L—lysosomes; AV—autophagic vacuole; CM—cell membrane; arrow—electron dense deposits in mitochondria.

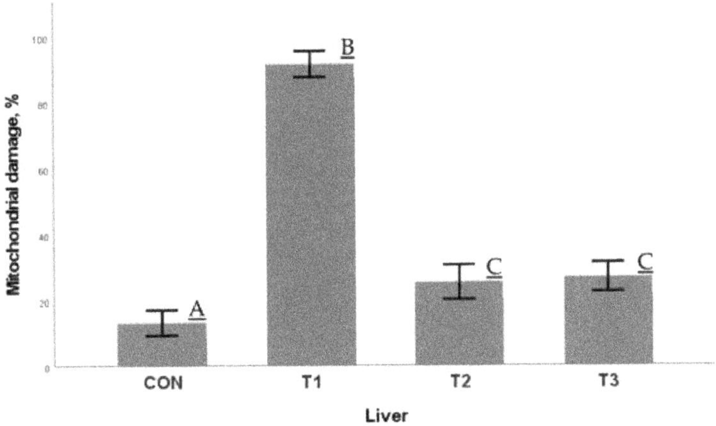

Figure 5. Mitochondrial damage in the liver. Data are given as mean ± SEM (n = 10); $^{A, B, C}$—values denoted with various letters differ significantly at $p \leq 0.01$; CON—control group; T1—diet with upper dose of inorganic Se (0.5 mg/kg feed); T2—diet with upper dose of Se in the organic form (0.5 mg/kg feed); T3—diet with upper dose of Se in the form of nanoparticles (0.5 mg/kg feed).

The use of inorganic Se (CON and T1) in the chickens' diet significantly ($p \leq 0.01$) decreased AcP in pectoral muscle tissue. At the same time, HEX activity increased significantly ($p \leq 0.05$) using SS in the chickens' diet at an increased dose. The activity of aminopeptidases including AlaAP, LeuAP, ArgAP decreased in the breast muscle ($p \leq 0.01$) when inorganic sources of Se were used (Table 5). The addition of other Se sources in the chickens' diet other than SS influenced ($p \leq 0.05$) the reduction of BGAL, BGLU and MAN activities. Nano-Se significantly ($p \leq 0.01$) influenced the increase in aminopeptidase activity. In addition, AcP, HEX and aGlu activities increased significantly ($p \leq 0.05$).

Exposure of chicken to a diet with the upper dose of inorganic Se (T1) significantly ($p \leq 0.01$) reduced the activity of AcP and HEX in the liver compared to the control group (CON). In contrast, BGRD and BGAL activities increased ($p \leq 0.01$). As for aminopeptidase activity, LeuAP activity increased significantly, while AlaAP and ArgAP activity decreased ($p \leq 0.05$). In general, in the liver, a decline in activity prevailed. Exposure of chicken to a diet with organic Se (T2) significantly increased AcP, BGAL and HEX activities and reduced BGRD activity ($p \leq 0.01$). As for the activity of aminopeptidases, there was an increase in the activity of all of them. In general, in the liver of chickens in T2, the increase in activity prevailed. In the group of chickens fed a diet with nano-Se (T3), the activities of AcP, BGAL, HEX and all three estimated aminopeptidases increased significantly ($p \leq 0.01$). Only BGRD activity decreased in this group of chickens ($p \leq 0.05$). Overall, an increase in degradative activity prevailed, and changes in enzyme activity followed a similar pattern to the previous group, T2, fed a diet with organic Se. In blood serum, of all the enzymes analysed, only the activity of aminopeptidases changed statistically significantly in all the groups of chickens studied ($p \leq 0.01$). A diet with the upper dose of inorganic Se significantly increased the activity of all aminopeptidases in serum, while feeding a diet with organic Se (T2) and nano-Se (T3) significantly decreased it.

Table 5. Enzymatic activity and antioxidant potential in selected tissues of male broiler chickens.

Indices	Group: Breast Muscle						Liver						Serum					
	CON	T1	T2	T3	SEM	*p* Value	CON	T1	T2	T3	SEM	*p* Value	CON	T1	T2	T3	SEM	*p* Value
AlaAP, nmol/mg protein/h	208.64 ^{A}	171.07 ^{A}	274.29 ^{B}	287.73 ^{B}	11.529	<0.001	265.91 ^{A}	267.32 ^{A}	406.13 ^{B}	467.25 ^{B}	23.211	0.001	40.12 ^{A}	46.32 ^{A}	22.17 ^{B}	28.91 ^{B}	2.0630	<0.001
LeuAP, nmol/mg protein/h	132.47 ^{A}	129.74 ^{A}	160.09 ^{AB}	189.48 ^{B}	6.335	<0.001	389.71 ^{A}	366.43 ^{A}	407.05 ^{AB}	524.87 ^{B}	19.878	0.013	48.38 ^{AB}	51.22 ^{A}	29.40 ^{C}	40.71 ^{B}	1.928	<0.001
ArgAP, nmol/mg protein/h	185.15 ^{A}	172.23 ^{A}	223.30 ^{B}	249.21 ^{C}	6.758	<0.001	319.19 ^{A}	321.11 ^{A}	369.28 ^{AB}	414.11 ^{B}	11.137	0.001	32.06 ^{A}	34.41 ^{A}	16.31 ^{B}	23.45 ^{B}	1.614	<0.001
AcP, nmol/mg protein/h	1330.75 ^{A}	1538.31 ^{A}	1899.71 ^{B}	2137.26 ^{B}	71.710	<0.001	1528.39 ^{A}	1240.78 ^{B}	1568.14 ^{A}	1759.12 ^{C}	42.830	<0.001	31.63	35.69	37.30	30.83	2.795	0.831
BGDR, nmol/mg protein/h	129.17	125.84	129.88	137.27	4.114	0.816	347.44 ^{A}	427.73 ^{B}	318.99 ^{A}	300.51 ^{A}	13.431	0.001	9.27	9.02	10.55	8.49	1.066	0.926
BGAL, nmol/mg protein/h	429.23 ^{b}	400.83 ^{ab}	372.47 ^{a}	391.54 ^{ab}	7.307	0.036	353.88 ^{A}	357.39 ^{A}	381.43 ^{A}	483.94 ^{B}	14.620	0.001	8.12	9.58	12.02	9.73	0.866	0.477
BGLU, nmol/mg protein/h	237.81 ^{AB}	264.24 ^{B}	211.54 ^{A}	237.58 ^{AB}	5.206	0.001	383.20	401.92	380.35	358.56	12.127	0.686	9.39	7.95	10.20	8.00	1.085	0.868
HEX, nmol/mg protein/h	617.80 ^{a}	692.00 ^{ab}	659.41 ^{ab}	745.95 ^{b}	17.003	0.042	1064.01 ^{AB}	941.48 ^{A}	1114.53 ^{AB}	1212.82 ^{B}	30.675	0.008	58.65	56.58	50.85	52.79	3.913	0.807
aGlu, nmol/mg protein/h	76.24 ^{A}	71.76 ^{A}	87.50 ^{AB}	102.46 ^{B}	3.328	0.001	323.77	240.78	304.10	294.70	12.844	0.093	7.26	6.65	8.22	6.45	0.746	0.852
MAN, nmol/mg protein/h	327.70 ^{A}	328.87 ^{A}	287.92 ^{B}	292.97 ^{B}	5.160	0.001	381.34	306.85	358.08	372.77	13.256	0.192	6.39	7.73	12.32	8.22	0.984	0.166
Vit. C, mg/100 mL	1.89 ^{A}	1.80 ^{A}	4.70 ^{B}	3.15 ^{AB}	0.323	0.001	2.57 ^{AB}	2.02 ^{A}	4.16 ^{BC}	4.36 ^{C}	0.287	0.001	4.46 ^{ab}	3.85 ^{b}	3.75 ^{b}	6.73 ^{a}	0.381	0.011
DPPH, %	52.55 ^{AB}	51.49 ^{A}	55.83 ^{BC}	56.12 ^{C}	0.588	0.002	85.53 ^{A}	85.71 ^{AB}	86.42 ^{BC}	86.56 ^{C}	0.132	0.004	82.82 ^{A}	81.84 ^{AB}	74.27 ^{BC}	71.20 ^{C}	1.369	0.002
GSH mM-SH	0.05 ^{A}	0.04 ^{A}	0.20 ^{B}	0.25 ^{B}	0.021	<0.001	0.85 ^{AB}	0.71 ^{A}	0.89 ^{B}	0.95 ^{B}	0.026	0.004	0.99 ^{A}	1.00 ^{A}	0.98 ^{A}	1.28 ^{B}	0.034	0.001

Data are given as mean ± SEM (n = 10); $^{A, B, C}$—values denoted with various letters differ significantly at $p \leq 0.01$; $^{a, b}$—values denoted with various letters differ significantly at $p \leq 0.05$; SEM—standard error of mean; CON—control group; T1—diet with upper dose of inorganic Se (0.5 mg/kg feed); T2—diet with upper dose of Se in the organic form (0.5 mg/kg feed); T3—diet with upper dose of Se in the form of nanoparticles (0.5 mg/kg feed). AlaAP—alanine aminopeptidase; LeuAP—leucine aminopeptidase; ArgAP—arginineaminopeptidase; AcP—acid phosphatase; BGRD—beta-glucuronidase; BGAL—beta-galactosidase; BGLU—beta-glucosidase; aGlu—alpha-glucosidase; MAN—mannosidase; HEX—N-acetyl-BD-hexosaminidase; vit. C—vitamin C; GSH—glutathione; DPPH—2;2-diphenyl-1-picrylhydrazyl.

The effect of Se type on the oxidative status in analysed tissues of broiler chickens was tested using the following determinations: vitamin C, glutathione (GSH), and 2,2-diphenyl-1-picrylhydrazyl (DPPH). In the pectoral muscle of chickens fed the upper dose of inorganic Se (T1 group), all indicators decreased, while in fed organic Se (T2 group) and nano-Se (T3), all indicators increased significantly. In the liver, both groups of chickens: T2 and T3, obtained significantly higher vitamin C, DPPH and glutathione results than the group considered a control. In contrast, the T1 group had statistically significantly lower vitamin C and glutathione levels and a higher result for DPPH than the CON group. Analysis of serum antioxidant potential in the T1 group showed a statistically significant reduction in vitamin C and DPPH levels ($p \leq 0.05$), while glutathione level increased ($p \leq 0.05$). Serum levels of all analysed factors decreased in chickens fed a diet supplemented with organic Se. The addition of nano-Se to the diet decreased the levels of vitamin C and GSH and increased the level of DPPH ($p \leq 0.05$).

4. Discussion

Food can be supplemented with selenium (Se) in various forms, such as inorganic, organic, and nanoparticle forms. The metabolism of these forms is different in birds. The chemical form and concentration of Se have significant roles in the rate of absorption, retention, and metabolism. Most inorganic Se is excreted in the urine, while nano-Se particles are excreted in faeces [46]. It has been shown that the inorganic form of Se has low bioavailability, accelerates oxidation processes and can be toxic, especially at high concentrations. In contrast, organic-Se and nano-Se exhibit low toxicity, high adsorption capacity, high bioavailability and high catalytic efficiency in chickens, sheep and goats [46–48]. Selenium is an important trace element that upregulates a vital component of the antioxidant defiance mechanism by controlling the body's glutathione pool and Se-dependent antioxidant enzymes such as superoxide dismutase and glutathione peroxidase [49,50]. These enzymes can help in reducing the concentration of hydrogen peroxide and lipid peroxides and enhance the immune response in numerous species of animals [51]. Several studies have illustrated that the dietary Se form influences growth performance, meat quality characteristics and antioxidative properties in chickens [52,53].

Based on the study, the different forms of Se in the feed at an increased dose (0.5 mg/kg) were not found to negatively affect the growth performance of the chickens. Our results showed that the addition of nanoparticle Se did not adversely impact the growth performance of chickens, indicating that nano-Se has high bioavailability and low toxicity compared to inorganic forms of Se such as SS [54]. The results obtained in this study are consistent with those of Yoon et al. [55], which found that the source of Se in the feed did not affect the growth performance of chickens. In contrast, groups of chickens fed a lower dose of 0.2 mg/kg of dietary organic Se or nano-Se had similar growth performance compared to a group supplemented with the same level of Se in the form of selenite [56]. In a study by Ahmad et al. [36], it was found that the application of increased Se intake (0.4 mg/kg feed) in the SY form and nano-Se was able to cover the requirement of this element by the birds and provide optimal growth conditions with no differences between the application of SY and nano-Se. There are known reports where Se influences the growth performance of birds, which is related to the expression of selenoprotein P and type I selenoenzymes, which play a key role in the synthesis of thyroid hormones and Se transport, both of which contribute to the proper functioning and growth of avian organisms [57].

The use of different sources of Se at an increased dose, as in the case of chicken growth performance, had no significant effect ($p \geq 0.05$) on the slaughter analysis results of cockerels at day 42 of rearing. Higher slaughter and BM performance was found in broilers that received SY in the feed. In contrast, Payne and Southern [58], in an experiment conducted on chickens receiving SS and SY, found that slaughter performance was not dependent on the form of Se in the feed. Downs et al. [52,58] found a similar relationship in their study on the use of SY in broiler feeding. The non-significant difference in body weight and other slaughter parameters in the present study may be due to feeding a balanced diet with appropriate practice [59] and optimal microclimate conditions with the welfare of the birds unaffected by, among other things, heat stress or other factors that may negatively affect the rearing performance of the chickens.

The highest concentration of Se was found in the BM of T2 cockerels, whereas the highest concentration of Se was found in the liver of the T1 group (SS, 0.5 mg/kg). This result is different from what was reported in [60]. Studies have indicated that nano-Se has a higher bioavailability and a lower risk of toxicity than other forms of Se [61]. Based on the results of the present study, it seems reasonable to recommend that broilers take nano-Se at a dose of 0.3–0.5 mg/kg [22]. The organic form of Se is accumulated to a greater extent than the inorganic form [62,63]; presumably, this may be related to differences in metabolic pathways between the inorganic and organic forms of Se. Inorganic Se is absorbed through the intestine through a passive diffusion process, with organic Se through an active transport mechanism [62]. A final product with increased nutritional value can be achieved by adequate supplementation of nano-Se at an increased dose (0.5 mg/kg). Our study confirms that nano-Se and liver enriched with nano-Se in this way are safe and can provide a valuable source of Se [34].

The post-mortem transformation of chicken muscle leads to an accumulation of lactic acid, which directly leads to a reduction in tissue pH. The delayed reduction in pH post-mortem results in reduced protein denaturation, which translates into an improved ability of skeletal muscle to retain its own water [64]. In the present study, a significant effect of Se application at an increased dose on pH_{24} was found. BM of cockerels in which SS was used in the diet at a dose of 0.5 mg/kg feed had the lowest pH_{24} and thus higher drip loss (3.01) and WHC (3.11), while the differences between CON, SY and nano-Se were not significant. This helps to conclude that the use of other forms of Se in chicken feed versus SS at an increased ration has a positive effect on the physico-chemical properties related to water retention in BM. Meat colour is dependent on a number of factors, including pH, myoglobin concentration, nitrite, etc. [65]. Se can significantly improve serum glutathione peroxidase activity, increase the resistance to oxidation of myoglobin or oxymyoglobin, among others, and deepen the colour of meat [22,66]. In this study, an increased dose of Se SS increases lightness (L*) and yellowness (b*), which directly affect the consumer's visual assessment of the pectoral muscle (ΔE).

Ultrastructural studies conducted showed the most significant changes in the muscle fibres of broiler chickens after supplementation of the inorganic Se at a dose of 0.5 mg/kg diet (Figure 2C,D). We showed a change in the profile of myofibrils and damage to sarcomeres and mitochondria. Furthermore, measurements of sarcomere size showed a reduction in length to 1.64 μm, which may suggest reduced elasticity of muscle tissue. This may be due to the weaker uptake of Se, which is more difficult to bind to glutathione peroxidase, which is responsible for scavenging oxygen free radicals and preventing oxidative stress [67–69]. In contrast, chickens that were fed Se in organic form (T2) and Se in nano form (T3) showed a slightly different picture of ultrastructural changes (Figure 2E–H). In this case, a stable structure of myofibrils with sarcomeres with a regular profile of actin and myosin fibres was demonstrated. No significant changes in mitochondrial structure were shown. However, measurements of the size of the sarcomeres showed an increase in length to 1.79 μm after organic Se (T2), and an increase to 1.81 μm after nanoparticles, the opposite of the group receiving inorganic Se. According to data [70], the length of sarcomeres can influence the better elasticity of muscle tissue and thus the quality of broiler chicken meat. To reveal the effect of different sources of Se supplementation on the ultrastructure of hepatocytes, the ultra-thin sections of the liver of chicken from all experimental groups (CON, T1, T2, T3) were also analysed by transmission electron microscopy (TEM). Our research has shown that the primary effect of Se at the ultrastructural level of the liver was the appearance of clusters of electron-dense (osmiophilic) inclusions within the mitochondrial matrix. However, these pathological changes were mainly observed in the hepatocytes of the chickens from the T1 group (Figure 4C,D) and to a minor extent in the mitochondria of the chickens fed the recommended diet containing SS at the level of 0.3 mg/kg feed (Figure 4A,B). Inclusions were not observed in the hepatocytes of chickens (T1) of the other groups, T2 and T3. In addition, almost all mitochondria in hepatocytes of chickens from the T1 group exhibit varying degrees of hydropic degeneration. They were swollen and had damaged inner membranes, including a high degree of cristae disorganization. The inclusions mentioned early on were found in the mitochondrial matrix that was transparent due to the influx of water. Abundant, spherical shape electron-dense deposits in the mitochondrial matrix were seen in numerous mitochondria of chicken fed with the upper dose of inorganic Se (Figure 4D). Abnormal mitochondria represent 91.83% of the total amount (Figure 5). In our opinion, hepatocyte mitochondria of chicken from the T1 group show features of a toxic response to excess inorganic Se in the feed. The effect of changes in the matrix volume in the physiological range is to stimulate the electron transfer chain and oxidative phosphorylation in order to satisfy the metabolic requirements of the cell. However, excessive matrix swelling caused by the continued opening of mitochondrial permeability transition pores (*p*TP) and other PTP-independent mechanisms makes mitochondrial function and integrity less efficient and leads to cell death [71]. Similar observations regarding the accumulation of electron-dense inclusions in the mitochondrial matrix were made by Medina et al. [72], who studied the effects of Se (Na_2SeO_3; SS) on the growth of three cell lines cultured in vitro. They performed analyses by transmission electron microscopy to determine any ultrastructural changes which correlated with the effects of Se treatment. They observed that the vast majority of the mitochondrial matrices were filled with dense osmophilic deposits. The accumulation was dependent on the dose of Se in the culture medium, the time of exposure, and the cell line. However, Se did not alter cytoplasmic microtubules or intermediate filament networks because they obtained abundant desmosomes [72]. Furthermore, they examined the chemical nature of the electron-dense material in the mitochondrial matrix by X-ray microanalysis of sections of studied cells. They obtained, as a result, an increased level of calcium, iron, and Se in the mitochondrial matrix [72]. There was some speculation about the nature of the material accumulated, as the nature of the mitochondrial inclusion was generally unknown, although it could be protein complexed by Se since Se reacts easily with mercapto groups of organic compounds [72,73]. Medina et al. [72] suggested that whereas high doses of Se inhibit the growth of cells, low doses stimulate the growth of some cell populations. They speculated that one of the early reasons of Se-mediated growth inhibition may be a modulation of mitochondrial function [72]. Klug et al. [73] found that succinic dehydrogenase was inhibited in vivo by sodium selenite. Liu et al. [74] used TEM to detect the accumulation of lentinan (LNT)-functionalized nano-Se in the mitochondria of tumour cells, and furthermore, they investigated the mechanism of selene targeting mitochondria. Analysis of ultra-thin sections showed that Se enters the cells through the caveolae-mediated endocytosis pathway and then enters the mitochondria via mitochondrial membrane fusion [74]. Zahedi et al. [75] analysed the effects of Se tetrachloride ($SeCl_4$) on the mitochondria of lung A549 cells. Analysis of mitochondrial morphology showed a significant increase in the swollen phenotype and a decrease in mitochondrial motility in Se treated cells [75]. Furthermore, they performed double staining of the cells with autophagy marker GFP-LC3 and MitoTracker. The results of this staining showed that the swollen mitochondria co-labeled with autophagosomes, indicating their targeted degradation via mitophagy [75]. Noteworthy in our study is the simultaneous appearance of damaged mitochondria and autophagic vacuoles, as well as a reduction in the number of primary lysosomes in hepatocytes of chickens treated with a high dose of inorganic Se (T1). In hepatocytes from chickens fed the upper dose of organic Se (T2) and nano-Se (T3), in contrast to the T2 group, only some mitochondria were swollen and had damaged cristae (accounting for 25.43% and 26.96% of the total amount in the cell, respectively). In contrast, numerous primary lysosomes appeared (Figure 4F). Overall, in the groups of chickens (CON, T2 and T3), in contrast to the T1 group, no significant pathological changes were observed in the ultrastructure of hepatocytes, either with respect to mitochondria or other organelles.

Parallel to morphological analysis, the activity of degradative enzymes was estimated. Morphological images correlate with the results of degradative enzyme activity (Table 5). In the liver of chickens exposed to a diet with an upper dose of inorganic Se, a decrease in the activity of the analysed enzymes prevailed. This indicates that the enzymes were involved in intracellular digestion processes, as evidenced by the presence of autophagic vacuoles and depletion of primary lysosomes, which are a cellular reservoir of acid hydrolases (Figure 4C,D). In contrast, an increase in the activity of the analysed enzymes in the liver prevailed in the other two groups of chickens (exposed to a diet with organic Se—T2, and nano-Se—T3), indicating a reduction in the level of lysosomal degradation. The ultrastructure of hepatocytes of both groups showed numerous primary lysosomes, presumably due to low lysosome consumption. The decrease in enzyme activity in the T1 group may indicate an increase in the degradation of organelles damaged by an excessive dietary supply of inorganic Se. According to Marzella et al. [76], initially, the enzyme activity increases in parallel with the induction of degradation, but it later decreases with the advanced stage of the sequestered cell organelle degradation [77,78]. With more advanced stages of lysosomal degradation, the activity seems to decline due to enzyme reserves consumption [76,77] and possibly insufficient synthesis. According to Kalamida et al. [79], the increase in lysosome accumulation, presumably due to a low rate of lysosome consumption, is accompanied by an increase in cathepsin D expression. Zahedi et al. [75], analysing the effects of Se tetrachloride on the mitochondria of lung A549 cells, obtained a significant dose-dependent increase in mitochondrial protein oxidation [75]. Several research teams confirm that Se (SS, selene-lentinan-functionalized, nano-Se) has been implicated in mitochondrial toxicity, with reactive oxygen species playing an important role [74,75,80,81]. They confirmed that after Se treatment, mitochondria become the main organelles of ROS production. Sun et al. [81] showed that excessive Se can enhance the toxicity of other metals (such as arsenic) by reacting with S–adenosylmethionine and glutathione. Mitochondria are very dynamic structures and have been used to study a variety of biological problems such as stress and drug response [75]. Environmental toxicants can have a negative impact on mitochondria and alter their morphology [75,82]. It is well known that mitochondrial dysfunction is a prominent phenomenon in the pathogenesis of a variety of diseases. In summary, mitochondria appear to be the main targets of the upper dose of Se, and changes in the structure and function of this organelle are commonly observed in Se-treated cells [72,74,75,83].

The influence of Se on the oxidative status in the analysed tissues of broiler chickens was tested using the following determinations: vitamin C, glutathione (GSH), and 2,2-diphenyl-1-picrylhydrazyl (DPPH). Both groups of chickens in T2 and T3 obtained significantly higher vitamin C, DPPH and glutathione results in muscle and liver tissues than in the group considered a control. In contrast, group T1 obtained significantly lower results in these tissues. The increase in the levels of antioxidant indicators in the muscle and liver tissues of chickens from the T2 and T3 groups indicates better oxidative stability of the organism, in contrast to the T1 group, where we obtained lower parameters. Visha et al. [84] obtained very similar results; namely, birds supplemented with the nano- and organic Se showed higher total antioxidant capacity in the serum and tissues compared to the inorganic Se supplemented group. Se is known to play an essential role in protecting cells from oxidative damage by affecting the antioxidant levels and activities of selenoenzymes [85]. Nano-Se is capable of scavenging free radicals by improving the activity of seleno-enzymes [85,86] and growth improvement along with the status of serum oxidants and retention of selenium in vivo. In comparison with other selenium species such as SeMet, SeCys, and SY, nano-Se showed lower acute toxicity but increased the activities of selenoenzymes. The antioxidant effect of nano-Se is mainly associated with glutathione peroxidase family (GPXs) and thioredoxin reductase (TR). The GPXs have a capacity to detoxify an extended range of peroxides, such as H_2O_2, phospholipid hydroperoxide, fatty acid hydroperoxides, and hydroperoxyl groups of thymine [87]. In conclusion, the study concluded that the use of nano-Se in the diet can be effective in increasing the antioxidant potential in the organisms of chickens, thereby increasing the ability to detoxify a wide range of peroxides.

5. Conclusions

Based on this study, the use of different forms of Se in broiler chicken diet besides commonly used sodium selenate had a positive effect on the physico-chemical properties of the breast muscle and on the health status, antioxidant potential and ultrastructure of the breast muscle and liver. It also improved quality parameters of the pectoral muscle, such as the length of the sarcomeres, which resulted in better elasticity of the muscle tissue and thus in meat quality. In addition, the use of nano-Se at an increased dose was confirmed to protect against mitochondrial damage in hepatocytes and increased antioxidant potential. Nano-Se did not show degenerative and toxic effects similar to sodium selenite at a dose of 0.5 mg/kg feed. The use of nano-Se at a dose higher than the recommended dose had high bioavailability and low toxicity without negatively affecting the growth performance of chickens.

Author Contributions: D.B., M.M. and P.K. conceived the project idea, designed the study and supervised the project. D.B., M.M. and A.M. performed the experiments. D.B., M.M., M.Ł.-G., A.J., A.W., A.M., M.K. and P.K. performed the laboratory work. D.B., M.M., M.Ł.-G., A.W. and P.K. analysed the data and wrote the manuscript. All authors have read and agreed to the published version of the manuscript.

Funding: This work was supported by the National Science Centre, grant no. 2018/29/B/NZ9/01351 "Bioactivity of cannabidiol and nano-selenium in the maintenance of gut immunological and integrity potential in chickens".

Institutional Review Board Statement: All procedures in the present study were performed in accordance with the principles of the European Union and Polish Law on Animal Protection. No action involving pain or suffering was practiced. The study was carried out in compliance with the Animal Research: Reporting of in Vivo Experiments (ARRIVE) guidelines. All applicable institutional guidelines for the care and use of animals were followed. The experimental procedures carried out in this study were approved by the 2nd Local Ethical Committee for Animal Experiments at the Warsaw University of Life Sciences (Ciszewskiego Str. 8, 02-786 Warsaw, Poland). Ethical review and approval were waived for this study; as a result, the activities undertaken on the animals in the project are typical zootechnical activities (evaluation of the effects of different forms of selenium as a feed additive in the feeding of broiler chickens), which do not fulfil the definition of a procedure (the animals will be maintained under standard environmental conditions and will not be subjected to any vital procedures). Considering the above, it must be considered that the planned research does not go beyond the standard husbandry and rearing of animals for agricultural operations. On this basis, the planned study design was considered not to require ethics committee approval for the experiment (1 September 2021, the 2nd Local Ethical Committee for Animal Experiments at the Warsaw University of Life Sciences).

Informed Consent Statement: Not applicable.

Data Availability Statement: All data are contained within the article.

Acknowledgments: The manuscript is a part of the thesis of Damian Bień.

Conflicts of Interest: The authors declare no conflict of interest.

References

1. Yildiz, A.; Kaya, Y.; Tanriverdi, O. Effect of the interaction between selenium and zinc on dna repair in association with cancer prevention. *J. Cancer Prev.* **2019**, *24*, 146–154. [CrossRef] [PubMed]
2. Castets, P.; Bertrand, A.T.; Beuvin, M.; Ferry, A.; Le Grand, F.; Castets, M.; Chazot, G.; Rederstorff, M.; Krol, A.; Lescure, A.; et al. Satellite cell loss and impaired muscle regeneration in selenoprotein N deficiency. *Hum. Mol. Genet.* **2011**, *20*, 694–704. [CrossRef] [PubMed]
3. Dudkiewicz, M.; Szczepińska, T.; Grynberg, M.; Pawłowski, K. A novel protein kinase-like domain in a selenoprotein, widespread in the tree of life. *PLoS ONE* **2012**, *7*, 32138. [CrossRef] [PubMed]

4. Sun, L.H.; Huang, J.Q.; Deng, J.; Lei, X.G. Avian selenogenome: Response to dietary Se and Vitamin E deficiency and supplementation. *Poult. Sci.* **2019**, *98*, 4247–4254. [CrossRef]
5. Surai, P.F.; Kochish, I.I.; Fisinin, V.I.; Velichko, O.A. Selenium in poultry nutrition: From sodium selenite to organic selenium sources. *J. Poult. Sci.* **2018**, *55*, 79–93. [CrossRef] [PubMed]
6. LeBlanc, K.L.; Kumkrong, P.; Mercier, P.H.J.; Mester, Z. Selenium analysis in waters. Part 2: Speciation methods. *Sci. Total Environ.* **2018**, *640–641*, 1635–1651. [CrossRef] [PubMed]
7. Di Marzio, A.; Lambertucci, S.A.; Fernandez, A.J.G.; Martínez-López, E. From Mexico to the Beagle Channel: A review of metal and metalloid pollution studies on wildlife species in Latin America. *Environ. Res.* **2019**, *176*, 108462. [CrossRef] [PubMed]
8. PAUL, T.; SAHA, N.C. Environmental Arsenic and Selenium Contamination and Approaches Towards Its Bioremediation Through the Exploration of Microbial Adaptations: A Review. *Pedosphere* **2019**, *29*, 554–568. [CrossRef]
9. Etteieb, S.; Magdouli, S.; Zolfaghari, M.; Brar, S.K. Monitoring and analysis of selenium as an emerging contaminant in mining industry: A critical review. *Sci. Total Environ.* **2020**, *698*, 134339. [CrossRef] [PubMed]
10. Minuye, A.; Zewge, F.; Chandravanshi, B.S. Levels of selenium in selected river, well and bottled waters in Ethiopia by microwave plasma-atomic emission spectrometry. *Bull. Chem. Soc. Ethiop.* **2020**, *34*, 215–226. [CrossRef]
11. Gammelgaard, B.; Rasmussen, L.H.; Gabel-Jensen, C.; Steffansen, B. Estimating intestinal absorption of inorganic and organic selenium compounds by in vitro flux and biotransformation studies in Caco-2 cells and ICP-MS detection. *Biol. Trace Elem. Res.* **2012**, *145*, 248–256. [CrossRef] [PubMed]
12. Han, X.J.; Qin, P.; Li, W.X.; Ma, Q.G.; Ji, C.; Zhang, J.Y.; Zhao, L.H. Effect of sodium selenite and selenium yeast on performance, egg quality, antioxidant capacity, and selenium deposition of laying hens. *Poult. Sci.* **2017**, *96*, 3973–3980. [CrossRef] [PubMed]
13. Schweizer, U.; Streckfuß, F.; Pelt, P.; Carlson, B.A.; Hatfield, D.L.; Köhrle, J.; Schomburg, L. Hepatically derived selenoprotein P is a key factor for kidney but not for brain selenium supply. *Biochem. J.* **2005**, *386*, 221–226. [CrossRef] [PubMed]
14. Yu, L.L.; Wang, R.L.; Zhang, Y.Z.; Kleemann, D.O.; Zhu, X.P.; Jia, Z.H. Effects of selenium supplementation on polyunsaturated fatty acid concentrations and antioxidant status in plasma and liver of lambs fed linseed oil or sunflower oil diets. *Anim. Feed Sci. Technol.* **2008**, *140*, 39–51. [CrossRef]
15. Schäfer, K.; Kyriakopoulos, A.; Gessner, H.; Grune, T.; Behne, D. Effects of selenium deficiency on fatty acid metabolism in rats fed fish oil-enriched diets. *J. Trace Elem. Med. Biol.* **2004**, *18*, 89–97. [CrossRef] [PubMed]
16. Zanini, S.F.; Torres, C.A.A.; Bragagnolo, N.; Turatti, J.M.; Silva, M.G.; Zanini, M.S. Effect of oil sources and vitamin E levels in the diet on the composition of fatty acids in rooster thigh and chest meat. *J. Sci. Food Agric.* **2004**, *84*, 672–682. [CrossRef]
17. González, E.; Tejeda, J.F. Effects of dietary incorporation of different antioxidant extracts and free-range rearing on fatty acid composition and lipid oxidation of Iberian pig meat. *Animal* **2007**, *1*, 1060–1067. [CrossRef] [PubMed]
18. Netto, A.S.; Zanetti, M.A.; Del Claro, G.R.; De Melo, M.P.; Vilela, F.G.; Correa, L.B. Effects of copper and selenium supplementation on performance and lipid metabolism in confined brangus bulls. *Asian Australasian J. Anim. Sci.* **2014**, *27*, 488–494. [CrossRef]
19. Pereira, A.S.C.; dos Santos, M.V.; Aferri, G.; Corte, R.R.P.D.S.; Silva, S.D.L.; Freitas Júnior, J.E.D.; Leme, P.R.; Rennó, F.P. Lipid and selenium sources on fatty acid composition of intramuscular fat and muscle selenium concentration of Nellore steers. *Rev. Bras. Zootec.* **2012**, *41*, 2357–2363. [CrossRef]
20. Ali, A.A. Use of nano-calciumand phosphors in broiler feeding. *Egypt. Poult. Sci. J.* **2017**, *37*, 637–650.
21. Habibian, M.; Sadeghi, G.; Ghazi, S.; Moeini, M.M. Selenium as a Feed Supplement for Heat-Stressed Poultry: A Review. *Biol. Trace Elem. Res.* **2015**, *165*, 183–193. [CrossRef] [PubMed]
22. Cai, S.J.; Wu, C.X.; Gong, L.M.; Song, T.; Wu, H.; Zhang, L.Y. Effects of nano-selenium on performance, meat quality, immune function, oxidation resistance, and tissue selenium content in broilers. *Poult. Sci.* **2012**, *91*, 2532–2539. [CrossRef] [PubMed]
23. Cemin, H.S.; Vieira, S.L.; Stefanello, C.; Kindlein, L.; Ferreira, T.Z.; Fireman, A.K. Broiler responses to increasing selenium supplementation using Zn-L-selenomethionine with special attention to breast myopathies. *Poult. Sci.* **2018**, *97*, 1832–1840. [CrossRef] [PubMed]
24. Ramiah, S.K.; Awad, E.A.; Mookiah, S.; Idrus, Z. Effects of zinc oxide nanoparticles on growth performance and concentrations of malondialdehyde, zinc in tissues, and corticosterone in broiler chickens under heat stress conditions. *Poult. Sci.* **2019**, *98*, 3828–3838. [CrossRef]
25. Patra, A.; Lalhriatpuii, M. Progress and Prospect of Essential Mineral Nanoparticles in Poultry Nutrition and Feeding-a Review. *Biol. Trace Elem. Res.* **2020**, *197*, 233–253. [CrossRef] [PubMed]
26. Khurana, A.; Tekula, S.; Saifi, M.A.; Venkatesh, P.; Godugu, C. Therapeutic applications of selenium nanoparticles. *Biomed. Pharmacother.* **2019**, *111*, 802–812. [CrossRef] [PubMed]
27. Guan, B.; Yan, R.; Li, R.; Zhang, X. Selenium as a pleiotropic agent for medical discovery and drug delivery. *Int. J. Nanomed.* **2018**, *13*, 7473–7490. [CrossRef] [PubMed]
28. Hosnedlova, B.; Kepinska, M.; Skalickova, S.; Fernandez, C.; Ruttkay-Nedecky, B.; Peng, Q.; Baron, M.; Melcova, M.; Opatrilova, R.; Zidkova, J.; et al. Nano-selenium and its nanomedicine applications: A critical review. *Int. J. Nanomed.* **2018**, *13*, 2107–2128. [CrossRef] [PubMed]
29. Skalickova, S.; Milosavljevic, V.; Cihalova, K.; Horky, P.; Richtera, L.; Adam, V. Selenium nanoparticles as a nutritional supplement. *Nutrition* **2017**, *33*, 83–90. [CrossRef] [PubMed]
30. Griffin, S.; Masood, M.I.; Nasim, M.J.; Sarfraz, M.; Ebokaiwe, A.P.; Schäfer, K.H.; Keck, C.M.; Jacob, C. Natural nanoparticles: A particular matter inspired by nature. *Antioxidants* **2018**, *7*, 3. [CrossRef] [PubMed]

31. Pelyhe, C.; Mézes, M. Myths and Facts About the Effects of Nano- Selenium in Farm Animals—Mini-Review. *Eur. Chem. Bull.* **2013**, *2*, 1049–1052. [CrossRef]
32. Mohapatra, P.; Swain, R.K.; Mishra, S.K.; Behera, T.; Swain, P.; Behura, N.C.; Sahoo, G.; Sethy, K.; Bhol, B.P.; Dhama, K. Effects of dietary nano-selenium supplementation on the performance of layer grower birds. *Asian J. Anim. Vet. Adv.* **2014**, *9*, 641–652. [CrossRef]
33. Hu, C.H.; Li, Y.L.; Xiong, L.; Zhang, H.M.; Song, J.; Xia, M.S. Comparative effects of nano elemental selenium and sodium selenite on selenium retention in broiler chickens. *Anim. Feed Sci. Technol.* **2012**, *177*, 204–210. [CrossRef]
34. Bień, D.; Michalczuk, M.; Szkopek, D.; Kinsner, M.; Konieczka, P. Changes in lipids metabolism indices as a result of different form of selenium supplementation in chickens. *Sci. Rep.* **2022**, *12*, 13817. [CrossRef]
35. Surai, P.F.; Kochish, I.I. Food for thought: Nano-selenium in poultry nutrition and health. *Anim. Health Res. Rev.* **2020**, *21*, 103–107. [CrossRef] [PubMed]
36. Ahmad, H.; Tian, J.; Wang, J.; Khan, M.A.; Wang, Y.; Zhang, L.; Wang, T. Effects of dietary sodium selenite and selenium yeast on antioxidant enzyme activities and oxidative stability of chicken breast meat. *J. Agric. Food Chem.* **2012**, *60*, 7111–7120. [CrossRef] [PubMed]
37. Michalczuk, M.; Jóźwik, A.; Damaziak, K.; Zdanowska-Sasiadek, Z.; Marzec, A.; Gozdowski, D.; Strzałkowska, N. Age-related changes in the growth performance, meat quality, and oxidative processes in breast muscles of three chicken genotypes. *Turkish J. Vet. Anim. Sci.* **2016**, *40*, 389–398. [CrossRef]
38. Michalczuk, M.; Zdanowska-Sąsiadek, Ż.; Damaziak, K.; Niemiec, J. Influence of indoor and outdoor systems on meat quality of slow-growingchickens. *CYTA J. Food* **2017**, *15*, 15–20. [CrossRef]
39. Bendowski, W.; Michalczuk, M.; Jóźwik, A.; Kareem, K.Y.; Łozicki, A.; Karwacki, J.; Bień, D. Using Milk Thistle (Silybum marianum) Extract to Improve the Welfare, Growth Performance and Meat Quality of Broiler Chicken. *Animals* **2022**, *12*, 1085. [CrossRef]
40. McDonald, J.K.; Barrett, A.J. Mammalian Proteinases: A Glossary and Bibliography Volume 2 Exopeptidases. *Biochem. Biol. Plasma Lipoproteins* **1987**, *224*, 235–236.
41. Barrett, A.J.; Heath, M.F. Lysosomal enzymes. In *Lysosomes. A Laboratory Handbook*; Dingle, J.T., Ed.; North-Holland Publ. Co.: Amsterdam, The Netherlands, 1972; pp. 19–135.
42. Michalczuk, M.; Holl, E.; Möddel, A.; Jóźwik, A.; Slósarz, J.; Bień, D.; Ząbek, K.; Konieczka, P. Phytogenic ingredients from hops and organic acids improve selected indices of welfare, health status markers, and bacteria composition in the caeca of broiler chickens. *Animals* **2021**, *11*, 3249. [CrossRef]
43. Brand-Williams, W.; Cuvelier, M.E.; Berset, C. Use of a free radical method to evaluate antioxidant activity. *LWT Food Sci. Technol.* **1995**, *28*, 25–30. [CrossRef]
44. Marzella, L.; Glaumann, H. Increased degradation in rat liver induced by vinblastine. I. Biochemical characterization. *Lab. Investig.* **1980**, *42*, 8–17.
45. *Aviagen Ross 308: Broiler Performance Objectives*; Aviagen Inc.: Huntsville, AL, USA, 2019; pp. 1–15.
46. Saleh, A.A.; Ebeid, T.A. Feeding sodium selenite and nano-selenium stimulates growth and oxidation resistance in broilers. *S. African J. Anim. Sci.* **2019**, *49*, 176–183. [CrossRef]
47. Toś-Luty, S.; Obuchowska-Przebirowska, D.; Latuszyńska, J.; Musik, I.; Tokarska-Rodak, M. Comparison of histological and ultrastructural changes in mice organs after supplementation with inorganic and organic selenium. *Ann. Agric. Environ. Med.* **2003**, *10*, 87–91. [PubMed]
48. Pedro, D.; Rosa, A.P.; Londero, A.; Forgiarini, J.; Branco, T.; De Freitas, H.M.; Giacomini, C.B.S.; Pilecco, M. Performance of broiler breeders supplemented with organic and inorganic minerals. *An. Acad. Bras. Cienc.* **2021**, *93*, e20191048. [CrossRef] [PubMed]
49. Albanes, D.; Till, C.; Klein, E.A.; Goodman, P.J.; Mondul, A.M.; Weinstein, S.J.; Taylor, P.R.; Parnes, H.L.; Gaziano, J.M.; Song, X.; et al. Plasma tocopherols and risk of prostate cancer in the Selenium and Vitamin e Cancer Prevention Trial (SELECT). *Cancer Prev. Res.* **2014**, *7*, 886–895. [CrossRef]
50. Shaheen, S.M.; Rinklebe, J.; Frohne, T.; White, J.R.; DeLaune, R.D. Biogeochemical Factors Governing Cobalt, Nickel, Selenium, and Vanadium Dynamics in Periodically Flooded Egyptian North Nile Delta Rice Soils. *Soil Sci. Soc. Am. J.* **2014**, *78*, 1065–1078. [CrossRef]
51. Li, J.L.; Zhang, L.; Yang, Z.Y.; Zhang, Z.Y.; Jiang, Y.; Gao, F.; Zhou, G.H. Effects of Different Selenium Sources on Growth Performance, Antioxidant Capacity and Meat Quality of Local Chinese Subei Chickens. *Biol. Trace Elem. Res.* **2018**, *181*, 340–346. [CrossRef]
52. Downs, K.M.; Hess, J.B.; Bilgili, S.F. Selenium Source Effect on Broiler Carcass Characteristics, Meat Quality and Drip Loss. *J. Appl. Anim. Res.* **2000**, *18*, 61–72. [CrossRef]
53. Marković, R.; Glišić, M.; Bošković, M.; Baltić, M. New scientific challenges—The possibilities of using selenium in poultry nutrition and impact on meat quality. *IOP Conf. Ser. Earth Environ. Sci.* **2017**, *85*, 12032. [CrossRef]
54. Ibrahim, D.; Kishawy, A.T.Y.; Khater, S.I.; Arisha, A.H.; Mohammed, H.A.; Abdelaziz, A.S.; Abd El-Rahman, G.I.; Elabbasy, M.T. Effect of dietary modulation of selenium form and level on performance, tissue retention, quality of frozen stored meat and gene expression of antioxidant status in ross broiler chickens. *Animals* **2019**, *9*, 342. [CrossRef]
55. Yoon, I.; Werner, T.M.; Butler, J.M. Effect of source and concentration of selenium on growth performance and selenium retention in broiler chickens. *Poult. Sci.* **2007**, *86*, 727–730. [CrossRef]

56. Couloigner, F.; Jlali, M.; Briens, M.; Rouffineau, F.; Geraert, P.A.; Mercier, Y. Selenium deposition kinetics of different selenium sources in muscle and feathers of broilers. *Poult. Sci.* **2015**, *94*, 2708–2714. [CrossRef] [PubMed]
57. Zhan, X.A.; Wang, H.F.; Yuan, D.; Wang, Y.; Zhu, F. Comparison of different forms of dietary selenium supplementation on gene expression of cytoplasmic thioredoxin reductase, selenoprotein P, and selenoprotein W in broilers. *Czech J. Anim. Sci.* **2014**, *59*, 571–578. [CrossRef]
58. Payne, R.L.; Southern, L.L. Comparison of inorganic and organic selenium sources for broilers. *Poult. Sci.* **2005**, *84*, 898–902. [CrossRef]
59. Rama Rao, S.V.; Prakash, B.; Raju, M.V.L.N.; Panda, A.K.; Poonam, S.; Murthy, O.K. Effect of supplementing organic selenium on performance, carcass traits, oxidative parameters and immune responses in commercial broiler chickens. *Asian-Australas. J. Anim. Sci.* **2013**, *26*, 247–252. [CrossRef]
60. Chen, K.; Fang, J.; Peng, X.; Cui, H.; Chen, J.; Wang, F.; Chen, Z.; Zuo, Z.; Deng, J.; Lai, W.; et al. Effect of selenium supplementation on aflatoxin B1-induced histopathological lesions and apoptosis in bursa of Fabricius in broilers. *Food Chem. Toxicol.* **2014**, *74*, 91–97. [CrossRef]
61. Zhan, X.A.; Wang, M.; Zhao, R.Q.; Li, W.F.; Xu, Z.R. Effects of different selenium source on selenium distribution, loin quality and antioxidant status in finishing pigs. *Anim. Feed Sci. Technol.* **2007**, *132*, 202–211. [CrossRef]
62. Skřivan, M.; Marounek, M.; Dlouhá, G.; Ševčíková, S. Dietary selenium increases vitamin E contents of egg yolk and chicken meat. *Br. Poult. Sci.* **2008**, *49*, 482–486. [CrossRef]
63. Sevcikova, S.; Skřivan, M.; Dlouhá, G.; Koucký, M. The effect of selenium source on the performance and meat quality of broiler chickens. *Czech J. Anim. Sci.* **2006**, *51*, 449–457. [CrossRef]
64. Berri, C.; Le Bihan-Duval, E.; Debut, M.; Santé-Lhoutellier, V.; Baéza, E.; Gigaud, V.; Jégo, Y.; Duclos, M.J. Consequence of muscle hypertrophy on characteristics of Pectoralis major muscle and breast meat quality of broiler chickens. *J. Anim. Sci.* **2007**, *85*, 2005–2011. [CrossRef] [PubMed]
65. King, N.J.; Whyte, R. Does it look cooked? A review of factors that influence cooked meat color. *J. Food Sci.* **2006**, *71*, R31–R40. [CrossRef]
66. Oliveira, T.F.B.; Rivera, D.F.R.; Mesquita, F.R.; Braga, H.; Ramos, E.M.; Bertechini, A.G. Effect of different sources and levels of selenium on performance, meat quality, and tissue characteristics of broilers. *J. Appl. Poult. Res.* **2014**, *1*, 15–22. [CrossRef]
67. Zhang, S.; Xie, Y.; Li, M.; Yang, H.; Li, S.; Li, J.; Xu, Q.; Yang, W.; Jiang, S. Effects of different selenium sources on meat quality and shelf life of fattening pigs. *Animals* **2020**, *10*, 615. [CrossRef] [PubMed]
68. Liu, B.; Xiong, Y.L.; Jiang, J.; Yu, D.; Lin, G. Cellular antioxidant mechanism of selenium-enriched yeast diets in the protection of meat quality of heat-stressed hens. *Food Biosci.* **2021**, *39*, 100798. [CrossRef]
69. Huang, J.C.; Huang, M.; Yang, J.; Wang, P.; Xu, X.L.; Zhou, G.H. The effects of electrical stunning methods on broiler meat quality: Effect on stress, glycolysis, water distribution, and myofibrillar ultrastructures. *Poult. Sci.* **2014**, *93*, 2087–2095. [CrossRef]
70. Ertbjerg, P.; Puolanne, E. Muscle structure, sarcomere length and influences on meat quality: A review. *Meat Sci.* **2017**, *132*, 139–152. [CrossRef]
71. Javadov, S.; Chapa-Dubocq, X.; Makarov, V. Different approaches to modeling analysis of mitochondrial swelling. *Mitochondrion* **2018**, *38*, 58–70. [CrossRef] [PubMed]
72. Medina, D.; Asch, B.B.; Oborn, C.J. Mitochondrial Inclusions in Selenium-Treated Mouse Mammary Epithelial Cell Lines. *Cancer Res.* **1983**, *43*, 2100–2105.
73. Klug, H.L.; Moxon, A.L.; Petersen, D.F.; Painter, E.P. Inhibition of rat liver succinic dehydrogenase by selenium compounds. *J. Pharmacol. Exp. Ther.* **1953**, *108*, 437–441.
74. Liu, H.J.; Qin, Y.; Zhao, Z.H.; Zhang, Y.; Yang, J.H.; Zhai, D.H.; Cui, F.; Luo, C.; Lu, M.X.; Liu, P.P.; et al. Lentinan-functionalized selenium nanoparticles target tumor cell mitochondria via TLR4/TRAF3/MFN1 pathway. *Theranostics* **2020**, *10*, 9083–9099. [CrossRef]
75. Zahedi, A.; On, V.; Phandthong, R.; Chaili, A.; Remark, G.; Bhanu, B.; Talbot, P. Deep Analysis of Mitochondria and Cell Health Using Machine Learning. *Sci. Rep.* **2018**, *8*, 16354. [CrossRef]
76. Marzella, L.; Ahlberg, J.; Glaumann, H. Isolation of autophagic vacuoles from rat liver: Morphological and biochemical characterization. *J. Cell Biol.* **1982**, *93*, 44–154. [CrossRef] [PubMed]
77. Wieczorek, A.; Lysek-Gladysinska, M.; Walaszczyk, A.; Jelonek, K.; Smolarz, M.; Pietrowska, M.; Gabrys, D.; Kulik, R.; Widlak, P. Changes in activity and structure of lysosomes from liver of mouse irradiated in vivo. *Int. J. Radiat. Biol.* **2018**, *94*, 443–453. [CrossRef] [PubMed]
78. Lysek-Gladysinska, M.; Wieczorek, A.; Walaszczyk, A.; Jelonek, K.; Jozwik, A.; Pietrowska, M.; Dörr, W.; Gabrys, D.; Widlak, P. Long-term effects of low-dose mouse liver irradiation involve ultrastructural and biochemical changes in hepatocytes that depend on lipid metabolism. *Radiat. Environ. Biophys.* **2018**, *57*, 123–132. [CrossRef]
79. Kalamida, D.; Karagounis, I.V.; Giatromanolaki, A.; Koukourakis, M.I. Important role of autophagy in endothelial cell response to ionizing radiation. *PLoS ONE* **2014**, *9*, e102408. [CrossRef] [PubMed]
80. Selvaraj, V.; Tomblin, J.; Yeager Armistead, M.; Murray, E. Selenium (sodium selenite) causes cytotoxicity and apoptotic mediated cell death in PLHC-1 fish cell line through DNA and mitochondrial membrane potential damage. *Ecotoxicol. Environ. Saf.* **2013**, *87*, 80–88. [CrossRef]

81. Sun, H.; Rathinasabapathi, B.; Wu, B.; Luo, J.; Pu, L.P.; Ma, L.Q. Arsenic and selenium toxicity and their interactive effects in humans. *Environ. Int.* **2014**, *69*, 148–158. [CrossRef]
82. Meyer, J.N.; Leung, M.C.K.; Rooney, J.P.; Sendoel, A.; Hengartner, M.O.; Kisby, G.E.; Bess, A.S. Mitochondria as a target of environmental toxicants. *Toxicol. Sci.* **2013**, *134*, 1–17. [CrossRef] [PubMed]
83. Wang, G.; Guo, Y.; Yang, G.; Yang, L.; Ma, X.; Wang, K.; Zhu, L.; Sun, J.; Wang, X.; Zhang, H. Mitochondria-Mediated Protein Regulation Mechanism of Polymorphs-Dependent Inhibition of Nanoselenium on Cancer Cells. *Sci. Rep.* **2016**, *6*, 31427. [CrossRef] [PubMed]
84. Visha, P.; Nanjappan, K.; Selvaraj, P.; Jayachandran, S.; Thavasiappan, V. Influence of Dietary Nanoselenium Supplementation on the Meat Characteristics of Broiler Chickens. *Int. J. Curr. Microbiol. Appl. Sci.* **2017**, *6*, 340–347. [CrossRef]
85. Bai, K.; Hong, B.; Hong, Z.; Sun, J.; Wang, C. Selenium nanoparticles-loaded chitosan/citrate complex and its protection against oxidative stress in d-galactose-induced aging mice. *J. Nanobiotechnol.* **2017**, *15*, 92. [CrossRef]
86. Sarkar, J.; Dey, P.; Saha, S.; Acharya, K. Mycosynthesis of selenium nanoparticles. *Micro Nano Lett.* **2011**, *6*, 599–602. [CrossRef]
87. Kondaparthi, P.; Flora, S.; Naqvi, S. Selenium nanoparticles: An insight on its Pro-oxidant and antioxidant properties. *Front. Nanosci. Nanotechnol.* **2019**, *6*, 1–5. [CrossRef]

MDPI
St. Alban-Anlage 66
4052 Basel
Switzerland
www.mdpi.com

Antioxidants Editorial Office
E-mail: antioxidants@mdpi.com
www.mdpi.com/journal/antioxidants

www.ingramcontent.com/pod-product-compliance
Lightning Source LLC
Chambersburg PA
CBHW061832210926
51850CB00006B/867

* 9 7 8 3 7 2 5 8 1 1 5 8 8 *